AF393361

Gian Luigi Chierici

2 Principles of Petroleum Reservoir Engineering

Translated from the Italian by

Peter J. Westaway

With 199 Figures and 42 Tables

Springer-Verlag

Berlin Heidelberg New York London Paris

Tokyo Hong Kong Barcelona Budapest

Professor GIAN LUIGI CHIERICI
University of Bologna
Chair of Petroleum Reservoir Engineering
Faculty of Engineering
Viale del Risorgimento, 2
I-40136 Bologna
Italy

Translator

PETER J. WESTAWAY
9 Nile Grove
Edinburgh EH10 4RE, Scotland
Great Britain

Title of the Original Italian edition:
Principi di ingegneria dei giacimenti petroliferi, Vol. 2

ISBN-13: 978-3-642-78245-9   e-ISBN-13: 978-3-642-78243-5
DOI: 10.1007/ 978-3-642-78243-5

Library of Congress Cataloging-in-Publication Data (Revised for vol. 2). Chierici, Gian Luigi. Principles of petroleum reservoir engineering. Includes bibliographical references and indexes. 1. Oil reservoir engineering. 2. Oil fields. 3. Gas reservoirs. I. Title. TN871.C494413 1995 622'.3382 93-26019

This work is subject to copyright. All rights are reserved, whether the whole or part of the material is concerned, specifically the rights of translation, reprinting, reuse of illustrations, recitation, broadcasting, reproduction on microfilm or in any other way, and storage in data banks. Duplication of this publication or parts thereof is permitted only under the provisions of the German Copyright Law of September 9, 1965, in its current version, and permission for use must always be obtained from Springer-Verlag. Violations are liable for prosecution under the German Copyright Law.

© Springer-Verlag Berlin Heidelberg 1995
Softcover reprint of the hardcover 1st edition 1995

The use of general descriptive names, registered names, trademarks, etc. in this publication does not imply, even in the absence of a specific statement, that such names are exempt from the relevant protective laws and regulations and therefore free for general use.

Typesetting: Macmillan India Ltd., Bangalore 25
SPIN: 10039904   32/3130/SPS – 5 4 3 2 1 0 – Printed on acid-free paper

Preface

Volume 1 of this book dealt with the techniques behind the acquisition, processing and interpretation of basic reservoir data. This second volume is devoted to the study, verification and prediction of reservoir behaviour, and methods of increasing productivity and oil recovery.

I should like to bring a few points to the reader's attention.

Firstly, the treatment of immiscible displacement by the method of characteristics. The advantage of this approach is that it brings into evidence the various physical aspects of the process, especially its dependence on the properties of the fluids concerned, and on the velocity of displacement. It was not until after the publication of the first, Italian, edition of this book (February 1990) that I discovered a similar treatment in the book *Enhanced Oil Recovery*, by Larry W. Lake, published in 1989.

Another topic that I should like to bring to the reader's attention is the forecasting of reservoir behaviour by the method of identified models. This original contribution to reservoir engineering is based on systems theory – a science which should, in my opinion, find far wider application, in view of the "black box" nature of reservoirs and their responses to production processes.

Finally, the improvement of productivity and oil recovery factor, to which the last chapter of this volume is devoted. With today's low oil prices (which are expected to persist for several years yet), the use of enhanced oil recovery methods (EOR) is restricted by economic considerations to only the most favourable situations, which tend to be few and far between. I have tried to point out how "improved" recovery methods (IOR) – water or non-miscible gas injection – can, on the other hand, if executed using advanced reservoir management techniques, achieve substantial increases in productivity and recovery within a free market framework.

It is really only in the last decade that reservoir engineering has recognised that truly effective field development demands a full and detailed knowledge of the reservoir's internal structure (its "architecture"). The first phase in building a picture of this internal structure is the construction of the geological (or "static") model, using information from reservoir rock outcrops, and studying the sedimentology, petrography and petrophysics of the reservoir through the synergetic analysis of cores and well logs, and its stratigraphy and lithology from 3D seismic surveys.

But it should not be forgotten that another – and perhaps more important phase – involves the transformation of the static model to a dynamic one, capable of replicating the observed response of the reservoir to the production process. The vital information necessary for this transformation can come only from a proper program of reservoir monitoring and surveillance aimed at determining the distribution of the fluid-dynamic characteristics of the reservoir rock in the interwell regions.

The design of these tests, the interpretation of their results, the modification of the static geological model to optimise the dynamic engineering model and, subsequently, using numerical models to forecast the behaviour of the reservoir under different production scenarios, are all typical activities that occupy the engineer in the domain of reservoir management.

Reservoir management is not, however, a task to be performed in "splendid isolation". Only an approach which combines the expertise of geologists, geophysicists, petrophysicists, geostatisticians, and production and reservoir engineers can have any hope of achieving the level of understanding of internal reservoir structure that is essential for reservoir management to be effective.

Hence the need, already being met in the more enlightened exploration and production companies, of multidisciplinary teams of engineers dedicated to all aspects of the management of the reservoir, from discovery to abandonment.

It would not be proper to finish my introduction without expressing my gratitude to those who have collaborated with me in the preparation of this second volume. Firstly, my thanks to the translator, my friend Peter J. Westaway, who, in addition to producing an impeccable translation, has drawn on his own reservoir engineering experience to suggest improvements to the text. More than a translator, I almost consider him to be the co-author of this book! Once again, thank you Peter.

My thanks also go to Prof. Guido Gottardi, Professor of the Faculty of Engineering at the University of Bologna, and co-lecturer of my course there, whose suggestions for modifications and improvements were so helpful in the preparation of the Italian edition of this book. Prof. Gottardi also provided the computer program that appears in Exercise 12.4. I should also like to thank Mrs. Irma Piacentini and Mr. Carlo Pedrazzini for their meticulous, patient and highly professional work in drawing all of the figures appearing in this volume.

Once again, thank you everybody!

Bologna   GIAN LUIGI CHIERICI
October 1994

Contents

11.1 Calculation of $S_{w,f}$ and $E_{R,o}$ at water breakthrough for three different water/oil viscosity ratios, p. 41. – 11.2 Calculation of $S_{w,f}$, $(f_w)_{BT}$ and E_D for the vertical displacement of oil by water, at three different frontal velocities, p. 44. – 11.3 Calculation of the critical velocity for the displacement of oil by gas, for five different sets of rock and fluid properties. p. 47. – 11.4 Calculation of the oil recovery factor for vertical displacement by gas, for five different velocities of the gas front. p. 48.

12.1 Calculation of f_w as a function of $E_{R,o}$, and $E_{R,o}(t)$ for the injection of water in a direct line drive pattern. p. 98. –
12.2 Calculation of the areal sweep efficiency $E_A(f_w)$ for water injection in direct and staggered line drive patterns, at different values of M_{WO}. p. 101. – 12.3 Calculation of the pseudo-relative permeability curves for a multilayered reservoir, with vertical interlayer communication. p. 104. – 12.4 Program to calculate the pseudo-curves of relative permeability and capillary pressure in segregated and dispersed flow regimes for reservoirs consisting of several layers in vertical communication. p. 107. –
12.5 Calculation of the vertical invasion efficiency curve $E_I(f_w)$ as a function of the fraction of water produced, for a reservoir consisting of several layers not in vertical communication. p. 114. – 12.6 Calculation of the critical oil flow rate for water coning for different penetration ratio and reservoir anisotropy. p. 118. –
12.7 Calculation of the time to water breakthrough, and the subsequent behaviour of f_w in a well producing oil above the critical rate for water coning. p. 119.

13 The Simulation of Reservoir Behaviour Using Numerical Modelling

13.1 Derivation of the continuity equation in radial coordinates.
p. 213. – 13.2 Derivation of an expression for the coefficient α in
Eq. (13.16b) for one- and two-dimensional geometries. p. 214. –
13.3 Derivation of equations to describe the behaviour of a
reservoir containing a moderately volatile oil, and a black oil,
from the generalised compositional equation (Eq. (13.41)] for
polyphasic flow. p. 217. – 13.4 Derivation of Eq. (13.77), for the
z-axis component of the transport term. p. 220. – 13.5 Derivation
of the finite difference equations for three-dimensional polyphasic
flow. p. 222. – 13.6 Expression of the general transport term in
fully implicit form, eliminating non-linearity. p. 226.

14   Forecasting Well and Reservoir Performance
         Through the Use of Decline Curves and Identified Models

15   Techniques for Improving the Oil Recovery

              15.1 Derivation of the equations relating oil recovery at
              breakthrough and N_{gv}, and the Corey n-exponent,
              for vertical non-miscible gas displacement. p. 369.

Most Commonly Used Symbols

Letter Symbol	Quantity	Dimensions
A	area	L^2
$\mathbf{A}$	matrix of coefficients	
b	well penetration from top of reservoir, also: h_w/h_t	L
b_D	dimensionless well penetration	
B_f	formation volume factor of fluid "f" ($f = g, o, w$)	
B_g	gas formation volume factor	
B_{gi}	gas formation volume factor, under initial reservoir conditions	
B_o	oil formation volume factor	
B_{od}	oil formation volume factor, for differential gas liberation	
B_{of}	oil formation volume factor, for flash gas liberation	
B_{oi}	oil formation volume factor, under initial reservoir conditions	
B_w	water (brine) formation volume factor	
c	compressibility	$m^{-1}Lt^2$
c_f	coefficient of porosity variation with pressure (pore compressibility)	$m^{-1}Lt^2$
c_g	gas compressibility	$m^{-1}Lt^2$
c_o	oil compressibility	$m^{-1}Lt^2$
c_t	total compressibility (pore volume + fluids)	$m^{-1}Lt^2$
c_w	water (brine) compressibility	$m^{-1}Lt^2$
C	tracer concentration in the injected water, also:	mL^{-3}
	aquifer constant [Eq. (9.25)]	$m^{-1}L^4t^2$
$C_{i,g}$	mass fraction of component "i" in the gaseous hydrocarbon phase	
$C_{i,o}$	mass fraction of component "i" in the oil phase	
$C_{i,w}$	mass fraction of component "i" in the water phase	
C_o	tracer concentration in the oil phase	mL^{-3}
C_r	specific heat capacity of the rock	$L^2t^{-2}T^{-1}$
C_s	solvent concentration in the miscible phase (volume ratio)	
C_w	tracer concentration in the water phase	mL^{-3}
d	diameter, also:	L
	distance	L
$\mathbf{d}$	column vector of constants	various
D	depth	L
$D(t)$	rate of oil production decline [Eq. (14.1)]	t^{-1}
D_1	longitudinal dispersion coefficient	L^2t^{-1}
D_o	molecular diffusion coefficient	L^2t^{-1}

Letter Symbol	Quantity	Dimensions
E_A	areal sweep efficiency	
E_D	microscopic displacement efficiency	
$(E_D)_{BT}$	microscopic displacement efficiency at breakthrough of displacing fluid	
E_I	vertical (invasion) efficiency	
E_R	recovery factor	
$E_{R,o}$	oil recovery factor	
$E_{R,a}$	final recovery factor at reservoir abandonment	
E_V	volumetric (invasion) efficiency	
$\bar{f}$	average number of shale lenses, per unit thickness of reservoir rock	L^{-1}
$f^*(t)$	signal sampled at time t, in a discrete system	various
f_D	displacing fluid fraction	
f_g	gas fraction	
f_L	Larmor frequency of proton-free precession	t^{-1}
f_o	oil fraction	
f_w	water fraction	
$f_{w,a}$	water fraction in produced liquids, at reservoir abandonment	
F	force	$m\,L\,t^{-2}$
F_R	retardation factor [Eq. (15.4)]	
F_s	fraction of reservoir rock volume consisting of impermeable lenses	
g	acceleration due to gravity	$L\,t^{-2}$
G	gas in place in the reservoir, total initial, under standard conditions	L^3
$G(z)$	transfer function in a discrete system	
GOC	gas/oil contact	L
GOR	gas/oil ratio	
G_p	gas produced, cumulative, under standard conditions	L^3
GWC	gas/water contact	L
GWR	gas/water ratio	
h	thickness, also:	L
	Planck's constant, h $(6.625 \cdot 10^{-34}$ J·s$)$	$m\,L^2\,t^{-1}$
h_c	thickness of capillary transition zone	L
$h_{D,V}$	dimensionless height of water cone	
h_o	oil column thickness at the well	L
h_t	total layer thickness	L
h_V	height of water cone, above static water/oil contact	L
h_w	water column thickness at the well	L
ΔH	specific enthalpy of steam, relative to original reservoir temperature	$L^2\,t^{-2}$
$H_{D,V}$	correlation parameter [Eq. (12.52a)]	
$(H_{D,V})_{BT}$	$H_{D,V}$ at water breakthrough by coning [Eq. (12.53b)]	
$\mathbf{H}_e$	Earth's magnetic field, tesla	$m\,t^{-1}q^{-1}$
H_K	normalized iteration parameter [Eq. (13.96b)]	$m^{-1}L^3t$
$\mathbf{H}_p$	polarizing magnetic field, tesla	$m\,t^{-1}q^{-1}$
$\mathbf{i}$	unit x-direction Cartesian vector	
i_w	injected water rate	$L^3\,t^{-1}$

Letter Symbol	Quantity	Dimensions
$\mathbf{j}$	unit y-direction Cartesian vector	
$J(S)$	Leverett J-function	
J_g	gas productivity index	$m^{-1} L^4 t$
$J_{go}(S_g)$	Leverett J-function, gas–oil systems	
J_1	productivity index of fluid "1"	$m^{-1} L^4 t$
J_o	oil productivity index	$m^{-1} L^4 t$
J_w	water productivity index	$m^{-1} L^4 t$
$J_{wo}(S_w)$	Leverett J-function, water–oil systems	
k	permeability	L^2
$\bar{k}$	average permeability	L^2
$\mathbf{k}$	unit z-direction Cartesian vector	
$[k]$	permeability tensor	L^2
k (superscript)	iteration counter: p^k = pressure value computed at the k-th iteration in the iterative ADI (IADI) method	
k_g	permeability, effective, to gas	L^2
k_h	horizontal permeability	L^2
k_o	permeability, effective, to oil	L^2
k_r	radial permeability, also: relative permeability	L^2
k_{rg}	permeability, relative, to gas	
k_{rg}^*	normalized relative permeability to gas [Eq. (13.124b)]	
k_{ro}	permeability, relative, to oil	
k_{ro}^*	normalized relative permeability to oil [Eqs. (13.123a) and (13.124a)]	
k_{rw}	permeability, relative, to water	
k_{rw}^*	normalized relative permeability to water [Eq. (13.123b)]	
$\bar{k}_{ro}$	pseudo-relative permeability to oil, in a multilayer pay zone	
$\bar{k}_{rw}$	pseudo-relative permeability to water, in a multilayer pay zone	
$k_{rg,or}$	endpoint relative permeability to gas ($S_g = 1 - S_{iw} - S_{or}$)	
$k_{ro,iw}$	endpoint relative permeability to oil ($S_o = 1 - S_{iw}$)	
$k_{rw,or}$	endpoint relative permeability to water ($S_w = 1 - S_{or}$)	
$\bar{k}_{ro,iw}$	endpoint pseudo-relative permeability to oil in a multilayer pay zone	
k_v	vertical permeability	L^2
$\bar{k}_x$	average permeability in the Cartesian x-direction	L^2
$\bar{k}_y$	average permeability in the Cartesian y-direction	L^2
k_w	permeability, effective, to water	L^2
K	thermal diffusivity, also: tracer partition equilibrium constant [Eq. (15.2)]	$L^2 t^{-1}$
K_{ani}	permeability anisotropy coefficient	
K_e	dispersion coefficient in miscible displacement	$L^2 t^{-1}$
K_i	equilibrium flash vaporization ratio for component "i"	
$K_{i,go}$	mass partition coefficient between gas and hydrocarbon liquid phase, for component "i"	
$K_{i,gw}$	mass partition coefficient between gas and water phase, for component "i"	
$\mathbf{l}$	unit radial-direction vector	
L	length	L
$\mathbf{L}$	lower triangular matrix	

Letter Symbol	Quantity	Dimensions
m	mass, also:	m
	real gas pseudo-pressure function, also:	$m\,L^{-1}t^{-3}$
	cementation factor [Archie's Eq. (15.8a)], also:	
	ratio of the initial hydrocarbon pore volume of the gas cap to that of the oil, both under reservoir conditions, also:	
	system order, in a discrete, sampled-data system	
m_{dis}	tracer mass flow rate resulting from dispersion	$m\,t^{-1}$
m_g	mass, gas phase	m
m_i	mass of component "i"	m
m_o	mass, liquid hydrocarbon phase, also:	m
	tracer mass flow rate in absence of dispersion	$m\,t^{-1}$
m_s	mass of injected steam	m
m_{tot}	total mass flow rate of tracer $= m_o + m_{dis}$	$m\,t^{-1}$
m_w	mass of water (brine) phase	m
M	molecular weight (mass)	m
$\mathbf{M}$	nuclear magnetization	$L^{-1}t^{-1}q$
M_{go}	mobility ratio, gas/oil	
M_s	mobility ratio, diffuse-front approximation [Eq. (11.33)]	
M_{wo}	mobility ratio, water/oil	
MMP	minimum miscibility pressure	$m\,L^{-1}t^{-2}$
n	number of components in a system, also:	
	saturation exponent [Archie's Eq. (15.8a)], also:	
	reciprocal of decline-curve exponent [Eq. (14.2a)]	
n (superscript)	time level (p^n means "pressure at time t_n")	
N	oil in place in the reservoir, initial, under standard conditions	L^3
N_{gv}	gravity-to-viscous force ratio [Eq. (15.11)]	
N_p	cumulative oil production, under standard conditions	L^3
N_p^*	cumulative oil production, fraction of mobile oil in place	
$N_{Pé}$	Péclet number [Eq. (12.64)]	
N_{pu}	mobile oil in place $[\,= V_R\,\varphi(1 - S_{iw} - S_{or})]$	L^3
N_r	volume of remaining oil in the reservoir, under standard conditions	L^3
N_{vc}	capillary number (viscous-to-capillary force ratio)	
N_{vg}	gravity number (viscous-to-gravity force ratio)	
N_{vg}^*	modified gravity number [Eq. (15.27)]	
N_x	number of grid blocks in the Cartesian x-direction	
N_y	number of grid blocks in the Cartesian y-direction	
N_z	number of grid blocks in the Cartesian z-direction	
p	pressure	$m\,L^{-1}t^{-2}$
$\bar{p}$	average pressure	$m\,L^{-1}t^{-2}$
p_b	pressure, bubble point (saturation)	$m\,L^{-1}t^{-2}$
p_g	pressure, in the gas phase	$m\,L^{-1}t^{-2}$
p_i	initial pressure, static	$m\,L^{-1}t^{-2}$
$p_{i,j,k}^n$	pressure in model block (i, j, k), at time t^n	$m\,L^{-1}t^{-2}$
$\bar{p}_K$	average pressure in model block K, at well bore	$m\,L^{-1}t^{-2}$
p_o	pressure, in the oil phase	$m\,L^{-1}t^{-2}$
p_{tf}	tubing head pressure, flowing	$m\,L^{-1}t^{-2}$
p_w	pressure, in the water (brine) phase	$m\,L^{-1}t^{-2}$
p_{wf}	bottomhole pressure, flowing	$m\,L^{-1}t^{-2}$
$p_{wf,U}$	bottomhole pressure, flowing, in the uppermost model block flowing into the well	$m\,L^{-1}t^{-2}$

Letter Symbol	Quantity	Dimensions
p_{woc}	pressure, at water/oil contact	$m\,L^{-1}t^{-2}$
p_{ws}	bottomhole pressure, static	$m\,L^{-1}t^{-2}$
P_c	capillary pressure	$m\,L^{-1}t^{-2}$
$\bar{P}_c$	average capillary pressure	$m\,L^{-1}t^{-2}$
P_c^{-1}	inverse function of P_c	
$P_{c,go}$	capillary pressure between oil and gas phases	$m\,L^{-1}t^{-2}$
$P_{c,ow}$	capillary pressure between oil and water phases	$m\,L^{-1}t^{-2}$
q_{crit}	critical (maximum) volumetric flow rate for front stabilization, reservoir conditions	L^3t^{-1}
q_g	volumetric gas flow rate, reservoir conditions	L^3t^{-1}
$q_{g,sc}$	volumetric gas flow rate, standard conditions	L^3t^{-1}
q_i	initial volumetric flow rate	L^3t^{-1}
q_l	volumetric flow rate of fluid "l" ($l = g, o, w$) produced from a model block	L^3t^{-1}
$\tilde{q}_m$	mass rate (source/skin) term for single-phase flow	$m\,t^{-1}$
$\tilde{q}_{m,g}$	mass rate (source/skin) term for flow of gas in equilibrium under standard conditions with oil	$m\,t^{-1}$
$\tilde{q}_{m,o}$	mass rate (source/skin) term for flow of oil in equilibrium under standard conditions with gas	$m\,t^{-1}$
$\tilde{q}_{m,w}$	mass rate (source/skin) term for flow of water	$m\,t^{-1}$
q_o	volumetric oil flow rate, reservoir conditions	L^3t^{-1}
$q_{o,crit}$	critical (maximum) volumetric flow rate of oil, standard conditions, for water coning	L^3t^{-1}
$q_{o,sc}$	volumetric flow rate of oil, standard conditions	L^3t^{-1}
q_{sc}	flow rate, standard conditions	L^3t^{-1}
q_t	volumetric flow rate of total fluid produced, also: volumetric flow rate at time "t"	L^3t^{-1} L^3t^{-1}
$\tilde{q}_v$	volumetric flow rate of fluids produced	L^3t^{-1}
$\tilde{q}_{v,g}$	volumetric flow rate of gas produced, standard conditions	L^3t^{-1}
$\tilde{q}_{v,o}$	volumetric flow rate of oil produced, standard conditions	L^3t^{-1}
$\tilde{q}_{v,w}$	volumetric flow rate of water produced, standard conditions	L^3t^{-1}
q_w	volumetric flow rate of water	L^3t^{-1}
r	radius	L
r_h	radius of reservoir rock heated by the injected steam	L
R	gas constant	$m\,L^2t^{-2}T^{-1}$
R^{k+1}	average of the absolute value of residuals after the $(k+1)$th iteration [Eq. (13.90a)]	
$R_{i,j,k}$	residual in the material balance equation for model block (i, j, k) [Eqs. (13.98a) and (13.111a)]	
$R_{o,i}$	resistivity, formation 100% saturated with injected water of resistivity $R_{w,i}$	$m\,L^3t^{-1}q^{-2}$
ROS	remaining oil saturation	
R_s	(also R_{so}) gas solubility in oil	
R_{sd}	gas solubility in oil, from differential liberation	
R_{sf}	gas solubility in oil, from flash liberation	
R_{si}	gas solubility in oil, at initial pressure	
R_{sw}	gas solubility in water	
$R_{t,R}$	true resistivity of reservoir rock (external to the flushed zone)	$m\,L^3t^{-1}q^{-2}$

Letter Symbol	Quantity	Dimensions
R_T	summation of the absolute value of material balance residuals of all model blocks [Eq. (13.113)]	
R_v	oil solubility in the gas phase	
$R_{w,i}$	resistivity, injected water	$m\,L^3\,t^{-1}\,q^{-2}$
$R_{w,R}$	resistivity, reservoir water	$m\,L^3\,t^{-1}\,q^{-2}$
s	spin quantic number	
S	cross-sectional area, also: skin factor	L^2
S_D	saturation, displacing fluid	
S_f	saturation, fluid "f" ($f = g, o, w$)	
S_g	saturation, gas	
S_g^*	normalized saturation, gas [Eq. (13.124c)]	
S_{gr}	saturation, residual gas	
S_i	saturation, fluid "i"	
S_{iw}	saturation, irreducible (critical) water	
$\bar{S}_{iw}$	average irreducible water saturation	
$\bar{\bar{S}}_{iw}$	average irreducible water saturation in a multilayer pay zone	
S_o	saturation, oil	
S_o^*	normalized saturation, oil [Eq. (15.13b)]	
S_{or}	saturation, residual oil	
$\bar{S}_{or}$	average residual oil saturation	
$\bar{\bar{S}}_{or}$	average residual oil saturation in a multilayer pay zone	
$S_{or,g}$	residual oil saturation after oil displacement by gas	
SOR	steam/oil ratio	
S_w	saturation, water	
S_w^*	normalized saturation, water [Eq. (13.123c)]	
$\bar{S}_w$	average saturation, water	
$\bar{\bar{S}}_w$	average water saturation in a multilayer pay zone	
S_{we}	water saturation at the outlet end of the porous medium	
$S_{w,f}$	water saturation at the displacing fluid front	
S_{wi}	saturation, initial, water	
t	time	t
Δt	time step	t
t_{BT}	time to breakthrough	t
t_D	dimensionless time	
$(t_D)_{BT}$	dimensionless time to water breakthrough in a well, owing to water coning	
T	temperature, also: transmissibility	T $m^{-1}\,L^4\,t$
T_c	temperature, critical	T
$T_{f,s}$	transmissibility to fluid "f" ($f = g, o, w$) in the "s" direction ($s = x, y, z$) in multiphase flow	$m^{-1}\,L^4\,t$
T_r	temperature, reduced	
T_R	temperature, reservoir	T
T_s	single-phase transmissibility in the "s" direction ($s = x, y, z$)	$m^{-1}\,L^4\,t$
T_1	longitudinal spin-lattice relaxation time constant	t
T_2	transversal spin–spin relaxation time constant	t
T_2^*	free-precession signal decay time constant	t

Letter Symbol	Quantity	Dimensions
$\mathbf{u}$	column vector of unknowns	various
$u(t)$	input signal, in a discrete system	various
u_g	Darcy velocity of gas	$L\,t^{-1}$
$\mathbf{u}_i$	vector "Darcy velocity", fluid "i"	$L\,t^{-1}$
u_{ix}	Cartesian x-direction component of vector $\mathbf{u}_i$	$L\,t^{-1}$
u_{iy}	Cartesian y-direction component of vector $\mathbf{u}_i$	$L\,t^{-1}$
u_{iz}	Cartesian z-direction component of vector $\mathbf{u}_i$	$L\,t^{-1}$
u_o	Darcy velocity of oil	$L\,t^{-1}$
$\mathbf{u}_p$	vector "Darcy velocity" of primary tracer (Sect. 15.3.2.2)	$L\,t^{-1}$
u_{st}	critical (maximum) Darcy velocity for miscible displacement under stabilized front conditions	$L\,t^{-1}$
u_t	Darcy velocity of total fluid $= q_t/A$	$L\,t^{-1}$
$(u_t)_{crit}$	critical (maximum) Darcy velocity for front stabilization, reservoir conditions	$L\,t^{-1}$
u_w	Darcy velocity of water (brine)	$L\,t^{-1}$
$\mathbf{u}_w$	vector "Darcy velocity" of water (brine)	$L\,t^{-1}$
u_z	Cartesian z-direction component of Darcy velocity	$L\,t^{-1}$
U	upper triangular matrix	
$U(z)$	z-transform of the input signal, in a discrete system	
v	velocity	$L\,t^{-1}$
$v(S)$	velocity of fluid saturation S moving through the porous medium	$L\,t^{-1}$
v_f	front velocity	$L\,t^{-1}$
$\mathbf{v}_w$	vector displacing water velocity, at microscopic level	$L\,t^{-1}$
V	volume	L^3
V_b	model block volume $\Delta x_i \Delta y_j \Delta z_k$	L^3
V_g	volume of gas phase, reservoir conditions	L^3
V_m	Darcy mass velocity	$m\,L^{-2}\,t^{-1}$
V_o	volume of oil phase, reservoir conditions	L^3
V_p	pore volume	L^3
$\Delta V_{p,i,j,k}$	pore volume in model block (i, j, k)	L^3
V_R	volume, reservoir rock	L^3
V_v	reservoir rock volume flooded by the injected fluid	L^3
V_w	volume, connate water, reservoir conditions	L^3
W	width, also:	L^2
	reservoir water volume, standard conditions	L^3
W_e	encroached water volume, standard conditions	L^3
W_i	cumulative volume of injected water, standard conditions	L^3
W_p	cumulative volume of produced water, standard conditions	L^3
WAR	water/air ratio, in wet combustion	
WC	water cut in well stream	
$(WC)_{lim}$	limiting value for the water cut, well perforated throughout the oil-bearing and water-bearing intervals of reservoir rock (Eq. (12.58)]	
WOC	water-oil contact	L
WOR	water/oil ratio	
$y(t)$	output signal, in a discrete system	various
$Y(z)$	z-transform of the output signal, in a discrete system	

Letter Symbol	Quantity	Dimensions
z	vertical depth below datum, also: gas compressibility factor	L
$\bar{z}$	gas compressibility factor at average pressure $\bar{p}$	
z_D	dimensionless height above the static water/oil contact [Eq. (12.45b)]	
z_{GOC}	vertical depth of the gas/oil contact, below datum	L
z_i	gas compressibility factor at initial reservoir pressure p_i	
$z_{i,j,k}$	vertical depth below datum of the centre of model block (i, j, k)	L
z_{WOC}	vertical depth below datum of the water/oil contact	L

Greek

α	geometrical factor [Eq. (13.16b)]	various
α_1	dispersivity, longitudinal	L
α_{MA}	dispersivity, at macroscopic scale	L
α_{ME}	dispersivity, at megascopic scale	L
β	geometrical factor [Eq. (13.30)]	various
γ	proton gyromagnetic ratio	$m^{-1}q$
$\dot{\gamma}$	shear rate [Eq. (15.34b)]	t^{-1}
γ_g	gas relative density (air = 1.00), standard conditions	
$\delta(t - k\Delta t)$	unit step function, in sampled-data systems	
ε	an infinitesimal, or very small, quantity	various
θ	dip angle of rock layer	
Θ_c	contact angle	
λ_g	gas mobility $= kk_{rg}/\mu_g$	$m^{-1}L^3t$
λ_o	oil mobility $= kk_{ro}/\mu_o$	$m^{-1}L^3t$
λ_w	water mobility $= kk_{rw}/\mu_w$	$m^{-1}L^3t$
μ	viscosity	$mL^{-1}t^{-1}$
μ_{app}	viscosity, apparent, non-Newtonian fluid	$mL^{-1}t^{-1}$
μ_g	viscosity, gas	$mL^{-1}t^{-1}$
μ_o	viscosity, oil	$mL^{-1}t^{-1}$
μ_s	viscosity, solvent, in miscible displacement	$mL^{-1}t^{-1}$
μ_w	viscosity, water	$mL^{-1}t^{-1}$
ρ_b	rock bulk density	mL^{-3}
ρ_f	density, pore fluid	mL^{-3}
ρ_g	density, gas, reservoir conditions	mL^{-3}
$\rho_{g,sc}$	density, gas, standard conditions	mL^{-3}
ρ_i	density, fluid "i"	mL^{-3}
ρ_o	density, oil, reservoir conditions	mL^{-3}
$\rho_{o,sc}$	density, oil, standard conditions	mL^{-3}
ρ_r	density, rock grains	mL^{-3}
ρ_s	density, solvent, in miscible displacement	mL^{-3}
ρ_w	density, water, reservoir conditions	mL^{-3}
$\rho_{w,sc}$	density, water, standard conditions	mL^{-3}

Letter Symbol	Quantity	Dimensions
σ	interfacial tension	$\mathrm{m\,t^{-2}}$
σ_k	iteration parameter, in the iterative alternating direction implicit (IADI) method [Eq. (13.96b)]	
σ_{mo}	interfacial tension, microemulsion/oil	$\mathrm{m\,t^{-2}}$
σ_{mw}	interfacial tension, microemulsion/water	$\mathrm{m\,t^{-2}}$
σ_{og}	interfacial tension, oil/gas	$\mathrm{m\,t^{-2}}$
σ_{vo}	interfacial tension, microemulsion at the invariant point (type III systems)/oil	$\mathrm{m\,t^{-2}}$
σ_{vw}	interfacial tension, microemulsion at the invariant point (type III systems)/water	$\mathrm{m\,t^{-2}}$
σ_{wo}	interfacial tension, water/oil	$\mathrm{m\,t^{-2}}$
$\sum$	neutron capture cross section, per unit volume	$\mathrm{L^{-1}}$
τ	tortuosity, hydraulic, also: shear stress	$\mathrm{m\,L^{-1}t^{-2}}$
τ_D	Coats' vertical equilibrium parameter [Eq. (11.49)]	
ϕ	porosity	
ϕ_f	free fluid porosity, NML	
Φ_D	dimensionless fluid potential [Eq. (12.45a)]	
$\Phi_{D,V}$	Φ_D value at top of water cone	
Φ_e	fluid potential per unit mass, at the external boundary of the drainage area	$\mathrm{L^2 t^{-2}}$
Φ_g	gas potential per unit mass	$\mathrm{L^2 t^{-2}}$
Φ_o	oil potential per unit mass	$\mathrm{L^2 t^{-2}}$
$\Phi_{o,V}$	oil potential per unit mass at top of water cone	$\mathrm{L^2 t^{-2}}$
Φ_w	water potential per unit mass	$\mathrm{L^2 t^{-2}}$
Φ_{wf}	bottomhole flowing potential, per unit mass	$\mathrm{L^2 t^{-2}}$
χ_o	static nuclear spin susceptibility	$\mathrm{m^{-1}L^{-1}q^2}$
ω	iteration parameter, SOR	
Ω	function [Eq. (12.47)] used in calculating the critical oil rate in water coning	

Mathematical Symbols, and Operators

div	divergence: $\operatorname{div}\mathbf{v} = \dfrac{\partial v_x}{\partial x} + \dfrac{\partial v_y}{\partial y} + \dfrac{\partial v_z}{\partial z}$	
F(s)	Laplace transform of $f(t)$	
F(z)	z-transform of sampled signal $f^*(t)$, in a discrete system	
grad	gradient: $\operatorname{grad}\Phi = \dfrac{\partial \Phi}{\partial x}\mathbf{i} + \dfrac{\partial \Phi}{\partial y}\mathbf{j} + \dfrac{\partial \Phi}{\partial z}\mathbf{k}$	
$\mathcal{L}[f(t)]$	Laplace transform of $f(t)$	
z^{-m}	delay operator for m time steps, in a sampled-signal discrete system	
∇	del, or nabla: $\nabla\Phi = \operatorname{grad}\Phi$	
$\nabla\cdot$	divergence: $\nabla\cdot\mathbf{v} = \operatorname{div}\mathbf{v}$	
∇^2	Laplacian: $\nabla^2\Phi = \dfrac{\partial^2 \Phi}{\partial x^2} + \dfrac{\partial^2 \Phi}{\partial y^2} + \dfrac{\partial^2 \Phi}{\partial z^2}$	
$\Delta_t u$	change of quantity "u" over a time step	
$\Delta_s T_s \Delta_s \Phi$	finite difference operator for the net inflow in the s-direction ($s = x, y, z$), single-phase flow (transport term)	

Letter Symbol	Quantity	Dimensions

$\Sigma_s \Delta_s T_s \Delta_s \Phi$	finite difference operator for the net inflow into a model block, single phase flow (transport term)
$\Delta_s T_{f,s}\,(\Delta_s p_f - g\,\rho_f\,\Delta_s D)$	finite difference operator for the net inflow of fluid "f" ($f = o,\,g,\,w$) in the s-direction ($s = x,\,y,\,z$)
$\Sigma_s \Delta_s T_{f,s}(\Delta_s p_f - g\,\rho_f\,\Delta_s D)$	finite difference operator for the net inflow of fluid "f" ($f = g,\,o,\,w$) into a model block, in multiphase flow
$\Delta_s R_s T_{o,s}(\Delta_s p_o - g\,\rho_o\,\Delta_s D)$	finite difference operator for the net inflow in the s-direction ($s = x,\,y,\,z$) of gas in solution in reservoir oil
$\Sigma_s \Delta_s R_s T_{o,s}(\Delta_s p_o - g\,\rho_o\,\Delta_s D)$	finite difference operator for the net inflow into a model block of gas in solution in reservoir oil
Σ_N	summation over all model blocks in a zone, in a lease or in the entire reservoir

Conversion Factors
for the Most Commonly Used Quantities

Quantity	SI Unit (1)	Practical unit (2)	Conversion factor, F (1) = F × (2)	
Space, time, speed				
Length	m	in	2.54*	E − 02
		ft	3.048*	E − 01
		mile (intl)	1.609344	E + 03
Area	m^2	sq in	6.4516*	E − 04
		sq ft	9.290304*	E − 02
		acre	4.046856	E + 03
		ha	1.0*	E + 04
Volume, capacity	m^3	cu ft	2.831685	E − 02
		US gal	3.785412	E − 03
		bbl (42 US gal)	1.589873	E − 01
		acre-ft	1.233489	E + 03
Time	s	min	6.0*	E + 01
		h	3.6*	E + 03
		d	8.64*	E + 04
		yr	3.15576*	E + 07
Speed	m/s	ft/min	5.08*	E − 03
		ft/h	8.466666	E − 05
		ft/day	3.527777	E − 06
		mile/year	5.099703	E − 05
		km/year	3.168809	E − 05
Mass, density, concentration				
Mass	kg	lbm (pound mass)	4.535924	E − 01
		US ton (short)	9.071847	E + 02
		US cwt	4.535924	E + 01
Density	kg/m^3	lbm/ft^3	1.601846	E + 01
		lbm/US gal	1.198264	E + 02
		lbm/bbl	2.853010	E + 00
		°API	(1.415 E + 05)/ (131.5 + API)	
Concentration	kg/m^3	lbm/bbl	2.853010	E + 00
	kg/kg	ppm (wt)	1.0*	E − 06
Pressure, compressibility, temperature				
Pressure	Pa	psi	6.894757	E + 03
		kg/cm^2	9.80665*	E + 04
		atm	1.01325*	E + 05
		mm Hg = torr	1.333224	E + 02

* Exact conversion factor

Quantity	SI Unit (1)	Practical unit (2)	Conversion factor, F (1) = F × (2)	
Pseudo-pressure of real gases	Pa/s	$(psi)^2/cP$	4.753767	E + 10
		$(kg/cm^2)^2/cP$	9.617038	E + 12
Pressure gradient	Pa/m	psi/ft	2.262059	E + 04
		$kg/cm^2 \times m$	9.80665*	E + 04
Compressibility	Pa^{-1}	psi^{-1}	1.450377	E − 04
		cm^2/kg	1.019716	E − 05
Temperature	K	°F	(°F + 459.7)/ 1.8	
		°C	°C + 273.2	
Temperature gradient	K/m	°F/ft	1.822689	E + 00
		°C/m	1.0*	E + 00

Flow rate, productivity index

Quantity	SI Unit (1)	Practical unit (2)	Conversion factor, F (1) = F × (2)	
Flow rate (volume basis)	m^3/s	ft^3/h	7.865791	E − 06
		ft^3/d	3.277413	E − 07
		bbl/h	4.416314	E − 05
		bbl/d	1.840131	E − 06
		m^3/h	2.777778	E − 04
		m^3/d	1.157407	E − 05
Mass flow rate	kg/s	lbm/h	1.259979	E − 04
		lbm/d	5.249912	E − 06
		US ton/h	2.519958	E − 01
		US ton/d	1.049982	E − 02
Gas/oil ratio	m^3/m^3	cu ft/bbl	1.781076	E − 01
Productivity index	$m^3/(s \cdot Pa)$	$bbl/(d \cdot psi)$	2.668884	E − 10
		$m^3/(d \cdot kg.cm^{-2})$	1.180227	E − 10

Transport properties

Quantity	SI Unit (1)	Practical unit (2)	Conversion factor, F (1) = F × (2)	
Permeability	m^2	md	9.869233	E − 16
		D	9.869233	E − 13
Viscosity	Pa·s	μP	1.0*	E − 07
		cP	1.0*	E − 03
Mobility (k k_r/μ)	$m^2/Pa{\cdot}s$	md/cp	9.869233	E − 13
		D/cp	9.869233	E − 10
Diffusivity	m^2/s	ft^2/h	2.58064*	E − 05
		cm^2/s	1.0*	E − 04
Surface tension	N/m	dyne/cm	1.0*	E − 03

11 Immiscible Displacement in Homogeneous Porous Media

11.1 Introduction

This chapter deals with the physics of the displacement of oil in a porous medium by fluids – water or gas – which are not miscible with the oil. In a reservoir engineering context, the term "displacement" describes the process wherein a fluid drives out and replaces the mobile oil contained in the pore spaces of the reservoir rock. The mobile oil in the case of immiscible displacement is that part of the oil which will not remain trapped in the pores as a residual oil saturation S_{or} at the end of the process.

In particular, the displacing fluid can be water (originating from the aquifer, injected into the aquifer, or injected into the reservoir) or gas (from the gas cap, which expands as a result of pressure depletion; or injected into the gas cap or reservoir).

In contrast to what was said in Chap. 10 (Vol. 1) about the material balance equation, the analysis and description of the immiscible displacement process by *analytical* methods are still of fundamental importance for the prediction of reservoir behaviour, despite the ready availability of numerical modelling methods today. More precisely, before a computerised numerical simulation can proceed, the engineer still needs to perform a critical analysis, by non-numerical means, of the displacement processes that are observed, or are considered to be likely, in the reservoir.

The reason for this is quite simple: numerical models can only be run when the reservoir has been discretised — that is, subdivided into blocks. These blocks will have sides of a few tens or, more commonly, a few hundreds, of metres, and are invariably much larger than the scale over which displacement phenomena occur, this being more of the order of pore size, or perhaps centimetres or, at most, a few metres.

An analysis of these processes on their true scale is therefore a prerequisite (especially, as we shall see, to obtain the so-called pseudo-curves of capillary pressures and relative permeabilities). Only in this way can the displacement mechanisms within each grid block of the numerical model be correctly described. This is the well-known problem of "upscaling" the petrophysical properties of the porous medium.

Reservoirs are, of course, three-dimensional. Thickness h, the dimension measured in the z-axis direction, is almost always much smaller than the areal dimensions in the x- and y-directions. Therefore, in the case of *homogeneous reservoir rock*, the study of displacement processes can be quite easily reduced to a two-dimensional one, with the reservoir divided into vertical slices in the x–z, or y–z, plane (Fig. 11.1a).

If the thickness h is extremely small, and is less than the capillary transition zone between the oil and the displacing fluid (see Sect. 3.4.4.4), the system can

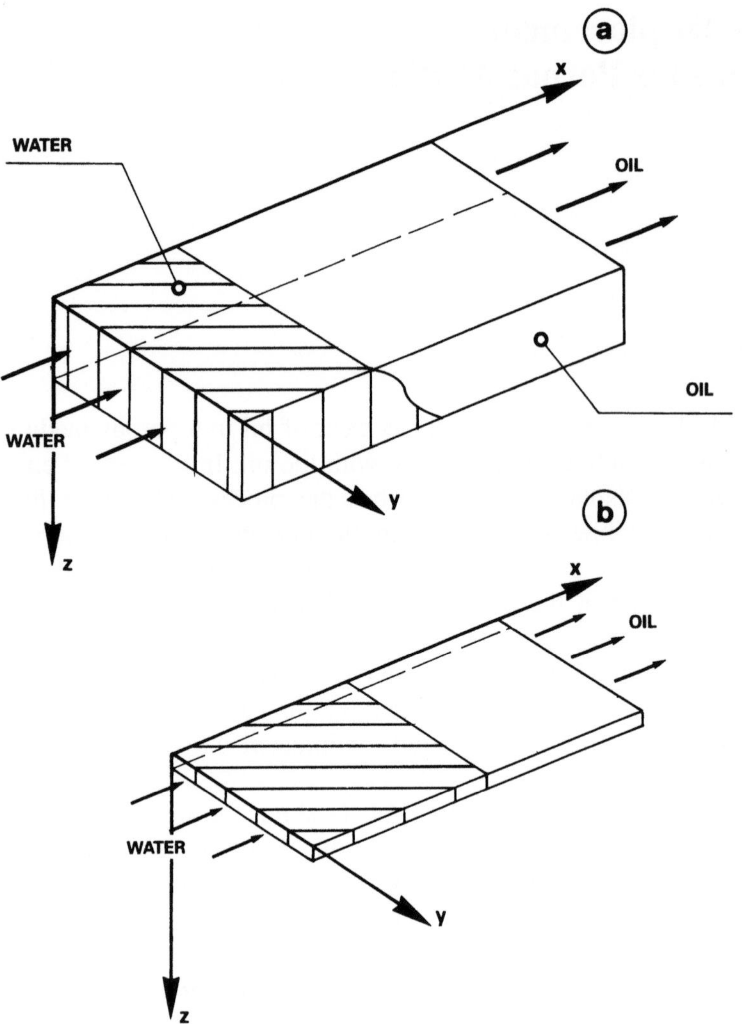

Fig. 11.1. *Displacement of oil by water in a homogeneous isotropic reservoir.* **a** *Two-dimensional and* **b** *One-dimensional representations*

be further reduced to a uni-dimensional one, along the x- or y-axis (Fig. 11.1b). In the following sections we will first look at uni-dimensional systems, and then two-dimensional with the reservoir cut into vertical slices.

Real reservoirs are usually three-dimensional, with petrophysical properties that vary from place to place. These can only be handled through numerical modelling (Chap. 13), but always after having evaluated the displacement process on a local scale by means of equivalent single and two-dimensional systems.

In Chap. 12, we will deal with the special cases among the heterogeneous reservoir types (including multilayered sedimentary systems with different rock properties in each layer) for which an analytical treatment of displacement is still feasible.

11.2 Basic Assumptions

The basic assumptions made when attempting to describe displacement processes are the same for one- and two-dimensional systems:

1. The oil and the displacing fluid are the only mobile fluids in the porous medium, and move in the same direction (unidirectional flow). If a third fluid is present (S_{gr} or S_{iw}), it is not mobile.
2. In the case of the displacement of oil by water, the rock is assumed to be preferentially water wet, so that the displacement is an imbibition process (Sect. 3.4.4.3). The movement of the two fluids is described by their relative permeability curves (Sect. 3.5.2.4), which are functions only of the fluid saturations.
3. In the case of the displacement of oil by gas, the gas is the non-wetting phase, and the displacement is a drainage process (Sect. 3.4.4.3). Once again, the movement of the two fluids is described by their relative permeability curves (Sect. 3.5.2.5), which are functions only of the fluid saturations.
4. The fluids are assumed to be incompressible. In other words, any variation in pressure along the path of the fluids is negligible relative to the average reservoir pressure, with the consequence that ρ_o, ρ_w, and ρ_g can be considered constant.
5. Vertical Equilibrium (VE) is maintained throughout the displacement. This assumes that the effective vertical permeability to the fluids is high enough to allow the instantaneous establishment of gravity/capillary equilibrium between the oil and the displacing fluid. The vertical saturation profile of the oil and displacing fluid will therefore conform to their equilibrium capillary pressure curve (Fig. 11.2) at all times and at all points along the fluid path.

This concept will be dealt with in more detail shortly.

The first of the listed assumptions intentionally excludes the displacement of oil *by gas liberated from the oil itself*, following a decrease in reservoir pressure (solution gas or depletion drive).

This kind of displacement is, in fact, considerably more complex than one driven by gas from an expanding gas cap. Firstly, owing to gravitational forces, the gas migrates towards the top of the reservoir. This is *in the opposite direction to the oil*, which, under the influence of the same gravitational forces, tends to move down towards the lower parts of the reservoir. Relative permeability curves describing this kind of countercurrent flow do not exist.

Secondly, the gravitational segregation of gas from the oil is dependent on the value of the critical gas saturation S_{gr} (Sect. 3.5.2.5), the vertical permeability of the reservoir rock, and the rate at which the gas evolves from solution. This latter in turn depends on the rate at which the reservoir pressure declines.

Solution gas drive can only be described (and even then only approximately) with a three-dimensional numerical model, or at least two-dimensional in the x–z plane.

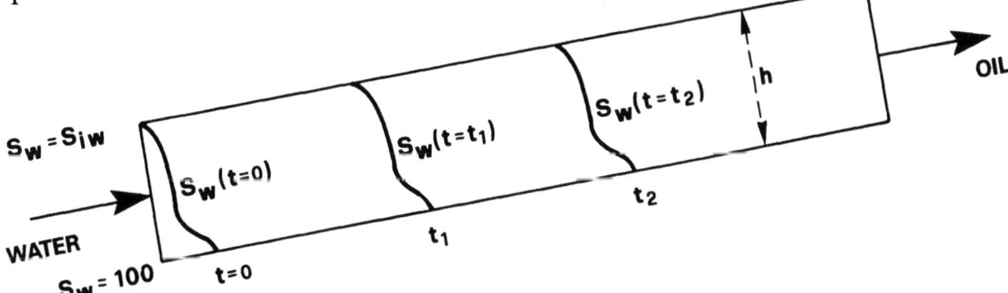

Fig. 11.2. *Movement of the vertical water saturation profile with time in a porous medium in which displacement occurs under conditions of vertical equilibrium (VE)*

11.3 Displacement in One-Dimensional Systems

11.3.1 The Fractional Flow Equation

We will consider a porous medium which is very thin relative to the capillary transition zone (so that it can be represented as a one-dimensional system), in which the oil is being displaced by, for example, water injected at a constant rate into the bottom end.

Adopting the usual convention of z increasing downwards, the x, z coordinate system that conforms to the porous medium will be as shown in Fig. 11.3. For convenience in the analysis that follows, we will define a second coordinate system (X, Z), with the same origin as the (x, z) system but with the X-axis horizontal. θ is the angle between the X- and x-axes.

Starting with the potential, as defined in Eq. (3.47), and the assumption of incompressible formation fluids stated in point 4 of Sect. 11.2, we have:

$$\Phi_o = \frac{p_o}{\rho_o} - gZ = \frac{p_o}{\rho_o} - g \times \sin\theta, \tag{11.1a}$$

$$\Phi_w = \frac{p_w}{\rho_w} - gZ = \frac{p_w}{\rho_w} - g \times \sin\theta. \tag{11.1b}$$

Bringing together Eqs. (3.49) describing polyphasic flow in a porous medium, and the capillary pressure defined in Eq. (3.14), we can write:

$$|u_o| = \frac{q_o}{A} = \frac{kk_{ro}}{\mu_o}\rho_o \,\mathrm{grad}\,\Phi_o$$
$$= \lambda_o \left[\frac{\partial}{\partial x}(p_w + P_c) - \rho_o g \sin\theta \right], \tag{11.2a}$$

$$|u_w| = \frac{q_w}{A} = \frac{kk_{rw}}{\mu_w}\rho_w \,\mathrm{grad}\,\Phi_w$$
$$= \lambda_w \left[\frac{\partial p_w}{\partial x} - \rho_w g \sin\theta \right], \tag{11.2b}$$

where λ is the mobility kk_r/μ [Eq. (3.52)], A is the cross-sectional area of the medium, u_o and u_w are the Darcy velocities of the oil and water phases, and q_o and q_w are their respective flow rates *under reservoir conditions*.

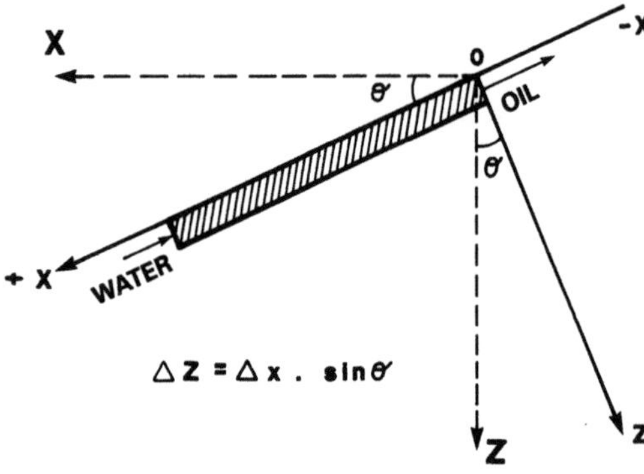

Fig. 11.3. *Coordinate system used to describe the displacement process in a one-dimensional porous medium*

Note that with the configuration shown in Fig. 11.3, under flowing conditions the potential *increases* in the positive x-direction, while the velocities have a *negative sign* because fluid movement is towards the minus x-direction. Hence the use of absolute values of velocity in Eqs. (11.2).

Subtracting[5] Eq. (11.2b) from (11.2a) we have:

$$\frac{|u_o|}{\lambda_o} - \frac{|u_w|}{\lambda_w} = \frac{\partial P_{c,ow}}{\partial x} + (\rho_w - \rho_o)g\sin\theta. \tag{11.3}$$

The total fluid flow rate through the porous medium, q_t, is constant in time because of the assumption of the incompressibility of the oil and water, and is equal to:

$$q_t = q_o + q_w = \text{const.}, \tag{11.4a}$$

so that:

$$|u_t| = \frac{q_t}{A} = |u_o| + |u_w|. \tag{11.4b}$$

We can substitute Eq. (11.4b) into Eq. (11.3) to get:

$$\frac{\lambda_w + \lambda_o}{\lambda_w}|u_w| = |u_t| - \lambda_o\left[\frac{\partial P_{c,ow}}{\partial x} + (\rho_w - \rho_o)g\sin\theta\right] \tag{11.5}$$

We now define f_w as the fraction of water in the total fluids (water + oil) flowing at any point in the porous medium:

$$f_w = \frac{q_w}{q_t} = \frac{q_w}{q_w + q_o} = \frac{u_w}{u_t}, \tag{11.6}$$

which is, from Eq. (11.5),

$$f_w = \frac{q_w}{q_t} = \frac{\lambda_w}{\lambda_w + \lambda_o}\left\{1 - \frac{\lambda_o}{|u_t|}\left[\frac{\partial P_{c,ow}}{\partial x} + (\rho_w - \rho_o)g\sin\theta\right]\right\} \tag{11.7}$$

Equation (11.7) is the *fractional flow equation for water in a porous medium.* The terms $\lambda_w = kk_{rw}/\mu_w$, $\lambda_o = kk_{ro}/\mu_o$ and $\partial P_{c,ow}/\partial x$ are functions of the local water saturation (or oil saturation, since $S_w + S_o = 1$). Consequently, of course, f_w is also function $f_w(S_w)$ *of the local value of* S_w. According to its definition [Eq. (11.6)], *it varies between zero (100% oil) and one (100% water).* A typical $f_w(S_w)$ curve for a medium viscosity oil is shown in Fig. 11.4.

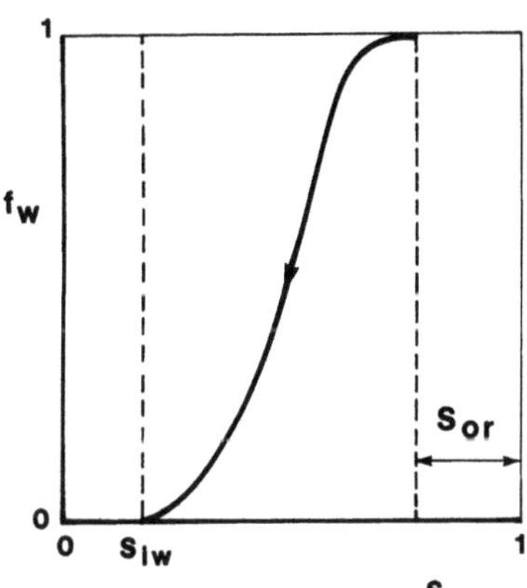

Fig. 11.4. *A typical fractional flow curve* $f_w = f_w(S_w)$

Note that the term $\partial P_{c,ow}/\partial x$ in Eq. (11.7) must be negative. In fact,

$$\frac{\partial P_{c,ow}}{\partial x} = \frac{dP_{c,ow}}{dS_w}\frac{\partial S_w}{\partial x}, \tag{11.8a}$$

where $dP_{c,ow}/dS_w$ is negative, since P_c decreases as S_w increases (Sect. 3.4.4.4), and $\partial S_w/\partial x$ is positive because S_w increases with x (Fig. 11.3). To emphasise this point, Eq. (11.8a) can also be expressed as:

$$\frac{\partial P_{c,ow}}{\partial x} = -\left|\frac{dP_{c,ow}}{dS_w}\right|\frac{\partial S_w}{\partial x}. \tag{11.8b}$$

The Leverett J-function, $J(S_w)$ is defined as:

$$J(S_w) = \frac{P_{c,ow}(S_w)}{\sigma\cos\Theta_c}\sqrt{\frac{k}{\phi}}, \tag{3.22}$$

so that:

$$\left|\frac{dP_{c,ow}}{dS_w}\right| = \sigma\cos\Theta_c\sqrt{\frac{\phi}{k}}\left|\frac{dJ}{dS_w}\right|. \tag{11.9}$$

Now since:

$$\frac{\lambda_w}{\lambda_w + \lambda_o} = \frac{\dfrac{k_{rw}}{\mu_w}}{\dfrac{k_{rw}}{\mu_w} + \dfrac{k_{ro}}{\mu_o}} = \frac{k_{rw}}{k_{rw} + \dfrac{\mu_w}{\mu_o}k_{ro}}, \tag{11.10}$$

from Eqs. (11.7) and (11.10) we get:

$$f_w = \frac{k_{rw}}{k_{rw} + \dfrac{\mu_w}{\mu_o}k_{ro}}\left\{1 - \frac{\lambda_o}{|u_t|}(\rho_w - \rho_o)g\sin\theta + \sqrt{\frac{\varphi}{k}\frac{\lambda_o}{|u_t|}}\sigma\cos\Theta_c\left|\frac{dJ}{dS_w}\right|\frac{\partial S_w}{\partial x}\right\}. \tag{11.11}$$

Equation (11.11) is the fractional flow equation in its most general form. The terms appearing in the brackets on the right-hand side have the following significance:

$$\frac{\lambda_o}{|u_t|}(\rho_w - \rho_o)g\sin\theta$$

Gravitational contribution. This dimensionless term is always positive (since $\rho_w > \rho_o$), and reduces the value of f_w for a given S_w. Note that its magnitude increases the smaller the total velocity u_t, the larger the difference in water and oil densities ($\rho_w - \rho_o$), or the larger the dip angle θ of the strata.

$$\sqrt{\frac{\varphi}{k}\frac{\lambda_o}{u_t}}\sigma\cos\Theta_c\left|\frac{dJ}{dS_w}\right|\frac{\partial S_w}{\partial x}$$

Capillarity contribution. This dimensionless term is always positive, and causes an increase in f_w for any given S_w. It is smaller in magnitude the larger the total velocity u_t, and the smaller the effective interfacial tension $\sigma\cos\Theta_c$. It is not an easy term to quantify, owing to the impossibility of calculating $\partial S_w/\partial x$ analytically. It is common practice to ignore it, on the understanding that there will be an error in the resultant f_w estimate.

If we ignore the gravity and capillary terms in Eq. (11.11) altogether, we are left with the simple expression:

$$f_w = \frac{\lambda_w}{\lambda_w + \lambda_o}$$

$$= \frac{k_{rw}}{k_{rw} + \dfrac{\mu_w}{\mu_o}k_{ro}} = \frac{1}{1 + \dfrac{k_{ro}/\mu_o}{k_{rw}/\mu_w}}. \tag{11.12}$$

Now, since by definition:

$$f_w + f_o = 1 \tag{11.13}$$

substituting for f_w from Eq. (11.11) we have:

$$f_o = 1 - f_w = \frac{k_{ro}}{k_{ro} + \dfrac{\mu_o}{\mu_w}k_{rw}} \left\{ 1 + \frac{\lambda_w}{|u_t|}(\rho_w - \rho_o)g \sin\theta \right.$$

$$\left. - \sqrt{\frac{\varphi}{k}} \frac{\lambda_w}{|u_t|}\sigma \cos\Theta_c \left| \frac{dJ}{dS_w} \right| \frac{\partial S_w}{\partial x} \right\}, \tag{11.14}$$

which, ignoring the gravity and capillary terms, reduces to:

$$f_o = \frac{\lambda_o}{\lambda_w + \lambda_o} = \frac{k_{ro}}{k_{ro} + \dfrac{\mu_o}{\mu_w}k_{rw}} = \frac{1}{1 + \dfrac{k_{rw}/\mu_w}{k_{ro}/\mu_o}}. \tag{11.15}$$

The fractional flow equations for gas (f_g) and oil (f_o), in the case of oil displacement by gas from a gas cap or gas injection, can be derived following the above procedure.

11.3.2 The Buckley-Leverett Displacement Equation

Suppose we have a linear porous medium in which oil, for example, is being displaced by water. The porous medium has a cross-sectional area A and a porosity ϕ. We will consider an element of length dx within the medium, with a local saturation $S_w = S_w(x, t)$.

If q_w is the flow rate of water entering the element, then there will be $[q_w + (\partial q_w/\partial x)dx]$ leaving from the other end. The difference between entry and exit rates accounts for any change in the volume of the water contained in the element.

Assuming incompressible fluids, volumetric balance applied to the element of porous medium over a time interval dt requires that:

(volume of water entering in time dt) $-$ (volume of water leaving in time dt)

$\qquad$ = (change in volume of water in time dt)

This is stated mathematically as:

$$q_w\, dt - \left(q_w + \frac{\partial q_w}{\partial x}dx \right) dt = \frac{\partial}{\partial t}(\phi A\, dx\, S_w)dt. \tag{11.16}$$

As ϕ and A are constant, and $q_w/A = u_w$, so that $u_w/u_t = f_w$, then:

$$-\frac{u_t}{\phi}\frac{\partial f_w}{\partial x} = \frac{\partial S_w}{\partial t}. \tag{11.17}$$

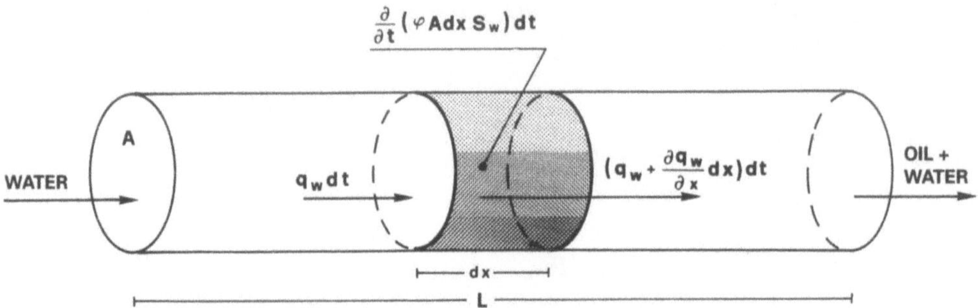

Fig. 11.5. *The displacement, or Buckley-Leverett, equation: volumetric material balance*

Since $f_w = f_w(S_w)$,

$$\frac{\partial f_w}{\partial x} = \frac{\mathrm{d} f_w}{\mathrm{d} S_w} \frac{\partial S_w}{\partial x}. \tag{11.18}$$

Substituting Eq. (11.18) into Eq. (11.17) gives:

$$\frac{u_t}{\phi} \frac{\mathrm{d} f_w}{\mathrm{d} S_w} \frac{\partial S_w}{\partial x} + \frac{\partial S_w}{\partial t} = 0. \tag{11.19a}$$

Equation (11.19a) is the *displacement*, or *Buckley-Leverett*[1] *equation* (Fig. 11.5). This is a first-order partial derivative equation which, when integrated, will give the distribution of water saturation $S_w(x, t)$ in space and time.

Some typical initial and boundary conditions that might be imposed are:

$$S_w = S_{iw} \qquad \text{for } 0 < x \le L, \quad t = 0, \tag{11.19b}$$

$$S_w = 1 - S_{or} \quad \text{for } x = 0, \quad t \ge 0. \tag{11.19c}$$

Equation (11.19) can be solved by the *method of characteristics*[10,11,13], which has the advantage of rendering the physical significance of the solution immediately obvious.

Since $S_w = S_w(x, t)$, the total differential $\mathrm{d}S_w$ will be given by

$$\mathrm{d}S_w = \frac{\partial S_w}{\partial x}\mathrm{d}x + \frac{\partial S_w}{\partial t}\mathrm{d}t. \tag{11.20}$$

We now want to describe the advance with time through the porous medium of a surface defined by a water saturation S_w.

In other words, we want a particular solution $[x(t)_{S_w}]$ to Eq. (11.19) which corresponds to a specified value of S_w.

Because $S_w = $ const., so that $\mathrm{d}S_w = 0$, we can rearrange Eq. (11.20) to:

$$\left(\frac{\mathrm{d}x}{\mathrm{d}t}\right)_{S_w} \frac{\partial S_w}{\partial x} + \frac{\partial S_w}{\partial t} = 0 \tag{11.21}$$

and, combining Eqs. (11.19) and (11.21), we get:

$$\left(\frac{\mathrm{d}x}{\mathrm{d}t}\right)_{S_w} = \frac{u_t}{\phi}\left(\frac{\mathrm{d} f_w}{\mathrm{d} S_w}\right)_{S_w}. \tag{11.22a}$$

Equation (11.22a) is the *equation of characteristics* (or *characteristic equation*) of the displacement Eq. (11.19). It tells us that *any saturation front S_w moves*

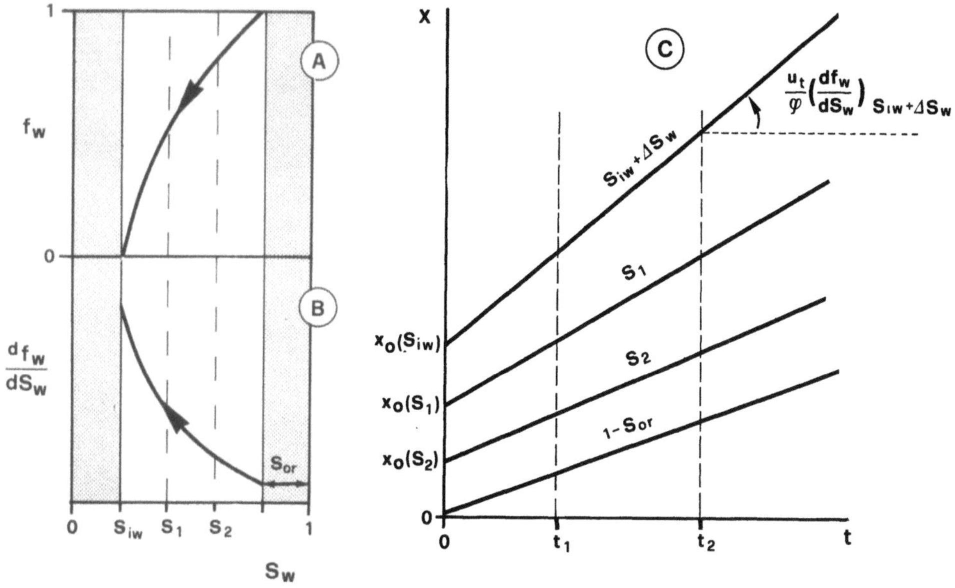

Fig. 11.6. *Displacement of oil by water. Case of a porous medium with an $f_w(S_w)$ curve which is concave downwards (curve **A**), showing the variation of the velocity of frontal advance with water saturation (curve **B**), and characteristics (**C**) which describe the advance of individual saturations S_w with time*

through a porous medium with a constant velocity $v(S_w)$, given by:

$$v(S_w) = \frac{u_t}{\phi}\left(\frac{df_w}{dS_w}\right)_{S_w}. \tag{11.22b}$$

Of course, the initial conditions $x_0(S_w)$ at $t = 0$ should be applied to Eq. (11.22). Therefore, after integrating Eq. (11.22a), we have:

$$[x(S_w)]_t = x_0(S_w) + \frac{u_t}{\phi}\left(\frac{df_w}{dS_w}\right)_{S_w} t. \tag{11.23}$$

The evolution with time of the water saturation in the porous medium is therefore described by a series of straight lines (*characteristics*) with an initial ordinate value $x_0(S_w)$, and slope $(u_t/\phi)(df_w/dS_w)_{S_w}$, as shown in Fig. 11.6.

Equation (11.23) is therefore the general solution to the displacement Eq. (11.19), inasmuch as it permits us to calculate the evolution of S_w in terms of x and t. The behaviour of this function $S_w(x, t)$ depends in turn on the behaviour of $f'(S_w) = df_w(S_w)/dS_w$.

We will now look in detail at three cases which illustrate some interesting aspects of the displacement process.

11.3.2.1 Case 1 – Fractional Flow Curve $f_w(S_w)$ Concave Downwards

Firstly, we shall look at the case where the curve $f_w(S_w)$ is concave downwards, as in Fig. 11.6A: a situation commonly encountered when the oil is very viscous ($\mu_o \gg \mu_w$) and in the absence of gravity effects (horizontal strata, $\theta = 0$).

The gradient $f'(S_w)$ increases progressively from its value at $S_w = 1 - S_{or}$ to a maximum at $S_w = S_{iw} + \Delta S_{iw}$ (Fig. 11.6B). Remember that the initial condition,

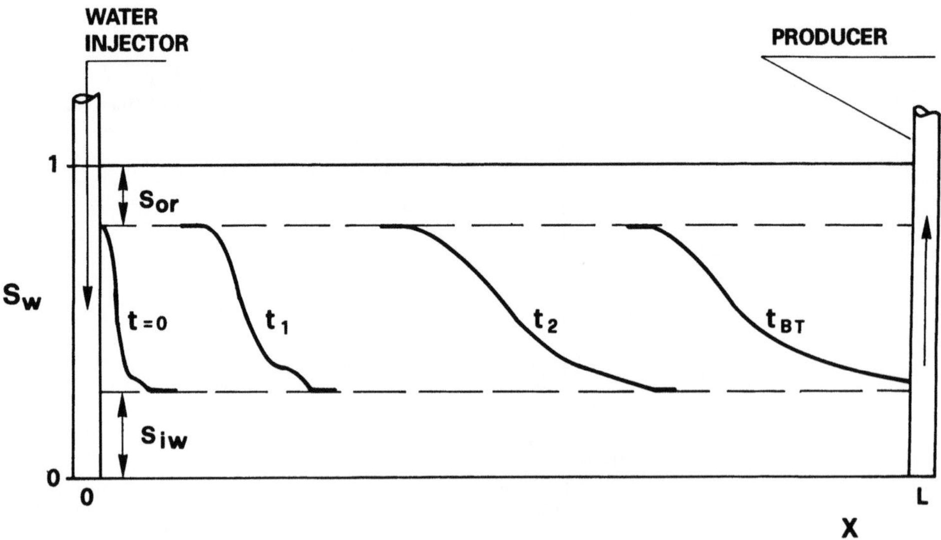

Fig. 11.7. *Evolution of the water saturation profile with time through a porous medium, in the case where the $f_w(S_w)$ curve is concave downwards. Water breakthrough is observed at time t_{BT}, after which an increasing fraction of water is produced*

given by Eq. (11.19c), is that $S_w = 1 - S_{or}$ at $x = 0$ when $t = 0$. The velocity of advance is therefore a maximum where S_w is just greater than S_{iw} (Fig. 11.6C), and decreases steadily towards a minimum where $S_w = 1 - S_{or}$. The progress of the water saturation profile at different times is as shown in Fig. 11.7: S_w at the exit face of the porous medium (producing well) will increase progressively with time, *and with it the fraction f_w of water in the produced fluids*, as soon as $S_w > S_{iw}$. This explains why, in wells producing heavy oil, the produced water cut increases steadily until the well has to be shut down, with, as a consequence, poor oil recovery.

11.3.2.2   Case 2 - Fractional Flow Curve $f_w(S_w)$ Concave Upwards

$f_w(S_w)$ is concave upwards (Fig. 11.8A) when the oil is very light ($\mu_o \ll \mu_w$) and a large gravitational effect causes a significant reduction in f_w [highly dipping strata, very low velocity u_t, high ($\rho_w - \rho_o$) – see Eq. (11.11)].

In this case, the gradient $f'(S_w)$ decreases steadily from its value at $S_w = 1 - S_{or}$ towards $S_w = S_{iw}$ (Fig. 11.8B); in other words, *the highest water saturations move with the highest velocity, with the maximum velocity at $S_w = 1 - S_{or}$. This means that saturations where $S_{iw} < S_w < (1 - S_{or})$ cannot exist in the porous medium, because they would be overtaken by the faster moving $S_w = 1 - S_{or}$* (Fig. 11.8C). Consequently, after a rapid initial buildup of S_w, a *shock front of water saturation $S_w = 1 - S_{or}$ traverses the porous medium, with $S_w = S_{wi}$ ahead of the front and $S_w = 1 - S_{or}$ in the swept region behind it* (Fig. 11.9). The shock front represents a discontinuity in the water saturation. This sort of discontinuity is, incidentally, a common feature in all phenomena involving shock waves.

At the exit face of the porous medium (the producing well), water-free oil will be produced until the shock front arrives. From then on, 100% water will be produced, and the well will have to be shut down. The oil remaining in the formation at the moment of breakthrough will be at the residual oil saturation S_{or}.

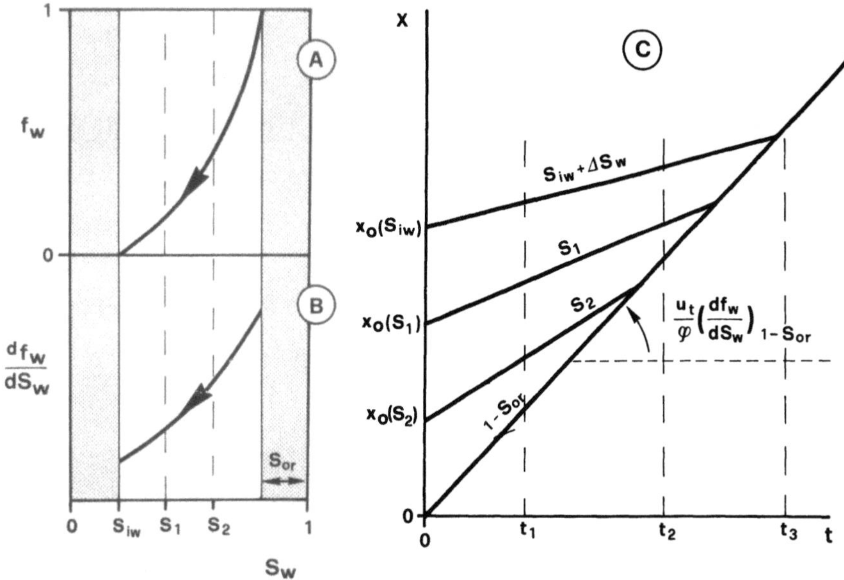

Fig. 11.8. *Displacement of oil by water. Case of a porous medium with an $f_w(S_w)$ curve which is concave upwards (curve **A**), showing the variation of the velocity of frontal advance with water saturation (curve **B**), and characteristics (**C**) which describe the formation of a saturation shock front*

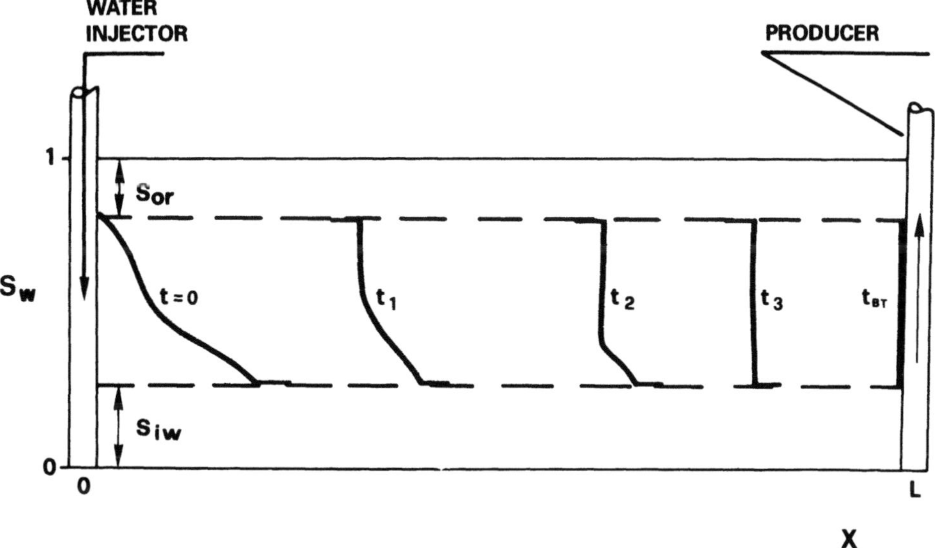

Fig. 11.9. *Evolution of the water saturation profile with time through a porous medium, in the case where the $f_w(S_w)$ curve is concave upwards, showing the formation of a shock front in the water saturation*

This explains the excellent recovery factors obtained from light oil reservoirs, especially where the strata are highly dipping, and where production is at a low enough rate to allow the front to stabilise (see Sect. 11.4.2). Once the front has formed, its velocity can be calculated quite simply by considering the material balance from one side of the front to the other.

If the velocity of the front is $v_f = dx_f/dt$, then, using the subscript L to indicate "behind the front", and R for "ahead", we can write (Fig. 11.10):

$$q_{w,L}\, dt - q_{w,R}\, dt = A\phi(S_{w,L} - S_{w,R})\, dx_f, \tag{11.24}$$

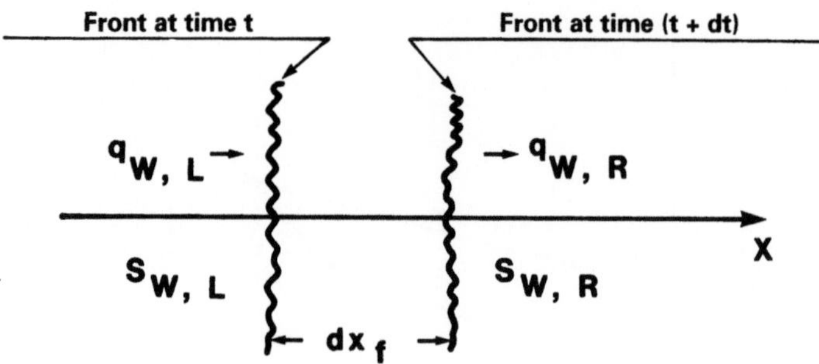

Fig. 11.10. *Water flow rate and saturation ahead of (subscript R) and behind (subscript L) the water front*

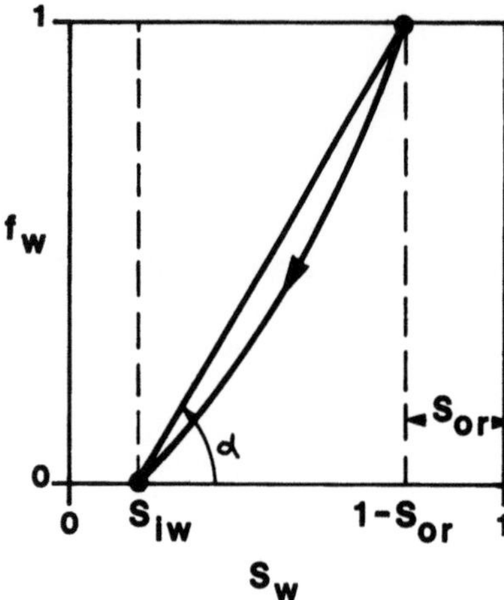

Fig. 11.11. *Calculation of the frontal velocity using the Rankine-Hugoniot condition $v_f = (u_t \tan\alpha)/\phi$*

or, since:

$$q_w = Au_w = Au_t f_w, \tag{11.25}$$

$$v_f = \frac{dx_f}{dt} = \frac{u_t}{\phi} \frac{f_{w,L} - f_{w,R}}{S_{w,L} - S_{w,R}}. \tag{11.26}$$

Equation (11.26) is the well-known *Rankine-Hugoniot condition* governing the frontal velocity in physical situations where shock fronts are observed.

In our case, we have (Fig. 11.11):

$$S_{w,L} = 1 - S_{or},$$

$$f_{w,L} = 1,$$

$$S_{w,R} = S_{iw},$$

$$f_{w,R} = 0.$$

Therefore,

$$\frac{f_{w,L} - f_{w,R}}{S_{w,L} - S_{w,R}} = \frac{1}{1 - S_{or} - S_{iw}} = \tan\alpha, \tag{11.27a}$$

where α is the angle shown in Fig. 11.11 between the S_w axis and the line joining $(S_{iw}, 0)$ and $(1 - S_{or}, 1)$. Consequently,

$$v_f = \frac{u_t}{\phi} \tan\alpha. \tag{11.27b}$$

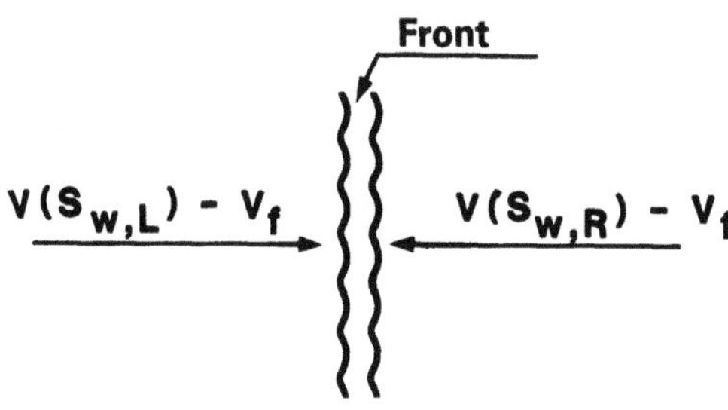

Fig. 11.12. *Schematic of the entropy condition at the front: the velocity of S_w behind the front is greater than its velocity ahead of the front*

The value of v_f obtained from Eq. (11.26) also satisfies the so-called *entropy condition*. In physical terms, this means that in its advance, the front must remain infinitesimally thin, with no tendency to spread out – introducing, in our case, a more gradual change in saturation.

The entropy condition is met if the velocity of the water saturation behind the front, $v(S_{w,L})$, is greater than the velocity v_f of the front itself, and if this in turn is greater than the velocity $v(S_{W,R})$ ahead of the front (see Fig. 11.12):

$$v(S_{w,L}) > v_f > v(S_{w,R}). \tag{11.28a}$$

Substituting from Eq. (11.22b), we have in our particular case (Fig. 11.11):

$$\left(\frac{\mathrm{d}f_w}{\mathrm{d}S_w}\right)_{1-S_{or}} > \tan\alpha > \left(\frac{\mathrm{d}f_w}{\mathrm{d}S_w}\right)_{S_{iw}}, \tag{11.28b}$$

so the entropy condition is automatically satisfied.

The ratio of the water mobility behind the front to the oil mobility ahead of it is called the *mobility ratio*, M_{wo}, defined in Chap. 3 as:

$$M_{wo} = \frac{\lambda_{w,S_{or}}}{\lambda_{o,S_{iw}}} = \frac{k_{rw,or}}{\mu_w} \bigg/ \frac{k_{ro,iw}}{\mu_o}. \tag{3.52}$$

From a mathematical viewpoint, the existence of a discontinuity (the front) in the solution to the displacement equation, as illustrated above, makes this a so-called weak solution.

11.3.2.3 Case 3 – Fractional Flow Curve $f_w(S_w)$ S-Shaped

The S-shaped $f_w(S_w)$ curve (Fig. 11.13A) is typical of an oil of medium density and viscosity, at displacement velocities usually existing in oil reservoirs. The S-shaped form, in other words, is the most diffuse type encountered in reservoir engineering. Because of the double curvature in $f_w(S_w)$, with downward concavity at high S_w, and upward concavity at low S_w, the slope $f'_w(S_w)$ increases and then decreases with decreasing S_w (Fig. 11.13B).

Consequently, moving forward through the porous medium from the point where the water first entered (where the saturation is now $1 - S_{or}$), we would initially observe a steady increase in the velocities of the various saturations as S_w decreases. However, we would observe a maximum velocity at a certain saturation $S_{w,f}$, where $S_{iw} < S_{w,f} < (1 - S_{or})$, beyond which the velocities would decrease with decreasing S_w. The characteristics are of the form illustrated in Fig. 11.13C.

This means that a shock front develops in this case, too. The saturation $S_{w,f}$ at the front moves faster than the saturations $S_w < S_{w,f}$ which lie ahead, and leaves saturations $S_w > S_{w,f}$ in its wake. The evolution of the saturation profile with time

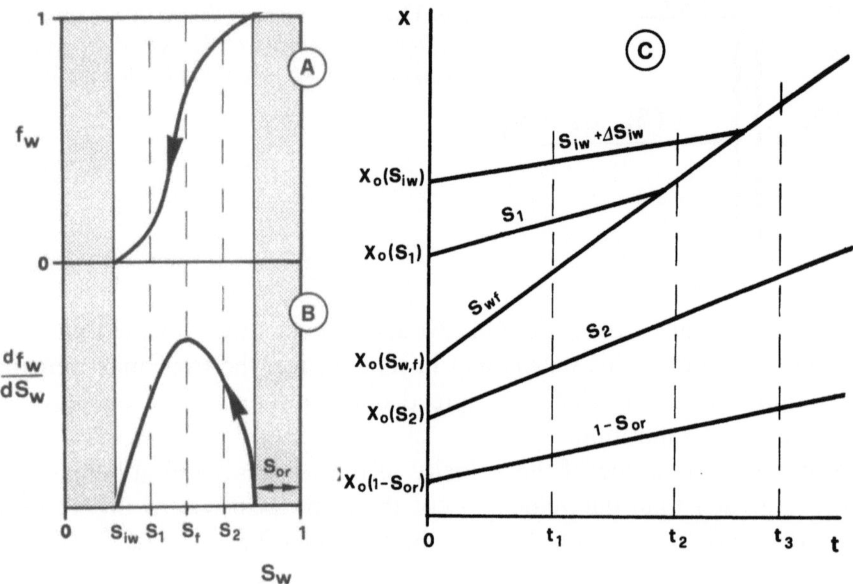

Fig. 11.13. *Displacement of oil by water. Case of a porous medium with an S-shaped $f_w(S_w)$ curve (**A**), showing the variation of the velocity of frontal advance with water saturation (curve **B**), and characteristics (**C**) which describe the advance of individual saturations S_w with time*

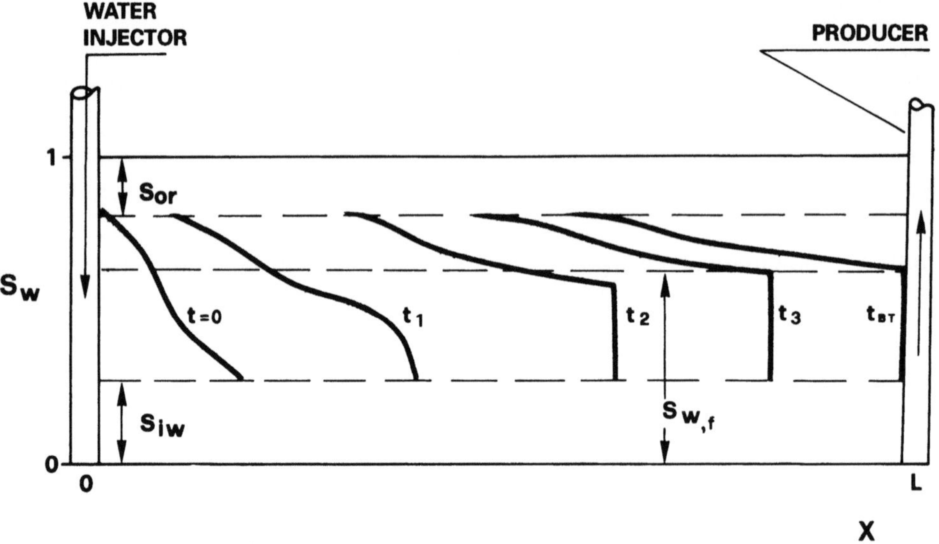

Fig. 11.14. *Evolution of the water saturation profile with time through a porous medium, in the case where the $f_w(S_w)$ curve is S-shaped. Formation of a saturation front followed by a continuous variation in saturation*

through the medium is shown in Fig. 11.14: after an initial transition period, the saturation front $S_{w,f}$ begins to form, travelling at a velocity:

$$v_f = v(S_{w,f}) = \frac{u_t}{\phi}\left(\frac{\mathrm{d}f_w}{\mathrm{d}S_w}\right)_{S_{w,f}}. \tag{11.29}$$

We apply the Rankine-Hugoniot condition [Eq. (11.26)] with:

$$S_{w,L} = S_{w,f},$$

$$f_{w,L} = f_w(S_{w,f}),$$

$$S_{w,R} = S_{iw},$$

$$f_{w,R} = 0.$$

This gives us:

$$v_f = \frac{u_t}{\phi} \frac{f_{w,L} - f_{w,R}}{S_{w,L} - S_{w,R}} = \frac{u_t}{\phi} \frac{f_w(S_{w,f})}{S_{w,f} - S_{iw}}. \tag{11.30}$$

From Eqs. (11.29) and (11.30), we deduce that the front should satisfy the relationship

$$\left(\frac{\mathrm{d}f_w}{\mathrm{d}S_w}\right)_{S_{w,f}} = \frac{f_w(S_{w,f})}{S_{w,f} - S_{iw}}. \tag{11.31a}$$

Equation (11.31a) provides us with a convenient graphic method for determining the water saturation $S_{w,f}$ at the front. This is achieved by *drawing a line through the point* $(S_{iw}, 0)$, *tangential to the curve* $f_w(S_w)$. This can be understood by looking at the triangle AFB in Fig. 11.15, where

$$\tan \beta = \frac{f_w(S_{w,f})}{S_{w,f} - S_{iw}} = \left(\frac{\mathrm{d}f_w}{\mathrm{d}S_w}\right)_{S_{w,f}} \tag{11.31b}$$

Note that Eq. (11.31a) is satisfied at, and only at, $S_{w,f}$. $S_{w,f}$ therefore represents the water saturation at the displacement front.

Ahead of the front, we will find oil in the presence of water at its irreducible saturation S_{iw}. Behind the front, on the other hand, water saturation increases gradually from $S_{w,f}$ towards $(1 - S_{or})$. Because $\partial S_w/\partial x$ is very small here, the capillary term in Eq. (11.11) can be ignored behind the front.

An observer positioned at the exit face of the medium would see only oil produced initially. After a time t_{BT}, corresponding to the time it takes the front to traverse the entire length L of the medium:

$$t_{BT} = \frac{L}{v_f} = \frac{L\phi}{u_t} \frac{S_{w,f} - S_{iw}}{f_w(S_{w,f})}, \tag{11.32}$$

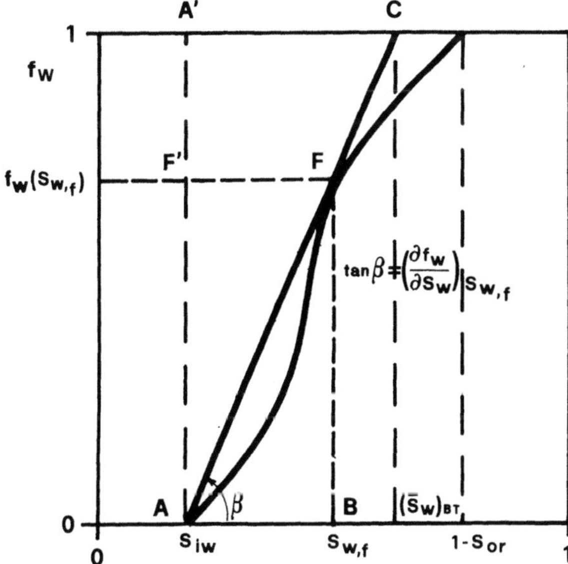

Fig. 11.15. *Graphical method for determining the value of the water saturation* $S_{w,f}$ *at the front, in the case of a porous medium where the* $f_w(S_w)$ *curve is S-shaped*

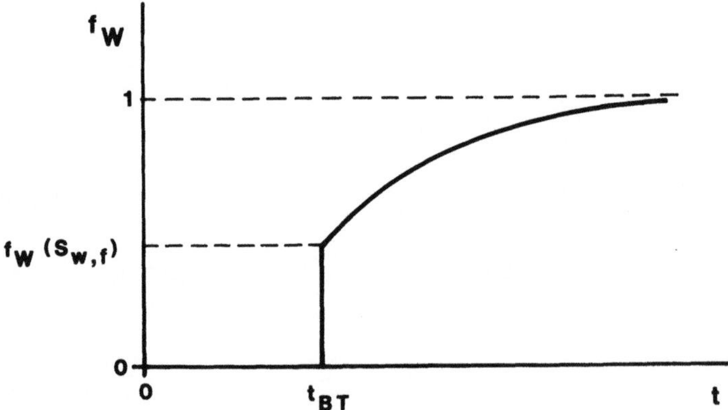

Fig. 11.16. *Variation of the produced water fraction f_w with time in the case of a porous medium with an S-shaped $f_w(S_w)$ curve*

a sudden increase (breakthrough) in water would be seen. The water fraction would then gradually increase with time (Fig. 11.16) as the saturation at the exit face builds up gradually from $S_{w,f}$ to $(1 - S_{or})$.

In the case of an S-shaped curve $f_w(S_w)$, a mobility ratio M_s is defined[6] at the front as:

$$M_s = \frac{k_{ro}(S_{w,f})}{k_{ro,iw}} + \frac{\mu_o}{\mu_w} \frac{k_{rw}(S_{w,f})}{k_{ro,iw}}. \tag{11.33}$$

The way in which the value of M_s determines the degree of instability at the front will be explained later in this chapter.

11.3.3 Calculation of the Average Saturation Behind the Front: Welge's Equation[16]

We will consider a one-dimensional porous medium of length L and cross-sectional area A (Fig. 11.17), with constant petrophysical properties throughout, initially saturated with oil in the presence of water at its irreducible saturation ($S_o = 1 - S_{iw}$), and with an S-shaped fractional flow curve $f_w(S_w)$.

Water is injected at a rate q_w into one end of the porous medium. At the other end, only oil will exit up to time t_{BT}, after which there will be a mixture of oil and an increasing percentage of water produced (Sect. 11.3.2).

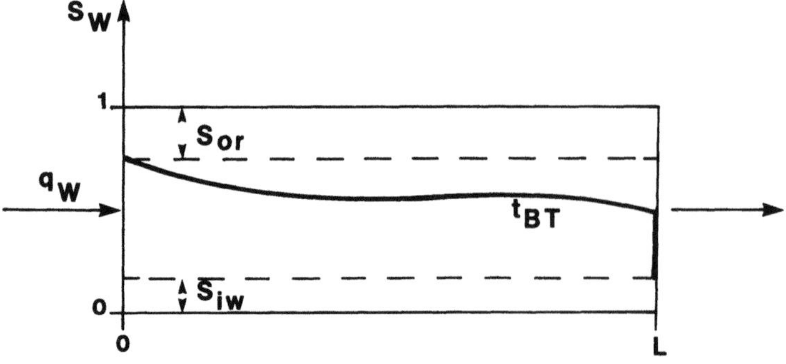

Fig. 11.17. *Water saturation profile along the porous medium at time t_{BT}, when the front breaks through at the exit face*

Since $u_t = q_w/A$, we can write Eq. (11.32) as:

$$t_{BT} = \frac{L}{v_f} = \frac{AL\phi}{q_w} \frac{S_{w,f} - S_{iw}}{f_w(S_{w,f})}.$$

(11.34)

Until water appears at the exit face, the volume N_p of oil produced at time t will equal the volume of water $W_i = q_w t$ injected. Furthermore, at breakthrough, the cumulative volume of oil produced can be expressed quite simply as the difference between the initial volume of oil $AL\phi(1 - S_{iw})$ and the volume currently remaining, $AL\phi[1 - (\bar{S}_w)_{BT}]$, where $(\bar{S}_w)_{BT}$ is the average water saturation in the porous medium at breakthrough. Therefore, we will have:

$$(N_p)_{BT} = q_w t_{BT} = AL\phi \frac{S_{w,f} - S_{iw}}{f_w(S_{w,f})} = AL\phi \left[(\bar{S}_w)_{BT} - S_{iw}\right],$$

(11.35a)

from which:

$$(\bar{S}_w)_{BT} - S_{iw} = \frac{S_{w,f} - S_{iw}}{f_w(S_{w,f})}.$$

(11.35b)

Substituting from Eq. (11.31a), this can also be written as:

$$(\bar{S}_w)_{BT} - S_{iw} = \frac{1}{\left(\dfrac{\mathrm{d}f_w}{\mathrm{d}S_w}\right)_{S_{w,f}}}.$$

(11.35c)

Combining Eqs. (11.35b) and (11.35c):

$$1 - f_w(S_{w,f}) = 1 - \frac{S_{w,f} - S_{iw}}{(\bar{S}_w)_{BT} - S_{iw}}$$

$$= \frac{(\bar{S}_w)_{BT} - S_{w,f}}{(\bar{S}_w)_{BT} - S_{iw}} = \left[(\bar{S}_w)_{BT} - S_{w,f}\right] \left(\frac{\mathrm{d}f_w}{\mathrm{d}S_w}\right)_{S_{w,f}},$$

(11.36a)

which, rearranging, becomes:

$$(\bar{S}_w)_{BT} = S_{w,f} + \frac{1 - f_w(S_{w,f})}{\left(\dfrac{\mathrm{d}f_w}{\mathrm{d}S_w}\right)_{S_{w,f}}}.$$

(11.36b)

This is Welge's equation[16] for the average water saturation at breakthrough.

Note that Eq. (11.35b) suggests a very simple graphic method for evaluating $(\bar{S}_w)_{BT}$.

Referring to Fig. 11.15, since AA'C and AF'F are similar triangles we have:

$$A'C = F'F \frac{AA'}{AF'}$$

In other words,

$$A'C = (S_{w,f} - S_{iw}) \frac{1}{f_w(S_{w,f})} = (\bar{S}_w)_{BT} - S_{iw},$$

(11.35d)

which means that *the tangent to the curve $f_w(S_w)$, drawn from the point $(S_{iw}, 0)$, cuts the line $f_w = 1$ at an x-axis value of $S_w = (\bar{S}_w)_{BT}$.* Thus, $(\bar{S}_w)_{BT}$ can be determined graphically. After breakthrough, S_w at the exit face will increase steadily from $S_{w,f}$

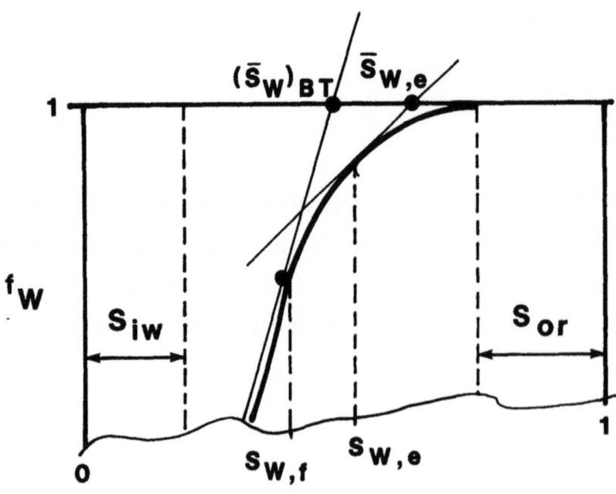

Fig. 11.18. *Graphic method for the determination of the average water saturation $\bar{S}_{w,e}$ in a porous medium when the water saturation $S_{w,e}$ has reached the exit face*

towards a maximum value of $(1 - S_{or})$. Correspondingly, f_w will increase from $f_w(S_{w,f})$ to its maximum $f_w = 1$, from which moment only water will be produced.

Welge[16] demonstrated that the average water saturation $\bar{S}_{w,e}$ in the porous medium when the saturation at the exit face is $S_{w,e}$ is given by an equation identical in form to Eq. (11.36b):

$$\bar{S}_{w,e} = S_{w,e} + \frac{1 - f_w(S_{w,e})}{\left(\dfrac{\mathrm{d}f_w}{\mathrm{d}S_w}\right)_{S_{w,e}}}. \tag{11.37}$$

In addition, the same graphic method can be used to determine $\bar{S}_{w,e}$. Simply draw the tangent to the $f_w(S_w)$ curve at the desired saturation $S_{w,e}$, and read the value of $\bar{S}_{w,e}$ from its intersection with the line $f_w = 1$ (Fig. 11.18).

11.3.4 Calculation of Oil Recovery as a Function of Time, and the Fraction of Displacing Fluid in the Production Stream

Once again we will consider the case of the displacement of oil by water, when the fractional flow curve $f_w(S_w)$ is S-shaped. A similar treatment can be used for the displacement of oil by gas. Obviously, prior to the arrival of the saturation front $S_{w,f}$ at the exit face of the porous medium, only oil will be produced.

The time t_{BT} at which water breakthrough occurs is given by Eq. (11.34), and the cumulative volume $(N_p)_{BT}$ of oil produced at breakthrough (equal to the injected water volume $(W_i)_{BT} = q_i t_{BT}$) by Eq. (11.35).

The fraction $(E_D)_{BT}$ of the initial oil recovered at breakthrough is:

$$(E_D)_{BT} = \frac{(N_p)_{BT}}{AL\phi(1 - S_{iw})} = \frac{(\bar{S}_w)_{BT} - S_{iw}}{1 - S_{iw}}. \tag{11.38}$$

If flushing of the porous medium continues, after a certain time all the mobile oil will have been displaced, and only a residual saturation S_{or} of oil will remain. From this moment on, only water will be produced ($f_w = 1$).

The fraction of oil $(E_D)_{ult}$ recovered at the time that $f_w = 1$ will be:

$$(E_D)_{ult} = \frac{1 - S_{iw} - S_{or}}{1 - S_{iw}} = 1 - \frac{S_{or}}{1 - S_{iw}}. \tag{11.39}$$

Up to this time, obviously $(E_D)_{BT} < E_D < (E_D)_{ult}$ and $0 < f_w < 1$. We will next look at the evolution of f_w and E_D from the moment of breakthrough up to the end of displacement ($f_w = 1$).

Recalling that

$$u_t = \frac{q_i}{A}, \quad q_i = \text{const.}, \tag{11.40}$$

we can use Eq. (11.22) to derive the time $(t)_{S_{w,e}}$ required by the saturation $S_{w,e}$ to travel the entire length of the porous medium:

$$(t)_{S_{w,e}} = \frac{L}{v(S_{w,e})} = \frac{AL\phi}{q_i} \frac{1}{\left(\dfrac{df_w}{dS_w}\right)_{S_{w,e}}}. \tag{11.41}$$

The cumulative injected volume of water $(W_i)_{S_{w,e}}$ when the saturation $S_{w,e}$ reaches the exit face of the medium will be:

$$(W_i)_{S_{w,e}} = q_i(t)_{S_{w,e}}, \tag{11.42}$$

so that, from Eqs. (11.41) and (11.42),

$$(W_i)_{S_{w,e}} = AL\phi \frac{1}{\left(\dfrac{df_w}{dS_w}\right)_{S_{w,e}}}. \tag{11.43}$$

Combining Eqs. (11.37) and (11.43), we have:

$$\bar{S}_{w,e} = S_{w,e} + \left[1 - f_w(S_{w,e})\right] \frac{(W_i)_{S_{w,e}}}{AL\phi}. \tag{11.44}$$

In addition,

$$(N_p)_{S_{w,e}} = AL\phi(\bar{S}_{w,e} - S_{iw}), \tag{11.45}$$

so that:

$$(E_D)_{S_{w,e}} = \frac{\bar{S}_{w,e} - S_{iw}}{1 - S_{iw}}. \tag{11.46}$$

Equations (11.41) to (11.46) constitute the sequence of calculations used to derive the evolution of f_w and E_D with time after water breakthrough, for a displacement process through water injection at a constant rate q_i. Note that, in these equations, the values of q_i, N_p and f_w *are expressed under reservoir conditions.*

In practice, the calculations are performed as follows:

1. Choose a value for $S_{w,e} > S_{w,f}$ (for the first pass, we usually choose $S_{w,e} = S_{w,f} + 0.01$) and read the corresponding values of $f_w(S_{w,e})$ and $(df_w/dS_w)_{S_{w,e}}$ from the curve $f_w(S_w)$.
2. Calculate the time $(t)_{S_{w,e}}$ required for the saturation $S_{w,e}$ to reach the exit face of the porous medium (when $f_w = f_w(S_{w,e})$ from step 1, using Eq. (11.41).
3. Calculate $(W_i)_{S_{w,e}}$ with Eq. (11.42), and then use Eq. (11.44) to estimate the average water saturation $\bar{S}_{w,e}$ in the porous medium at time $(t)_{S_{w,e}}$.
4. Equations. (11.45) and (11.46) will now give, respectively, the cumulative volume of oil produced, $(N_p)_{S_{w,e}}$, and the fractional oil recovery $(E_D)_{S_{w,e}}$, at time $(t)_{S_{w,e}}$.
5. Repeat steps 1 to 4 for increasing values of $S_{w,e}$ up to $S_{w,e} = 1 - S_{or}$.

Note that this procedure *is based on the analytical solution of the displacement equation,* and as such does not suffer from spatial discretisation errors that we will encounter in the case of the simulation of reservoir behaviour through numerical modelling (Chap. 13). As a consequence, values of $N_p(t)$ and $f_w(t)$ obtained by the method described above are used to evaluate the detailed behaviour of two-dimensional reservoir blocks, when the thickness of the blocks is small enough for vertical variations in S_w can be ignored.

11.3.5 The Effect of Oil Viscosity and Flow Rate on Displacement

It will have become apparent from the discussions in the previous sections that the recovery at breakthrough, and the production behaviour thereafter, depend to a very large extent on the shape of the curve $f_D(S_D)$, where the subscript D signifies the displacing fluid ($D = $ w for water, $D = $ g for gas). In the case of displacement by water, Eq. (11.11) is the full statement of $f_w(S_w)$.

As seen in Sect. 11.3.2.3, $\partial S_w/\partial x$ is zero ahead of the front $S_w = S_{w,f}$, because $S_w = S_{iw} = $ constant; and is extremely small behind it, owing to the very gradual variation of S_w with x. To a first approximation, therefore, we can ignore the capillary term in Eq. (11.11). In addition, the gravitational term is usually very small because $(\rho_w - \rho_o)$ is small. With the range of values of u_t typically encountered in the reservoir, this term is, for all practical purposes, negligible, and is in fact zero in a horizontal stratum ($\theta = 0$).

Equation (11.11) therefore can be simplified to Eq. (11.12):

$$f_w = \frac{k_{rw}}{k_{rw} + \dfrac{\mu_w}{\mu_o} k_{ro}} \tag{11.12}$$

This states that, for a porous medium with a given set of relative permeability curves $k_{rw}(S_w)$ and $k_{ro}(S_w)$, $f_w(S_w)$ depends on the viscosity ratio μ_o/μ_w. The form of this dependence is illustrated in Fig. 11.19, for a real porous medium.

As can be seen, as μ_o/μ_w increases (which means, in fact, as μ_o increases), the curvature of $f_w(S_w)$ changes from being strongly concave upwards (Case 2,

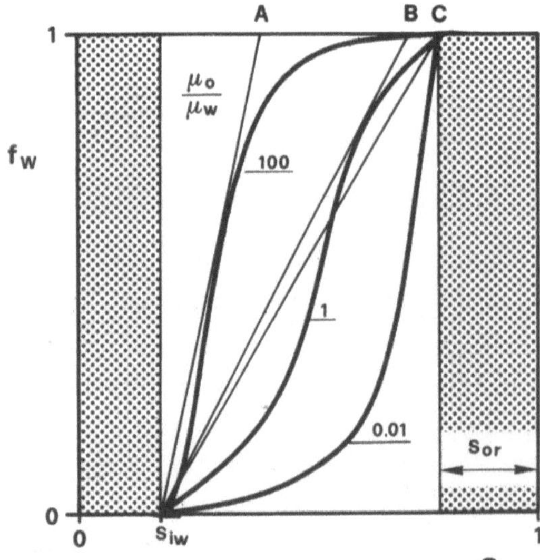

Fig. 11.19. *Displacement of oil by water. Variation of $f_w(S_w)$ and the recovery factor at water breakthrough, with the oil/water viscosity ratio μ_o/μ_w*

Sect. 11.3.2.2), to being S-shaped (Case 3, Sect. 11.3.2.3), and then concave downwards (Case 1, Sect. 11.3.2.1). Consequently, *the amount of oil recovered at breakthrough decreases as μ_o/μ_w increases, and the period during which oil is produced with a steadily increasing water fraction (f_w increasing) is extended in time.*

The separation of the water from the produced oil, and its disposal, are part of the costs of the operation. In the case of a very viscous oil, separating out the water content can be extremely difficult if, as sometimes occurs, a stable emulsion forms. Even disposal of the water itself may not be without problems; for example, all traces of oil must be removed if it is to be reinjected into the formation. In order to improve this situation, either with the objective of increased oil recovery at breakthrough, or reduced treatment and disposal costs, it will be necessary to reduce the ratio μ_o/μ_w in the reservoir.

This can be achieved in two ways:

1. *Very dense, viscous oil*: increasing the reservoir temperature by means of steam injection or partial combustion of the oil. The technology behind these methods is described in Chap. 15. Although both μ_o and μ_w decrease as a result of the heating, there is a net reduction in their ratio, owing to the greater sensitivity of $\mu_o(T)$ to temperature.
2. *Medium viscosity oil*: increasing the viscosity of the injected water by the addition of appropriate polymers. This technique, too, is described in Chap. 15.

In the case of the displacement of oil by gas, ignoring the capillary term, the function $f_g(S_g)$ is expressed as

$$f_g = \frac{k_{rg}}{k_{rg} + \dfrac{\mu_g}{\mu_o} k_{ro}} \left\{ 1 - \frac{\lambda_o}{u_t}(\rho_o - \rho_g)g \sin\theta \right\}. \tag{11.47a}$$

This is easily derived by following the procedure outlined in Sect. 11.3.1, with gas as the displacing fluid.

All the conclusions developed in Sects. 11.3.2 and 11.3.4 for oil/water apply equally to the oil/gas displacement process described by Eq. (11.47a).

Owing to the low viscosity of gas, we will always have $\mu_g \ll \mu_o$. If we make the reasonable approximation that $\mu_g/\mu_o \approx 0$, we can simplify Eq. (11.47a) to:

$$f_g = 1 - \frac{k}{u_t}\frac{k_{ro}(S_g)}{\mu_o}(\rho_o - \rho_g)g \sin\theta. \tag{11.47b}$$

For every value of u_t there will be an S_g value, S_g^*, such that:

$$\frac{k}{u_t}\frac{k_{ro}(S_g^*)}{\mu_o}(\rho_o - \rho_g)g \sin\theta = 1. \tag{11.48a}$$

Now, since k_{ro} increases as S_g decreases (Sect. 3.5.2.5), this means that:

$$f_g = 0 \quad \text{for } S_g \leq S_g^* \tag{11.48b}$$

(f_g cannot, of course, go negative).

In order to satisfy Eq. (11.48a), S_g^* will have to increase for smaller values of u_t. Consequently, as u_t decreases, the $f_g(S_g)$ curve moves further to the right, and changes from being concave downwards (Case 1, Sect. 11.3.2.1), to being S-shaped (Case 3, Sect. 11.3.2.3), to concave upwards (Case 2, Sect. 11.3.2.2).

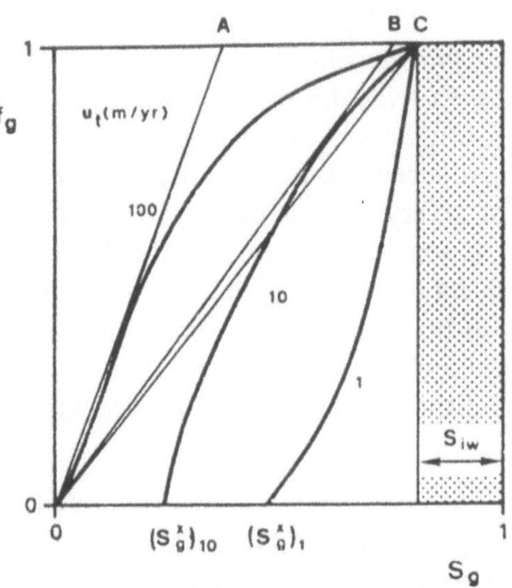

Fig. 11.20. *Displacement of oil by gas. Variation of the recovery factor at gas breakthrough with the displacement velocity u_t*

The percentage of oil recovered at gas breakthrough is therefore improved by reducing u_t, as illustrated in Fig. 11.20. When the $f_g(S_g)$ curve is concave upwards, all the oil is recovered before the gas breaks through.

In Eq. (11.48a), note that, for a given rock at a given value of S_g^* [and therefore of $k_{ro}(S_g^*)$], u_t will be higher, the larger the permeability k, the angle of dip θ of the strata, and the difference in oil and gas densities $(\rho_o - \rho_g)$, or the lower the oil viscosity μ_o. The combination of a velocity u_t (and, therefore, oil production rate) large enough to meet the economic requirements of the field, and a good recovery at breakthrough (high S_g^*), will only be attained when the oil is very light (low viscosity μ_o), in high permeability strata with vertical or subvertical displacement $(\theta \approx 90°)$. In such circumstances, the oil recovery can be very high[9], and almost as good as that obtained via miscible displacement (see Chap. 15).

The comments made about the effect of gravity forces are, strictly speaking, also valid for oil displacement by water, but with two fundamental differences:

1. $(\rho_w - \rho_o) \ll (\rho_o - \rho_g)$: therefore, for a given k and θ, the gravitational effect is much less where water is concerned.
2. in the imbibition process, S_{or} is always much higher than in drainage (where some consider it to be absolutely zero).

11.3.6 Frontal Instability: Fingering

Reservoir rock, although it may be homogeneous on a macroscopic scale, will always contain local heterogeneities of permeability on a scale of millimetres or centimetres. The rock will certainly be intrinsically heterogeneous on the scale of the pores.

In Fig. 11.21, we consider a rock which is globally homogeneous, but contains a thin streak of higher-than-average permeability. The porous medium is initially saturated with oil in the presence of water at S_{iw}, and the oil is to be displaced by water. When the water/oil front – which we will assume has already formed upstream – reaches the high permeability streak, it will tend to advance more rapidly along it than through the rest of the medium. There are two distinct cases to consider.

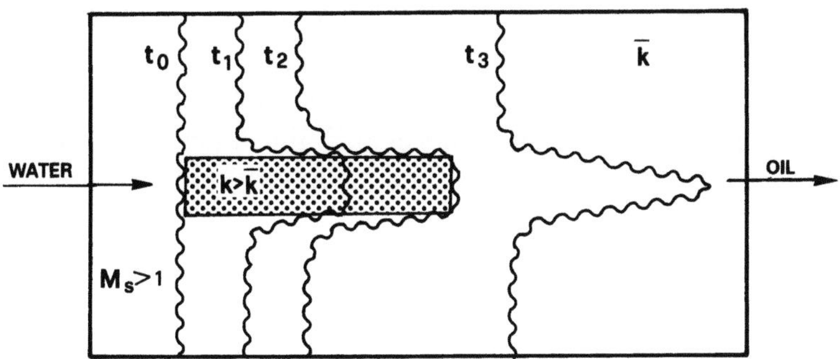

Fig. 11.21. *How water starts to finger along a thin layer of high permeability* ($M_s > 1$)

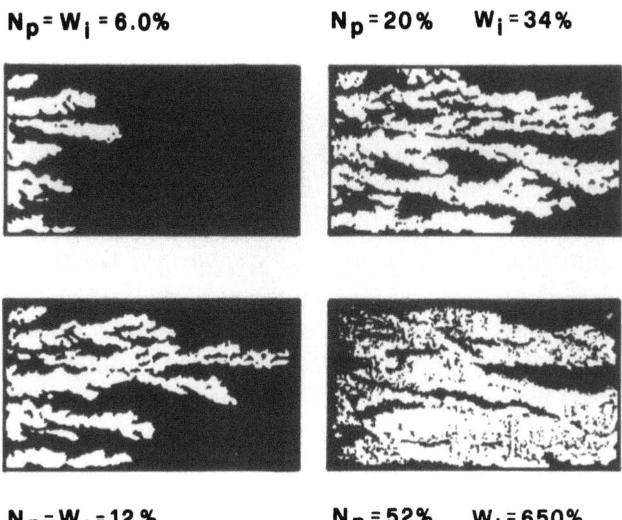

Fig. 11.22. *The fingering process observed in a laboratory sample of porous medium, apparently homogeneous, with $\mu_o/\mu_w = 80$. (From Ref. 15 1958, Society of Petroleum Engineers of AIME. Reprinted with permission of the SPE)*

1. The mobility $k_{rw}(S_{w,f})/\mu_w$ of the water behind the front is higher than that of the oil $k_{ro}(S_{iw})/\mu_o$ ahead. The displacement of oil by water in the high permeability streak steadily reduces the resistance to local frontal advance. The phenomenon tends to be self-enhancing, to the extent that when the water/oil front has travelled the entire length of the streak, it looks like a finger of water within the oil.
2. The mobility of the water behind the front is less than that of the oil ahead. In this case, resistance to frontal advance in the streak increases as the front advances. The process therefore tends to stifle itself, and no water fingering will occur within the oil.

Looking at the phenomenon of fingering on a more complex level, we would find that it is in fact an interplay between numerous parallel fingers of water which form along the front, spreading laterally and merging by diffusion and dispersion. Detailed treatments of these phenomena are to be found in the classic works of van Meurs et al[15], Chuoke et al.[2] and Rachford[12]. Hagoort[8] demonstrated that the presence or absence of fingering at the front depended on the local mobility ratio M_s, defined in Eq. (11.33) for oil/water.

If $M_s > 1$, fingering will occur; if $M_s < 1$, it will not.

Roughly speaking, this means that there will be a risk of fingering whenever the oil is more viscous than the fluid displacing it. The phenomenon is therefore very common when oil is being displaced by gas, because of its very low viscosity.

Figure 11.22, taken from the work of van Meurs et al.[15], illustrates the evolution of fingers of water during the displacement of oil in a laboratory sample of porous medium, where $\mu_o/\mu_w = 80$. Where fingering is extensive, it is no longer realistic to speak of a displacement "front" as such, but rather of a region where the *average* water saturation varies between S_{iw} and $(1 - S_{w,f})$, while the actual saturation at any point is *either* S_{iw} *or* $(1 - S_{w,f})$.

11.4　Displacement in a Two-Dimensional System (x, z)

We will now consider a two-dimensional system in the (x, z) plane, representing a vertical section of the reservoir.

Provided certain assumptions are valid, this two-dimensional system can be reduced to an equivalent one-dimensional system of identical behaviour. This has obvious advantages from a reservoir engineering point of view: the ability to model, in a single dimension, phenomena that actually occur over the vertical thickness $(x, z$ plane) of the reservoir, allows us to address the three-dimensional case of a real reservoir by means of an equivalent *two-dimensional system in a horizontal plane* (x, y).

The advantage of this is readily appreciated when reservoir behaviour is studied through numerical modelling (Chap. 13). Instead of a three-dimensional (3D) model of the reservoir, a much simpler 2D model can be used – as long as certain conditions (described shortly) are satisfied. This represents a significant saving in memory requirement and computing time. Of course, in the 2D model, every point in the (x, y) plane will need to have parameters and functions which define the flow behaviour in the third (z) dimension. The material in the rest of this chapter should be considered as an introduction to the numerical modelling of reservoir behaviour.

The treatment of a two-dimensional system (x, z), whose vertical thickness is a lot less than that of the capillary transition zone (Sect. 11.1), as a one-dimensional equivalent system has already been dealt with in Sect. 11.3. In this case, it is reasonable to assume that the saturation, relative permeability and capillary pressure at any point are constant over the thickness of the formation. At a position $x(t)$, the saturation profile will be a constant in the vertical direction. The flow regime here is referred to as *dispersed*.

On the other hand, when the formation thickness is greater than the capillary transition zone, we cannot ignore the effects of gravity and capillarity on the vertical distribution of the saturation, which will no longer be constant. At any position $x(t)$, there will be a vertical saturation profile $S(z, t)_x$.

The following sections will describe the conditions under which it is still possible to use a one-dimensional equivalent system to describe displacement in a vertical plane (x, z).

11.4.1　Segregated Flow – the Concept of Vertical Equilibrium (VE)

The notion of a *segregated flow regime*, although not expressed in so many words, was proposed for the first time by Terwilliger et al.[14] in their work on gravity displacement, published in 1951. It was developed as the underlying hypothesis in the classic work of Dietz[7] on stabilised flow, in which he states: "The water encroachment on a monoclinal flank can be studied in a representative cross section of a field and the problem is therefore reduced to one of two dimensions. A sharp interface, rather than a transition zone, is assumed between the oil-bearing and the flooded part of the formation. No pressure drop is assumed across the interface, although a constant one might have been introduced without affecting results." To generalise the discussion to include displacement by gas, we will refer to "the displacing fluid" (meaning water or gas), and indicate its characteristics with the subscript D.

In a segregated flow regime, then, we assume that, ahead of the interface between the displacing fluid and the oil, oil alone is flowing, in the presence of immobile water at its irreducible saturation. Behind the interface, only displacing fluid is flowing, in the presence of immobile oil at its residual saturation.

For example, in the case of displacement by water we will have:

	Saturations		Mobilities	
	Water	Oil	Water	Oil
Behind the front	$1 - S_{or}$	S_{or}	$kk_{rw,or}/\mu_w$	zero
Ahead of the front	S_{iw}	$1 - S_{iw}$	zero	$kk_{ro,iw}/\mu_o$

The process can therefore be regarded as a "piston" of displacing fluid pushing a quantity of oil $(1 - S_{iw} - S_{or})$ ahead of it, and leaving behind it a quantity S_{or}. For this reason, segregated flow is often referred to as "piston-like displacement". Physically, at the interface of the displacing fluid and the oil, there must be a capillary transition zone (in the present case assumed to be of small extent relative to the layer thickness). We can therefore visualise a vertical saturation profile, corresponding to capillary equilibrium, moving forward parallel to itself, with oil at its residual saturation S_{or} behind it and oil at saturation $(1 - S_{iw})$ in front. Figure 11.23 illustrates this for the case of oil displacement by water. Using the terminology introduced by Coats et al.[3] in 1967, we say that, in the case of segregated flow, *vertical equilibrium* (VE) exists.

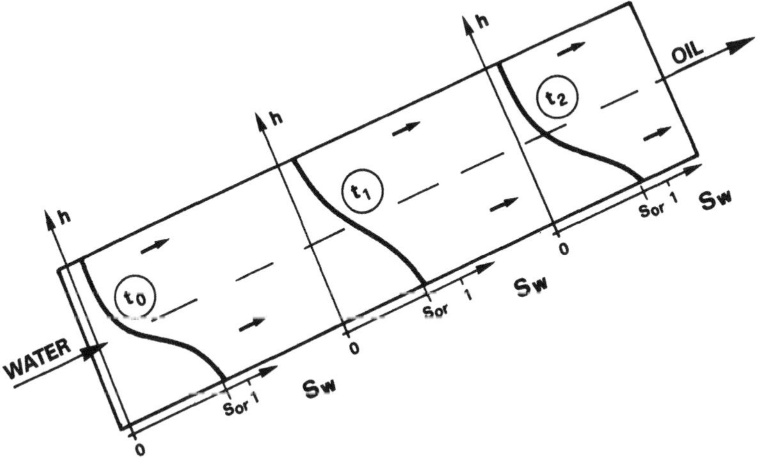

Fig. 11.23. *Piston-like displacement in the presence of a capillary transition zone*

Qualitatively, VE exists when the following conditions prevail:

- high vertical permeability k_v,
- large density contrast $|\rho_D - \rho_o|$ between the displacing fluid and oil.
  This condition is especially true when gas is the displacing fluid, where
  $(\rho_o - \rho_g) \gg (\rho_w - \rho_o)$
- strong capillary forces,
- low oil viscosity,
- layer thickness not too large,
- low fluid velocity, and therefore low viscous forces.

Under such conditions, the ratio of viscous and capillary forces:

$$N_{vc} = \frac{u\mu_o}{\sigma \cos \Theta_c} \tag{3.53b}$$

and of viscous and gravity forces:

$$N_{vg} = \frac{u\mu_o}{k_o g |\rho_D - \rho_o|} \tag{3.53a}$$

are both very small. The fact that capillary and gravity forces predominate over viscous means that a continuous, instantaneous, readjustment of the displacing fluid saturation (according to the capillary pressure curve) can occur at the front as it advances through the oil. As has already been suggested, this can also be thought of as a fixed saturation profile, corresponding to the initial equilibrium saturation distribution $S_D(z)$ across the thickness of the reservoir, moving forward through the porous medium as displacement proceeds.

Coats et al[3] proposed a criterion that must be satisfied for VE conditions to occur in the case of co-current immiscible displacement. This is based on the ratio of the "stabilisation time" (the time required for the displacing fluid saturation at the front to establish static equilibrium, as defined by the capillary pressure curve) and the time taken for the front to travel a distance equal to the thickness of the reservoir. This ratio is expressed in dimensionless form as:

$$\tau_D = \frac{q_D h}{A \left| \dfrac{dP_c}{dS_D} \right| \dfrac{\lambda_D \lambda_o}{\lambda_D + \lambda_o}}, \tag{11.49}$$

where q_D is the flow rate of the displacing fluid, h the thickness of the reservoir and A its cross-sectional area, and the terms $|dP_c/dS_D|$, λ_D and λ_o are evaluated at the mean saturation $\bar{S}_D$ at the front. The smaller τ_D is, the more rapid will be the readjustment of S_D at the front, and the more probable that VE will be established.

Coats et al[4] proposed a different VE criterion for the displacement of oil by a countercurrent movement of gas evolving from within the oil as the pressure declines in a reservoir under depletion drive.

The most commonly used, and most reliable, method of verifying whether the flow in a reservoir is taking place under VE conditions is to model the flow occurring in a representative vertical section of the reservoir using a numerical simulator (Chap. 13). In essence, the method involves comparing results obtained by two different numerical models.

The first model (Fig. 11.24A) is one-dimensional in the x-direction, there being no vertical discretisation, so that there is only one grid block in the z-direction. The fluid dynamic properties of the porous medium are modelled by means of

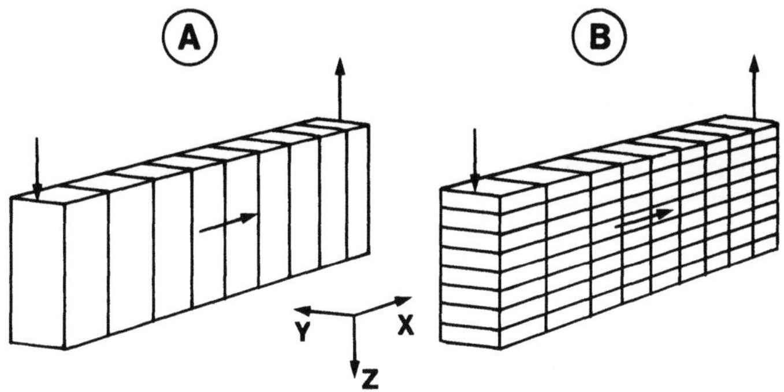

Fig. 11.24. *Schematics of the one-dimensional model (**A**), and two-dimensional model with discretisation in the vertical direction (**B**), used to verify whether displacement will occur under conditions of vertical equilibrium (VE). (From Ref. 6. Reprinted with permission of Elsevier Science Publishers and Prof. L.P. Dake)*

pseudo-curves of relative permeability and capillary pressure, calculated assuming VE by a procedure which will be explained in Sect. 11.5.

The second model (Fig. 11.24B) is two-dimensional in the (x, z) directions. The thickness of the layer is now discretised as a sufficiently large number of blocks, and the fluid dynamic properties are represented by the true relative permeability and capillary pressure curves of the reservoir rock, which are the same for all blocks.

If the two models give results (fluid rates as a function of time, percentage oil recovery versus volume of fluid injected, etc.) which agree closely, the displacement is in vertical equilibrium, and the oil/displacing fluid flow is segregated. If this is the case, the reservoir, which is in reality three-dimensional, can be studied by means of a 2D model in which, at every point (x, y), we assume that VE conditions exist. Pseudo-curves must of course be used to model the fluid dynamic properties of the rock.

If the results from the two models differ significantly, the displacement is not happening under VE conditions. Then an examination of the distribution of S_D with time across the thickness of the reservoir, computed with the 2D model (Fig. 11.24B), will enable us to ascertain if the flow regime is *dispersed* (Sect. 11.3), or perhaps only *partially segregated*.

In the former case, the reservoir can still be modelled in two dimensions, using the actual relative permeability and capillary pressure curves of the rock. In the latter case, the reservoir can only be studied effectively using a full 3D numerical model.

11.4.2 Gravity Stabilisation of the Front in Segregated Flow

We consider here a displacement process occurring under VE conditions – that is, with the displacement fluid/oil in the segregated flow regime. If the displacement velocity is low enough, every part of the front will advance at the same speed. Consequently, the inclination of the front relative to the direction of fluid movement will remain unchanged as it advances through the porous medium (Fig. 11.25a). The force of gravity due to the difference in fluid densities $\Delta\rho$ has time to stabilise the shape of the slow-moving front: the displacement in this case is referred to as *"gravity stabilised"*.

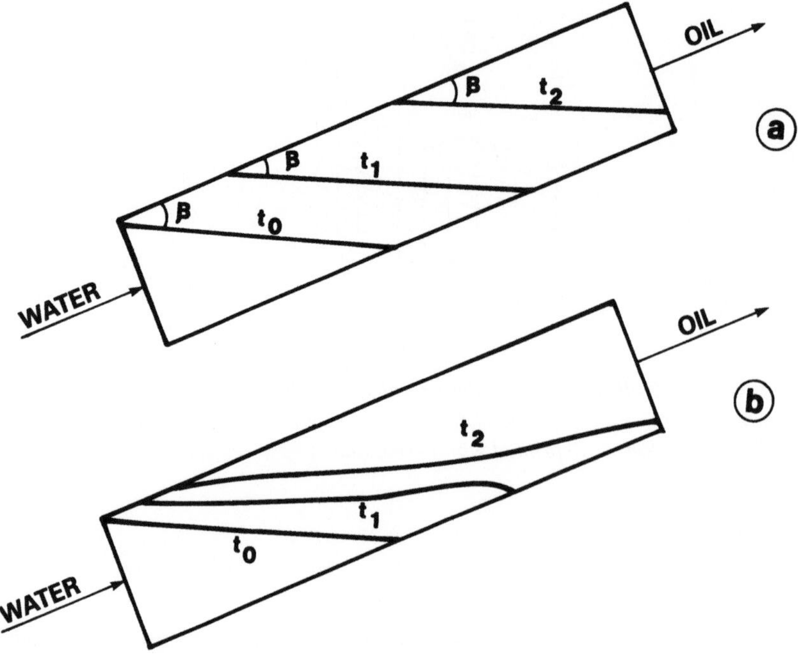

Fig. 11.25. *Advance of the water front in the case of displacement stabilised* (**a**) *and not stabilised* (**b**) *by gravity*

On the other hand, if the displacement velocity is high, the viscous forces associated with fluid movement are more important than gravitational forces, and cause the front to develop a tongue of water which advances beneath the oil, along the base of the layer (Fig. 11.25b), or a tongue of gas along the top of the layer, advancing over the oil. This sort of displacement is *"non-stabilised"*.

Dietz[7] defined the conditions for the gravity stabilisation of a displacement front in vertical equilibrium, and the following discussion has been adapted from his classic work on the subject, with reference to the "piston-like" displacement of oil by water.

Consider a vertical section (x, z) of a homogeneous and isotropic porous medium (Fig. 11.26), in which water is displacing oil in vertical equilibrium with a velocity $u_o = u_w = u_t$, the two fluids being assumed incompressible. θ is the angle between the direction of flow and the horizontal. The configuration of the cartesian axes is shown in Fig. 11.26, with the convention that z increases downwards. Note that the flow is in the *negative x-direction* (u_t is negative). If D is the depth (also increasing downwards), we have

$$dD = \sin\theta \, dx = \cos\theta \, dz \tag{11.50}$$

We now define a circuit ABCD as shown in the figure: A and C are situated right on the water/oil contact (WOC), B lies in the region where only water flows (in the presence of S_{or}), while D lies in the region where only oil flows (in the presence of S_{iw}).

Along path ABC:

$$p_C = p_A + (p_B - p_A) + (p_C - p_B) \tag{11.51a}$$

and, along path ADC:

$$p_C = p_A + (p_C - p_D) + (p_D - p_A). \tag{11.51b}$$

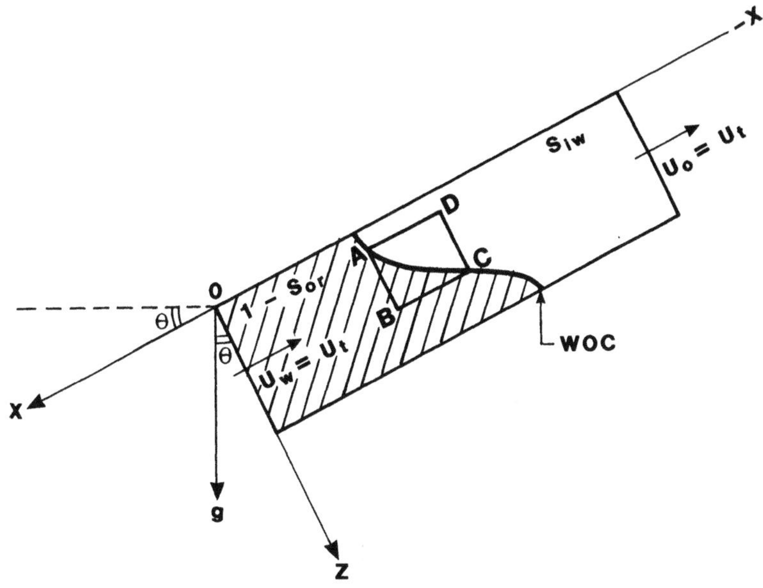

Fig. 11.26. *A gravity-stabilised displacement front, showing the geometry used to determine the conditions for stabilisation*

Referring to Eq. (11.50), we will therefore have:

$$p_B - p_A = \rho_w g(D_B - D_A) = \rho_w g \cos\theta(z_B - z_A) \tag{11.52a}$$

and

$$p_C - p_D = \rho_o g(D_C - D_D) = \rho_o g \cos\theta(z_C - z_D). \tag{11.52b}$$

The water and oil potentials are defined as:

$$\Phi_w = \frac{p_w}{\rho_w} - gD, \qquad \Phi_o = \frac{p_o}{\rho_o} - gD \tag{11.53a}$$

and the generalised Darcy equation for polyphasic flow is:

$$u = -\frac{kk_r}{\mu}\rho\frac{\partial\Phi}{\partial x}, \tag{11.53b}$$

Equations (11.50), (11.53a) and (11.53b) can be combined to give:

$$-u_t = \frac{kk_{rw,or}}{\mu_w}\left(\frac{\partial p_w}{\partial x} - g\rho_w\sin\theta\right)$$

$$= \frac{kk_{ro,iw}}{\mu_o}\left(\frac{\partial p_o}{\partial x} - g\rho_o\sin\theta\right), \tag{11.54a}$$

from which we derive:

$$\frac{\partial p_w}{\partial x} = g\rho_w\sin\theta - \frac{u_t\mu_w}{kk_{rw,or}} \tag{11.54b}$$

and

$$\frac{\partial p_o}{\partial x} = g\rho_o\sin\theta - \frac{u_t\mu_o}{kk_{ro,iw}}. \tag{11.54c}$$

Therefore,

$$p_C - p_B = \left(\rho_w g\sin\theta - \frac{u_t\mu_w}{kk_{rw,or}}\right)(x_C - x_B) \tag{11.54d}$$

and

$$p_{\mathrm{D}} - p_{\mathrm{A}} = \left(\rho_{\mathrm{o}} g \sin\theta - \frac{u_{\mathrm{t}}\mu_{\mathrm{o}}}{kk_{\mathrm{ro,iw}}} \right) (x_{\mathrm{D}} - x_{\mathrm{A}}). \tag{11.54e}$$

Now, because

$$x_{\mathrm{C}} - x_{\mathrm{B}} = x_{\mathrm{D}} - x_{\mathrm{A}}$$

and

$$z_{\mathrm{B}} - z_{\mathrm{A}} = z_{\mathrm{C}} - z_{\mathrm{D}}$$

we can, in a few easy steps, derive the following relationship from Eqs. (11.52) and (11.54):

$$\frac{k(\rho_{\mathrm{w}} - \rho_{\mathrm{o}})g \sin\theta}{u_{\mathrm{t}}} \left[\cot\theta \frac{z_{\mathrm{B}} - z_{\mathrm{A}}}{x_{\mathrm{C}} - x_{\mathrm{B}}} + 1 \right] = \frac{\mu_{\mathrm{w}}}{k_{\mathrm{rw,or}}} - \frac{\mu_{\mathrm{o}}}{k_{\mathrm{ro,iw}}}$$

$$= \frac{\mu_{\mathrm{w}}}{k_{\mathrm{rw,or}}}(1 - M_{\mathrm{wo}}), \tag{11.55}$$

remembering that:

$$M_{\mathrm{wo}} = \frac{k_{\mathrm{rw,or}}/\mu_{\mathrm{w}}}{k_{\mathrm{ro,iw}}/\mu_{\mathrm{o}}}.$$

If we now bring points A and C closer and closer together, until AC becomes infinitesimal, we have:

$$\lim_{\mathrm{AC}\to 0} \frac{z_{\mathrm{B}} - z_{\mathrm{A}}}{x_{\mathrm{C}} - x_{\mathrm{B}}} = \left(\frac{\mathrm{d}z}{\mathrm{d}x} \right)_{\mathrm{f}} = \tan\beta,$$

where $(\mathrm{d}z/\mathrm{d}x)_{\mathrm{f}}$ is the inclination of the WOC at the point under consideration $(\mathrm{A} \equiv \mathrm{C}$, Fig. 11.27), and β is the angle between the tangent at this point and the x-axis.

Note that *in order for the front to be stable*:

$$\left(\frac{\mathrm{d}z}{\mathrm{d}x} \right)_{\mathrm{f}} = \tan\beta \leq 0.$$

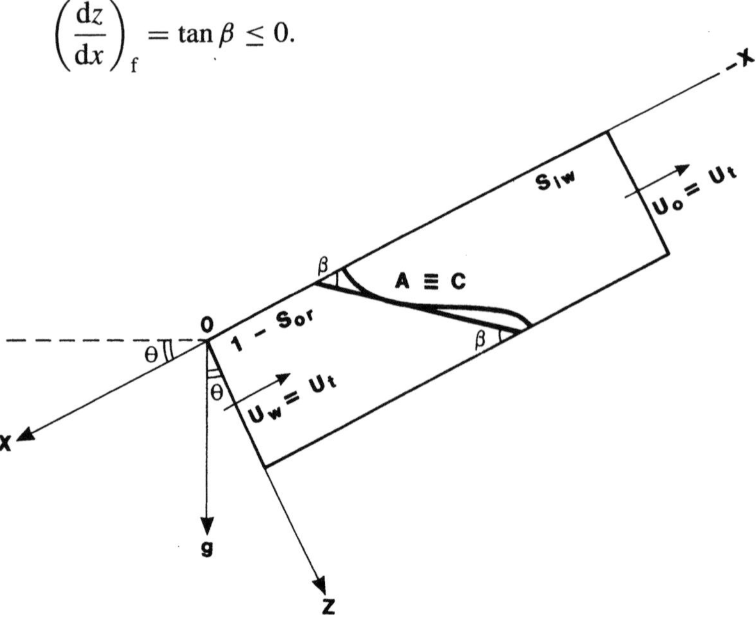

Fig. 11.27. *Relationship between the dip angle θ of the layer and the inclination β of the front at any point on the WOC, when the front is stabilised by gravity*

If this were not the case, the depth z of each point on the WOC would *decrease* as the front advanced (x becoming more negative), and an ever-thickening tongue of water would form along the base of the layer.

By introducing the notation:

$$G = -\frac{kk_{\text{rw,or}}}{\mu_{\text{w}}u_{\text{t}}}(\rho_{\text{w}} - \rho_{\text{o}})g \sin\theta, \tag{11.56}$$

where G *is positive* because u_{t} is negative and all the other parameters are positive, we can rewrite Eq. (11.55) as

$$M_{\text{wo}} - 1 = G\left[1 + \cot\theta\left(\frac{dz}{dx}\right)_{\text{f}}\right]. \tag{11.57a}$$

Alternatively, after rearranging:

$$\cot\theta\left(\frac{dz}{dx}\right)_{\text{f}} = \frac{M_{\text{wo}} - 1 - G}{G}, \tag{11.57b}$$

where both G and $\cot\theta$ are positive.

The condition for frontal stability:

$$\tan\beta = \left(\frac{dz}{dx}\right)_{\text{f}} < 0$$

can therefore be stated as:

$$[M_{\text{wo}} - 1 - G] < 0$$

or:

$$[G - (M_{\text{wo}} - 1)] > 0 \tag{11.58}$$

Equation (11.58) is an expression of the condition for frontal stability in the segregated displacement of oil by water.

The stability of the front therefore depends on M_{wo} and G.

We can distinguish three cases:

a) $M_{\text{wo}} < 1$ – *The front will always be stable, regardless of G.*
    From Eq. (11.57b) we have

$$\frac{\tan\beta}{\tan\theta} < -1$$

    so that $|\beta| > \theta$.
    This is the situation illustrated in Fig. 11.28a.

b) $M_{\text{wo}} = 1$ – *The front is also stable in this case, regardless of G.*
    From Eq. (11.57b) we have

$$\frac{\tan\beta}{\tan\theta} = -1$$

    so that $|\beta| = \theta$.
    The water/oil front will remain horizontal during the displacement process (Fig. 11.28b).

c) $M_{\text{wo}} > 1$ – *The front will only be stable if*

$$G > (M_{\text{wo}} - 1) \tag{11.59}$$

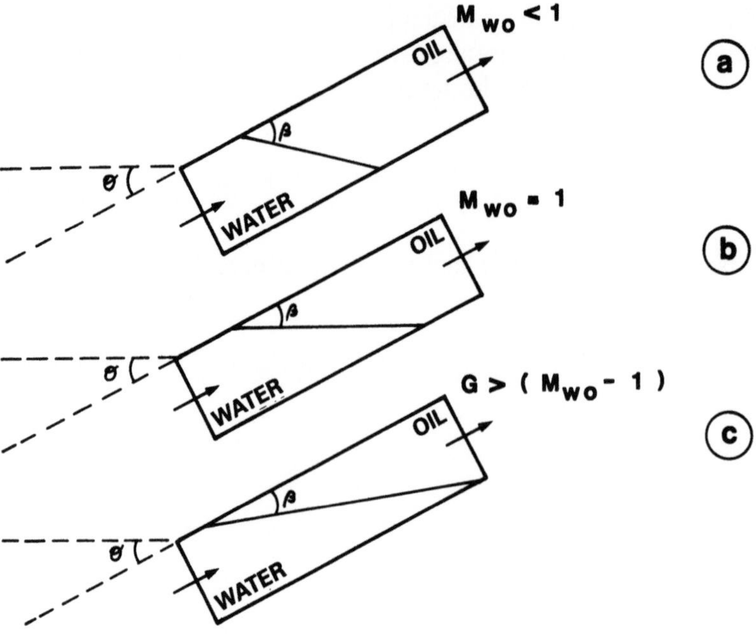

Fig. 11.28a–c. *Inclination of the water/oil front under three different conditions of stabilisation [Eq. (11.57b)]*

From Eqs. (11.56) and (11.59) we can deduce that the maximum frontal velocity (and therefore the flow rate) possible for gravity stabilisation still to occur is given by:

$$|(u_\mathrm{t})_\mathrm{crit}| = \frac{q_\mathrm{crit}}{A} = \frac{k k_\mathrm{rw,or}}{\mu_\mathrm{w}(M_\mathrm{wo} - 1)}(\rho_\mathrm{w} - \rho_o)g \sin \theta. \tag{11.60}$$

q_crit is defined as *"the critical flow rate for front stabilisation"*. When $q \geq q_\mathrm{crit}$, the front becomes unstable, and the water advances as a tongue beneath the oil (Fig. 11.25b).

To summarise, *when oil is displaced by water under conditions of vertical equilibrium (segregated flow), the front will always be stabilised by gravity when the water mobility is less than, or equal to, that of the oil ($M_\mathrm{wo} \leq 1$).*

When the water is more mobile than the oil, stabilisation can only occur if the flow rate is less than a critical value q_crit, defined in Eq. (11.60).

Equation (11.60) lends itself to an interesting physical interpretation when M_wo is expressed fully:

$$-\frac{(u_\mathrm{t})_\mathrm{crit}}{k\dfrac{k_\mathrm{rw,or}}{\mu_\mathrm{w}}}\left(\frac{\dfrac{k_\mathrm{rw,or}}{\mu_\mathrm{w}}}{\dfrac{k_\mathrm{ro,iw}}{\mu_\mathrm{o}}} - 1\right)$$

$$= -(u_\mathrm{t})_\mathrm{crit}\left(\frac{\mu_\mathrm{o}}{k k_\mathrm{ro,iw}} - \frac{\mu_\mathrm{w}}{k k_\mathrm{rw,or}}\right) = (\rho_\mathrm{w} - \rho_\mathrm{o})g \sin \theta. \tag{11.61a}$$

With reference to the generalised Darcy equation [Eq. (3.49)], Eq. (11.61a) tells us that at the critical flow rate q_{crit} the following applies:

$$\left(\frac{\partial p_{\text{o}}}{\partial x}\right) = \left(\frac{\partial p_{\text{w}}}{\partial x}\right). \tag{11.61b}$$

This explains the physical significance of Eq. (11.60): in the VE regime, regardless of the water/oil mobility ratio, the stability of the front can be maintained by gravitational forces as long as the pressure gradient in the oil phase is less than the gradient in the water.

In the case of displacement of oil by gas, which has a very low viscosity, the mobility ratio

$$M_{\text{go}} = \frac{k_{\text{rg,or}}}{\mu_{\text{g}}} \frac{\mu_{\text{o}}}{k_{\text{ro,iw}}} \tag{3.52c}$$

is always greater than 1. Consequently, the conditions $M_{\text{go}} < 1$ and $M_{\text{go}} = 1$ equivalent to those described above in points a and b for water and oil can never be satisfied.

The only possible situation where oil displacement by gas can occur under stabilised conditions is expressed in Eq. (11.60), which takes the form:

$$(u_{\text{t}})_{\text{crit}} = \frac{q_{\text{crit}}}{A} \frac{k k_{\text{rg,or}}}{\mu_{\text{g}}(M_{\text{go}} - 1)}(\rho_{\text{o}} - \rho_{\text{g}})g \sin \theta. \tag{11.62a}$$

Expanding M_{go}, this becomes:

$$(u_{\text{t}})_{\text{crit}} = \frac{q_{\text{crit}}}{A} = \frac{k k_{\text{rg,or}}}{\dfrac{k_{\text{rg,or}}}{k_{\text{ro,iw}}} - \mu_{\text{g}}}(\rho_{\text{o}} - \rho_{\text{g}})g \sin \theta. \tag{11.62b}$$

Since μ_{g} is very small, we can ignore it, and Eq. (11.62b) simplifies to:

$$q_{\text{crit}} = \frac{k k_{\text{ro,iw}} A}{\mu_{\text{o}}}(\rho_{\text{o}} - \rho_{\text{g}})g \sin \theta. \tag{11.63}$$

This is the equation commonly used to estimate the critical rate (or limiting rate) below which the displacement front will be stabilised by gravity.

When $q < q_{\text{crit}}$, the stable front retains its shape as it advances (Fig. 11.29A); when $q > q_{\text{crit}}$, the front is no longer stable and a tongue of gas will advance along the top of the layer, over the oil (Fig. 11.29B). If the velocity of the front is sufficiently low for gravitational forces to predominate so that the curve $f_{\text{g}}(S_{\text{g}})$ is concave upwards (Sect. 11.3.5), the residual oil saturation behind the front will be almost nil. This explains the high oil recovery factors achieved in some reservoirs[17] under immiscible gas displacement, sometimes on a par with those obtained by miscible gas displacement (Chap. 15).

In practical terms, the following conditions must be met in order for a high recovery factor and an economically viable oil production rate to be obtained:

- high reservoir rock permeability, k,
- vertical or subvertical displacement ($\sin \theta \cong 1$),
- a large contact surface area A between the gas and the oil,
- very low oil viscosity μ_{o}.

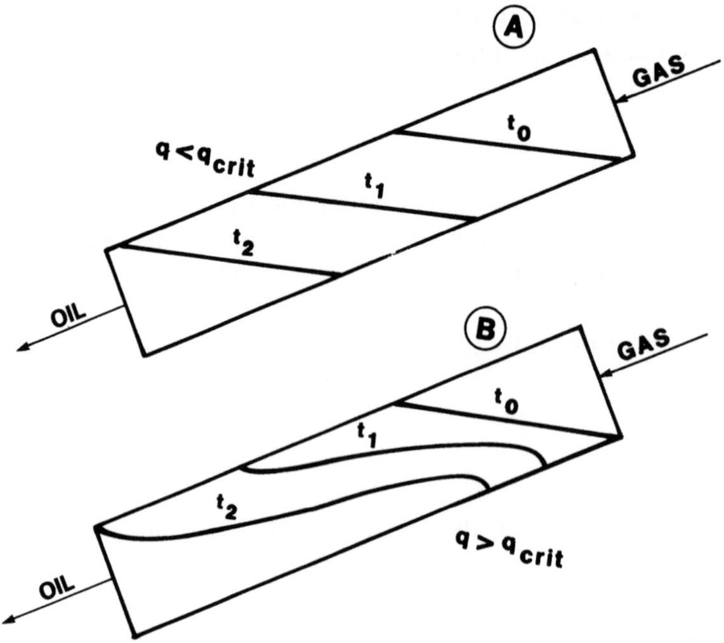

Fig. 11.29. *Displacement of oil by gas. Front stabilised by gravity (**A**) and not stabilised (**B**)*

11.5   Relative Permeability and Capillary Pressure Pseudo-Curves

Sections 11.1 and 11.4.1 described in some detail how the displacement processes which take place in the reservoir, although intrinsically three-dimensional in nature, can, under certain conditions, be represented by an equivalent two-dimensional system (x, y) by using relative permeability and capillary pressure pseudo-curves[4]. This offers an important advantage when using numerical simulation to model reservoir behaviour (Chap. 13). The absence of discretisation in the vertical direction results in a significant reduction in the number of grid blocks (and therefore in memory requirement and computing time).

When calculating pseudo-curves, there are three cases to consider, according to whether the reservoir thickness h_t is much less than, roughly equal to, or much greater than the height h_c of the capillary transition zone (Sect. 3.4.4.4). One basic assumption is that, *at any point*, the reservoir is vertically homogeneous. However, although its petrophysical properties are constant in the z-direction, they may vary areally in the (x, y) plane. Reservoirs consisting of overlying zones (or sedimentary units; Sect. 3.6) of different petrophysical properties will be dealt with in Chap. 12.

In this chapter, we will look at the most common case of oil displacement by water. The relative permeability and capillary pressure curves we shall require will be those measured in laboratory *imbibition* experiments, as described in Sects. 3.5.2.4 and 3.4.4.4. In order to derive pseudo-curves for the case of oil displacement by gas, it will of course be necessary to refer to the relative permeability and capillary pressure curves for the *drainage* process (Sects. 3.5.2.5 and 3.4.4.4).

11.5.1 Total Layer Thickness h_t Much Less Than the Height h_c of the Capillary Transition Zone

When the total thickness h_t of the oil-bearing interval is much less than the height h_c of the capillary transition zone, it is reasonable to assume that the saturation S_w is effectively constant in the z-direction [although it can of course vary in the (x, y) plane]. The displacement process can therefore be described as a one-dimensional system in the (x, z) plane [or (y, z)], using the techniques outlined in Sect. 11.3.

Most importantly, the relative permeability curves $k_{ro}(S_w)$ and $k_{rw}(S_w)$ and the capillary pressure curve $P_c(S_w)$ *will be the real curves measured on rock samples in the laboratory*, averaged, in the case of a statistically homogeneous reservoir, by the methods described in Sects. 3.5.4 and 3.4.4.7, respectively. So in thin strata, there is no need to talk of pseudo-curves.

The variation of S_w along the x- or y-axis is only dependent on the shape of the fractional flow curve $f_w(S_w)$ (see Sect. 11.3.2). When $f_w(S_w)$ is concave upwards (Sect. 11.3.2.2), there is a "piston-like" displacement of the oil by water, with the formation of a shock front, leaving oil at saturation S_{or} behind it. When $f_w(S_w)$ is S-shaped (Sect. 11.3.2.3), a shock front also develops, with oil at saturation $(1 - S_{iw})$ ahead of it and, in this case, leaving an oil saturation which varies between $(1 - S_{w,f})$ and S_{or} behind it. The recovery of oil with time, and the producing water fraction f_w, are calculated using Welge's equation following the procedure in Sect. 11.3.4.

11.5.2 Total Layer Thickness h_t of the Same Order as the Height h_c of the Capillary Transition Zone

When the total thickness h_t of the oil-bearing interval is of the same order as the height h_c of the capillary transition zone, the assumption made in Sect. 11.5.1 that the saturation S_w is constant in the z-direction is no longer valid. Consequently, the variation with height of S_w along the water/oil contact, and the resulting changes in $P_c[S_w(z)]$, $k_{ro}[S_w(z)]$ and $k_{rw}[S_w(z)]$, must all be considered in the displacement calculations as the front advances. In the following discussion, we assume that *the displacement occurs in vertical equilibrium*, which means that the vertical distribution of saturation always corresponds to gravity-capillary equilibrium.

For any layer of thickness h_t, it is a straightforward matter to derive the curves $P_{c,ow}(z)$ and $S_w(z)$, corresponding to successive positions of the WOC, by graphic or numerical means from the average $P_{c,ow}(S_w)$ curve measured in the laboratory (Fig. 11.30A). In the rest of this section, to simplify the treatment of the problem, we will define a slightly different coordinate system from that used previously. The x- and y-axes will now lie in a *horizontal* plane, the z-axis is *vertical*, with z increasing upwards.

If z is the height of an arbitrary point in the layer, measured from its base, and z_{woc} the height of the WOC [at the same position (x, y)], we have:

$$P_{c,ow}(z) = g(\rho_w - \rho_o)(z - z_{woc}) \tag{11.64a}$$

$$S_w(z) = S_w[P_{c,ow}(z)]$$

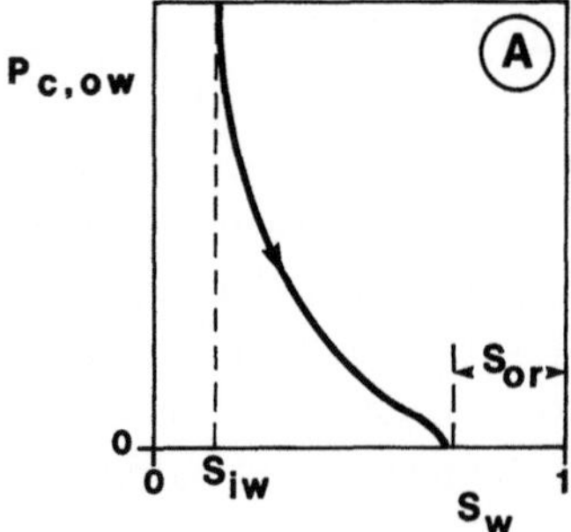
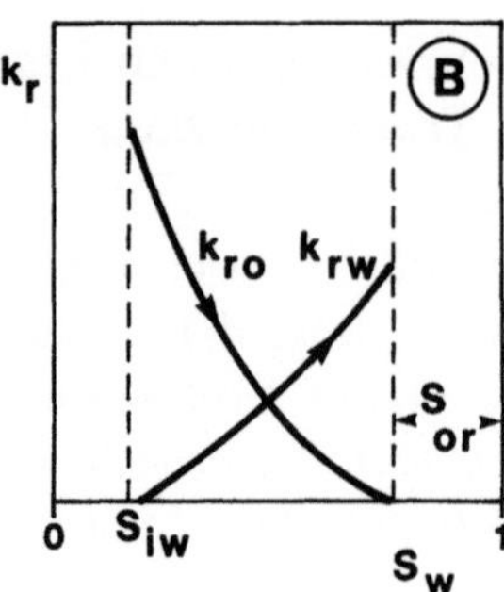

Fig. 11.30. *Laboratory imbibition capillary pressure (**A**) and water/oil relative permeability (**B**) curves used to calculate the pseudo-curve*

as shown in Fig. 11.30A, with the constraint that

$$S_w = 1 - S_{or} \tag{11.64b}$$

for all $z \le z_{woc}$.

Examples of $P_{c,ow}(z)$ and $S_w(z)$ curves derived from the $P_{c,ow}(S_w)$ curve in Fig. 11.30A, are displayed in the left-hand column of Fig. 11.31 for three different positions of the WOC within the layer.

To be precise, we have:

Fig. 11.31A – WOC at the base of the layer;

Fig. 11.31B – WOC in the middle of the layer;

Fig. 11.31C – WOC at the top of the layer.

The mean values of the water saturation, $\bar{S}_w$, and capillary pressure, $\bar{P}_{c,ow}$, as a function of the height z_{woc} of the WOC above the base of the layer can be calculated graphically or numerically from the equations:

$$\bar{S}_w(z_{woc}) = \frac{(1 - S_{or})z_{woc} + \int_{z_{woc}}^{h_t} S_w(z)\,dz}{h_t}, \tag{11.65a}$$

and

$$\bar{P}_{c,ow}(z_{woc}) = \frac{g(\rho_w - \rho_o)}{h_t} \int_0^{h_t} (z - z_{woc})\,dz$$

$$= g(\rho_w - \rho_o)\left(\frac{h_t}{2} - z_{woc}\right). \tag{11.65b}$$

As we might expect, we will find $\bar{P}_{c,ow} = 0$ when $z_{woc} = h_t/2$. From Eqs. (11.65) we can derive, point by point, the curve $\bar{P}_{c,ow} = \bar{P}_{c,ow}(\bar{S}_w)$ *which represents the pseudo-curve for capillary pressure in the layer under consideration.*

Starting from the curve $S_w(z)$ corresponding to a given height z_{woc} of the water/oil contact above the base of the layer, the corresponding curves $k_{ro}(z)$ and $k_{rw}(z)$ can be constructed quite easily. In fact, all we need do at each height z is read the value of $S_w(z)$ (Fig. 11.31, left-hand column), and plot the corresponding values of $k_{rw}(S_w)$ and $k_{ro}(S_w)$ read from the experimental curves obtained from core measurements (Fig. 11.30B). Examples of these relative permeability versus

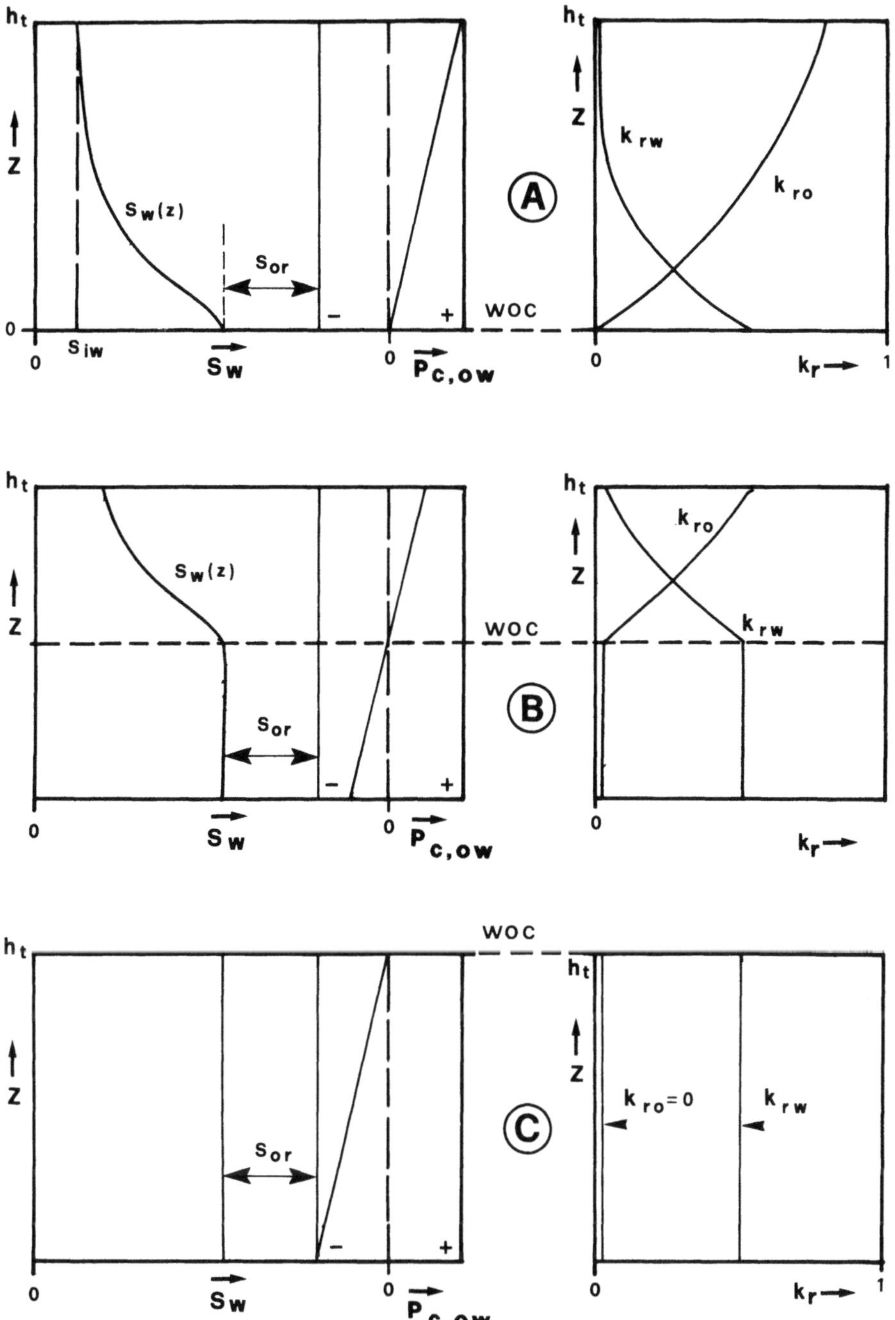

Fig. 11.31A-C. *Capillary pressure and relative permeability distributions for different positions of the* WOC *in a layer whose thickness exceeds the height of the capillary transition zone, and where displacement occurs under vertical equilibrium conditions. (From Ref. 6. Reprinted with permission of Elsevier Science Publishers and Prof. L.P. Dake)*

height diagrams appear on the right-hand side of Fig. 11.31 for the three positions of the WOC.

For each value of z_{woc} we then estimate, graphically or numerically, the following values:

$$\bar{k}_{ro}(z_{woc}) = \frac{\int_{z_{woc}}^{h_t} k_{ro}(z)\, dz}{h_t}, \tag{11.66a}$$

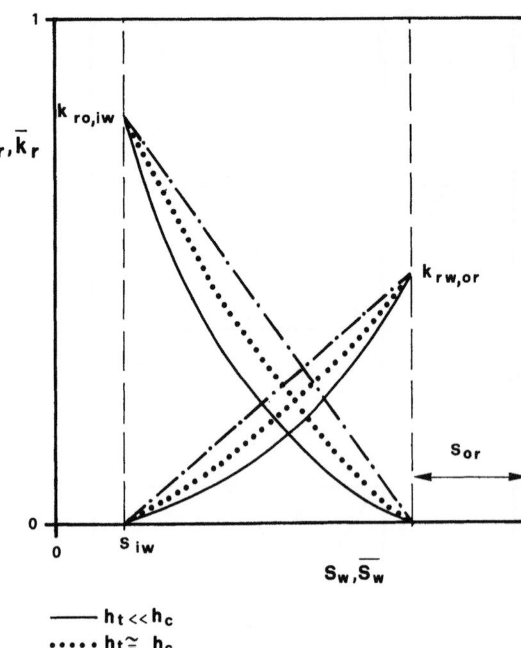

Fig. 11.32. *Comparison of relative permeability pseudo-curves for the cases where $h_t \cong h_c$ and $h_t \gg h_c$ with the curves measured on the core ($h_t \ll h_c$). (From Ref.6. Reprinted with permission of Elsevier Science Publishers and Prof. L.P. Dake)*

and

$$\bar{k}_{rw}(z_{woc}) = \frac{k_{rw,or}z_{woc} + \int_{z_{woc}}^{h_t} k_{rw}(z)\,dz}{h_t} \tag{11.66b}$$

Using these values, and $\bar{S}_w = \bar{S}_w(z_{woc})$ from Eq. (11.65a), we can build up the curves:

$$\bar{k}_{ro}(\bar{S}_w) \quad \text{and} \quad \bar{k}_{rw}(\bar{S}_w)$$

point by point, and thus obtain the *relative permeability pseudo-curves for the layer under study*. Note that these pseudo-curves always lie *above* the experimental curves $k_{ro}(S_w)$ and $k_{rw}(S_w)$ measured on cores (Fig. 11.32).

The use of relative permeability and capillary pressure pseudo-curves will allow us to model the displacement process with one-dimensional equations developed in Sect. 11.3. In other words, *assuming vertical equilibrium* (Sect. 11.4.1), we can reduce a two-dimensional system to an equivalent one-dimensional system. This offers a number of computational advantages, already described.

11.5.3 Total Layer Thickness h_t Much Greater Than the Height h_c of the Capillary Transition Zone

When the total thickness h_t of the oil-bearing interval is much greater than the height h_c of the capillary transition zone, we can quite reasonably ignore the influence of the latter on both the average water saturation $\bar{S}_w$, and the biphasic flow of oil and water. For flow occurring under VE conditions (Sect. 11.4.1), we can define the height h_w of the water table above the base of the layer at any point in the layer (Fig. 11.33). Beneath the water table we have:

$$S_w = 1 - S_{or}; \qquad k_{rw} = k_{rw,or}; \qquad k_{ro} = 0.$$

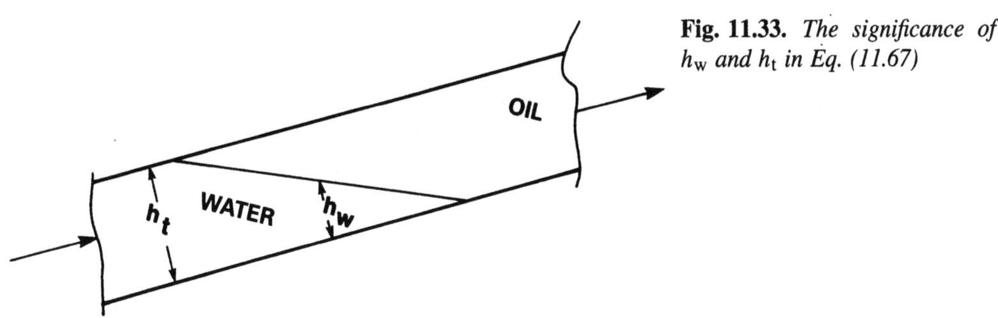

Fig. 11.33. *The significance of h_w and h_t in Eq. (11.67)*

Above the water table:

$$S_w = S_{iw}; \qquad k_{rw} = 0; k_{ro} = k_{ro,iw}.$$

Therefore, assuming ϕ and k to be constant in the vertical direction:

$$\bar{S}_w = \frac{h_w(1 - S_{or}) + (h_t - h_w)S_{iw}}{h_t}, \tag{11.67a}$$

$$\bar{k}_{rw} = \frac{h_w k_{rw,or}}{h_t}, \tag{11.67b}$$

$$\bar{k}_{ro} = \frac{(h_t - h_w)k_{ro,iw}}{h_t}. \tag{11.67c}$$

If we define:

$$b = \frac{h_w}{h_t}, \tag{11.68}$$

then from Eq. (11.67a):

$$\bar{S}_w = S_{iw} + b(1 - S_{iw} - S_{or}) \tag{11.69a}$$

so that:

$$b = \frac{\bar{S}_w - S_{iw}}{1 - S_{iw} - S_{or}}. \tag{11.69b}$$

Equations (11.67b) and (11.67c) therefore become:

$$\bar{k}_{rw}(\bar{S}_w) = \frac{\bar{S}_w - S_{iw}}{1 - S_{or} - S_{iw}} k_{rw,or}, \tag{11.70a}$$

and

$$\bar{k}_{ro}(\bar{S}_w) = \frac{1 - S_{or} - \bar{S}_w}{1 - S_{or} - S_{iw}} k_{ro,iw}. \tag{11.70b}$$

Equations (11.70) show that, in the case where $h_t \gg h_c$, *the pseudo-curves for relative permeability are linear functions of the mean water saturation $\bar{S}_w$* (Fig. 11.32). With these pseudo-curves, we can again describe the displacement process using the equations for a one-dimensional system (Sect. 11.3). In particular, since we are assuming $P_{c,ow} = 0$, and if we ignore gravitational effects, the following relationship can be derived quite simply from Eq. (11.12):

$$f_w(\bar{S}_w) = \frac{M_{wo}(\bar{S}_w - S_{iw})}{(M_{wo} - 1)\bar{S}_w + 1 - S_{or} - M_{wo}S_{iw}}, \tag{11.71}$$

so that:

$$\frac{\mathrm{d}f_\mathrm{w}}{\mathrm{d}\bar{S}_\mathrm{w}} = \frac{M_\mathrm{wo}(1 - S_\mathrm{or} - S_\mathrm{iw})}{[(M_\mathrm{wo} - 1)\bar{S}_\mathrm{w} + 1 - S_\mathrm{or} - M_\mathrm{wo}S_\mathrm{iw}]^2}, \tag{11.72}$$

Equation (11.72) tells us that when $M_\mathrm{wo} > 1$, the slope of the curve $f_\mathrm{w}(\bar{S}_\mathrm{w})$ decreases as $\bar{S}_\mathrm{w}$ increases. Therefore $f_\mathrm{w}(\bar{S}_\mathrm{w})$ is concave downwards and it is impossible for a displacement front to form (Sect. 11.3.2.1). When $M_\mathrm{wo} = 1$, $f_\mathrm{w}(\bar{S}_\mathrm{w})$ is a straight line. When $M_\mathrm{wo} < 1$, the slope of the $f_\mathrm{w}(\bar{S}_\mathrm{w})$ curve increases with $\bar{S}_\mathrm{w}$: it is therefore concave upwards and a shock front will be able to form (Sect. 11.3.2.2).

Referring to what has been said in Sect. 11.4.2, note that for $M_\mathrm{wo} < 1$ the front is stabilised with its inclination β greater than that of the layer, θ. If β tends towards $\pi/2$, the two-dimensional displacement in the layer takes on the form of a true shock front as described in Sect. 11.3.2.2 for a one-dimensional geometry.

References

1. Buckley SE, Leverett MC (1942) Mechanism of fluid displacement in sands. Trans, AIME 146: 107–116
2. Chuoke RL, van Meurs P, van der Poel C (1959) The instability of slow, immiscible, viscous liquid-liquid displacements in permeable media. Trans, AIME 216: 188–194
3. Coats KH, Nielsen RL, Terhune MH, Weber AG (1967) Simulation of three-dimensional, two-phase flow in oil and gas reservoirs. Soc Pet Eng J Dec. 377–388; Trans, AIME 240
4. Coats KH, Dempsey JR, Henderson JH (1971) The use of vertical equilibrium in two-dimensional simulation of three-dimensional reservoir performance. Soc Pet Eng J March: 63–71; Trans, AIME 251
5. Craft BC, Hawkins MF (1969) Applied petroleum reservoir engineering. Prentice-Hall, Englewood Cliffs
6. Dake LP (1978) Fundamentals of reservoir engineering. Elsevier, Amsterdam
7. Dietz DN (1953) A theoretical approach to the problem of encroaching and by-passing edge water. Proc Ned Akad Wet Ser B, 56(1): 83–92
8. Hagoort J (1974) Displacement stability of water drives in water wet connate water bearing reservoirs. Soc Pet Eng J Feb: 63–74; Trans, AIME 257
9. Hagoort J (1980) Oil recovery by gravity drainage. Soc Pet Eng J June: 139–150; Trans, AIME 269
10. Hildebrand FB (1962) Advanced calculus for applications. Prentice-Hall, Englewood Cliffs
11. Lee EH, Fayers FJ (1959) The use of the method of characteristics in determining boundary conditions for problems in reservoir analysis. Tran, AIME 216: 284–289
12. Rachford HH Jr (1964) Instability in water flooding oil from water-wet porous media containing connate water. Soc Pet Eng J June: 133–148; Trans, AIME 231
13. Sheldon JW, Zondek B, Cardwell WT Jr (1959) One-dimensional, incompressible, noncapillary two-phase fluid flow in porous media. Trans, AIME 216: 290–296
14. Terwilliger PL, Wilsey LE, Hall HN, Bridges PM, Morse RA (1951) An experimental and theoretical investigation of gravity drainage performance. Trans, AIME 192: 285–296
15. van Meurs P, van der Poel C (1958) A theoretical description of water-drive processes involving viscous fingering. Trans, AIME 213: 103–112
16. Welge HJ (1952) A simplified method for computing oil recovery by gas or water drive. Trans, AIME 195: 91–98
17. Ypma JGJ (1985) Analytical and numerical modeling of immiscible gravity-stable gas injection into stratified reservoirs. Soc Pet Eng J Aug: 554–564; Trans, AIME 279

EXERCISES

Exercise 11.1

We have a reservoir rock saturated with oil in the presence of interstitial water at an irreducible saturation $S_{iw} = 0.15$. The residual oil saturation after flushing with water is $S_{or} = 0.30$.

The endpoint relative permeabilities are $k_{ro,iw} = 0.90$ and $k_{rw,or} = 0.40$.

The oil/water relative permeability curves are described by Eq. (3.59) with

$$C = 2, \qquad N = 1,$$

$$D = 2, \qquad P = 1,$$

Assuming the formation to be horizontal ($\sin\theta = 0$) and, ignoring the capillary pressure gradient, calculate the $f_w(S_w)$ curve for each of the following cases:

a) $\mu_o/\mu_w = 0.1$,

b) $\mu_o/\mu_w = 1$,

c) $\mu_o/\mu_w = 10$.

For each of these cases, use graphic methods to evaluate the following:

1) the water saturation at the front at breakthrough

2) percentage oil recovery at water breakthrough.

Solution

Using Eqs. (3.59), we first calculate the values of $k_{ro}(S_w)$, $k_{rw}(S_w)$ and $k_{ro}/k_{rw} = f(S_w)$. The results are listed in Table E11/1.1, and k_{ro}/k_{rw} is plotted against S_w in Fig. E11/1.1.

Since the formation is horizontal and we are to ignore capillary pressures, we can use Eq. (11.12) to calculate $f_w(S_w)$:

$$f_w = \cfrac{1}{1 + \cfrac{k_{ro}}{k_{rw}}\cfrac{\mu_w}{\mu_o}}. \tag{11.12}$$

Table E11/1.2 lists the three sets of results, and the curves $f_w(S_w)$ corresponding to $\mu_o/\mu_w = 0.1$, 1 and 10 are presented in Fig. E11/1.2.

To calculate the saturation $S_{w,f}$ at the front and the percentage oil recovery at the time of water breakthrough, the graphic method described in Sects. 11.3.2.3 and 11.3.3 (Welge's equation) should be used.

From Fig. E11/1.2 we obtain

$\dfrac{\mu_o}{\mu_w}$	$S_{w,f}$	$(f_w)_{BT}$	$\bar{S}_w$	E_D
0.1	0.58	0.97	0.60	0.529
1.0	0.53	0.94	0.55	0.471
10	0.45	0.91	0.47	0.376

where $(f_w)_{BT}$ is the water fraction in the produced fluids, under reservoir conditions, when breakthrough occurs, and E_D is the cumulative fraction of oil recovered, *assuming the porous medium to be completely homogeneous, and the front to have no instability.*

Table E11/1.1. *Oil/water relative permeability curves*

S_w	R_w (Eq. 3.59(c))	k_{ro}^* (Eq. 3.59(a))	k_{rw}^* (Eq. 3.59(b))	k_{ro} (Eq. 3.54(b))	k_{rw} (Eq. 3.54(c))	k_{ro}/k_{rw}
0.15	0.000	1.000	0.000	0.900	0.000	∞
0.20	0.100	0.819	2.00×10^{-9}	0.737	8×10^{-10}	9.21×10^{8}
0.25	0.222	0.641	1.23×10^{-4}	0.577	5×10^{-5}	1.15×10^{4}
0.30	0.375	0.472	4.82×10^{-3}	0.425	2×10^{-3}	212.5
0.35	0.571	0.319	0.030	0.287	0.012	23.92
0.40	0.833	0.189	0.091	0.170	0.036	4.722
0.45	1.200	0.091	0.189	0.082	0.076	1.079
0.50	1.750	0.030	0.319	0.027	0.128	0.211
0.55	2.667	4.83×10^{-3}	0.472	4.35×10^{-3}	0.189	0.024
0.60	4.500	1.23×10^{-4}	0.641	1.11×10^{-4}	0.256	4.34×10^{-4}
0.65	10.000	2.06×10^{-9}	0.819	1.85×10^{-9}	0.328	5.65×10^{-9}
0.70	∞	0.000	1.000	0.000	0.400	zero

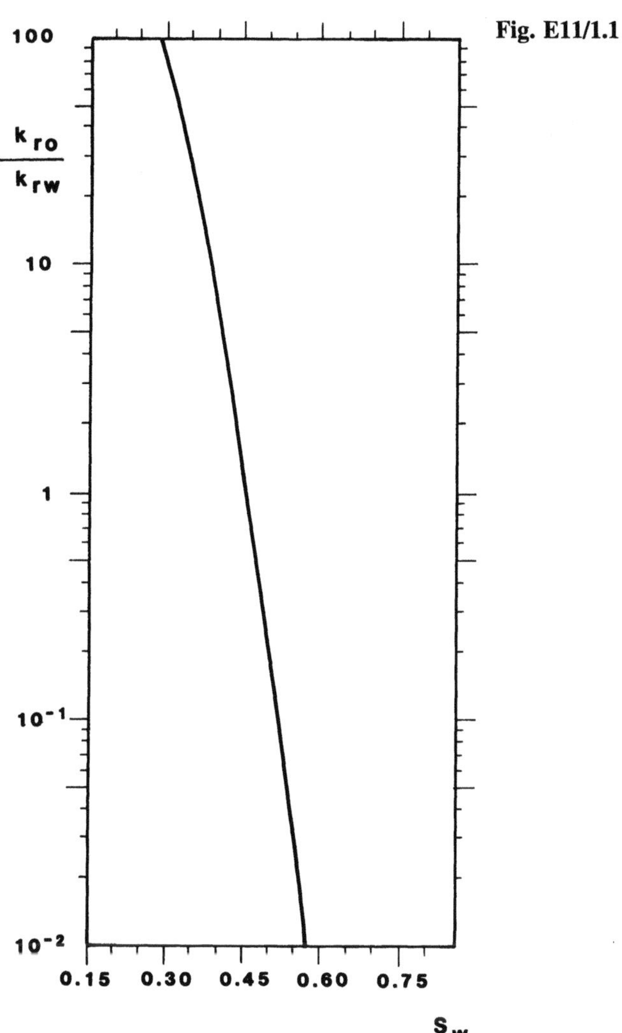

Fig. E11/1.1

Table E11/1.2. *Values of $f_w(S_w)$ for $\mu_o/\mu_w = 0.1$, 1 and 10*

S_w	$f_w(S_w)$ for		
	$\mu_o/\mu_w = 0.1$	$\mu_o/\mu_w = 1$	$\mu_o/\mu_w = 10$
0.15	0.000	0.000	0.000
0.20	1.1×10^{-10}	1.1×10^{-9}	1.1×10^{-8}
0.25	8.7×10^{-6}	8.7×10^{-5}	8.7×10^{-4}
0.30	4.7×10^{-4}	0.005	0.045
0.35	0.004	0.040	0.295
0.40	0.021	0.175	0.679
0.45	0.085	0.481	0.903
0.50	0.322	0.826	0.979
0.55	0.806	0.977	0.998
0.60	0.996	0.999	0.999
0.65	0.999	1.000	1.000
0.70	1.000	1.000	1.000

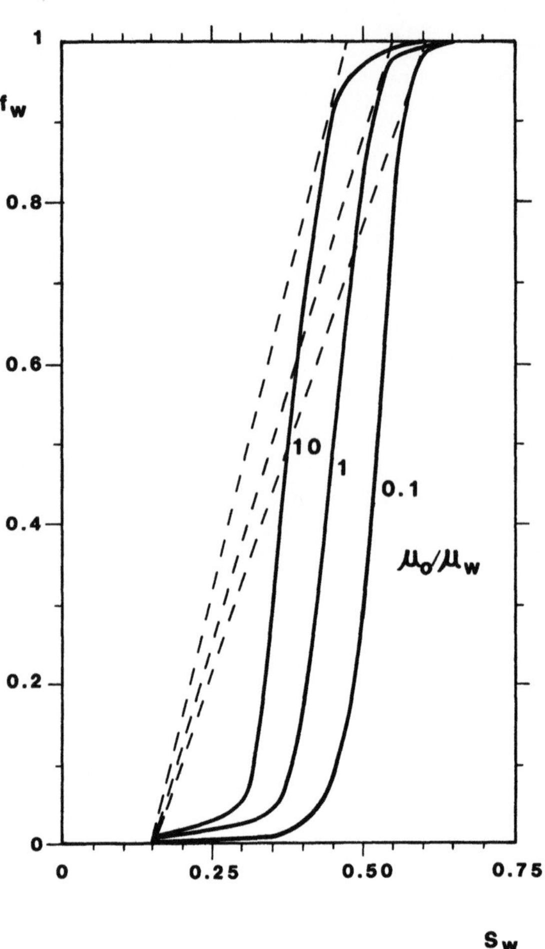

Fig. E11/1.2

Note that for $\mu_o/\mu_w = 10$ [$M_s = 0.935$, Eq. (11.33)] it is unlikely that fingering will be observed (Sect. 11.3.6).

$$\diamond \; \diamond \; \diamond$$

Exercise 11.2

A reservoir with the same oil/water relative permeability characteristics as those listed in Table E11/1.1 is producing by water displacement from an *underlying aquifer* ("bottom water drive"). Consequently, beyond a certain distance from each well, the water is displacing the oil *vertically*.

The reservoir rock and fluid properties are as follows:

- porosity : ϕ $= 0.25$
- permeability : k $= 200$ md
- oil viscosity : μ_o $= 5$ cP
- water viscosity : μ_w $= 0.5$ cP
- oil density : ρ_o $= 750$ kg/m^3
- water density : ρ_w $= 1030$ kg/m^3

Assuming a linear system, calculate, at water breakthrough:

- water saturation $S_{w,f}$ at the front
- produced water fraction, under reservoir conditions
- microscopic displacement efficiency E_D

for the following three different frontal velocities, measured in a vertical direction:

$$v_1 = 5 \text{ m/yr}$$

$$v_2 = 20 \text{ m/yr}$$

$$v_3 = 80 \text{ m/yr}$$

The vertical capillary pressure gradient can be considered negligible.

Solution

As has been explained, the relationship between the actual velocity v of a fluid in a porous medium and its corresponding Darcy velocity u is

$$v = \frac{u}{\phi}.\qquad(11/2.1)$$

In this example, where $\phi = 0.25$, the Darcy velocities u_t corresponding to the three different velocities of advance of the water front are

$$u_{t,1} = 5 \times 0.25 \;\; = 1.25 \text{ m/yr}$$
$$u_{t,2} = 20 \times 0.25 = 5 \text{ m/yr}$$
$$u_{t,3} = 80 \times 0.25 = 20 \text{ m/yr}.$$

From Eq. (11.11), ignoring the capillary pressure gradient, we have

$$f_w = \frac{1}{1 + \dfrac{k_{ro}}{k_{rw}}\dfrac{\mu_w}{\mu_o}} \left\{ 1 - \frac{kk_{ro}}{\mu_o u_t}(\rho_w - \rho_o)g\sin\theta \right\}$$

where all parameters are expressed in SI units.

The term outside the brackets was calculated in Exercise 11.1, for the case $\mu_o/\mu_w = 10$. If we define

$$N_{gv} = \frac{kk_{ro}}{\mu_o u_t}(\rho_w - \rho_o)g\sin\theta$$

and convert all the parameters to SI units, we get:

$$N_{gv} = \frac{k_{ro}k(md) \times 9.869 \times 10^{-16}(\rho_w - \rho_o)\left(\dfrac{\text{kg}}{\text{m}^3}\right) \times 9.806}{\mu_o(cP) \times 10^{-3}\dfrac{u_t(\text{m/yr})}{3.15576 \times 10^7}}$$

remembering that $\sin\theta = 1$ $(\theta = \pi/2)$.

Table E11/2.1

S_w	$\dfrac{1}{1 + \dfrac{k_{ro}}{k_{rw}}\dfrac{\mu_w}{\mu_o}}$	k_{ro}	f_w (%)		
			$u_t = 1.25$	$u_t = 5$	$u_t = 20$
0.15	0.000	0.900	0.000	0.000	0.000
0.20	1.1×10^{-8}	0.737	Negative	5.45×10^{-9}	9.61×10^{-9}
0.25	8.7×10^{-4}	0.577	Negative	5.27×10^{-4}	7.84×10^{-4}
0.30	0.045	0.425	Negative	0.032	0.042
0.35	0.295	0.287	0.063	0.237	0.281
0.40	0.679	0.170	0.363	0.600	0.659
0.45	0.903	0.082	0.700	0.852	0.890
0.50	0.979	0.027	0.907	0.961	0.974
0.55	0.998	4.35×10^{-3}	0.986	0.995	0.997
0.60	0.999	1.11×10^{-4}	0.998	0.999	0.999
0.65	1.000	1.85×10^{-9}	1.000	1.000	1.000
0.70	1.000	0.000	1.000	1.000	1.000

Therefore,

$$N_{gv} = 3.054 \times 10^{-4} \frac{k(\text{md})\,k_{ro}(\rho_w - \rho_o)\left(\dfrac{\text{kg}}{\text{m}^3}\right)}{\mu_o(\text{cP})\,u_t(\text{m/yr})},$$

which, inserting the appropriate values, becomes:

$$N_{gv} = 3.054 \times 10^{-4} \frac{200\,k_{ro}(1030 - 750)}{5u_t(\text{m/yr})} = 3.42\frac{k_{ro}}{u_t(\text{m/yr})}$$

Therefore,

v (m/yr)	u_t (m/yr)	N_{gv}
5	1.25	2.736 k_{ro}
20	5	0.684 k_{ro}
80	20	0.171 k_{ro}

With these values we can calculate the curve $f_w(S_w)$ (Table E11/2.1) corresponding to each of the three specified frontal velocities. The curves are presented in Fig. E11/2.1.

Now, using the graphic method described in Sects. 11.3.2.3 and 11.3.3, we obtain

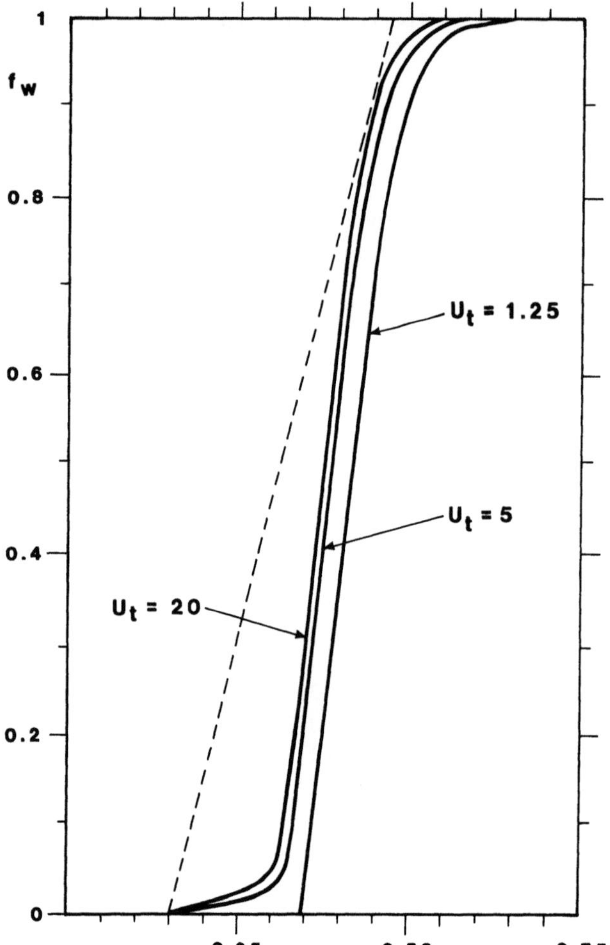

Fig. E11/2.1

v (m/yr)	$S_{w,f}$	$(f_w)_{BT}$	$(E_D)_{BT}$
5	0.50	0.91	45.3
20	0.45	0.85	40.6
80	0.43	0.84	38.2

It should be apparent that in this case (and, generally speaking, in most cases where oil is being displaced by water), both the microscopic displacement efficiency at breakthrough and $(f_w)_{BT}$ are hardly affected by the velocity of the front (although there is a trend towards better values at lower velocities).

For practical purposes, over the range of frontal velocities typically encountered in a producing reservoir (more than 10 m/yr), the gravitational term can usually be omitted from the fractional flow equation.

Exercise 11.3

Calculate the critical velocity u_{crit} (m/yr) of oil displacement by gas in a gravity-stabilised regime (Sect. 11.4.2), and the corresponding annual oil production rate Q_o per km^2 of gas/oil contact surface area, for each of the following sets of reservoir conditions:

Case	k (md)	μ_o (cP)	ρ_o (kg/m^3)	B_{of} (m^3/m^3)
a	1000	0.5	750	1.8
b	1000	5	800	1.3
c	1000	50	850	1.1
d	100	0.5	750	1.8
e	10	0.5	750	1.8

The formation is inclined at $\theta = 20°$ from the horizontal.

For all five cases, the following applies:

$$k_{ro,iw} = 1.00,$$

$$\rho_g = 120 \text{ kg/m}^3.$$

Solution

We shall use the following equation, expressed in SI units:

$$u_{crit} = \frac{q_{crit}}{A} = \frac{k k_{ro,iw}}{\mu_o}(\rho_o - \rho_g)g \sin\theta. \tag{11.63}$$

With

$$1 \text{ md} = 9.869233 \times 10^{-16} \text{ m}^2,$$

$$1 \text{ cP} = 10^{-3} \text{ Pa.s},$$

$$1 \text{ yr} = 3.15576 \times 10^7 \text{ s}$$

we have

$$\frac{u_{crit}(\text{m/yr})}{3.15576 \times 10^7}$$

$$= \frac{k(\text{md}) \times 9.869233 \times 10^{-16} k_{ro,iw}}{\mu_o(\text{cP}) \times 10^{-3}}(\rho_o - \rho_g)\left(\frac{\text{kg}}{\text{m}^3}\right) g \left(\frac{\text{m}}{\text{s}^2}\right) \sin\theta.$$

Therefore,

$$u_{crit}(\text{m/yr}) = 3.114 \times 10^{-5} \frac{k(\text{md})k_{ro,iw}}{\mu_o(\text{cP})} (\rho_o - \rho_g) \left(\frac{\text{kg}}{\text{m}^3}\right) g \left(\frac{\text{m}}{\text{s}^2}\right) \sin\theta. \qquad (11/3.1)$$

We can insert the following values:

$$k_{ro,iw} = 1.00$$

$$g = 9.806 \text{ m/s}^2$$

$$\sin 20° = 0.3420$$

$$1 \text{ km}^2 = 10^6 \text{ m}^2$$

so that

$$u_{crit}(\text{m/yr}) = 1.04455 \times 10^{-4} \frac{k(\text{md})}{\mu_o(\text{cP})} (\rho_o - \rho_g) \left(\frac{\text{kg}}{\text{m}^3}\right)$$

and

$$Q_{o,sc}(\text{m}^3/\text{yr}) = \frac{10^6 u_{crit}(\text{m/yr})}{B_{of}}.$$

For the five specified reservoir conditions, we then calculate:

Case	u_{crit} (m/yr)	$Q_{o,sc}$ (m³/yr)
a	131.6	73.1×10^6
b	14.2	10.9×10^6
c	1.53	1.4×10^6
d	13.16	7.3×10^6
e	1.32	0.7×10^6

Exercise 11.4

A reservoir overlain by a gas cap is producing a light oil. Reservoir pressure is maintained more or less constant (and therefore above the bubble point) by crestal injection of gas into the gas cap.

The displacement of the oil by gas can therefore be treated as a one-dimensional system.

The properties of the reservoir rock are:

− porosity :	ϕ	$= 0.35$
− vertical permeability:	k_v	$= 500$ md
− irreducible water saturation :	S_{iw}	$= 0.15$
− critical gas saturation :	S_{gc}	$= 0.08$
− residual oil saturation :	S_{or}	$= 0.05$

and the reservoir fluid properties are:

− oil volume factor :	$B_{of} = 1.8$ m³/m³
− oil density :	$\rho_o = 750$ kg/m³
− oil viscosity :	$\mu_o = 0.5$ cP
− gas density :	$\rho_g = 120$ kg/m³
− gas viscosity :	$\mu_g = 0.013$ cP

Calculate the critical displacement velocity of the oil, u_{crit}, assuming vertical equilibrium conditions, and the recovery factor of the oil at the time of gas breakthrough, for each of the following Darcy displacement velocities u_t:

Case	u
a	u_{crit}
b	$0.5\,u_{\mathrm{crit}}$
c	$0.1\,u_{\mathrm{crit}}$
d	$0.05\,u_{\mathrm{crit}}$
e	$0.01\,u_{\mathrm{crit}}$

Solution

First, we have to calculate the oil/gas relative permeability curves, which have not been provided in the reservoir data.

We shall use the Corey equations:

$$k_{\mathrm{ro}} = \left(\frac{S_{\mathrm{o}} - S_{\mathrm{or}}}{1 - S_{\mathrm{iw}} - S_{\mathrm{or}}}\right)^4,\tag{3.56a}$$

$$k_{\mathrm{rg}} = (S_{\mathrm{g}} - S_{\mathrm{gc}})^3 \frac{2(1 - S_{\mathrm{iw}} - S_{\mathrm{or}}) - (S_{\mathrm{g}} - S_{\mathrm{gc}})}{(1 - S_{\mathrm{iw}} - S_{\mathrm{or}})^4},$$

$$[k_{\mathrm{rg}} = 0 \text{ for } S_{\mathrm{g}} \leq S_{\mathrm{gc}}].$$

With the data supplied, these become:

$$k_{\mathrm{ro}} = 2.4414(0.8 - S_{\mathrm{g}})^4,$$

$$k_{\mathrm{rg}} = 2.4414(S_{\mathrm{g}} - 0.08)^3(1.68 - S_{\mathrm{g}}).$$

$$[S_{\mathrm{g}} \geq 0.08]$$

These equations are used to calculate the values of $k_{\mathrm{ro}}(S_{\mathrm{g}})$ and $k_{\mathrm{rg}}(S_{\mathrm{g}})$, listed in Table E11/4.1. u_{crit} is calculated from Eq. (11/3.1), with $\sin\theta = 1.00$ ($\theta = \pi/2$) and $k_{\mathrm{ro,iw}} = 1$. Then

$$u_{\mathrm{crit}} \text{ (m/yr)} = 3.114 \times 10^{-5}\frac{500}{0.5}(750 - 120) \times 9.806 = 192.4 \text{ m/yr}$$

Table E11/4.1

S_{g}	k_{ro}	k_{rg}	$k_{\mathrm{ro}}/k_{\mathrm{rg}}$
0.00	1.0000	0.0000	∞
0.05	0.7725	0.0000	∞
0.08	0.6561	0.0000	∞
0.10	0.5862	3.09×10^{-5}	1.90×10^4
0.15	0.4358	1.28×10^{-3}	340.1
0.20	0.3164	6.24×10^{-3}	50.71
0.25	0.2234	0.0172	13.0245
0.30	0.1526	0.0359	4.2507
0.35	0.1001	0.0639	1.5665
0.40	0.0625	0.1024	0.6104
0.45	0.0366	0.1521	0.2406
0.50	0.0198	0.2134	0.0928
0.55	9.54×10^{-3}	0.2864	0.0333
0.60	3.91×10^{-3}	0.3707	0.0105
0.65	1.24×10^{-3}	0.4657	2.66×10^{-3}
0.70	2.44×10^{-4}	0.5702	4.28×10^{-4}
0.75	1.53×10^{-5}	0.6829	2.24×10^{-5}
0.80	zero	0.8019	zero

To address the second part of the exercise – the calculation of the percentage oil recovery at breakthrough for each of the specified displacement velocities – we must first compute $f_g(S_g)$ for each value of u.

The equation to use is:

$$f_g = \frac{\lambda_g}{\lambda_g + \lambda_o} \left\{ 1 - \left[\frac{\lambda_o}{u}(\rho_o - \rho_g)g \sin\theta + \frac{\partial P_{c,go}}{\partial z} \right] \right\}.$$

If we ignore the capillary pressure gradient, write the λ terms in full, and rearrange, we arrive at

$$f_g(S_g) = \frac{1}{1 + \dfrac{k_{ro}}{k_{rg}}\dfrac{\mu_g}{\mu_o}} \left\{ 1 - \left[\frac{kk_{ro,iw}}{\mu_o u_{crit}}(\rho_o - \rho_g)g \sin\theta \right] \frac{k_{ro}}{k_{ro,iw}} \frac{u_{crit}}{u} \right\}. \tag{11/4.1}$$

Now, according to Eq. (11.63):

$$\frac{kk_{ro,iw}}{\mu_o u_{crit}}(\rho_o - \rho_g)g \sin\theta = 1. \tag{11.63a}$$

Substituting this into Eq. (11/4.1) we obtain:

$$f_g(S_g) = \frac{1}{1 + \dfrac{k_{ro}}{k_{rg}}\dfrac{\mu_g}{\mu_o}} \left[1 - \frac{k_{ro}}{k_{ro,iw}} \frac{1}{(u/u_{crit})} \right]. \tag{11/4.2}$$

From this, it is an easy matter to calculate $f_g(S_g)$ for any value of u.

The calculations are presented in Table E11/4.2, and plotted in Figs. E11/4.1 and 4.2.

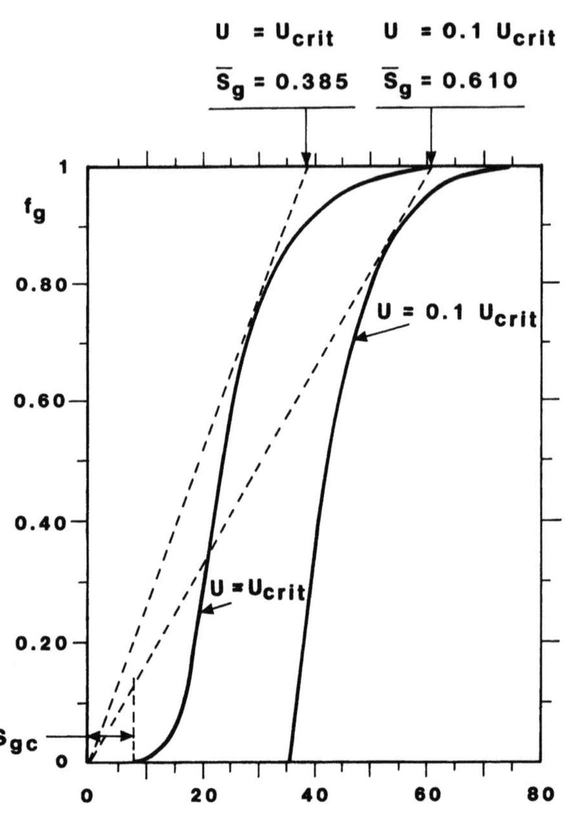

Fig. E11/4.1

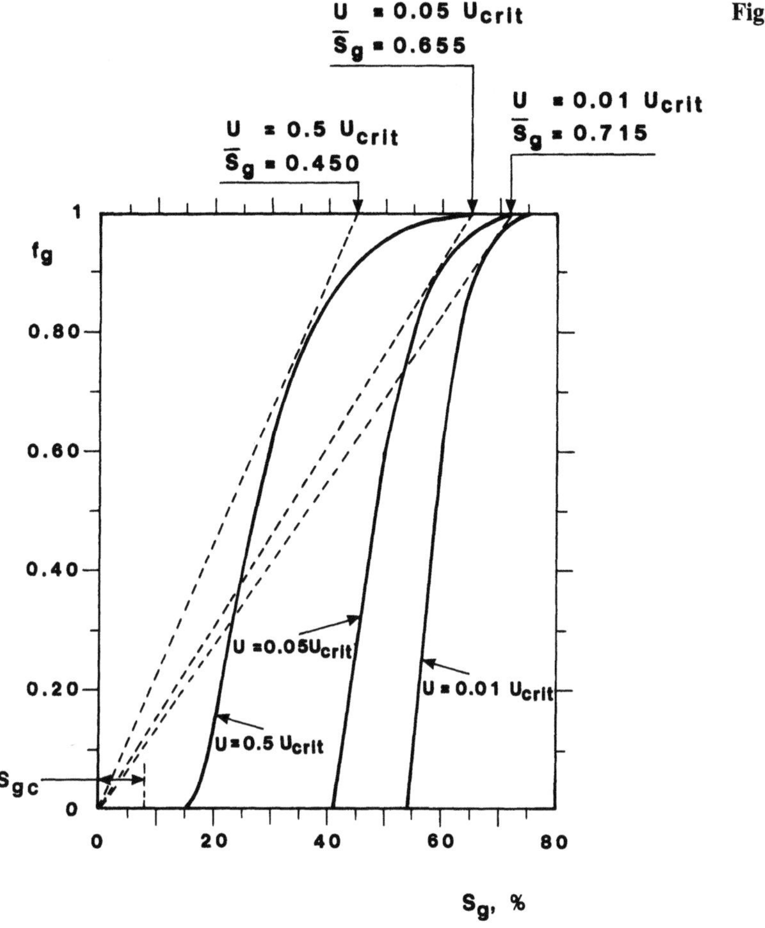

With the following definitions:

$(\bar{S}_g)_{BT}$: the average gas saturation behind the displacement front at the moment of breakthrough

$(\bar{E}_D)_{BT}$: average microscopic displacement efficiency (i.e. on a pore scale) in the rock swept by gas, at the moment of breakthrough

$Q_{o,sc}$: the annual production of oil, at surface conditions, per km^2 of gas/oil contact area

t_r : the duration of production from the reservoir (at specified u) ÷ the duration if $u = u_{crit}$.

we can apply Welge's method to the $f_g(S_g)$ curves in Figs. E11/4.1 and 2, with the following results:

$\dfrac{u}{u_{crit}}$	$(\bar{S}_g)_{BT}$	$(\bar{E}_D)_{BT}$	$Q_{o,sc}(m^3/yr.km^2)$	t_r
1	0.385	45.3%	106.9×10^6	1
0.5	0.450	52.9%	53.4×10^6	2
0.1	0.610	71.8%	10.7×10^6	10
0.05	0.655	77.1%	5.3×10^6	20
0.01	0.715	84.1%	1.1×10^6	100

In practical terms, in order to achieve any significant increase in the percentage of oil recovered at breakthrough, the oil displacement velocity must be at least one order of magnitude lower than the critical velocity for stabilised displacement (u_{crit}).

Any gain in recovery achieved by slower displacement must of course be evaluated against the net value of the oil, discounted forward over the increased time period required to produce it.

In any case, subcritical displacement is only economically viable in thick reservoirs with high vertical permeability, with a large gas/oil contact area and low viscosity oil.

Table E11/4.2

S_g	$\dfrac{1}{1 + \dfrac{k_{ro}}{k_{rg}} \dfrac{\mu_g}{\mu_o}}$	k_{ro}	$f_g(S_g)$ when $(u/u_{crit}) =$				
			1.00	0.50	0.10	0.05	0.01
0.10	2.020×10^{-3}	0.5862	8.28×10^{-4}	Negative	Negative	Negative	Negative
0.15	0.10160	0.4358	0.05732	0.01305	Negative	Negative	Negative
0.20	0.43132	0.3164	0.29485	0.15838	Negative	Negative	Negative
0.25	0.74703	0.2234	0.58014	0.41326	Negative	Negative	Negative
0.30	0.90048	0.1526	0.76307	0.62565	Negative	Negative	Negative
0.35	0.96086	0.1001	0.86468	0.76850	Negative	Negative	Negative
0.40	0.98438	0.0625	0.92286	0.86133	0.36914	Negative	Negative
0.45	0.99378	0.0366	0.95741	0.92104	0.63006	0.26633	Negative
0.50	0.99759	0.0198	0.97784	0.95809	0.80007	0.60254	Negative
0.55	0.99913	9.54×10^{-3}	0.98960	0.98007	0.90391	0.80850	0.04596
0.60	0.99973	3.91×10^{-3}	0.99582	0.99191	0.96064	0.92155	0.60884
0.65	0.99993	1.24×10^{-3}	0.99869	0.99745	0.98753	0.97513	0.87594
0.70	0.99998	2.44×10^{-4}	0.99974	0.99949	0.99754	0.99510	0.97558
0.75	0.99999	1.53×10^{-5}	0.99997	0.99996	0.99983	0.99968	0.99845
0.80	1.00000	zero	1.00000	1.00000	1.00000	1.00000	1.00000

An important distinction should be made at this point: the concept of a critical velocity (u_{crit}) pertains to the stability of the gas/oil front – the absence of gas fingering through or otherwise over-riding the oil; on the other hand, the dependence of the oil recovery factor on the frontal advance velocity is a direct consequence of the formulation of the Buckley-Leverett equation.

In the case of displacement by gas then, first and foremost, we must have a displacement velocity less than u_{crit} to establish a stable front. To improve oil recovery, we must then seek to reduce the velocity as far below u_{crit} as is economically feasible.

12 The Injection of Water into the Reservoir

12.1 Introduction

In an oil reservoir which is not in contact with an aquifer, or where the water influx (Chap. 9) is not adequate to replace the volume of oil extracted, the reservoir pressure will decline steadily, perhaps even rapidly. As a consequence, the productivity of the wells will also deteriorate until, eventually, it will no longer be economically viable to continue production.

Furthermore, the efficiency E_D (Sect. 11.3.4) with which oil is displaced from the pores by any gas liberated in the reservoir because of the declining pressure is usually very small. A large part of the energy that might otherwise be available to drive the oil from pore to pore towards the producing wells is lost to the surface because the high mobility of the liberated gas allows it to finger through the oil to the wells.

The technology commonly used in this situation involves the injection of fluid into the reservoir such that, by replacing the produced oil volume for volume, the reservoir pressure can be maintained at a level sufficient to keep the wells producing within economic limits. The choice of injection fluid, and the method of injection, should of course achieve a better global oil recovery factor E_R (Sect. 4.3.7) than would be obtained by simple extraction of the oil with no injection (i.e. *depletion drive*). The fluids normally used for injection are:

- natural gas, not miscible with the oil, which is injected into the upper region of the reservoir in order to enhance the contribution of gravity to the displacement mechanism (Sects. 11.3.5 and 11.4.2).
- water, injected into the aquifer through wells usually drilled around the periphery of the reservoir, or injected into the oil-bearing part of the reservoir itself through injection wells alternating with producers in a predetermined geometrical pattern.

Traditionally, water or gas injection was only instigated when the reservoir showed signs of approaching the end of its primary production phase as its natural drive energy (Sect. 10.5.2) became exhausted. Injection was therefore – and still is, to some extent – considered to be a *secondary recovery process*. These days, injection is often implemented as part of the primary production strategy, either during, or immediately following the start of, production: in this context it is more appropriately referred to as an *improved recovery process*[22].

Chapter 15 will deal with improved recovery processes based on the injection of more complex, or "exotic" fluids, such as steam, carbon dioxide, surfactants and polymers, or hydrocarbon gases miscible with oil. These methods, aimed at recovering at least some part of the residual oil left after waterflooding, were traditionally referred to as *tertiary recovery processes*. The term *enhanced recovery* is more appropriate for present-day applications, because these processes can be employed even during the primary production phase: in the extreme case of very heavy and viscous oils, in fact, steam injection is often the only means of achieving any production at all.

To summarise, the so-called enhanced recovery processes constitute a subgroup of the broader category of "improved recovery" methods. In this context, the injection of water or immiscible natural gas into oil fall into a second subgroup, "conventional improved recovery" processes. Of the conventional improved recovery methods, water injection has seen by far the longest use, for the simple reason that water is almost universally available (from surface or subterranean sources), and costs considerably less than natural gas, whose present-day commercial value is, per unit of energy, only slightly less than that of oil. In addition, the equipment costs and operating expenses for water injection are lower than those for gas.

Gas injection is usually only favoured in remote locations, where there is no local use for the gas produced from the reservoir, and where the cost of transporting it to industrialised areas (including the investment in pipelines, gas liquefaction facilities and tankers to carry it in the form of LNG) is too high. By injecting it back into the reservoir, too, the gas is effectively being stored, leaving the possibility of some future commercialisation should economic conditions change.

12.2   The Development of Water Injection Technology

It is commonly accepted that the first instance of water injection into an oil reservoir came about quite by accident, through a mechanical failure in a producing well in the Pithole City[1] region of Pennsylvania in 1865. As a result of the failure of a crude packer, water from a shallow aquifer leaked into the producing interval. Although this killed the well in question, the production rates of nearby wells increased – a fact which was attributed to the pushing effect of the influx of water into the reservoir. The injection of water whose source is a water-bearing formation in the same well was the very first example of what was to become the widely used technique of "dump-flooding".

John F. Carll[15] was the first to suggest, in 1880, that water injected into an oil-bearing rock might actually displace the oil towards producing wells, causing a direct increase in recovery. The practice of water injection was the subject of much initial controversy, and was slow to spread. It was, in fact, prohibited in some states of the USA because it was considered to endanger the reservoir.

To begin with, injection was started in a single well in the field. As successive producing wells cut water and flooded out, they were converted to injectors. This gradual expansion of the reservoir area thus swept by injected water became known as "circle-flooding[15]". Water injection was practised initially only in the Bradford field area in Pennsylvania. There is evidence of subsequent cases in Ontario, Canada, in 1913, and the Kern River field in California in 1917[25].

The Forrest Oil Co. was the first operator to use a regular geometrical well pattern, injecting simultaneously into lines of injection wells alternating with lines of producers ("line drive"; Fig. 12.1). The "five spot" pattern (Fig. 12.1) was used for the first time in 1924 in the Bradford field. This pattern of injection and producing wells, if repeated many times, could cover a substantial area of the reservoir. In Oklahoma, USA, injection was first tried in the Bartlesville formation, Nowata County, in 1931: within a few years, almost all reservoir units producing

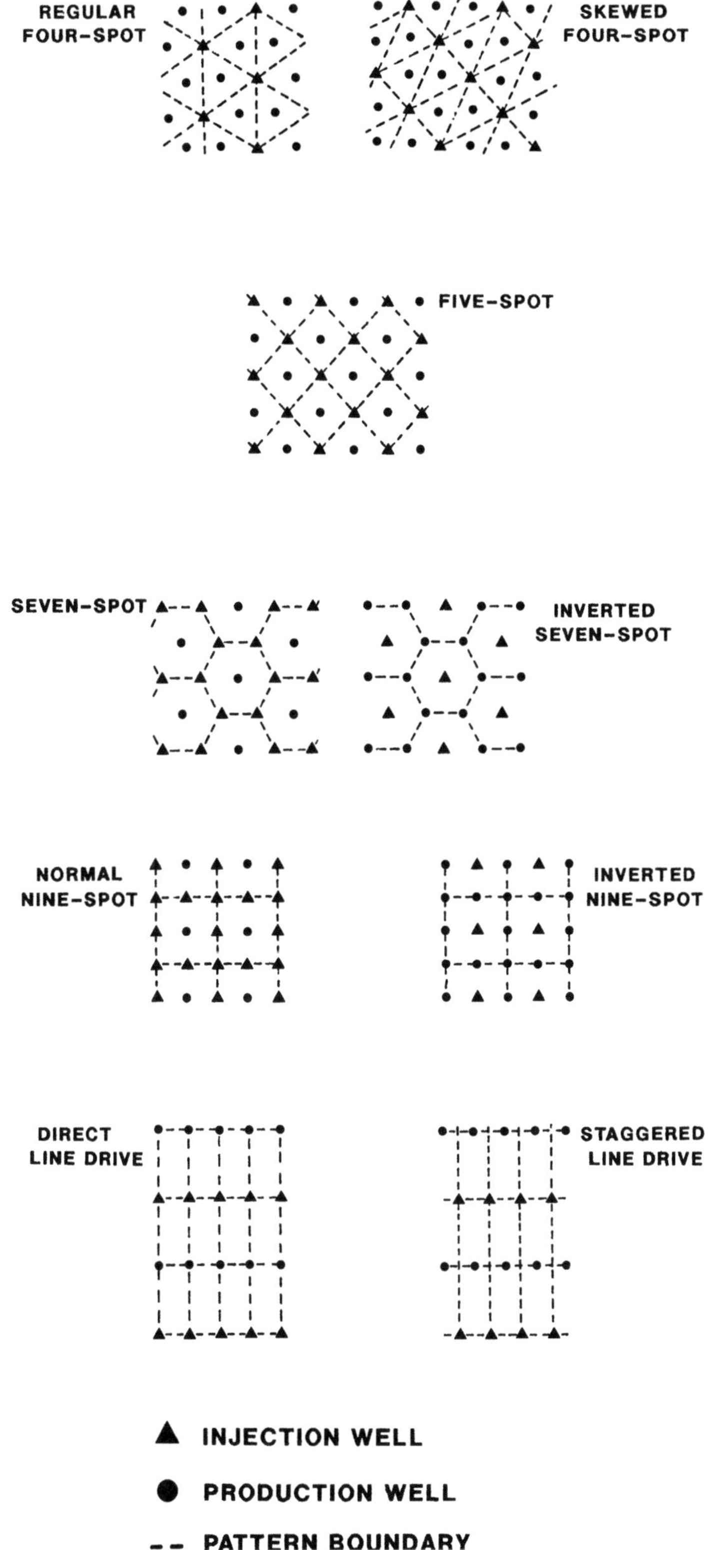

Fig. 12.1. *The most commonly used well patterns for water injection. (From Ref. 15, 1971, Society of Petroleum Engineers of AIME. Reprinted with permission of the SPE)*

from this formation were under water injection. The Fry formation, Brown County, became the first candidate for injection in Texas in 1936.

By World War II, water injection was quite widely used, especially throughout North America, but it was not until the 1950s that it gained recognition as an important and effective means of increasing oil recovery, in addition to maintaining reservoir pressure[15]. More than 50% of US oil production is currently from reservoirs undergoing water injection. It is now in use almost worldwide, especially in the so-called giant reservoirs (see Chap. 4) of the Middle East, North Africa and the CIS. In most cases, particularly in the USA, the well configurations tend to be of the "dispersed" type, the wells being distributed within the reservoir in regular geometrical patterns of alternating injectors and producers.

Four, five, seven and nine spot patterns are all in common use. As Fig. 12.1 illustrates, variations are possible within these configurations: *regular* and *skewed*, *normal* and *inverted*. In addition, line drive patterns can be *direct* or *staggered*. The chosen pattern is repeated so as to cover the area of the reservoir to be flooded. Patterns like these with close well spacings provide more immediate results from the producers, and allow higher injection rates than are possible with peripheral injection.

If the reservoir pressure has already fallen below the bubble point before water injection is initiated, the water must first replace as much as possible of the liberated gas in the pores. Little or no response will be observed at the producing wells until this "fill-up" period is well advanced. Some of the gas will redissolve, some will be pushed ahead of the water, or ahead of the oil being displaced by the water.

Water-flooding technology has made considerable advances over the years, to circumvent some of the problems often encountered during the process. Before injection, for instance, the water must be filtered to remove all solid particles from suspension. These would otherwise accumulate downhole in the injected zones, forming an impermeable barrier which could reduce or totally block injectivity.

For the same reason, the water must be treated with bactericides (usually chlorine or sodium hypochlorite) to inhibit the growth of bacteria in the reservoir. Bacterial colonies can develop into gelatinous masses which reduce permeability and, if of the sulphate-reducing type, convert sulphates in the water to H_2S. As well as altering the characteristics of the oil, this is also the potential cause of sometimes severe corrosion problems in the well completion and surface equipment. To further combat risks of corrosion, the water must also be totally deaerated (under vacuum, or by bubbling natural gas through it in packed columns), and treated with inhibitor. Water used for injection, therefore, is not just *any* water.

Chemical compatibility between the injection water, the reservoir rock, and the formation water is also a serious, and controversial, matter. The water must be of a salinity (or, more precisely, ionic strength) such that clay particles in the rock will not swell or flocculate. Both phenomena, although brought about by different mechanisms, would result in a loss of injectivity through a reduction in permeability. The mixing of the two different waters might result in the precipitation of insoluble salts. SO_4^{2-} ions in the injection water, for example, will react with any Ca^{2+}, Sr^{2+} or Ba^{2+} in the formation water to produce sulphates of calcium, strontium or barium. Such precipitation can lead to the plugging of pores – yet another potential loss of permeability.

Recent work[5] has shown that the consequences of precipitation in the reservoir itself are almost negligible. Serious problems are, however, often encountered at the producing wells. In fact, most of the precipitation takes place in the near-wellbore region of the producers, and, carried by the water, tends to deposit in the well itself as a hard and often highly radioactive "scale" on the inside walls of the casing and tubing. Scale build-up can be impeded by injection of inhibitor into the producing perforations, which covers the surface of each grain of precipitate with a thin film which discourages agglomeration. Otherwise, scaling may be so severe as to choke back the well, necessitating regular workover which may involve pulling and changing the tubing, and milling out the obstruction in the casing.

More recent studies of the internal structure of reservoir rock[11] have cast doubt on the universal usefulness of *regular* well patterns for water injection. In the case of highly heterogeneous reservoir rock, deposited in medium to high energy sedimentary environments, a regular well pattern gives no guarantee that every lens and pocket making up the reservoir is in communication with at least one injection and one producer. Ideally, then, it would be desirable to vary the well distribution on a local basis, so that all the oil is contacted by injected water and driven towards producing wells. This, however, would require an accurate and detailed knowledge of the internal structure of the reservoir.

12.3 Factors Influencing Oil Recovery by Water Injection

Generally speaking, in any process of displacement by fluid, either injected or, in the case of water or gas, emanating from an aquifer or gas cap, we indicate by $V_V(t)$ the volume of rock which has been flooded by the displacing fluid at time t. V_R is the total bulk volume of reservoir rock (Chap. 4).

$N_p(t)$ is the cumulative volume of oil produced at time t, and $\bar{S}_{or}(t)$ is the average residual oil saturation behind the displacement front (assuming that the reservoir pressure has not yet fallen below the bubble point; Sect. 2.2). Then:

$$N_p(t) = V_V(t)\phi \left\{ \frac{1 - S_{iw}}{B_{of,i}} - \frac{\bar{S}_{or}(t)}{B_{of}[p(t)]} \right\} \tag{12.1}$$

From Chap. 4 we have the relationship:

$$N = V_R\phi \frac{1 - S_{iw}}{B_{of,i}} \tag{12.2}$$

so that:

$$\frac{N_p(t)}{N} = \frac{V_V(t)}{V_R} \left\{ 1 - \frac{B_{of,i}}{B_{of}[p(t)]} \frac{\bar{S}_{or}(t)}{1 - S_{iw}} \right\} \tag{12.3}$$

The following nomenclature is commonly used:

$$E_{R,o}(t) = \frac{N_p(t)}{N} = \text{oil recovery factor at time } t \tag{12.4a}$$

$$E_V(t) = \frac{V_V(t)}{V_R} = \text{volumetric efficiency of the displacing fluid}$$

$$= \text{fraction of reservoir rock flooded by the fluid} \tag{12.4b}$$

and a term we have already encountered in Chap. 11:

$$E_D(t) = \frac{1 - S_{iw} - \bar{S}_{or}(t)}{1 - S_{iw}} = \text{microscopic displacement efficiency}$$

$$= \text{fraction of oil displaced in the pore within the volume of}$$
$$\text{reservoir that has been flooded.} \qquad (11.39)$$

With the above notation, if we assume B_{of} to be constant (i.e. the reservoir pressure does not vary significantly), we can derive the following from Eq. (12.3):

$$E_{R,o}(t) = E_V(t)E_D(t), \qquad (12.5)$$

which quite clearly means:

$$\text{(fraction of the reservoir oil volume produced at time } t)$$
$$= \text{(fraction of the reservoir volume flooded by the displacing fluid}$$
$$\text{at time } t) \times \text{(fraction of the oil displaced in the pores}$$
$$\text{within the flooded volume of the reservoir).}$$

At some moment in the producing life of the reservoir, the displacing fluid will break through at one or more wells. From then on, we will see a gradual increase in the fraction f_D of displacing fluid in the production stream.

In the case of oil displacement by water (injected or from an aquifer), we have

$$f_w = \frac{q_w}{q_t} = \frac{q_w}{q_o + q_w} \qquad (11.6)$$

In the field, f_w is referred to as the "water cut". A second term is also frequently used to define the water fraction – the "water/oil ratio" (WOR):

$$\text{WOR} = \frac{q_w}{q_o} \qquad (12.6)$$

From Eqs. (11.6) and (12.6) we have:

$$\text{WOR} = \frac{f_w}{1 - f_w} \qquad (12.7)$$

Provided there are no significant workovers affecting water production in the producing wells, f_w is a function $f_w(t)$ which is almost always uniquely dependent on time. Eq. (12.5) can therefore also be expressed as:

$$E_{R,o}(f_w) = E_V(f_w)E_D(f_w). \qquad (12.8a)$$

We are particularly interested in the values of E_D and E_V at the end of the producing life of the reservoir, when the water cut f_w will have reached the maximum level $f_{w,a}$ beyond which the costs of continued production would outweigh the revenue earned from the produced oil. What $f_{w,a}$ happens to be for a particular reservoir depends on local conditions (operating costs, royalties, taxation) and the current oil price. In the USA, values of $f_{w,a}$ as high as $0.97 \sim 0.98$ are acceptable.

From Eq. (12.8a) at abandonment we will have:

$$E_{R,o}(f_{w,a}) = E_V(f_{w,a})E_D(f_{w,a}), \qquad (12.8b)$$

where $E_{R,o}(f_{w,a}) = E_{R,a}$ is the *final recovery factor* for the reservoir oil.

With the present-day availability of numerical techniques to model reservoir behaviour (Chap. 13), the values of $N_p(t)$, $f_w(t)$ and, from them, $E_{R,o}(f_w)$, can be predicted directly by simulating the production/injection process for any desired distribution of wells and flow rates. $E_D(f_w)$ and $E_V(f_w)$, among others, will also be computed.

In the following pages we will look in some detail at how the functions $E_D(f_w)$ and $E_V(f_w)$ are influenced by various reservoir parameters. We will limit ourselves to some of the very simple displacement geometries encountered when the production and injection wells are arranged in regular patterns (Fig. 12.1). These cases do not require recourse to numerical simulation; in fact, we will see how the results obtained can be used to help in the detailed simulation of reservoir production by numerical methods.

12.4 E_D: Microscopic Displacement Efficiency

The relationship that exists between E_D, the *microscopic displacement efficiency* (on the scale of pores) and the water fraction in the production stream can be easily derived from the discussion in Chap. 11.

If $S_{w,e}$ is the water saturation at the exit face of the porous medium (the sand face of the producing well), and $f_{w,e}$ is the produced water cut, Eqs. (11.46) and (11.37) can be combined, for the case of linear flow, to give:

$$E_D(f_{w,e}) = \frac{1}{1 - S_{iw}} \left[S_{w,e} - S_{iw} + \frac{1 - f_{w,e}}{\left(\dfrac{df_w}{dS_w}\right)_{S_{w,e}}} \right] \tag{12.9}$$

Equation (12.9) is only valid for $f_{w,e} > 0$, which means $S_{w,e} > S_{iw}$.

For any given value of $f_{w,e}$, we read the corresponding $S_{w,e}$ and $(df_w/dS_w)_{S_{w,e}}$ from the $f_w(S_w)$ diagram (Fig. 12.2), and then compute $E_D(f_{w,e})$ from Eq. (12.9). The entire $E_D(f_w)$ curve is constructed point by point up as far as $f_{w,a}$, corresponding to the depletion of the reservoir. The $f_w(S_w)$ curve is calculated from Eq. (11.11) or, if we can ignore the effects of gravity and capillarity, from Eq. (11.12). Both of these equations contain the relative permeabilities $k_{rw}(S_w)$ and $k_{ro}(S_w)$ and the ratio μ_o/μ_w of the oil viscosity to that of the water displacing it.

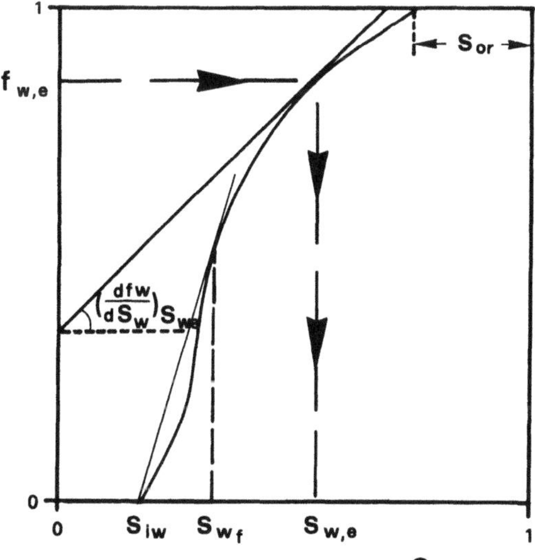

Fig. 12.2. *Graphic determination of $S_{w,e}$ and $(df_w/dS_w)_{S_{w,e}}$ for a given value of $f_{w,e}$*

In Sect. 11.5 we saw how in a reservoir, whose thickness h_t was less than the height h_c of the capillary transition zone, and which could therefore be treated as a one-dimensional system, the relative permeability curves could be taken directly from core measurements. On the other hand, in the case of two-dimensional (x, z) systems (Sect. 11.4), if we can assume displacement under vertical equilibrium conditions (Sect. 11.4.1), we use relative permeability pseudo-curves, determined as described in Sects. 11.5.2 or 11.5.3, depending on the ratio h_t / h_c.

The influence of the viscosity ratio μ_o / μ_w is expressed via the mobility ratio:

$$M_{wo} = \frac{k_{rw,or} \mu_o}{k_{ro,iw} \mu_w}. \tag{3.52a}$$

In Sect. 11.3.5 we saw that for a given f_w, E_D will be increased by a decrease in M_{wo}, such as might occur through a decrease in μ_o / μ_w. This explains why water is sometimes injected whose viscosity has been increased by means of "thickening" agents, such as certain polymers, to reduce μ_o / μ_w, thereby increasing the microscopic displacement efficiency.

12.5 E_V: Volumetric Efficiency

We will take the example of water injection in a five-spot pattern into a reservoir consisting of a number of layers of different permeabilities. Breakthrough will occur when the water injected in the central well first reaches the producing wells through the most permeable layer. At this time, the injection fronts in the other layers will have travelled only part of the distance to the producers – just how far depends on their respective permeabilities (Fig. 12.3). As injection continues, the water cut f_w will increase, and there will be a further frontal advance in all layers. The volumetric efficiency E_V (the fraction of the total volume of reservoir rock within the five-spot pattern that has been swept by water) therefore increases with f_w; in general, we can say that $E_V = E_V(f_w)$.

In order to make a distinction between the influence of geometrical factors (well positions and production rates) and reservoir heterogeneities on the displacement process, E_V can be expressed as a product of two terms:

$$E_V(f_w) = E_A(f_w) E_I(f_w), \tag{12.10}$$

where $E_A(f_w)$ = *areal sweep efficiency*: the ratio of the area swept by water in the most permeable layer, and the total area contained by the well pattern, when the producing water cut is f_w. $E_A(f_w)$ depends on the geometrical distribution of the wells, their production rates and the mobility ratio M_{wo}.
and $E_I(f_w)$ = *vertical invasion efficiency* or *conformance factor*: the ratio of the total volume of rock invaded by water, and the total volume of rock contained within a vertical surface passing through the water front in the most permeable layer, measured when the producing water cut is f_w. $E_I(f_w)$ depends on the vertical heterogeneity of the reservoir, and on M_{wo}.

We will now look at how $E_A(f_w)$ and $E_I(f_w)$ are affected by the mobility ratio M_{wo}. We will only consider the case of regular well patterns, because they are easier to quantify. Irregular well patterns must be studied with the help of numerical or analytical models[13].

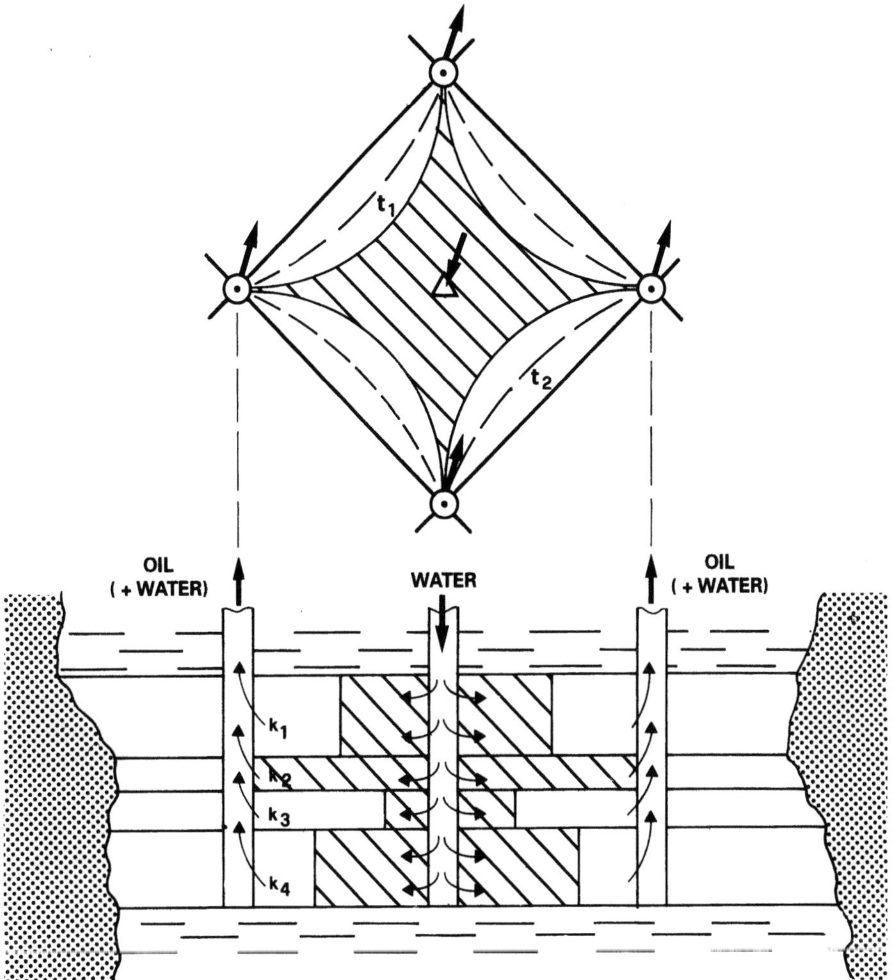

Fig. 12.3. *Representation of the volumetric efficiency E_V in a multilayered reservoir*

12.5.1 E_A: Areal Sweep Efficiency

The main research into areal sweep efficiency (E_A), its dependence on the mobility ratio M_{wo}, and its variation with the producing water cut f_w, was carried out in the 1950s using scale models of porous media[16,27,32] and electrical analogues[2].

Figures 12.4 and 12.5 have been reprinted from this work to demonstrate, respectively, the position of the water front at breakthrough for several different mobility ratios in one-quarter of a five-spot pattern, and the movement of the water front in a line drive (alternating lines of injection wells and producing wells).

Mainly through work on scale models[10], the behaviour of E_A as a function of f_w for different values of M_{wo} has been quantified for the principal well patterns (Fig. 12.1). An example of the results obtained[8] for a five-spot pattern is shown in Fig. 12.6. It is apparent that the higher M_{wo}, the smaller E_A for a given f_w.

We can conclude that for very viscous oils ($\mu_o/\mu_w > 1$; $M_{wo} > 1$), any increase in the injected water viscosity (achieved by the addition of polymers) to reduce the value of μ_o/μ_w and M_{wo} can only improve the areal sweep efficiency and, consequently, the oil recovery factor $E_{R,o}$.

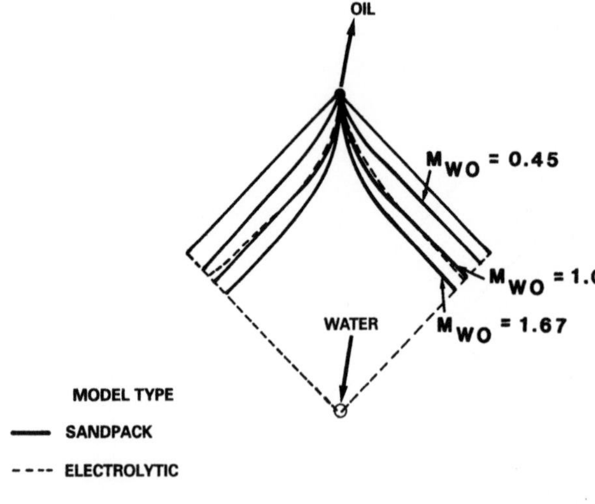

Fig. 12.4. *The position of the displacement front at breakthrough for different oil/water mobility ratios M_{wo} in a five-spot pattern. (From Ref. 27, 1952, Society of Petroleum Engineers of AIME. Reprinted with permission of the SPE)*

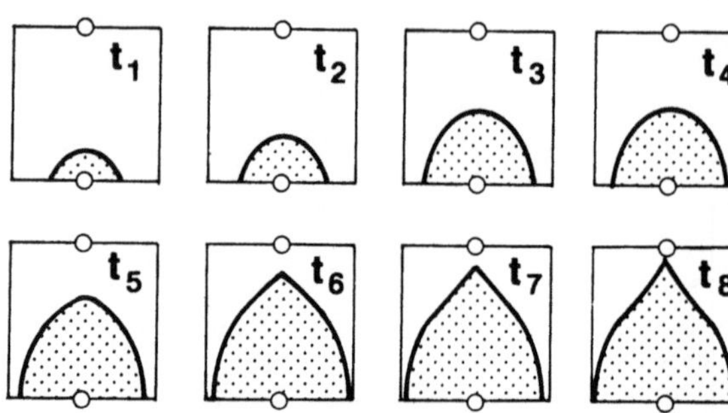

Fig. 12.5. *Advance of the water front with time in a line-drive configuration. (From Ref. 32, 1933, Society of Petroleum Engineers of AIME. Reprinted with permission of the SPE)*

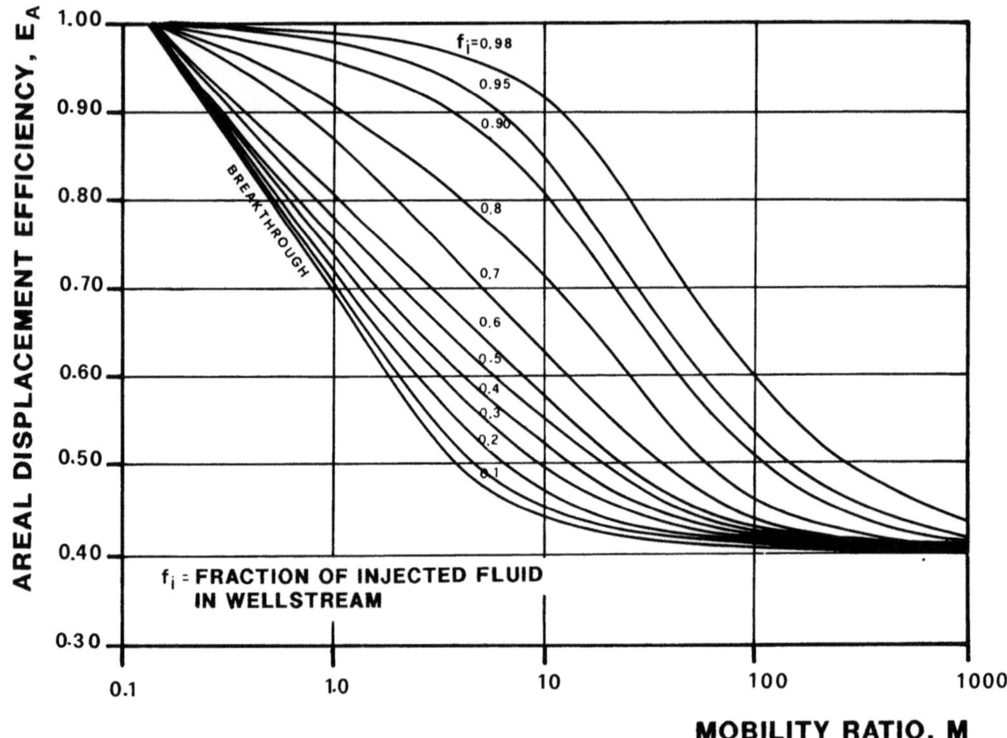

Fig. 12.6. *Five-spot injection pattern. Areal sweep efficiency E_A as a function of the mobility ratio M_{wo}, from the initial breakthrough of the displacing fluid until the fraction of displacing fluid in the production stream reaches 0.98. (From Ref. 8, 1967, Society of Petroleum Engineers of AIME. Reprinted with permission of the SPE)*

Table 12.1. *Coefficients of Eq. (12.11) for different well patterns*

| | | Line drive | |
	Five spot	Direct	Staggered
a_1	−0.2062	−0.3014	−0.2077
a_2	−0.0712	−0.1568	−0.1059
a_3	−0.511	−0.9402	−0.3526
a_4	0.3048	0.3714	0.2608
a_5	0.123	−0.0865	0.2444
a_6	0.4394	0.8805	0.3158

The relationship between E_A, f_w and M_{wo} for the more commonly used of the "dispersed" injection patterns is described by the following equation:

$$\frac{1 - E_A}{E_A} = [a_1 \ln(M_{wo} + a_2) + a_3] f_w + a_4 \ln(M_{wo} + a_5) + a_6 \tag{12.11}$$

This was proposed by Fassihi[19], who evaluated and interpolated the results of the major work on scale models. The coefficients appearing in Eq. (12.11) are listed in Table 12.1.

12.5.2 E_I: Vertical Invasion Efficiency

Although a few published references[31] deal with the tentative application of radial geometry to the determination of the vertical invasion efficiency E_I in reservoirs consisting of several layers (or "zones") with different petrophysical properties, it is more common to assume linear flow. Only linear flow will be covered in the following pages. There are two broad categories:

- no vertical communication between layers: k_v is therefore zero at the interfaces between layers. This is the most common situation.
  In Sect. 3.6 we saw that each layer represents a sedimentary unit corresponding to a different depositional history (conditions and material). Between successive cycles of sedimentation there is frequently a low energy period during which fine particles (silt, clay) are deposited. These effectively seal off the top of the unit from the overlying one formed in the next cycle.
- vertical communication between layers: k_v is not zero at the interfaces.

In both cases, each layer is internally homogeneous *in a statistical sense*, and is characterised by mean values of its petrophysical properties (ϕ, k, S_{iw}, S_{or}, etc.), as well as their standard deviations from the mean, as described in the context of permeability in Sect. 3.5.1.7.

12.5.2.1 *Layers Not in Vertical Communication*

To start, we shall consider a statistically homogeneous horizontal layer of thickness h, length L, and of unit width (Fig. 12.7). ϕ, k, S_{iw} are the mean values of the porosity, permeability and irreducible water saturation in the layer. It is assumed

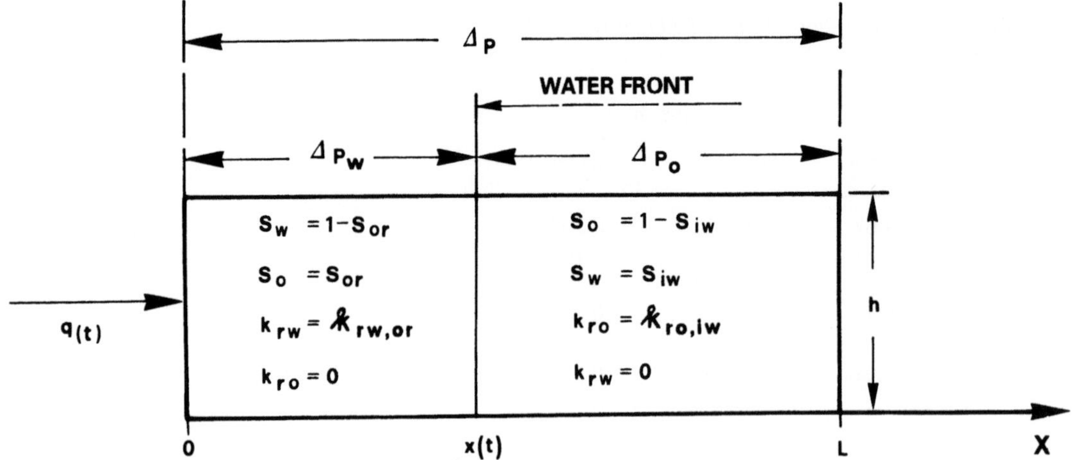

Fig. 12.7. *The pressure drops, saturations and relative permeabilities behind and ahead of the water front in a homogeneous medium*

initially to contain only oil in the presence of water at S_iw, so there is no free gas present, and therefore no "fill-up" period where the incoming water replaces gas. The reservoir oil and injected water will be assumed incompressible.

At $t = 0$, injection of water commences at one end of the layer. This initiates linear flow under a potential difference (pressure in this case, the layer being horizontal) which is constant and equal to Δp, between the inlet and outlet ends of the layer.

We will assume piston-like displacement under conditions of vertical equilibrium (Sect. 11.4.1), so that behind the front we will find oil at a saturation S_or with water at saturation $(1 - S_\text{or})$. Here, then, the effective permeability to oil will be zero, while that to water will be $kk_\text{rw,or}$. Ahead of the front will be oil at $S_\text{o} = (1 - S_\text{iw})$ and water at S_iw. The effective permeability to water is now zero, and to oil $kk_\text{ro,iw}$. At $t = 0$, the instant at which water injection starts, the flow rate $q(t = 0)$ of fluid (oil) leaving from the opposite end is:

$$q(t = 0) = \frac{khk_\text{ro,iw}}{\mu_\text{o}} \frac{\Delta p}{L} \tag{12.12}$$

If at any general time t, $q(t)$ is the fluid flow rate in the layer, $x(t)$ the position of the displacement front, and $\Delta p_\text{w}(t)$ and $\Delta p_\text{o}(t)$ the pressure drops across, respectively, the region flooded by water and the region ahead of the front (Fig. 12.7), then:

$$q(t) = \frac{khk_\text{rw,or}}{\mu_\text{w}} \frac{\Delta p_\text{w}(t)}{x(t)} = \frac{khk_\text{ro,iw}}{\mu_\text{o}} \frac{\Delta p_\text{o}(t)}{L - x(t)} \tag{12.13}$$

Now, we have stipulated that:

$$\Delta p_\text{w}(t) + \Delta p_\text{o}(t) = \Delta p = \text{ const.} \tag{12.14}$$

which, combined with Eqs. (12.12) and (12.13), gives:

$$\frac{q(t)}{q(t = 0)} = \frac{v_\text{f}(t)}{v_\text{f}(t = 0)} = \frac{M_\text{wo}}{M_\text{wo} + (1 - M_\text{wo})\dfrac{x(t)}{L}} \tag{12.15}$$

v_f being the instantaneous frontal velocity.

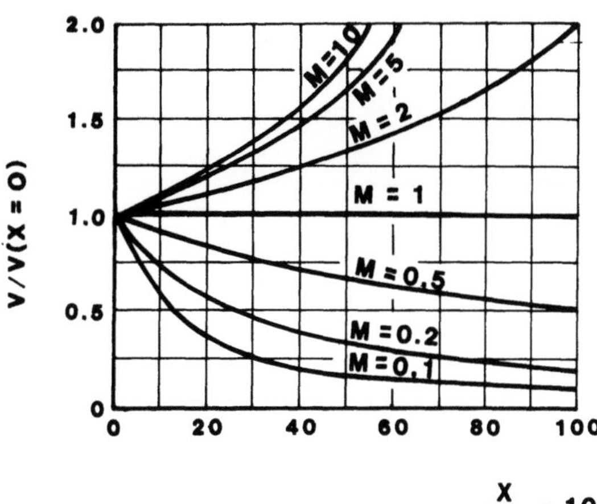

$$\frac{x}{L} \cdot 100$$

Fig. 12.8. *Relative frontal advance velocities during the displacement of oil by water, for different mobility ratios M. (From Ref. 8, 1967, Society of Petroleum Engineers of AIME. Reprinted with permission of the SPE)*

The following observations can be made about Eq. (12.15) (see Fig. 12.8):

for $M_{wo} = 1$:   the velocity of frontal advance remains constant (equal to its initial value) during the displacement.

for $M_{wo} > 1$:   the denominator in Eq. (12.15) decreases as x increases: so the front moves progressively faster as it traverses the porous medium.

for $M_{wo} < 1$:   the denominator in Eq. (12.15) increases as x increases: so the front moves progressively slower as it traverses the porous medium.

Since:

$$v_f = \frac{dx}{dt} \tag{12.16a}$$

and:

$$v_f(t) = \frac{q(t)}{\phi h (1 - S_{iw} - S_{or})} \tag{12.16b}$$

we obtain from Eq. (12.15):

$$L \left[M_{wo} + (1 - M_{wo})\frac{x}{L} \right] d\left(\frac{x}{L}\right) = v_f(t = 0) M_{wo} \, dt$$

and therefore:

$$\int_0^{x/L} \left[M_{wo} + (1 - M_{wo})\frac{x}{L} \right] d\left(\frac{x}{L}\right) = \frac{q(t = 0) M_{wo}}{\phi h L (1 - S_{iw} - S_{or})} t \tag{12.17a}$$

where:

$$\phi h L (1 - S_{iw} - S_{ow}) = N_{p,u} \tag{12.17b}$$

represents the volume of oil that will have been produced at the time of water breakthrough at the outlet end of the layer. Since this is piston-like displacement, this will also be the maximum volume recoverable from the layer.

If we integrate Eq. (12.17a) and rearrange the terms we get:

$$(1 - M_{wo})\left(\frac{x}{L}\right)^2 + 2M_{wo}\frac{x}{L} - \frac{2q(t = 0) M_{wo} t}{N_{p,u}} = 0 \tag{12.18a}$$

from which:

$$\frac{x}{L} = \frac{-M_{\mathrm{wo}} + \sqrt{M_{\mathrm{wo}}^2 + \dfrac{2q(t=0)M_{\mathrm{wo}}(1-M_{\mathrm{wo}})t}{N_{\mathrm{p,u}}}}}{1 - M_{\mathrm{wo}}} \qquad (12.18\mathrm{b})$$

Water will break through at the exit face ($x/L = 1$; $t = t_{\mathrm{BT}}$) when:

$$M_{\mathrm{wo}}^2 + \frac{2q(t=0)M_{\mathrm{wo}}(1-M_{\mathrm{wo}})t_{\mathrm{BT}}}{N_{\mathrm{p,u}}} = 1 \qquad (12.19\mathrm{a})$$

i.e., when:

$$t_{\mathrm{BT}} = \frac{1+M_{\mathrm{wo}}}{M_{\mathrm{wo}}} \frac{N_{\mathrm{p,u}}}{2q(t=0)} \qquad (12.19\mathrm{b})$$

Equation (12.18b) is undefined as zero/zero when $M_{\mathrm{wo}} = 1$, and becomes meaningless. In this case, $q(t) = q(t=0) = $ const., so that from Eq. (12.16):

$$\frac{x}{L} = \frac{q(t=0)t}{\phi h(1 - S_{\mathrm{iw}} - S_{\mathrm{or}})L} = \frac{q(t=0)t}{N_{\mathrm{p,u}}} \qquad (12.20)$$

The value of t_{BT} calculated from Eq. (12.20) will be the same as that obtained from Eq. (12.19b) with $M_{\mathrm{wo}} = 1$.

We will now consider the multilayered reservoir depicted in Fig. 12.9, consisting of n horizontal strata containing oil with a viscosity μ_{o}, and not in vertical communication. Each layer has its own, probably different, set of values for h, ϕ, k, S_{iw}, S_{or}, $k_{\mathrm{ro,iw}}$, $k_{\mathrm{rw,or}}$. We wish to calculate E_{I} as a function of f_{w} in a block of the reservoir L in length and of unit width. The procedure to follow is summarised as follows:

1. Assume an arbitrary value for Δp, and calculate the initial flow rate $q(t=0)$ in each layer using Eq. (12.12). Calculate M_{wo} and $N_{\mathrm{p,u}}$ [Eq. (12.17b)] for each layer as well.
2. Compute t_{BT} for each layer using Eq. (12.19b). $(t_{\mathrm{BT}})_{\mathrm{max}}$, the maximum value of t_{BT}, will correspond, of course, to the layer with the lowest permeability.
3. Divide the time interval $[0-(t_{\mathrm{BT}})_{\mathrm{max}}]$ into a large number of times t, to include all of the t_{BT} values calculated in step 2 for the other layers.

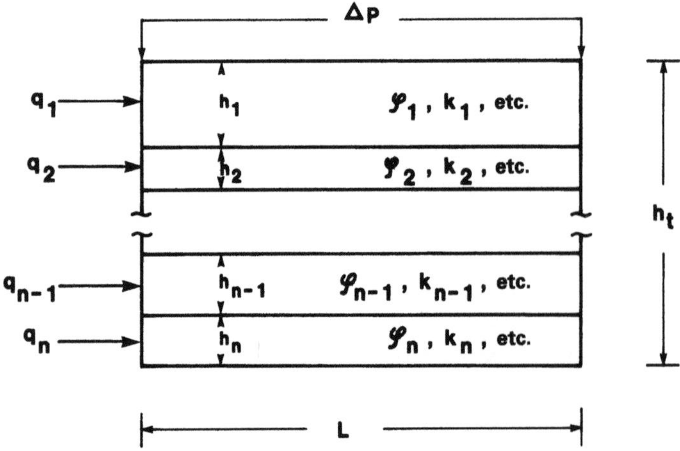

Fig. 12.9. *Representation of a multilayered reservoir*

4. Use Eq. (12.18b) to calculate x/L in each layer at each of the times t. For $t \geq t_{BT}$, the layer produces only water, so the calculation can be terminated in any layer when $t = t_{BT}$, i.e. when $x/L = 1$.

5. Calculate $q(t)$ in each layer at each time t (i.e. at each x/L) with Eq. (12.15). Note that as long as $x/L < 1$, the produced fluid is oil; if $x/L = 1$, the produced fluid is water.

6. At any time t, the total oil flow rate $q_o(t)$ in this section of reservoir is:

$$q_o(t) = \Sigma_{(x/L)<1}\, q(t) \tag{12.21a}$$

where the summation is performed in all layers where $x/L < 1$. The total water flow rate $q_w(t)$ is:

$$q_w(t) = \Sigma_{(x/L)=1}\, q(t) \tag{12.21b}$$

where, here, the summation is performed in all layers where $(x/L) = 1$.
We can now get $f_w(t)$ from:

$$f_w(t) = \frac{q_w(t)}{q_w(t) + q_o(t)} \tag{12.22}$$

7. When computing the vertical invasion efficiency $E_I(t)$ at time t, the contribution from each layer is weighted according to its ϕ, S_{iw} and S_{or}:

$$E_I(t) = \frac{\displaystyle\sum_{1}^{n}{}_i\left[N_{p,u}\frac{x(t)}{L}\right]_i}{\displaystyle\sum_{1}^{n}{}_i(N_{p,u})_i} \tag{12.23}$$

8. The curve:

$$E_I = E_I(f_w) \tag{12.24}$$

can then readily be constructed from the curves $E_I(t)$ and $f_w(t)$, and the objective of the exercise is met.

This procedure, which can be performed with a calculator, is based, with a few modifications, on the methods proposed by Dykstra and Parsons[17] for $M_{wo} \neq 1$ and Stiles[30] for $M_{wo} = 1$.

Dykstra and Parsons go on to present a series of diagrams of E_I as a function of M_{wo} for various values of f_w. These were calculated for statistically homogeneous reservoir rock, with the degree of heterogeneity characterised by the *permeability variation* defined by Law in Eq. (3.45a). One of these diagrams has been reproduced in Fig. 12.10.

These diagrams were based on the rather tenuous assumption that the reservoir rock consists of a series of strata which are *areally continuous over the entire reservoir*, with no communication between them, and with permeabilities distributed in a vertical sense as defined by the permeability variation. However, we know from Sect. 3.6 that the permeability "lenses" in any statistically homogeneous reservoir rock are likely to have limited areal extent. In addition, even if they are indeed isolated from one another vertically, they may well be in communication laterally. This renders Dykstra and Parsons' hypothesis physically questionable. Chierici et al.[14] have extended the method described above to cover the case where displacement in each layer is *dispersed* rather than *piston-like* (Sect. 11.4).

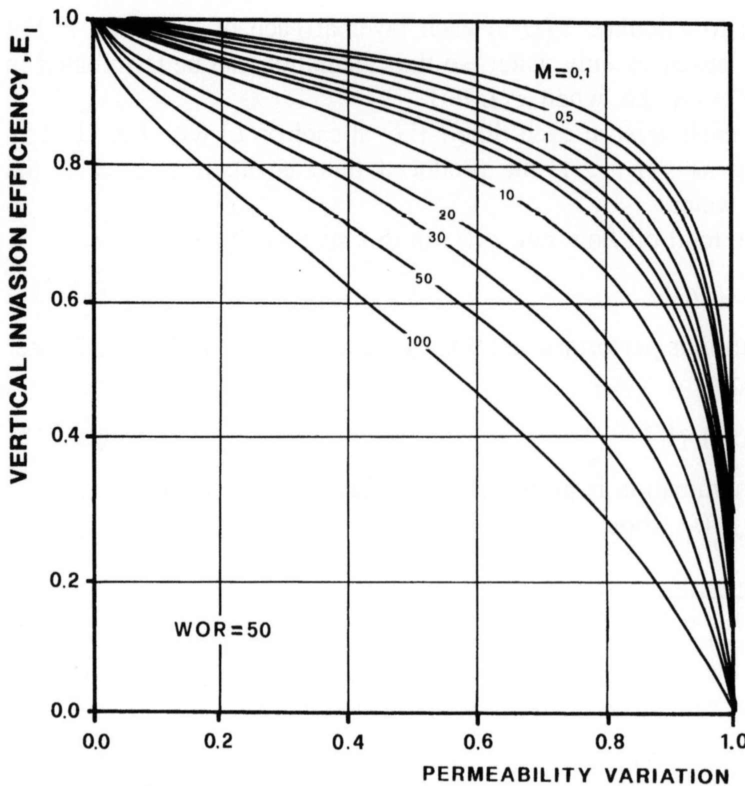

Fig. 12.10. *Vertical invasion efficiency E_I as a function of the permeability variation and the mobility ratio, for a WOR of 50. (From Ref. 19, 1986, Society of Petroleum Engineers of AIME. Reprinted with permission of the SPE)*

A totally different approach can be used to model the displacement of oil by water in a linear system made up of a number of layers not in vertical communication. This consists of defining an equivalent *homogeneous* system and using relative permeability pseudo-curves (Sect. 11.5). The methodology will be explained briefly.

The permeability $\bar{k}$ of the homogeneous porous medium with the same hydrodynamic behaviour as the multilayered system will be

$$\bar{k} = \frac{\sum_{1}^{n}{}_{i}(hk)_i}{h_t},\tag{12.25}$$

where h_t is the total thickness of the reservoir. The irreducible water saturation $\bar{\bar{S}}_{iw}$ is

$$\bar{\bar{S}}_{iw} = \frac{\sum_{1}^{n}{}_{i}(\phi h S_{iw})_i}{\sum_{1}^{n}{}_{i}(\phi h)_i},\tag{12.26}$$

while the "pseudo" relative permeability to oil $\bar{\bar{k}}_{ro,iw}$ at $\bar{\bar{S}}_{iw}$ is:

$$(\bar{\bar{k}}_{ro})_{\bar{\bar{S}}_{iw}} = \frac{\sum_{1}^{n}{}_{i}[hkk_{ro,iw}]_i}{\sum_{1}^{n}{}_{i}(hk)_i} = \frac{\sum_{1}^{n}{}_{i}[hkk_{ro,iw}]_i}{h_t\bar{k}}.\tag{12.27}$$

From Eqs. (12.12) and (12.27) we get:

$$\frac{\bar{k}h_t\bar{\bar{k}}_{\mathrm{ro,iw}}}{\mu_o}\frac{\Delta p}{L} = \sum_1^n {}_i[q(t=0)]_i. \tag{12.28}$$

At any time t, the mean water saturation $(\bar{\bar{S}}_w)_t$ in the porous medium, assuming piston-like displacement of oil by water, will be:

$$(\bar{\bar{S}}_w)_t = \frac{\sum\limits_1^n {}_i \left\{ \phi h \left[S_{iw} + \frac{x(t)}{L}(1 - S_{or} - S_{iw}) \right] \right\}_i}{\sum\limits_1^n {}_i(\phi h)_i}, \tag{12.29}$$

where $x(t)/L$ is calculated through Eq. (12.18b) for each layer.

If $\bar{\bar{k}}_{\mathrm{ro}}(\bar{\bar{S}}_w)_t$ and $\bar{\bar{k}}_{\mathrm{rw}}(\bar{\bar{S}}_w)_t$ are the pseudo-relative permeabilities to oil and water respectively at saturation $(\bar{\bar{S}}_w)_t$ in the equivalent homogeneous porous medium, then from Eq. (12.21) we have:

$$\frac{\bar{k}h_t\bar{\bar{k}}_{\mathrm{ro}}(\bar{\bar{S}}_w)_t}{\mu_o}\frac{\Delta p}{L} = \sum_{(x/L)<1} q(t) \tag{12.30a}$$

and

$$\frac{\bar{k}h_t\bar{\bar{k}}_{\mathrm{rw}}(\bar{\bar{S}}_w)_t}{\mu_w}\frac{\Delta p}{L} = \sum_{(x/L)=1} q(t). \tag{12.30b}$$

Dividing Eqs. (12.30a) and (12.30b) by Eq. (12.28) we get:

$$\bar{\bar{k}}_{\mathrm{ro}}(\bar{\bar{S}}_w)_t = \frac{\sum_{(x/L)<1} q(t)}{\sum\limits_1^n {}_i[q(t=0)]_i}\,\bar{\bar{k}}_{\mathrm{ro,iw}} \tag{12.31a}$$

and

$$\bar{\bar{k}}_{\mathrm{rw}}(\bar{\bar{S}}_w)_t = \frac{\mu_w}{\mu_o}\frac{\sum_{(x/L)=1} q(t)}{\sum\limits_1^n {}_i[q(t=0)]_i}\,\bar{\bar{k}}_{\mathrm{ro,iw}} \tag{12.31b}$$

Now by calculating $(\bar{\bar{S}}_w)_t$, $\bar{\bar{k}}_{\mathrm{ro}}(\bar{\bar{S}}_w)_t$, and $\bar{\bar{k}}_{\mathrm{rw}}(\bar{\bar{S}}_w)_t$ for increasing values of $t \leq (t_{\mathrm{BT}})_{\max}$, we obtain the pseudo-curves for relative permeability which will allow us to describe the behaviour of the real multilayered porous medium by means of the equivalent homogeneous system. In this way, the influence of the vertical invasion efficiency E_I is automatically incorporated in the relative permeability pseudo-curves. The procedure described above is particularly useful in the numerical simulation of layered reservoir behaviour.

12.5.2.2 Layers with Vertical Communication

When the layers in the reservoir are in vertical communication, $E_I(f_w)$ is conveniently calculated using a method which is in some respects the opposite to that used for non-communicating layers. The relative permeability pseudo-curves are calculated first. From these we derive the fractional flow equation $f_w(S_w)$, and, from this, $E_I(f_w)$.

As far as what has been written in Sect. 11.5 about the pseudo-curves for relative permeability and capillary pressure is concerned, there are two cases to consider: the displacement *in each layer* of the reservoir may be dispersed (when the layer is of about the same thickness as the thickness of the capillary transition zone), or segregated (layer much thicker than the capillary transition zone).

In the case of a *dispersed flow regime* in each layer, we can apply the procedure outlined in Sect. 11.5.1 for homogeneous layers.

While:

$$P_{c,ow}(h) = g(\rho_w - \rho_o)h \tag{3.17}$$

(where h is the height above the free water level) is continuous over the thickness of the reservoir, the curve $S_w(P_{c,ow})$ [$S_w(h)$ in other words] may exhibit discontinuities at the interface of two adjacent layers with different petrophysical properties (Fig. 3.14).

If $\bar{\bar{S}}_w(t)$ is the mean water saturation at time t over the thickness h_t of a layer, and $S_w(z, t)$ is the water saturation at time t at a height z above the base of the layer, we have:

$$\bar{\bar{S}}_w(t) = \frac{\int_0^{h_t} \phi(z) S_w(z, t)\, dz}{\int_0^{h_t} \phi(z)\, dz}. \tag{12.32}$$

In other words, if there are n layers, and we assign the subscript i to denote the properties of the ith layer:

$$\bar{\bar{S}}_w(t) = \frac{\sum_{i}^{n} h_i \phi_i \bar{S}_{w,i}(t)}{\sum_{i}^{n} h_i \phi_i} \tag{12.33a}$$

with:

$$\bar{S}_{w,i}(t) = \frac{\int_0^{h_i} S_w(z, t)\, dz}{h_i}. \tag{12.33b}$$

The relative permeability to water at $\bar{\bar{S}}_w(t)$ will be:

$$\bar{\bar{k}}_{rw}[\bar{\bar{S}}_w(t)] = \frac{\int_0^{h_t} k(z) k_{rw}[S_w(z, t)]\, dz}{\int_0^{h_t} k(z)\, dz} \tag{12.34a}$$

and to oil:

$$\bar{\bar{k}}_{ro}[\bar{\bar{S}}_w(t)] = \frac{\int_0^{h_t} k(z) k_{ro}[S_w(z, t)]\, dz}{\int_0^{h_t} k(z)\, dz}. \tag{12.34b}$$

The integrals can be expressed as summations:

$$\bar{\bar{k}}_{rw}[\bar{\bar{S}}_w(t)] = \frac{\sum_{i}^{n} h_i k_i \bar{k}_{rw,i}[\bar{S}_{w,i}(t)]}{\sum_{i}^{n} h_i k_i} \tag{12.35a}$$

and

$$\bar{\bar{k}}_{ro}[\bar{\bar{S}}_w(t)] = \frac{\sum_{1}^{n} {}_i h_i k_i \bar{k}_{ro,i}[\bar{S}_{w,i}(t)]}{\sum_{1}^{n} {}_i h_i k_i} \tag{12.35b}$$

with:

$$\bar{k}_{r,i}[\bar{S}_{w,i}(t)] = \frac{\int_0^{h_i} k_r[S_w(z,t)]\,dz}{h_i} \tag{12.36}$$

For any value of z, $S_w(t)$ increases with t as the water-oil contact (where $S_w = 1 - S_{or}$) rises. The height of the contact above the base of the reservoir at time t is denoted by $z_{S_{or}}(t)$.

Finally, using a computer, calculate $\bar{\bar{S}}_w$ from Eq. (12.32) or (12.33), and the corresponding values of $\bar{\bar{k}}_{rw}$ and $\bar{\bar{k}}_{ro}$ from Eqs. (12.34) or (12.35), for different heights $z_{S_{or}}$ so as to define, point by point, the curves $\bar{\bar{k}}_{rw}(\bar{\bar{S}}_w)$ and $\bar{\bar{k}}_{ro}(\bar{\bar{S}}_w)$.

In the case of *segregated flow*, where the displacement of oil by water in each layer is piston-like, we calculate the relative permeability pseudo-curves at terms of the times t_j at which each of the n layers becomes fully flooded by the rising water–oil contact.

When the height $z_{S_{or}}$ of the contact coincides with the top of the jth layer (j increases from the bottom of the reservoir upwards), we have:

$$\bar{\bar{S}}_{w,j} = \frac{\sum_{1}^{j} {}_i h_i \phi_i (1 - S_{or,i}) + \sum_{j+1}^{n} {}_i h_i \phi_i S_{iw,i}}{\sum_{1}^{n} {}_i h_i \phi_i}; \tag{12.37a}$$

$$\bar{\bar{k}}_{rw}(\bar{\bar{S}}_{w,j}) = \frac{\sum_{1}^{j} {}_i h_i k_i k_{rw,or}}{\sum_{1}^{n} {}_i h_i k_i}; \tag{12.37b}$$

$$\bar{\bar{k}}_{ro}(\bar{\bar{S}}_{w,j}) = \frac{\sum_{j+1}^{n} {}_i h_i k_i k_{ro,iw}}{\sum_{1}^{n} {}_i h_i k_i}. \tag{12.37c}$$

If we consider all values of j from 1 to n, we will obtain a series of points defining the relative permeability pseudo-curves for oil and water, derived from the true relative permeability curves. Note that the displacement process and the shape of the pseudo-curves do not depend solely on the petrophysical properties and thickness of each layer, but also on *the order of the layering in the reservoir.*

For example, if the permeability of the layers increases from bottom to top, the preferential advance of water in the upper, more permeable, layers is counterbalanced by gravitational forces. This tends to produce a fairly uniform vertical distribution of saturation, as illustrated in Fig. 12.11. Such a sequence of permeabilities might have resulted from the deposition of successively coarser sediments during a marine transgression ("fining downwards").

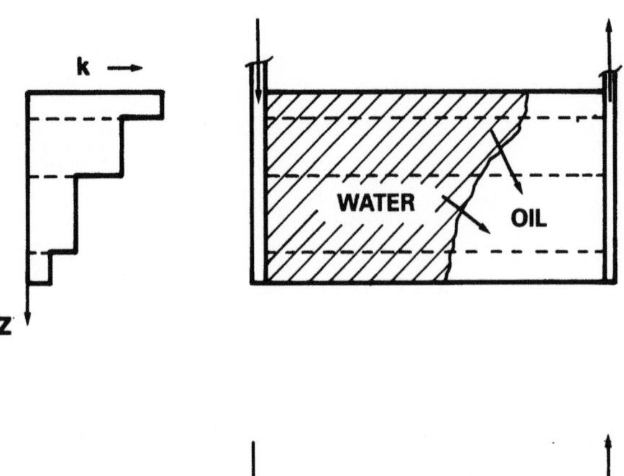

Fig. 12.11. *Shape of the water front in a reservoir consisting of a number of layers in vertical communication, with permeability increasing upwards*

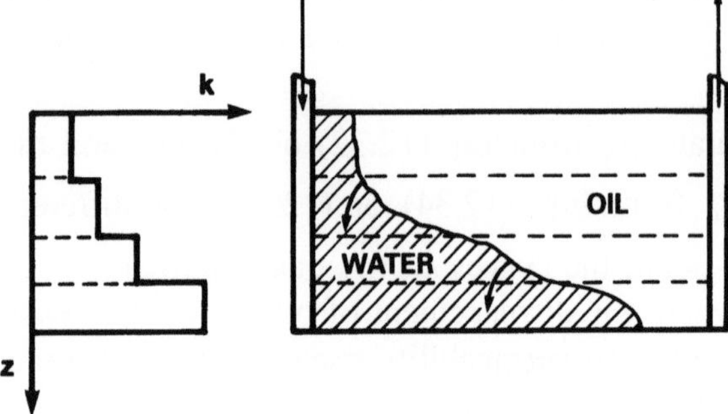

Fig. 12.12. *Shape of the water front in a reservoir consisting of a number of layers in vertical communication, with permeability decreasing upwards*

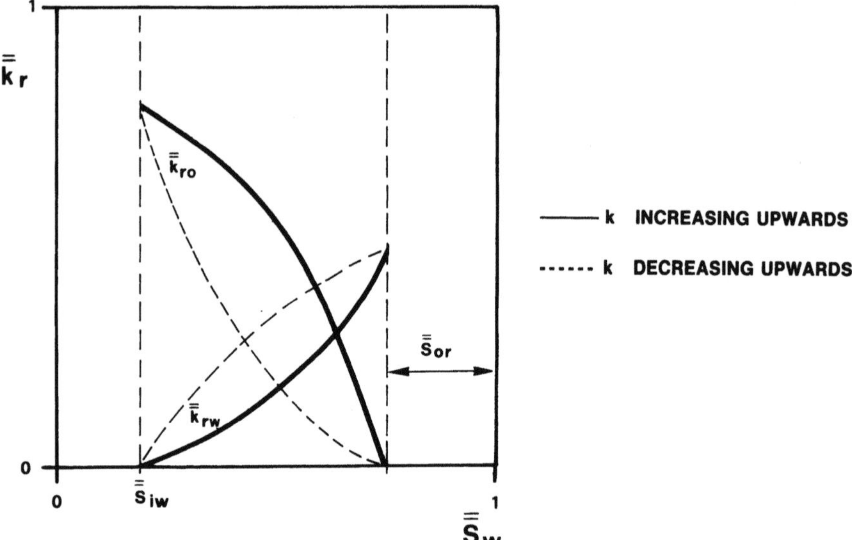

Fig. 12.13. *Relative permeability pseudo-curves for two sedimentary sequences, one with permeability increasing upwards, the other decreasing upwards, but otherwise identical*

The opposite occurs when the permeability *decreases* from the bottom up: the tendency for the water to advance preferentially along the lower layers is exacerbated by gravitational forces (Fig. 12.12). This sort of permeability distribution might have been caused by marine ingression, depositing successively smaller sediments ("fining upwards"). The behaviour of the pseudo-curves in these two cases is illustrated qualitatively in Fig. 12.13.

As has already been amply discussed in Sects. 11.1, 11.4.1 and 11.5, the pseudo-curves obtained by the methods described above allow us to model the displacement process in real reservoirs – two-dimensional in the vertical plane, and heterogeneous in the sense of being multilayered – by means of an equivalent system, one-dimensional in the vertical plane and homogeneous. In particular, these pseudo-curves, coupled with the oil and water viscosities, provide the means of calculating the fractional flow curve $f_w(\bar{S}_w)$ (Sect. 11.3.1). From this, using Welge's equation [Eq. (11.37)], the mean water saturation in the porous medium, $\bar{S}_{w,e}$, can be estimated as a function of the produced water fraction f_w.

The $\bar{\bar{S}}_w(f_w)$ curve obtained in this way can easily be interpreted in terms of the vertical invasion efficiency E_I.

Consider a volume V_R of reservoir rock consisting of a number of overlying layers which are in vertical communication. The vertical invasion efficiency when the produced water fraction is f_w is $E_I(f_w)$.

Assuming piston-like displacement, so that the oil behind the front is at its residual saturation S_{or}, the volume of produced oil $N_p(f_w)$ will be:

$$N_p(f_w) = \bar{\bar{\phi}} V_R E_I(f_w)(1 - \bar{\bar{S}}_{iw} - \bar{\bar{S}}_{or}) \tag{12.38a}$$

In terms of the equivalent homogeneous porous medium, this is:

$$N_p(f_w) = \bar{\bar{\phi}} V_R [\bar{\bar{S}}_w(f_w) - \bar{\bar{S}}_{iw}] \tag{12.38b}$$

In Eqs. (12.38), the double bar over each term indicates that the weighted mean value has been taken over all layers. In particular, $\bar{\bar{S}}_w(f_w)$ is calculated by means of Welge's fractional flow equation using the relative permeability pseudo-curves.

Combining Eqs. (12.38) we get:

$$E_I(f_w) = \frac{\bar{\bar{S}}_w(f_w) - \bar{\bar{S}}_{iw}}{1 - \bar{\bar{S}}_{iw} - \bar{\bar{S}}_{or}} \tag{12.39}$$

This expresses the relationship between $E_I(f_w)$ for the multilayered system and $\bar{\bar{S}}_w(f_w)$ for the equivalent homogeneous porous medium.

12.6 Deformation of the Water/Oil Contact in the Near-Wellbore Region: "Water Coning"

Non-uniform advance of a water front, characterised by the areal sweep efficiency E_A and the vertical invasion efficiency E_I introduced in the preceding sections, is caused by distortions in the potential field in the regions of flow convergence at producing and injection wells, and in heterogeneities in the reservoir rock.

If the oil-bearing section is underlain by water (the original aquifer, or water that has moved in from injectors or from an edge aquifer and has either segregated beneath the oil through gravitational forces, or has advanced preferentially along high permeability strata at the base of the reservoir), it is common practice to limit the perforations to a safe distance above the water/oil contact. In this way, the onset of water production can be avoided completely, or at least delayed. However, partial penetration of the reservoir interval leads to an essentially hemispherical potential distribution beneath the open interval (Fig. 12.14), with a *vertical* potential gradient $\partial \Phi_o / \partial z$ across the water/oil contact. This causes a deformation of the

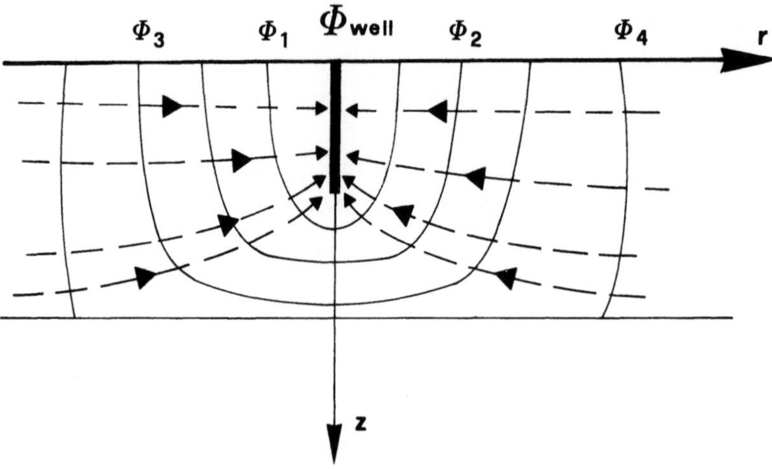

Fig. 12.14. *Equipotentials and lines of flow around a partially penetrating well or completion*

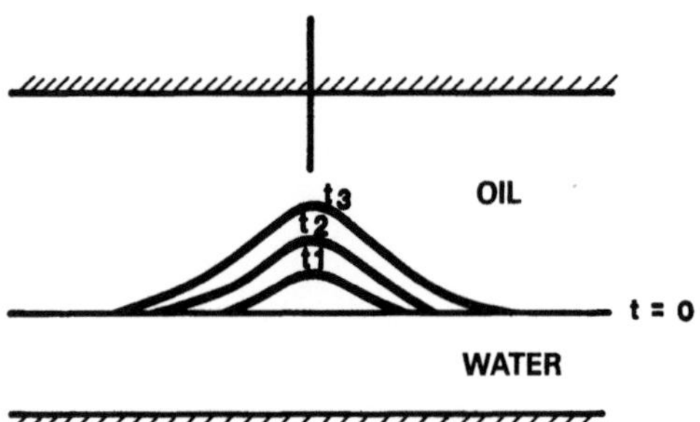

Fig. 12.15. *Stages in the local rise of the water/oil contact beneath the open interval, when the oil production rate is less than critical*

contact surface, which reaches a maximum at the axis of the well, where $\partial\Phi_o/\partial z$ is largest. The water/oil contact therefore takes the shape of a cone – this is the phenomenon of "water coning".

In principle, this deformation of the water/oil contact in the vertical plane is analogous to the deformation of a water displacement front in the horizontal plane. Both are caused by the distortion of the potential field resulting from the convergence of flow towards the well. Both, too, affect the volumetric efficiency.

Coning of a gas/oil contact can also occur in wells where there is a gas cap. Since this chapter is devoted to water injection, we will only discuss water coning here. The cone develops as soon as the well is put on production, and the tip rises at a velocity which depends on the local value of $\partial\Phi_o/\partial z$ and on the vertical permeability of the rock. The cone rises until the gravitational force due to the density difference $(\rho_w - \rho_o)$ balances the vertical pressure gradient (Fig. 12.15). The tip of the cone will not extend to the base of the open interval if the production rate is less than a certain critical value $q_{o,crit}$. The cone is said to be stabilised, and as long as $q_o < q_{o,crit}$, only oil is produced.

When the oil production rate, q_o, exceeds the critical rate, $q_{o,crit}$, and/or the water/oil contact rises as a consequence of oil production, the cone becomes unstable and reaches the bottom of the well (Fig. 12.16); as a consequence, the well starts producing water and oil ("wet oil").

In order to plan the development of the reservoir, the following factors must be determined for each well:

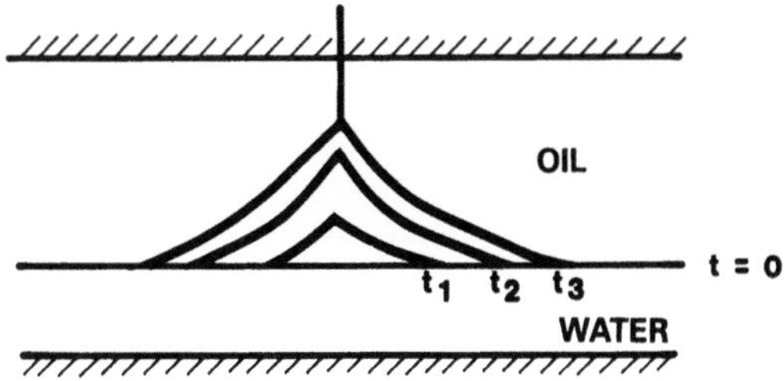

Fig. 12.16. *Stages in the local rise of the water/oil contact beyond the base of the open interval, when the oil production rate is greater than critical*

- the critical flow rate $q_{o,crit}$ for production without water;
- the time at which water will arrive at the well, and the variation of f_w with time when $q_o > q_{o,crit}$.

In the following sections, we shall see how to estimate these important terms.

12.6.1 Calculation of the Critical Flow Rate for Water Coning

Muskat and Wyckoff[26] were the first to present a study of the phenomenon of water coning. They also proposed a means of calculating $q_{o,crit}$. Successively more sophisticated methods were later presented by Arthur[3], Chaney et al.[9] and Chierici et al.[12] The method developed by Chierici et al. from Muskat and Wyckoff's original work will be presented here.

The oil-bearing section of a formation is of thickness h_o (Fig. 12.17), and it is underlain by an aquifer of indefinite thickness. A producing well is open over an interval b at the top of the formation. The bottom of the open interval is therefore a distance $(h_o - b)$ above the initial water/oil contact. The drainage radius of the well (Sect. 5.6.2) is r_e. The well has been producing at the critical rate $q_{o,crit}$ for a sufficient time for the cone to stabilise, at a height h_v above the initial contact.

Since the water within the cone is immobile, at any point (r, z) below the surface Γ of the cone we have:

$$\Phi_w(r, z) = \text{const.} \tag{12.40a}$$

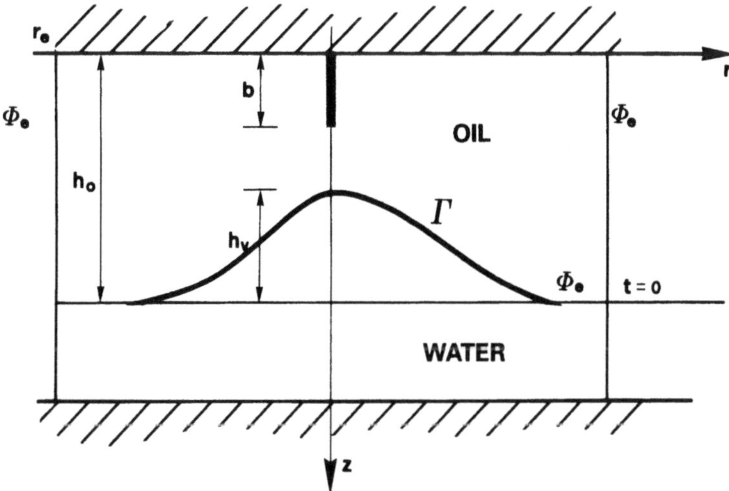

Fig. 12.17. *Geometry and nomenclature used in the discussion of water coning*

and, more importantly:

$$\frac{\partial \Phi_w}{\partial z} . = 0 \qquad (12.40b)$$

In addition, when $S_o = S_{or}$ the capillary pressure is, by definition, zero, so that inside the cone and along the surface Γ we have:

$$p_o = p_w = p. \qquad (12.41)$$

Introducing Φ_o and Φ_w [defined in Eqs. (3.47)], and ignoring the compressibility of the oil and water, we can write:

$$\Phi_w(r, z) = \frac{\rho_o}{\rho_w}[\Phi_o(r, z) + gz] - gz = \text{ const.}, \qquad (12.42)$$

from which we can derive the following relationship, *valid only within the cone and along its surface*:

$$\frac{\partial \Phi_o}{\partial z} = g\frac{\rho_w - \rho_o}{\rho_o} = \text{ const..} \qquad (12.43)$$

If Φ_e is the potential in the oil phase at the drainage boundary ($r = r_e$), it will be the same at the water/oil contact, where the oil is immobile. Therefore,

$$\frac{\Phi_e - \Phi_{o,v}}{h_v} = \left(\frac{\partial \Phi_o}{\partial z}\right)_v = g\frac{\rho_w - \rho_o}{\rho_o}, \qquad (12.44a)$$

where $\Phi_{o,v}$ is the potential in the oil phase at the tip of the cone, and $(\partial \Phi_o/\partial z)_v$ is its derivative in the vertical direction [constant inside the cone according to Eq. (12.43)].

If $\Delta\Phi$ is the constant difference between Φ_e and the potential in the oil phase at the well, we can define the following dimensionless terms:

$$\Phi_D(r, z) = \frac{\Phi_e - \Phi_o(r, z)}{\Delta\Phi} \qquad (12.45a)$$

and

$$z_D = \frac{z}{h_o}, \qquad (12.45b)$$

where:

$$0 \le \Phi_D \le 1$$

$$0 \le z_D \le 1$$

and, in particular,

$$h_{D,v} = \frac{h_v}{h_o}.$$

Equation (12.44a) can now be expressed as:

$$\left(\frac{\partial \Phi_D}{\partial z_D}\right)_v = -\frac{\Phi_{D,v}}{h_{D,v}}. \qquad (12.44b)$$

Equation (12.44b) suggests a handy graphic method for determining the critical height of the tip of the cone above the initial water/oil contact. Starting at the point $(z_D = 1, \Phi_D = 0)$, simply draw the tangent to the curve $\Phi_D = \Phi_D(z_D)$ as measured along the z-axis (corresponding to the axis of the well). According to Eq. (12.44b),

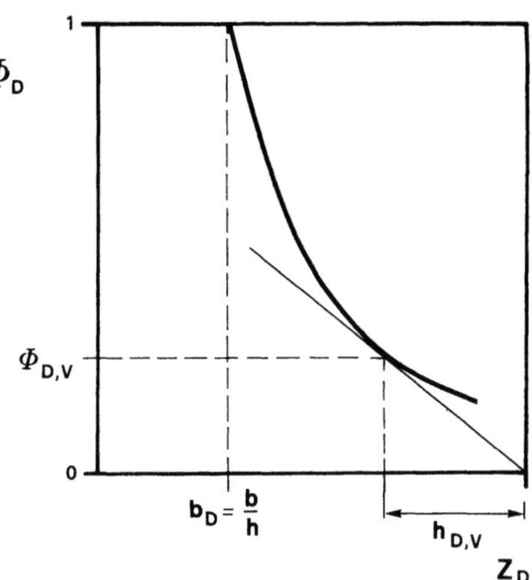

Fig. 12.18. *Graphic determination of the critical cone tip height above the water/oil contact*

the tangent touches the curve at $z_D = h_{D,v}$, the dimensionless height of the critical cone tip (Fig. 12.18).

To calculate $q_{o,crit}$ itself, we refer to the potential gradient at the outer limit of the drainage area rather than, as might seem more intuitive, the gradient at the well.

If k_r is the permeability of the reservoir rock in the radial direction, we have:

$$q_{o,crit} = -2\pi \frac{k_r k_{ro,iw}}{B_{of}\mu_o}\rho_o \int_0^{h_o} r_e \left(\frac{\partial \Phi_o}{\partial r}\right)_{r_e} dz. \tag{12.46a}$$

If we define the new dimensionless term:

$$r_D = \frac{r}{h_o}\sqrt{\frac{k_v}{k_r}}, \tag{12.45c}$$

and bring in those already defined in Eq. (12.45a,b), Eq. (12.46a) becomes:

$$q_{o,crit} = 2\pi h_o \frac{k_r k_{ro,iw}}{B_{of}\mu_o}\rho_o \Delta\Phi \int_0^1 r_{De} \left(\frac{\partial \Phi_D}{\partial r_D}\right)_{r_{De}} dz_D. \tag{12.46b}$$

Equation (12.44a) can be expressed in dimensionless form as:

$$\frac{\Phi_{D,v}}{h_{D,v}} = g\frac{\rho_w - \rho_o}{\rho_o}\frac{h_o}{\Delta\Phi}. \tag{12.44c}$$

Rearranging, we get:

$$\rho_o \Delta\Phi = g(\rho_w - \rho_o)h_o \frac{h_{D,v}}{\Phi_{D,v}}. \tag{12.44d}$$

Substituting this into Eq. (12.46b), we obtain the final equation:

$$q_{o,crit} = 2\pi g h_o^2(\rho_w - \rho_o)\frac{k_r k_{ro,iw}}{B_{of}\mu_o}\Omega(r_{De}, b_D) \tag{12.47a}$$

with:

$$\Omega(r_{De}, b_D) = \frac{h_{D,v}}{\Phi_{D,v}}\int_0^1 r_{De}\left(\frac{\partial \Phi_D}{\partial r_D}\right)_{r_{De}} dz_D. \tag{12.47b}$$

The function $\Omega(r_{De}, b_D)$ was determined by Chierici et al.[12] by simulation using equivalent electrical models, over the following range of conditions:

$$5 \le r_{De} \le 80$$

$$0.1 \le b_D \le 0.75.$$

The results they obtained will be found in Fig. 12.19. Note that the use of electrical analogues (valid as long as the fluid flow is in steady state) to calculate the Ω function has the advantage over numerical modelling that spatial discretisation of the porous medium is not required. An electrical field model eliminates the errors associated with numerical dispersion and truncation.

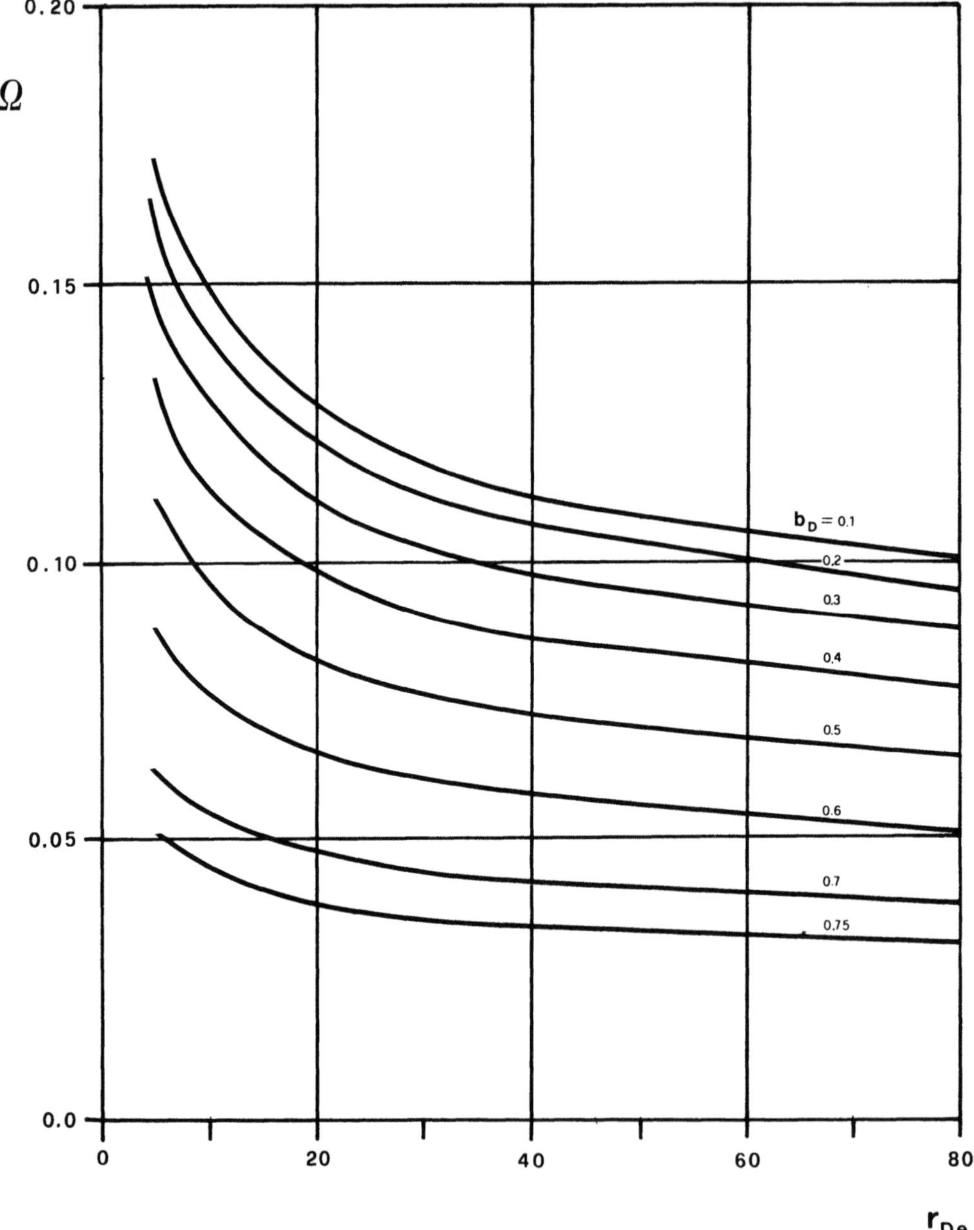

Fig. 12.19. *The function $\Omega(r_{De}, b_D)$ used in the calculation of the critical flow rate for water coning*

The Ω function presented in Fig. 12.19 is expressed by the equation:

$$\Omega(r_{De}, b_D) = A(b_D)\frac{1}{B(b_D) + C(b_D)\ln r_{De}}, \tag{12.48a}$$

where:

$$A(b_D) = 0.993 + \frac{0.00119}{0.769 - b_D}, \tag{12.48b}$$

$$B(b_D) = \frac{1}{0.302 - 0.053b_D - 0.336b_D^2}, \tag{12.48c}$$

and

$$C(b_D) = 1.459\exp[0.803b_D^2\,\ln(5.664b_D^{2.85} + 9)]. \tag{12.48d}$$

In practical metric units, Eq. (12.47a) becomes:

$$q_{o,crit} = 5.256 \times 10^{-6}h_o^2(\rho_w - \rho_o)\frac{k_r k_{ro,iw}}{B_{of}\mu_o}\Omega(r_{De}, b_D) \tag{12.49}$$

with:

$q_{o,crit}$: m^3/day

h_o : m

ρ_w, ρ_o : kg/m^3

k_r : md

B_{of} : m^3/m^3

μ_o : cP.

In US oil-field units, Eq. (12.47a) is

$$q_{o,crit} = 3.073 \times 10^{-3}h_o^2(\rho_w - \rho_o)\frac{k_r k_{ro,iw}}{B_{of}\mu_o}\Omega(r_{De}, b_D), \tag{12.50}$$

with:

$q_{o,crit}$: bbl/day

h_o : ft

ρ_w, ρ_o : g/cm^3

k_r : md

B_{of} : bbl/STB

μ_o : cP.

The dimensionless function $\Omega(r_{De}, b_D)$ appearing in Eqs. (12.49) and (12.50) is defined by Eq. (12.48) and presented in Fig. 12.19.

Correct evaluation of the vertical and radial permeabilities, k_v and k_r respectively, whose ratio appears in Eq. (12.45c) defining the dimensionless radius, is extremely important for the calculation of $q_{o,crit}$. For example, the presence of an impermeable barrier of even moderate areal extent between the base of the open interval and the water/oil contact will lead to a distortion of the potential field in the oil phase that will invalidate the equations cited above.

In the case of a reservoir consisting of a number layers in vertical communication, k_r should be calculated using the equation for strata in parallel:

$$k_r = \frac{\sum_{i=1}^{n} h_i k_{r,i}}{\sum_{i=1}^{n} h_i}$$
(12.51a)

and k_v using the equation for strata in series:

$$k_v = \frac{\sum_{i=1}^{n} h_i}{\sum_{i=1}^{n} \dfrac{h_i}{k_{v,i}}}.$$
(12.51b)

For completeness, it should be mentioned that Chierici et al.[12] also determined the Ω functions for the downward coning of a gas cap, and for the simultaneous coning of water and gas.

In the latter case, the authors presented a procedure for determining the optimum length of the open section and its position within the oil-bearing interval to maximise oil production without simultaneous production of water or free gas. Guo and Lee[33] have published a purely analytical method for the optimisation of b_D and $q_{o,crit}$ when only water coning is present.

12.6.2 Production Above the Critical Rate

It often happens that, from the outset of field development, the critical flow rate for coning is so low that the well cannot produce dry oil economically. Even if this is not the case initially, it will be found that as the water/oil contact rises with ongoing production, $q_{o,crit}$ will eventually become so low (as a consequence of the diminishing h_o) that a similar situation has to be confronted.

In both cases, the most likely decision would be to produce the well at a higher rate than the critical. The water produced as a consequence must then be separated out in the surface production facilities – not necessarily an easy process in the case of viscous, heavy oils.

To dewater (and, at the same time, desalinate) the oil, it must be heated, demulsifiers added, and the water left to segregate (coagulation of water droplets is sometimes encouraged by the application of an electric field). Where the oil is very viscous, the separation process may take several days, in which case the equipment needed is quite sizeable and costly, with significant running costs.

When production of a well above the critical coning rate is unavoidable, the following factors must be assessed;

- the time t_{BT} at which the cone will break through at the base of the open interval (this marks the onset of water production with the oil),
- the change in the water cut with time when $t > t_{BT}$.

Sobocinski and Cornelius[29] and Bournazel and Jeanson[6] published some fundamental work on the evaluation of t_{BT}. Both groups defined a dimensionless cone tip height $H_{D,v}$:

$$H_{D,v} = \frac{2\pi g h_o^2 (\rho_w - \rho_o) k_r}{q_o B_{of} \mu_o} h_{D,v},$$
(12.52a)

where:

$$h_{D,v} = \frac{h_v}{h_o}$$
(12.45b)

and a dimensionless time t_D:

$$t_D = \frac{g(\rho_w - \rho_o) k_v (1 + M_{wo}^\alpha)}{2\mu_o \phi h_o} t,$$
(12.52b)

where:

$\alpha = 0.5$ for $M_{wo} \leq 1$

$\alpha = 0.6$ for $1 < M_{wo} \leq 10$.

When the tip of the cone reaches the bottom of the open interval, i.e. when

$$(h_{D,v})_{BT} = 1 - b_D = 1 - \frac{b}{h_o}$$

water will appear in the production stream. The breakthrough time t_{BT} can be expressed in dimensionless form as:

$$(t_D)_{BT} = \frac{g(\rho_w - \rho_o) k_v (1 + M_{wo}^\alpha)}{2\mu_o \phi h_o} t_{BT},$$
(12.53a)

at which time:

$$(H_{D,v})_{BT} = \frac{2\pi g h_o^2 (\rho_w - \rho_o) k_r}{q_o B_{of} \mu_o} (1 - b_D).$$
(12.53b)

Sobocinski and Cornelius derived an experimental curve relating $(t_D)_{BT}$ and $(H_{D,v})_{BT}$, using scale models and numerical simulation.

The equation of their curve is[23]

$$(t_D)_{BT}^{SOBO} = \frac{(H_{D,v})_{BT}}{4} \frac{16 + 7(H_{D,v})_{BT} - 3(H_{D,v})_{BT}^2}{7 - 2(H_{D,v})_{BT}}.$$
(12.54a)

Bournazel and Jeanson's equation[23] for the same relationship had the form

$$(t_D)_{BT}^{BOUR} = \frac{(H_{D,v})_{BT}}{3 - 0.7(H_{D,v})_{BT}}.$$
(12.54b)

Note that the Sobocinski and Cornelius correlation has $(t_D)_{BT}$ going to infinity (i.e. stabilised cone) at $H_{D,v} = 3.5$, while for Bournazel and Jeanson this happens at $H_{D,v} = 4.286$. Table 12.2 contains a comparison of the results of the two correlations.

Sobocinski and Cornelius predict values of $(t_D)_{BT}$ which are more than twice as large as those of Bournazel and Jeanson. It would be preferable to err on the side of caution by using the Bournazel and Jeanson correlation.

Once $(t_D)_{BT}$ has been obtained, t_{BT} can be calculated directly from Eq. (12.53a). In more convenient practical metric units:

$q_o \quad$: m^3/day

$t_{BT} \quad$: days

Table 12.2. *Comparison of the Sobocinski et al. and Bournazel et al. correlations*

$(H_{D,v})_{BT}$	$(t_D)_{BT}$	
	Sobocinski *et al.*[29]	Bournazel *et al.*[6]
0.5	0.391	0.188
1.0	1.000	0.435
1.5	1.852	0.769
2.0	3.000	1.250
2.5	4.609	2.000
3.0	7.500	3.333
3.5	∞	6.364
4.286	–	∞

ϕ : decimal fraction

k_r, k_v : md

h_o : m

ρ_w, ρ_o : kg/m^3

μ_o : cP

B_{of} : m^3/m^3,

the equations are:

$$(t_D)_{BT} = 4.181 \times 10^{-7} \frac{(\rho_w - \rho_o)k_v(1 + M_{wo}^\alpha)}{\mu_o \phi h_o} t_{BT} \tag{12.55a}$$

and

$$(H_{D,v})_{BT} = 5.256 \times 10^{-6} \frac{h_o^2(\rho_w - \rho_o)k_r}{q_o B_{of} \mu_o}(1 - b_D). \tag{12.55b}$$

In US oil-field units:

q_o : bbl/day

t_{BT} : days

ϕ : decimal fraction

k_r, k_v : md

h_o : ft

ρ_w, ρ_o : g/cm^3

μ_o : cP

B_{of} : bbl/bbl,

the equations are:

$$(t_D)_{BT} = 1.372 \times 10^{-3} \frac{(\rho_w - \rho_o)k_v(1 + M_{wo}^\alpha)}{\mu_o \phi h_o} t_{BT} \tag{12.56a}$$

and

$$(H_{D,v})_{BT} = 3.073 \times 10^{-3} \frac{h_o^2(\rho_w - \rho_o)k_r}{q_o B_{of} \mu_o}(1 - b_D). \tag{12.56b}$$

To predict the development of the water cut in the production stream at any time after breakthrough, we can use the correlation derived by Kuo and Des Brisay[23], which they based on the results of a series of numerical simulations using circular symmetry (r, z). These authors defined a "limiting water cut", $(WC)_{lim}$ – the water cut that would be obtained *at any given time* if the entire pay section, including that part already invaded by water, were open to production.

If, at time t, h_o is the thickness of the current oil-bearing section of formation, and h_w the total thickness of the water-bearing section, including that part of the pay zone already invaded by water, then, if the interval $(h_o + h_w)$ were completely open to production, we would have:

$$\frac{q_w}{q_o} = \frac{h_w}{h_o}\frac{k_{rw,or}\mu_o}{k_{ro,iw}\mu_w}\frac{B_{of}}{B_w} = \frac{h_w}{h_o}\frac{B_{of}}{B_w}M_{wo},$$ (12.57)

assuming k to be uniform over the entire formation. In Eq. (12.57), h_o and h_w are the values at the moment in time, t, at which we wish to calculate the water cut WC, and therefore allow for the fact that the water/oil contact rises as production continues.

From Eq. (12.57) we get:

$$(WC)_{lim} = \frac{q_w}{q_w + q_o} = \frac{h_w M_{wo}}{h_w M_{wo} + \dfrac{B_w}{B_{of}}h_o}.$$ (12.58)

Kuo and Des Brisay's correlation, presented in Fig. 12.20, can be expressed as follows, denoting, as usual, the time at which water breakthrough occurs at the

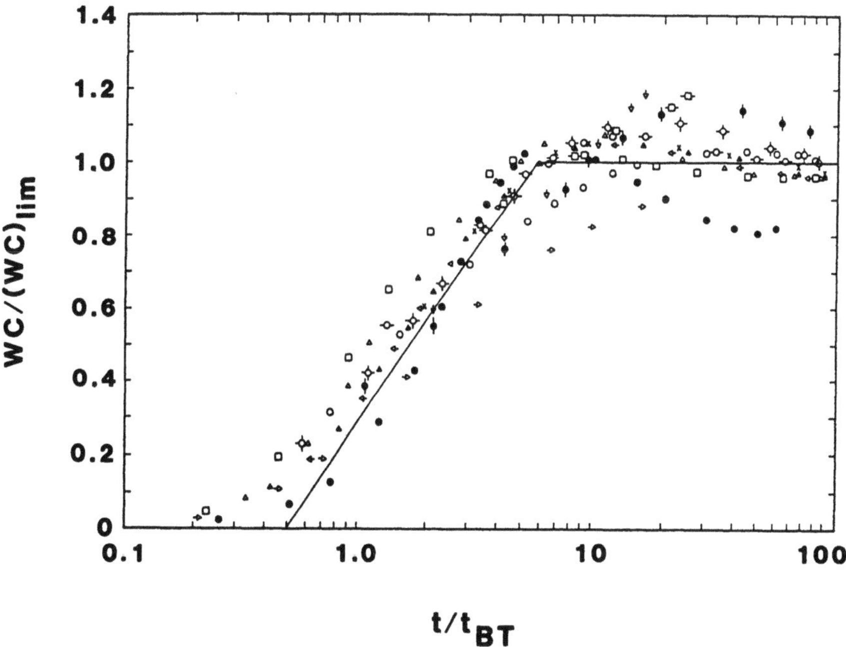

Fig. 12.20. *Kuo and Des Brisay's correlation for the calculation of the water cut at any time after breakthrough, when the well is being produced above the critical water coning rate. (From Ref. 23, 1983, Society of Petroleum Engineers of AIME. Reprinted with permission of the SPE)*

well by t_{BT}:

$$\text{WC} = 0 \qquad \text{for } \frac{t}{t_{BT}} < 0.56$$

$$\left[\frac{\text{WC}}{(\text{WC})_{\lim}}\right]_t = 0.94 \log\left(\frac{t}{t_{BT}}\right) + 0.24$$

$$\text{for } 0.56 \leq \left(\frac{t}{t_{BT}}\right) \leq 5.7$$

$$\left[\frac{\text{WC}}{(\text{WC})_{\lim}}\right]_t = 1.0 \quad \text{for } \left(\frac{t}{t_{BT}}\right) > 5.7$$

One unsatisfactory aspect of this correlation is that it predicts a non-zero water cut even when $t < t_{BT}$.

12.7 Surveys and Tests in Reservoirs Produced Under Water Injection

In present-day reservoir engineering technology, it is quite common to plan the implementation of water injection at a very early stage, as soon as a suitable number of development wells are available. By studying the reservoir behaviour during the initial water injection phase, the reservoir engineer will endeavour to deduce the information he will need to plan the further development of the field. In particular, he will have to decide:

- the drilling of infill wells within the area already under waterflood, in order to improve the volumetric efficiency (Sect. 12.5) – in terms of either the areal sweep or vertical invasion,
- the possible use, during or shortly after the current water injection phase, of enhanced recovery methods with the injection of specialised fluids (see Chap. 15).

A knowledge of the response of the reservoir at any time to the injection of water, and especially of the movement of the water front, is essential to enable sensible decisions to be made on these key points. Other than this, the bulk of the information available was obtained during drilling and is *static* in nature (the distribution of lithology and petrophysical properties over the hydrocarbon-bearing interval, at the very localised point of investigation that the wellbore represents).

The injection phase, on the other hand, provides the opportunity to obtain *dynamic* data which relate to the hydraulic continuity between wells of any layers in the pay section, areal variations in their properties, the presence and position of permeability barriers, etc. This sort of information can be obtained through tests in individual wells or among a number of wells.

Well testing is a very large and highly specialised topic. In the following pages it will only be possible to provide an overview of the technical aspects of the various tests and equipment used. For more details, the reader is referred to the references at the end of this chapter.

12.7.1   Monitoring the Advance of the Water Front

The measurements themselves have already been described in Chap. 8, so in this section we will just look at the philosophy underlying their application. Firstly, a regular check of the injection profile over the pay section in each injector is essential. Continuous fluid velocity measurement with a spinner flowmeter, or stationary measurements using converging flowmeters or radioactive tracer ejection (Sect. 8.3.1), and temperature or differential temperature logging (Sect. 8.3.5) are commonly used methods.

If the measurements show a loss of injectivity in part of the interval (for example, as a result of the presence of fine solids or bacteria in the water), it may be necessary to prescribe selective stimulation (e.g. acidisation) to reestablish a uniform injection profile. A zone of very high injectivity (a thief zone) is particularly undesirable, as it will almost certainly lead to a short-circuit between the injector and producing well or wells, which results in a serious loss of overall volumetric efficiency. At the producing wells, the production stream must be checked continually, for the presence of water. The measurement of water cut is part of routine field procedure.

As soon as a well starts producing water, the following questions have to be answered: at what point or points is the water entering the well? Is it entering at the bottom of the open interval? (This might mean water coning if the production rate is greater than $q_{o,crit}$; see Sect. 12.6.1) Or does the entry correspond to layers of high permeability and, if so, where?

A conventional production logging survey will normally be adequate to resolve the problem (Sect. 8.3), through the following measurements:

- continuous profile of the fluid velocity (FBS, CFS);
- fluid density (GMS, PTS, NFD) or capacitance (HUM);
- water content (XFT).

But what if the water reaches the well via a layer that has not been perforated for production in that well?

In order to monitor this sort of eventuality, it is necessary to run periodic surveys using pulsed neutron logging tools which can respond to the presence of water behind casing [most commonly, the thermal decay tools (TDT, PDK) described in Sect. 8.7.2].

From the information gathered, it should be possible for the reservoir engineer to map the disposition of the water front at any particular time, both areally and vertically. He can then program any well interventions that he may consider necessary (stimulations, isolation of thief zones) in order to keep the frontal advance as uniform as possible and maximise the volumetric displacement efficiency.

Should the measurements indicate that part of the reservoir is in poor communication with the rest of the field, it may be necessary to consider drilling infill wells – producers and/or injectors. Adherence to this sort of systematic approach to reservoir management will increase the probability of obtaining a very high volumetric displacement efficiency, perhaps obviating the need for an enhanced recovery phase of dubious economic viability.

Figures 12.21 and 12.22 illustrate[18] the results obtained in a "giant" reservoir (Sect. 4.5) where the philosophy of regular monitoring described above was applied. Note the fairly uniform advance of the water/oil contact – despite the heterogeneous

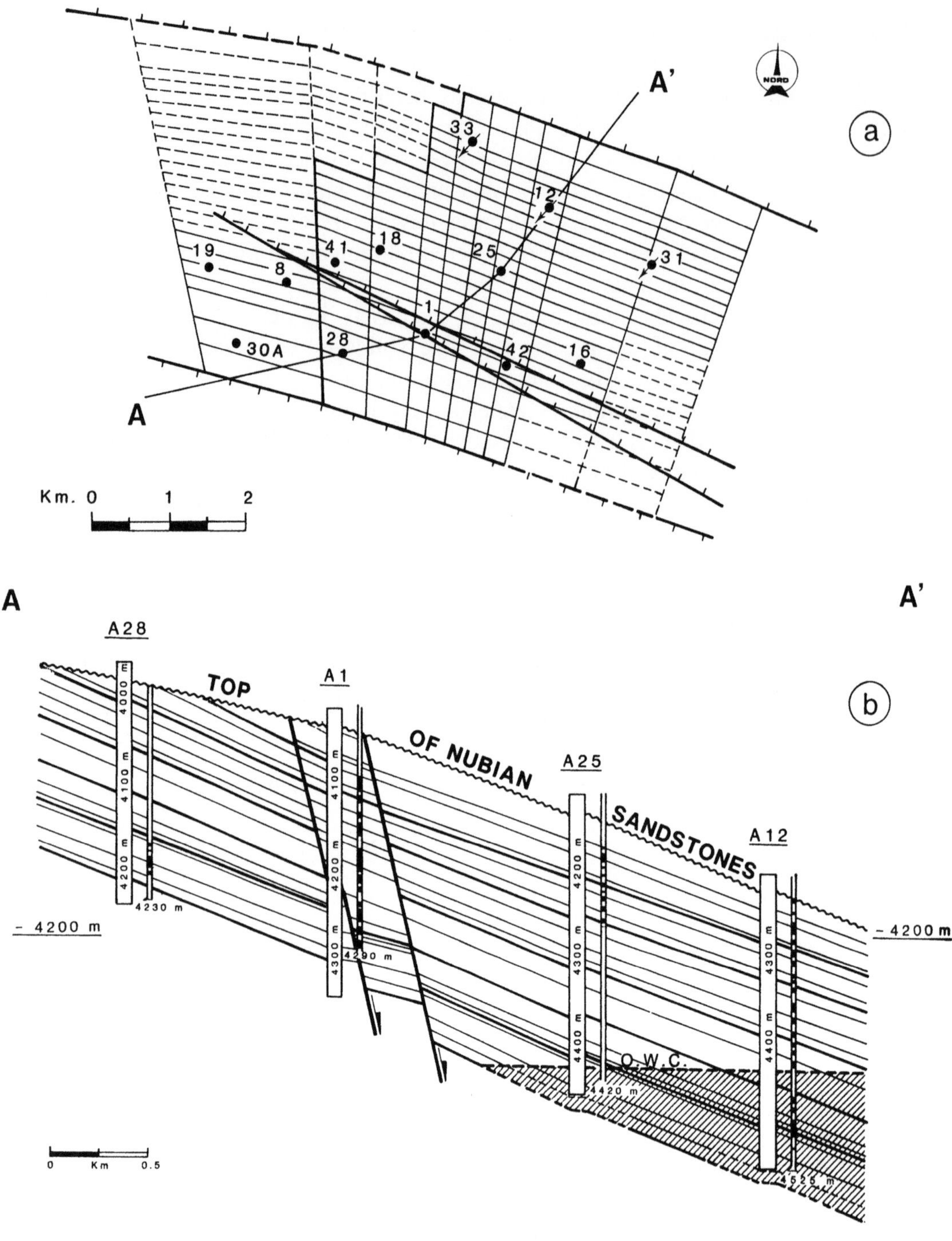

Fig. 12.21a,b. *The Bu Attifel reservoir (SPLAJ).* **a** *Grid-block model used to optimise water injection in part of the field;* **b** *reservoir cross section, showing zonation and initial WOC. (Ref. 18)*

nature of the reservoir rock – achieved through appropriate well interventions and recompletions based on the results of the logging surveys.

The effect on the movement of the water/oil contact of any planned well intervention was examined by means of numerical simulation before going ahead with the actual operation. The grid-block model (Chap. 13) is shown in Fig. 12.21 (top).

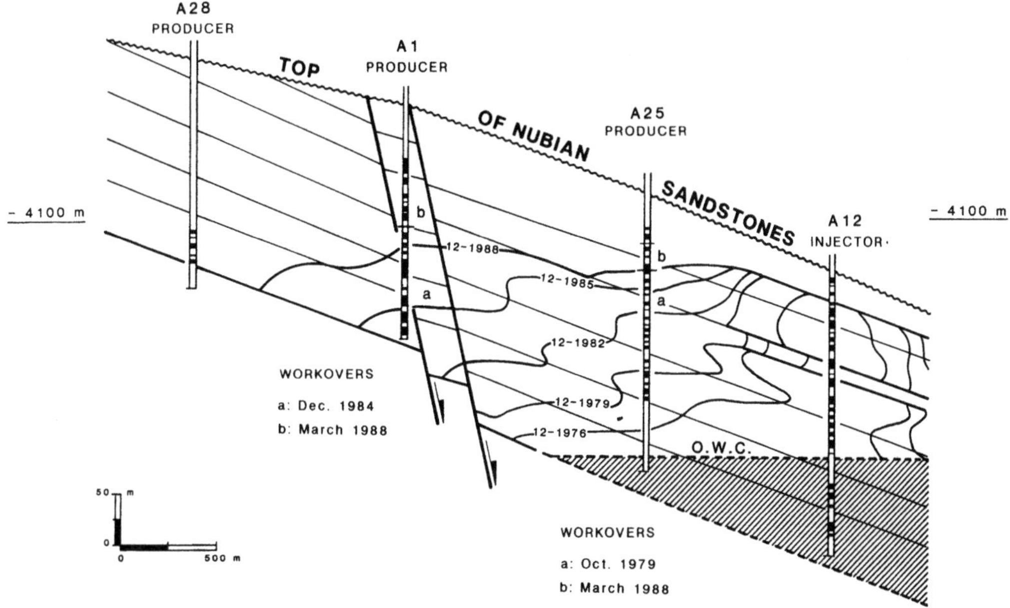

Fig. 12.22. *The Bu Attifel reservoir (SPLAJ). Cross section, showing the movement of the water front with time, and the well interventions performed to correct for preferential advance of the water. (Ref. 18)*

12.7.2 Interwell Testing

Interwell testing provides a means of verifying the presence of permeability barriers in the region between neighbouring wells and, in their absence, of estimating the average permeability and porosity-compressibility product. This is true both in reservoirs being produced under natural depletion and under water injection. An understanding of the anisotropy of reservoir permeability derived on a "megascopic" scale in this way will enable the directions of preferential water movement to be identified.

In reservoirs under water injection, the following interwell tests are commonly performed:

- interference tests (Sect. 6.11);
- pulse tests (Sect. 6.12)
- chemical or radioactive tracer tests.

Interference and pulse tests have already been covered in Chap. 6. The important point to remember about such tests is that the flow rates of all wells must be carefully monitored during the test period if a correct interpretation is to be made of the pressure data. Ideally, the injection or production rates of all wells in the region of investigation should be held constant long enough for the pressure to stabilise.

The injectors are then raised successively to a high injection rate (interference testing), or pulsed at alternating high and low (or zero) rates (pulse testing), and the bottom-hole pressure behaviour is recorded in the producers.

All wells outside the area of investigation must be kept at rigorously constant rates during the test in order to avoid the superposition of extraneous pressure signals which would complicate the analysis.

As seen in Chap. 6, these tests allow us to evaluate the *average* permeability of the reservoir rock between an injector and each of the surrounding producers. It is also possible to identify the presence of permeability barriers and their approximate position. However, we cannot determine anything about the nature or degree of the reservoir heterogeneity.

The execution of interference and pulse tests will necessitate shutting in producing wells in the area under investigation for a significant period of time (which can sometimes run into weeks in low permeability reservoirs). The undesirable economic consequences of well closure are avoided by some operators who prefer to keep the wells on production at a constant rate. There is, however, a trade-off in terms of data quality because additional flowing transients appear in the measured pressure responses which, as a consequence, become more difficult to interpret unambiguously.

Tracer tests, on the other hand, do not require production to be interrupted, either before or during the test. A fairly small pad of tracer fluid is simply pumped into the injector or injectors. The tracer, either a chemical product or a radioactive isotope, is prepared as an aqueous solution of appropriate concentration, a few tens to a few hundred cubic metres in volume. If the produced water contains any of the tracer, its presence can be detected easily and its concentration measured. The tracer should not, of course, react with the reservoir fluids or rock minerals, nor be adsorbed onto the grain surfaces. It should simply travel through the reservoir at the same speed as the water in which it is dissolved.

Some typical chemical tracers[20] are sodium, potassium or ammonium thiocyanates and nitrates. If the reservoir water contains no bromine or iodine, these ions too may be used to "tag" the injection water. The use of organic tracers (fluorescein, EDTA, picric or salicylic acid) is not recommended, as they are readily adsorbed; neither are bivalent or trivalent ions, which tend to undergo an exchange with potassium and sodium ions present on the surface of clay platelets. The most commonly used radioactive tracer is "heavy water", containing a small quantity of tritium oxide ($T_2O = H_2^3O$). Tritium is an isotope of hydrogen ($T = H^3$) which emits β-particles and has a radioactive half-life of 12.3 years (see Sect. 8.5.2).

If several injectors are involved in the test, then a different tracer can be used for each (the so-called "multicoloured tracer" technique). Each tracer is easily distinguishable from the others, so the origin of the water appearing at a producer can be identified.

Through these interwell tests (interference, pulse or tracer), the degree of continuity, if any, between each injector and the producers surrounding it can be assessed, as well as the average permeability in the interwell region.

It is very important to recognise areal anisotropy in the permeability ($k_x \neq k_y$). This leads to preferential directions of flow, deforming the water front relative to the isotropic case, with a consequent loss of areal sweep efficiency (Fig. 12.23). Corrective action should be taken, such as the drilling of infill wells (Fig. 12.23) or, if the field is still being developed, an appropriate change in the placement of future wells so as to increase the interwell spacings in the direction of higher permeability (Fig. 12.24).

In the mid-1980s, a more sophisticated technique was introduced for the interpretation of tracer tests. Based on the variations in tracer concentration in the produced water, this method made it possible in many cases to determine both the nature and degree of heterogeneity of the reservoir rock in the interwell region.

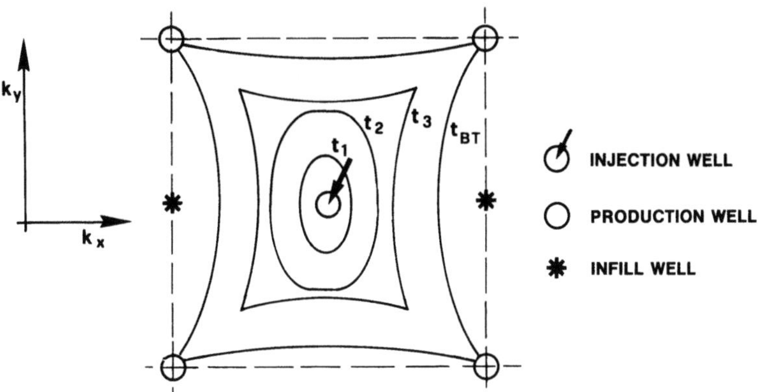

Fig. 12.23. *The use of infill wells to increase the areal sweep efficiency E_A in a regular five-spot waterflood pattern where the reservoir permeability exhibits areal anisotropy ($k_y > k_x$)*

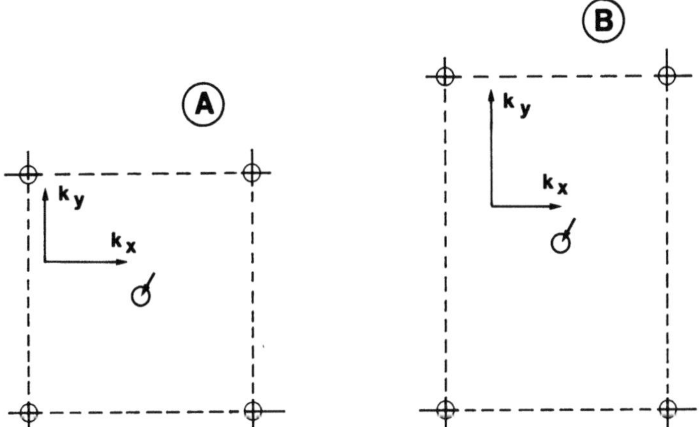

Fig. 12.24. **A** *Water injection in a regular five-spot pattern into an isotropic reservoir;* **B** *use of an elongated five-spot pattern in a reservoir with anisotropic permeability ($k_y > k_x$)*

Refer to the list of references for a detailed discussion of the technique. Briefly, this technique is based on a correlation between rock heterogeneity and the dispersion coefficient of the tracer moving through it[4,28].

Suppose we have a *linear* aquifer ($S_w = 1$) of uniform cross-sectional area A, length L, and porosity ϕ, into which water is being injected at a constant rate i_w. At $t = t_0$, we inject a pad of water of volume $\Delta q_w = i_w \Delta t$ for a time Δt into one end face of the aquifer. This water contains tracer at a concentration C_0. Following the pad, injection of clear water is continued at the same rate i_w. If Δt is small relative to the time taken for the water to traverse the length of the aquifer, the injection of the tracer can be represented as a "disturbance" described by a Dirac function.

If no dispersion or diffusion occurs in the porous medium, the tracer will move with a uniform velocity, equal to the Darcy velocity i_w/A of the water carrying it, so that, after a time $\phi AL/i_w$, it will appear as an undisturbed pad at the exit face of the aquifer (Fig. 12.25). In fact, while travelling through the aquifer, it will undergo the phenomenon of diffusion (due to the concentration gradients across the leading and trailing edges of the pad) and, more importantly, various *mechanical dispersion* effects.

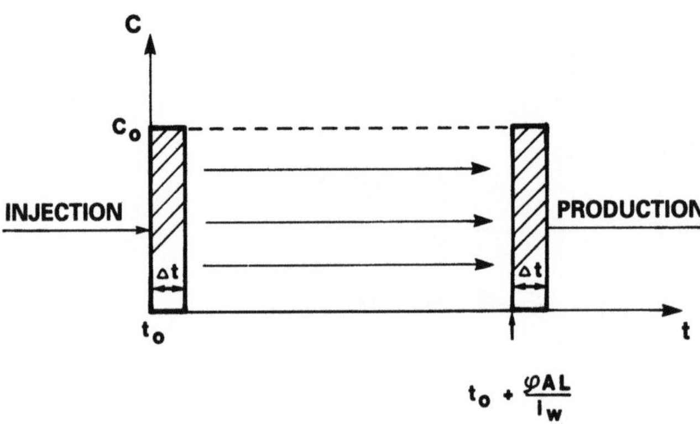

Fig. 12.25. *Piston-like advance of the tracer pad in the absence of dispersion and diffusion*

Heterogeneities in the porous medium create water flow streams of varying paths and lengths so that, although the Darcy velocity is constant, *local velocity vector v_w is by no means constant, either in magnitude or direction.*

The continual branching and rejoining of flow streams, each having travelled a different distance, results in mixing of tagged and untagged water to varying degrees along the displacement front, and a mechanical, or "dynamic", dispersion of the tracer.

If m_o is the mass flow rate of the tracer in the absence of dispersion, and C is its concentration, we have:

$$m_o = C i_w. \tag{12.59a}$$

The mass flow rate of the tracer transported by dispersion, m_{dis}, is proportional to the effective cross section to flow, $A\phi$, to the Darcy velocity, and to the concentration gradient. This is expressed as follows:

$$m_{disp} = \alpha_1 (A\phi) \left(\frac{i_w}{A} \right) \frac{\partial C}{\partial x}, \tag{12.59b}$$

where:

$\alpha_1 = $ the longitudinal dispersivity.

α_l has the dimensions of length. Note that $\partial C / \partial x$ is negative at the leading edge of the pad, and positive at the trailing edge.

There are in fact two components[4] to α_1: a *macroscopic dispersivity* α_{MA} and a *megascopic dispersivity* α_{ME}. The former is related to heterogeneities on a scale of one to several tens of centimetres, typical of the local variations in depositional conditions that are always found in reservoir rock. The second is related to more substantial heterogeneities resulting from stratification, variations in facies, the presence of barriers, etc. which are usually encountered on a scale of metres or larger.

If m_{tot} is the total mass flow rate of the tracer (being the sum of the rate m_o associated with transport at constant concentration and the rate m_{disp} associated with dispersion), we have from Eqs. (12.59):

$$m_{tot} = m_o - m_{disp} = i_w \left(C - \alpha_1 \phi \frac{\partial C}{\partial x} \right). \tag{12.60}$$

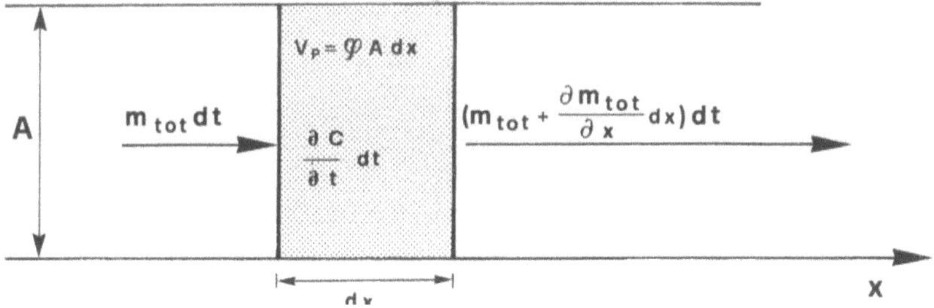

Fig. 12.26. *Diagram showing the mass balance for the tracer between two sections a distance* dx *apart in the direction of flow*

For conservation of mass of the tracer between two sections dx apart in the direction of flow (Fig. 12.26), we have:

$$\left[m_{\text{tot}} - \left(m_{\text{tot}} + \frac{\partial m_{\text{tot}}}{\partial x}\,\mathrm{d}x \right) \right]\,\mathrm{d}t = \phi A\,\mathrm{d}x\,\frac{\partial C}{\partial t}\,\mathrm{d}t. \tag{12.61}$$

In other words:

$$\alpha_1 \phi u_{\mathrm{w}} \frac{\partial^2 C}{\partial x^2} - u_{\mathrm{w}} \frac{\partial C}{\partial x} - \phi \frac{\partial C}{\partial t} = 0, \tag{12.62a}$$

where

$$u_{\mathrm{w}} = i_{\mathrm{w}}/A.$$

If we now define

$$\alpha_1 = \frac{D_1}{u_{\mathrm{w}}} \tag{12.63a}$$

Equation (12.62a) becomes:

$$\phi D_1 \frac{\partial^2 C}{\partial x^2} - u_{\mathrm{w}} \frac{\partial C}{\partial x} - \phi \frac{\partial C}{\partial t}. = 0 \tag{12.62b}$$

D_1 is the *longitudinal dispersion coefficient.* In the SI system, it is expressed as m²/s.

If we wish to include molecular diffusion due to the concentration gradient (Fick's law), Eq. (12.63a) takes the form:

$$D_1 = \frac{D_0}{\sqrt{\tau}} + \alpha_1 u_{\mathrm{w}} \tag{12.63b}$$

with

$D_0 = $ molecular diffusion coefficient $(u_{\mathrm{w}} = 0)$, m²/s;

$\tau \;\; = $ tortuosity of the porous medium.

Equations (12.62) describe the distribution in time and space of the tracer concentration $C(x, t)$ in the porous medium. There will, of course, be initial and boundary conditions as well.

For the injection of a pad of tracer fluid of concentration C_0 they are:

$$\begin{aligned}
C &= 0; & 0 &\le x \le L, & t &\le t_0 \\
C &= C_0; & x &= 0, & t_0 &< t \le (t_0 + \Delta t) \\
u_{\mathrm{w}} &= \text{const}; & 0 &\le x \le L, & t. &
\end{aligned} \tag{12.62c}$$

In the study of transport phenomena, the diffusion/dispersion properties are usually characterised by the Péclet number N_{Pe}:

$$N_{Pe} = \frac{u_w L}{\phi D_l} = \frac{L}{\phi \alpha_l}. \tag{12.64}$$

The smaller the ratio of dispersivity α_l and length L for the porous medium, the larger N_{Pe} becomes. Measurements made in statistically homogeneous reservoirs suggest the relationship[4]:

$$\frac{\alpha_l}{L} \cong 0.1, \tag{12.65a}$$

so that

$$N_{Pe} \cong \frac{10}{\phi}. \tag{12.65b}$$

Expressing the volume of water injected after the introduction of the tracer pad as a fraction of the total pore volume:

$$W_{i,t}^* = \frac{u_w(t - t_o - \Delta t)}{\phi L}. \tag{12.66}$$

Hagoort[21] furnished the solution to Eq. (12.62b) at $x = L$, subject to the boundary conditions listed in Eq. (12.62c), as

$$\frac{C(L,t)}{C_o} = \frac{\Delta t}{2(t - t_o)} \sqrt{\frac{N_{Pe}}{\pi W_{i,t}^*}} \exp\left[-\left(\frac{1 - W_{i,t}^*}{2}\right)^2 \frac{N_{Pe}}{W_{i,t}^*}\right]. \tag{12.67}$$

From this solution, Hagoort also determined the following relationships (Fig. 12.27):

$$\frac{C_{peak}}{C_o} = \frac{W_{tr}^*}{2} \sqrt{\frac{N_{Pe}}{\pi}}; \tag{12.68a}$$

$$W_{i,peak}^* = 1 - \frac{3}{N_{Pe}}; \tag{12.68b}$$

$$\Delta W_{i,peak}^* = 2\sqrt{\frac{2}{N_{Pe}}}; \tag{12.68c}$$

where

C_{peak}:   maximum concentration of tracer measured in the produced fluid;

W_{tr}^*:   volume of tracer (i.e. of tagged water) injected, expressed as a fraction of the total pore volume;

$W_{i,peak}^*$:   volume (expressed as a fraction of total pore volume) of water injected from the moment of introduction of the tracer pad until the appearance of the tracer concentration peak at the exit face of the porous medium;

$\Delta W_{i,peak}^*$:   "width" of the tracer peak, expressed as pore volumes of water injected over the period between the appearance of the two inflexion points on the $C(t)$ curve at the exit face.

In dimensioned form, Eqs. (12.68) become:

$$\frac{C_{peak}}{C_o} = \frac{\Delta t}{2\sqrt{\pi L}} \frac{u_w}{\phi} \frac{1}{\sqrt{\phi \alpha_l}}; \tag{12.69a}$$

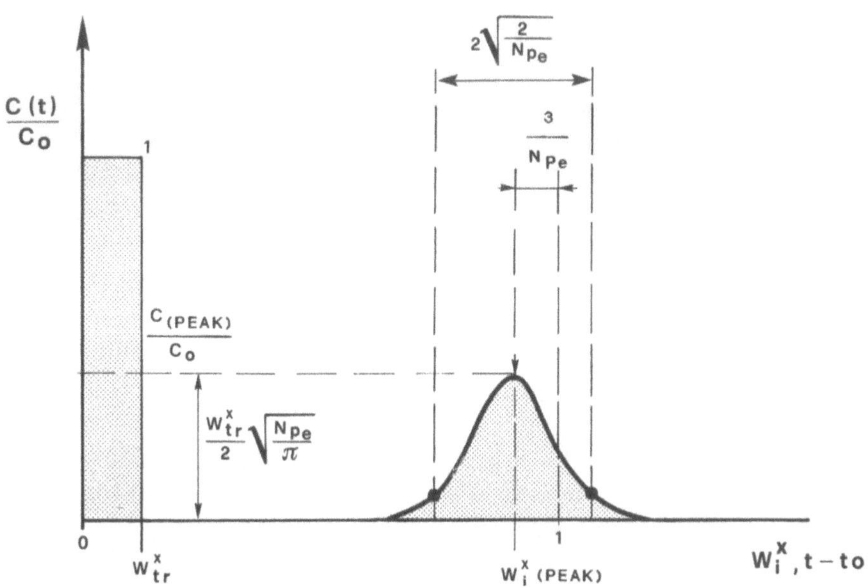

Fig. 12.27. *Dispersion of tracer in a porous medium. The diagram shows the relationships between the dimensions of the tracer peak and the relevant parameters*

$$t_{\text{peak}} = (t_{\text{peak}})_{\alpha_1=0} - \frac{3\phi^2\alpha_1}{u_{\text{w}}}; \qquad (12.69b)$$

$$(\Delta t)_{\text{peak}} = \sqrt{8L}\frac{\phi}{u_{\text{w}}}\sqrt{\phi\alpha_1}; \qquad (12.69c)$$

where

t_{peak}:   time at which the tracer peak arrives at the exit face of the porous medium;

$(t_{\text{peak}})_{\alpha_1=0}$: value of t_{peak} if there were no dispersion;

$(\Delta t)_{\text{peak}}$:   time elapsed between the appearance of the first and second inflexions on the curve $C(L, t)$ at the exit face.

Equation (12.69a) is only valid under dispersive conditions ($\alpha_1 \neq 0$); it has no meaning when $\alpha_1 = 0$.

It is apparent from Eqs. (12.69) (or, indeed, just from an intuitive analysis of what is happening) that the tracer concentration peak will arrive earlier: (1) the greater the part played by dispersion in the transport process, and (2) the longer the time actually spent by the tracer pad in the porous medium (u_{w}/ϕ small).

For a given volume of tracer (i.e. a given injection time Δt), the peak will be weaker, the greater the length of the porous medium, the stronger the dispersion effect, and the lower the Darcy velocity (i.e. the longer the time it spends traversing the medium). The same factors will determine the width of the peak.

Equations (12.69) show that, in a linear porous medium, the longitudinal dispersivity α_1 and the longitudinal dispersion coefficient D_1 can be determined from the shape of the $C(L, t)$ curve obtained in response to the injection of a pad of water tagged with a tracer. You will find detailed accounts of the relationship between α_1 and the heterogeneity of a porous medium in the works published by Hagoort[21], Smith and Brown[28] and Arya et al[4].

Unfortunately, flow in reservoirs is rarely linear, and their complex geometries are reflected in areal sweep efficiencies which are less than unity. Furthermore, the reservoir rock is often not statistically homogeneous but stratified, each layer representing a depositional unit which is itself homogeneous.

The effect of geometry and layering is to produce concentration response curves $C(L, t)$ which are much more complex in form than the simplistic bell shape described by Eq. (12.67). Figure 12.28 illustrates the $C(L, t)$ response that was measured during a test with a five-spot pattern in a reservoir known to consist of at least ten layers[7].

Interpretation of this sort of tracer response requires numerical modelling. Experience has shown[24] that there is no need to resort to miscible displacement models for this purpose. Suitable adjustments can be made to the relative permeability and capillary pressure curves to enable dispersive phenomena to be simulated with simpler immiscible displacement models. This makes the description of the dispersion process in each layer much easier.

It sometimes happens, however, that the same measured $C(L, t)$ curve can be interpreted satisfactorily by means of several different models, with different geometries and petrophysical characteristics. In other words, the solution cannot be determined uniquely. This is a common occurrence with "inverse" problems where we have to determine the internal properties of a system from its response to an external stimulus. Despite this, the interpretations of tracer tests should not be ignored. They are one more valuable source of information about the dynamic behaviour of the reservoir in the region between wells. Taken with all the other

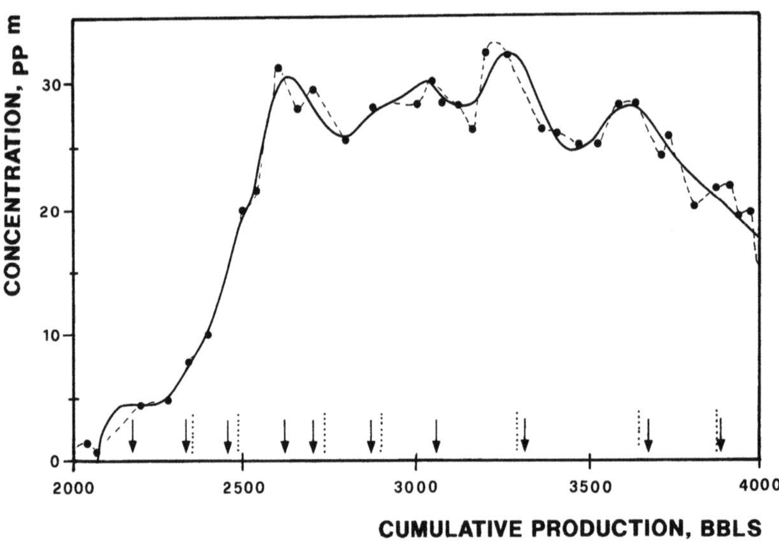

Fig. 12.28. *Measured variations in the concentration of tracer (points) in the produced fluid of a well in a layered reservoir. The solid curve is the interpretation, in terms of number of layers and their characteristics, simulated with a numerical model. (From Ref. 7, 1987, Society of Petroleum Engineers of AIME. Reprinted with permission of the SPE*

information available – especially that concerning the sedimentology – they will help to build the "most probable" reservoir model.

In addition to their application in the evaluation of the internal structure of the reservoir, the dispersivity estimates obtained from tracer tests are an extremely useful input to the planning of enhanced recovery projects that might well be instigated after the injection phase. In fact, one of the most undesirable occurrences in enhanced recovery is the dilution (or, worse still, the disappearance) of the pad of miscible fluid that has been injected, as a consequence of dispersion in the reservoir. The likelihood of such problems, and the probability of a successful enhanced recovery process, can be assessed by means of tracer tests to determine dispersivity.

References

1. API (1961) History of petroleum engineering. American Petroleum Institute, Dallas, Texas
2. Aronofsky JS, Ramey HJ (1956) Mobility ratio – its influence on injection or production histories in five-spot water flood. Trans, AIME 207: 205–210
3. Arthur MG (1944) Fingering and coning of water and gas in homogeneous oil sand. Trans, AIME 155: 184–199
4. Arya A, Hewett TA, Larson R, Lake LW (1988) Dispersion and reservoir heterogeneity. SPE Reservoir Eng Feb: 139–148
5. Bertero L, Chierici GL, Gottardi G, Mesini E, Mormino G (1988) Chemical equilibrium models: their use in simulating the injection of incompatible waters. SPE Reservoir Eng Feb: 288–294
6. Bournazel C, Jeanson B (1971) Fast water-coning evaluation method. Paper SPE 3628 Presented at the 47th Annu Fall Meet of the Society of Petroleum Engineers, New Orleans, Louisiana, 3–5 Oct
7. Brigham WE, Abbaszadeh-Degani M (1987) Tracer testing for reservoir description. J Pet Techol May: 519–527
8. Caudle BH (1967, 1968) Fundamentals of reservoir engineering – Part I and Part II. Society of Petroleum Engineers, Dallas, Texas
9. Chaney PE, Noble MD, Henson WL, Rice, TD (1956): How to perforate your well to prevent water and gas coning. Oil Gas J 55 May 7: 108–120
10. Chierici GL (1976) Modelli per la simulazione del comportamento dei giacimenti – una revisione critica. Boll Assoc Min Subalpina 13(3): 237–251
11. Chierici GL (1985) Petroleum reservoir engineering in the year 2000. Energy Explor Exploit 3: 173–193
12. Chierici GL, Ciucci GM, Pizzi G (1964) A systematic study of gas and water coning by potentiometric models. J Pet Techol Aug: 923–929; Trans, AIME, 231
13. Chierici GL, Ciucci, GM, Pizzi G (1965) Sweep efficiency optimization in injection projects – a potentiometric analyser study. Proc 5th Arab Petr Congr, Cairo, Egypt, March 16–23
14. Chierici GL, Ciucci GM, Pizzi G (1967) Recovery calculations in heterogeneous water-drive reservoirs containing undersaturated oil. J Inst Pet 53 (528): 394–401
15. Craig FF Jr (1971) The reservoir engineering aspects of waterflooding. SPE Monogr vol 3, Society of Petroleum Engineers, Dallas, Texas
16. Dyes AB, Caudle BH, Erickson RA (1954) Oil production after breakthrough as influenced by mobility ratio. Trans, AIME 201: 81–84
17. Dykstra H, Parsons HL (1950) The prediction of oil recovery by waterflooding, 2nd edn. Secondary recovery of oil in the US. API, New York, pp 160–174
18. Erba M, Rovere A, Turriani C, Safsaf S (1983) Bu Attifel field – a synergetic geological and engineering approach to reservoir management. Proc 11th World Pet Congr, vol 3. John Wiley & Sons, London, pp 89–99
19. Fassihi MR (1986) New correlations for calculation of vertical coverage and areal sweep efficiency. SPE Reservoir Eng Nov: 604–606

20. Greenkorn RA (1962) Experimental study of waterflood tracers. Trans, AIME 225: 87–92
21. Hagoort J (1982) The response of interwell tracer tests in watered-out reservoirs. Paper SPE 11131 presented at the 57th Annual Technical Conf, Society of Petroleum Engineers, New Orleans, Louisiana, Sept: 26–29
22. IOCC (1983) Improved oil recovery. Interstate Oil Compact Commission, Oklahoma City, Oklahoma
23. Kuo MCT, Des Brisay CL (1983). A simplified method for water coning predictions. Paper SPE 12067 presented at the 58th Annu Technical Conf, Society of Petroleum Engineers, San Francisco, California, Oct: 5–8
24. Lantz RB (1970) Rigorous calculation of miscible displacement using immiscibile reservoir simulators. Soc Pet Eng J June: 192–202; Trans, AIME 249
25. Lewis JO (1917) Methods for increasing the recovery from oil sands. US Bureau of Mines, Bull No 148
26. Muskat M, Wyckoff RD (1935) An approximate theory of water-coning in oil production. Trans, AIME 114: 144–163
27. Slobod RL, Caudle BH (1952) X-ray shadowgraph studies of areal sweepout efficiencies. Trans, AIME 195: 265–270
28. Smith PJ, Brown CE (1984) The interpretation of tracer test response from heterogeneous reservoirs. Paper SPE 13262 presented at the 59th Annu Technical Conf, Society of Petroleum Engineers, Houston, Texas, Sept: 16–19
29. Sobocinski DL, Cornelius AJ (1965) A correlation for predicting water coning time. J Pet Tech May: 594–600; Trans, AIME 234
30. Stiles WE (1949) Use of permeability distribution in waterflood calculations. Trans, AIME 186: 9–13
31. Suder FE, Calhoun JC, Jr. (1949) Water-flood calculations. API Drill Prod Pract pp 260–285
32. Wyckoff RD, Botset HG, Muskat M (1933) The mechanics of porous flow applied to waterflooding problems. Trans, AIME 103: 219–249
33. Guo B, Lee RL (1992) A simple approach to optimization of completion interval in oil/water coning systems. Paper SPE 23994, Society of Petroleum Engineers (submitted 19 June 1992)

EXERCISES

Exercise 12.1

A reservoir of predominantly undersaturated oil is being produced by waterflood using a line drive configuration (alternating parallel lines of production and injection wells). The reservoir pressure is kept above the bubble point throughout the life of the reservoir.

A line of injectors is 400 m from the next line of producers, and the wells in each line are 200 m apart.

The average injection rate per well, i_w, is 500 m^3/day, and can be assumed to be constant with time.

The reservoir rock is assumed to be homogeneous and isotropic. It has thickness $h = 15$ m, porosity $\phi = 0.25$, and an initial water saturation $S_{iw} = 0.15$.

The reservoir oil has a viscosity of ten times higher than the water ($\mu_o/\mu_w = 10$). It has a volume factor $B_{of} = 1.15$.

Define the geometry of an elementary component of the well network (an injector well plus a producer) by an equivalent linear system ($E_v = 1.00$).

Using the $f_w(S_w)$ curve calculated in Ex. 11.1, and ignoring the effects of gravity and fingering, calculate the following:

- the variation in the water cut f_w under reservoir conditions with the oil recovery factor E_R;
- the variation in the oil recovery factor with time.

Solution

The equivalent linear system has the following dimensions:

- length: $L = 400$ m
- width: $W = 200$ m
- thickness: $h = 15$ m.

Its pore volume is:

$$V_p = \phi L W h = 0.25 \times 400 \times 200 \times 15 = 300\,000 \text{ m}^3.$$

The initial volume of oil in the system is:

$$N B_{of} = V_p(1 - S_{iw}) = 300\,000 \times (1 - 0.15) = 255\,000 \text{ m}^3.$$

From the $f_w(S_w)$ curve calculated in Ex. 11.1, we have, for $\mu_o/\mu_w = 10$, *at water breakthrough at the producing well*:

- water saturation at the front: $\quad (S_{w,f})_{BT} = 0.450$
- water fraction in the production stream: $\quad (f_w)_{BT} = 0.903$
- oil recovery factor: $\quad (E)_R = 0.376$

for $S_w \geq (S_{w,f})_{BT}$ we have the data given in Table E12/1.1.

These data are reproduced on an amplified scale in Fig. E12/1.1.

To calculate the recovery factor $E_R(f_{w,e})$ as a function of the producing water cut, we can use Eq. (12.9):

$$E_R(f_{w,e}) = \frac{1}{1 - S_{iw}} \left[S_{w,e} - S_{iw} + \frac{1 - f_{w,e}}{\left(\dfrac{\mathrm{d}f_w}{\mathrm{d}S_w}\right)_{S_{w,e}}} \right], \tag{12.9}$$

where $S_{w,e}$ is the water saturation at the exit face of the porous medium and $(\mathrm{d}f_w/\mathrm{d}S_w)_{S_{w,e}}$ is determined graphically from Fig. E12/1.1.

The relevant calculations are listed in Table E12/1.2, and the results plotted in Fig. E12/1.2. These results indicate that by the time $f_w = 0.98$ – *a mere 2% oil in the production stream* – only 46% recovery has been achieved.

To summarise so far we have:

- at breakthrough: $\quad E_R = 0.3764 \quad f_w = 0.9030$
- at $f_w = 0.98$: $\quad E_R = 0.4600$
- at $f_w = 1.00$: $\quad E_R = 0.6471$

In other words, in order to reach high percentage recoveries, it will be necessary to continue production until high water cuts occur, especially in the case of medium-heavy and heavy oils. In the light of this, it is not surprising that many reservoirs in the USA are produced until as much as 98% water cut.

Table E12/1.1

S_w	f_w
0.45	0.903
0.50	0.979
0.55	0.998
0.60	0.999
0.65	~1.000
0.70	1.000

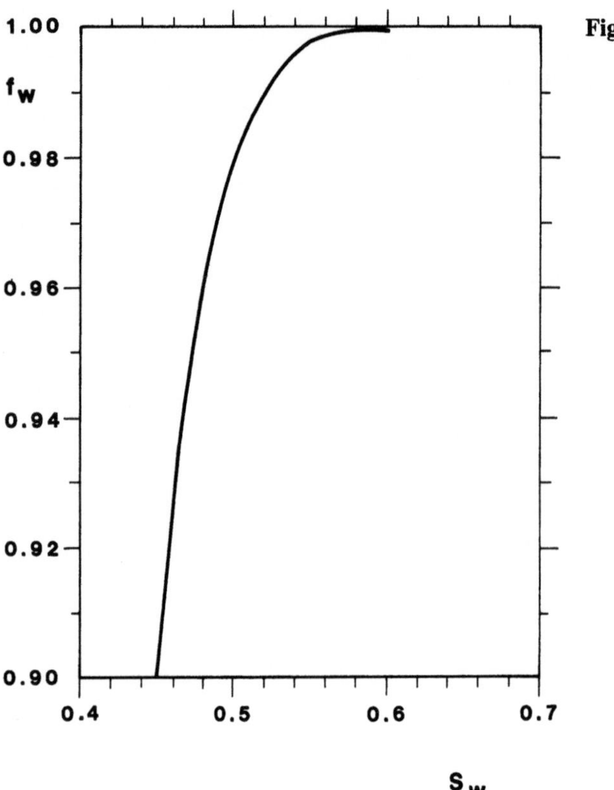

Fig. 12/1.1

Table E12/1.2

$S_{w,e}$	$f_{w,e}$	$\left(\dfrac{df_w}{dS_w}\right)_{S_{w,e}}$	E_R
0.45	0.9030	4.95	0.3764
0.46	0.9253	2.72	0.3970
0.48	0.9606	1.10	0.4304
0.50	0.9790	0.58	0.4544
0.52	0.9893	0.39	0.4676
0.54	0.9960	0.24	0.4784
0.56	0.9982	0.11	0.5016
0.58	0.9995	0.04	0.5206
0.70	1.0000	0.00	0.6471

To determine how the recovery factor, $E_R(t)$, varies with time, we proceed as follows. We start from the relationships:

$$N_p(t) = N E_R(t);\tag{12/1.1}$$

$$q_o = \frac{dN_p}{dt} = N\frac{dE_R}{dt} = \frac{1 - f_w}{B_{of}}i_w;\tag{12/1.2}$$

where q_o and N_p are, respectively, the total daily production rate and cumulative production of oil, *both measured at surface conditions.*

When we integrate Eq. (12/1.2) between 0 and t we get

$$t = \frac{N B_{of}}{i_w}\int_0^{E_R}\frac{dE_R}{1 - f_w},\tag{12/1.3}$$

where t is in days if i_w is in m^3/day.

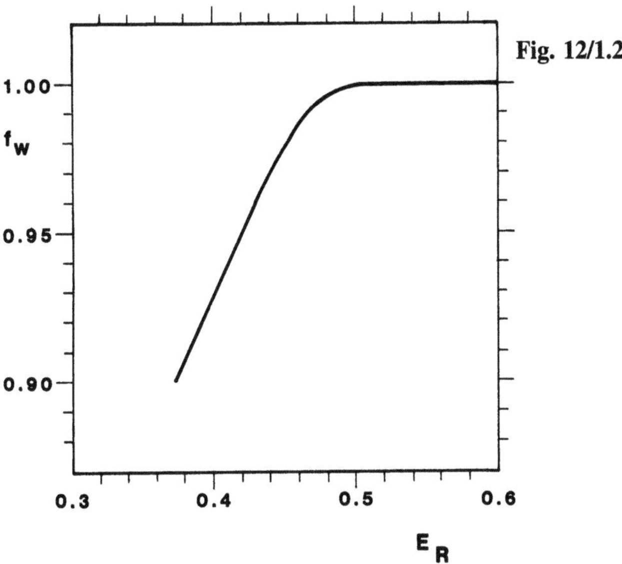

Fig. 12/1.2

Table E12/1.3

E_R	$f_o = 1 - f_w$	$\int_0^{E_R} \dfrac{dE_R}{1 - f_w}$	t	Production data	
				t	N_p
(frac)	(frac)	(dimensionless)	(days)	(years)	(stock tank m^3)
0.3764	0.0970	0.37640	192	0.526	83 463
0.3970	0.0747	0.62047	316	0.866	88 109
0.4304	0.0394	1.26789	647	1.770	96 299
0.4544	0.0210	2.14389	1 093	2.993	102 165
0.4676	0.0107	3.07500	1 568	4.294	105 438
0.4784	0.0040	4.92967	2 514	6.883	108 461
0.5016	0.0018	14.27411	7 280	19.929	114 470
0.5206	0.0005	38.5519	19 661	53.830	120 660

The integral in Eq. (12/1.3) is quite simple to evaluate numerically using the values plotted in Fig. E12/1.2, and the results are listed in Table E12/1.3.

Note how slowly oil is recovered once water breakthrough has occurred.

Water breaks through at the producers approximately 6 months after the start of injection, by which time the cumulative oil production is about 83 500 m^3 (stock tank). It then needs 2.5 years of continued production to recover a further 18 500 m^3 of oil, and, another 17 years to recover 12 000 m^3 more.

This behaviour is typical of medium-heavy to heavy oil reservoirs.

Exercise 12.2

Using Eq. (12.11) (proposed by Fassihi[19]), determine how the areal efficiency E_A varies with f_w under reservoir conditions for a waterflood using alternating lines of injectors and producers with the following two configurations:

a) direct line drive;
b) staggered line drive;

and for water/oil mobility ratios, M_{wo}, equal to:

0.5; 1; 5 and 10.

Solution

Fassihi's Eq. (12.11) can be written in the following form:

$$E_A(f_w) = \frac{1}{A(M_{wo})f_w + B(M_{wo}) + 1} \qquad (12/2.1)$$

Table E12/2.1

M_{wo}	Direct line drive		Staggered line drive	
	$A(M_{wo})$	$B(M_{wo})$	$A(M_{wo})$	$B(M_{wo})$
0.5	−0.61787	0.55252	−0.15920	0.23882
1	−0.88880	0.84690	−0.32935	0.37282
5	−1.41568	1.47176	−0.68243	0.74799
10	−1.62944	1.73245	−0.82864	0.92261

Table E12/2.2. *Direct line drive*

f_w	E_A			
	$M_{wo} = 0.5$	$M_{wo} = 1$	$M_{wo} = 5$	$M_{wo} = 10$
0.1	0.6708	0.5688	0.4291	0.3892
0.2	0.6998	0.5991	0.4569	0.4155
0.3	0.7314	0.6328	0.4885	0.4457
0.4	0.7661	0.6705	0.5248	0.4806
0.5	0.8041	0.7130	0.5669	0.5214
0.6	0.8462	0.7613	0.6164	0.5699
0.7	0.8928	0.8165	0.6753	0.6282
0.8	0.9450	0.8804	0.7467	0.6998
0.9	0.9980	0.9551	0.8350	0.7898

Table E12/2.3. *Staggered line drive*

f_w	E_A			
	$M_{wo} = 0.5$	$M_{wo} = 1$	$M_{wo} = 5$	$M_{wo} = 10$
0.1	0.8177	0.7463	0.5953	0.5436
0.2	0.8285	0.7651	0.6205	0.5692
0.3	0.8396	0.7849	0.6480	0.5974
0.4	0.8510	0.8057	0.6780	0.6285
0.5	0.8626	0.8277	0.7108	0.6630
0.6	0.8747	0.8509	0.7471	0.7015
0.7	0.8870	0.8754	0.7872	0.7448
0.8	0.8997	0.9014	0.8319	0.7938
0.9	0.9128	0.9290	0.8820	0.8497

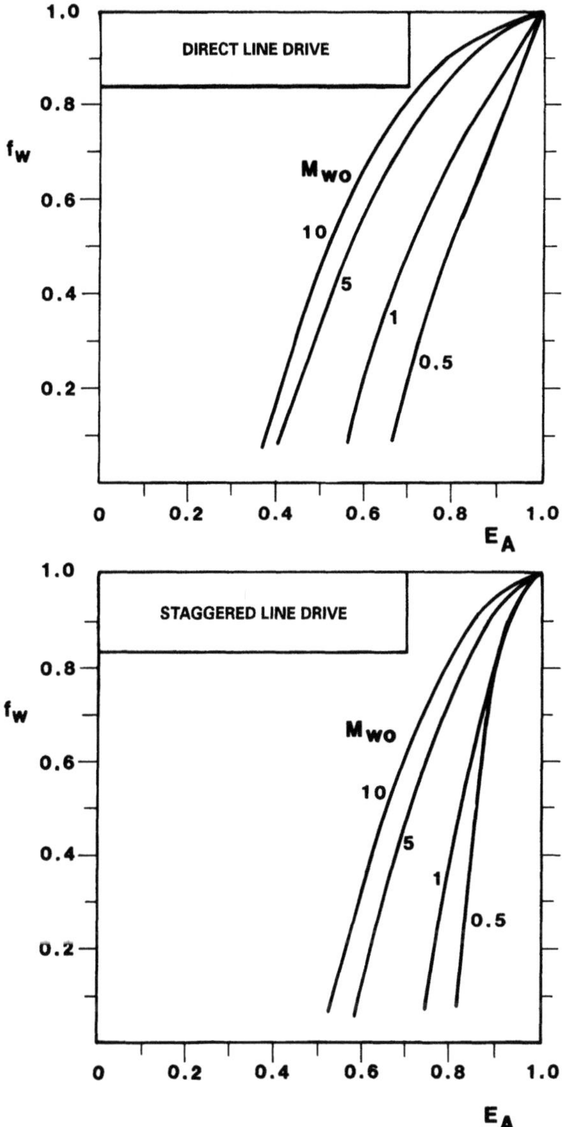

Fig. E12/2.1

where

$$A(M_{wo}) = a_1 \ln(M_{wo} + a_2) + a_3 \tag{12/2.2a}$$

and

$$B(M_{wo}) = a_4 \ln(M_{wo} + a_5) + a_6. \tag{12/2.2b}$$

Table 12.1 lists the coefficients a_1, a_2, a_3, a_4, a_5, and a_6 for both direct and staggered line drive configurations.

Values of $A(M_{wo})$ and $B(M_{wo})$ for each of the cases specified are listed in Table E12/2.1.

The values of $E_A(f_w)$ calculated using A and B from the above table are listed in Tables E12/2.2 and E12/2.3, and are plotted in Fig. E12/2.1.

It is evident from these results that, for a given f_w, the line drive geometry with the producing wells staggered by a half-spacing from the injectors (Fig. 12.1) achieves a better areal efficiency than with the wells directly facing one another.

In both these cases, note that E_A is poorer when $M_w > 1$, again for a given f_w.

$$\Diamond \ \Diamond \ \Diamond$$

Exercise 12.3

A reservoir consists of 5 layers (or "sedimentary units"), *all in vertical communication*. The layer properties are, from top to bottom:

Layer	h (m)	ϕ (frac)	k (md)
1	1	0.40	1000
2	3	0.30	200
3	5	0.25	100
4	4	0.28	150
5	2	0.20	50

The following relationships apply to all five layers:

$$S_{iw} = 0.25 - 2 \times 10^{-3}\sqrt{k/\phi}; \tag{12/3.1a}$$

$$S_{or} = 0.45 - 3 \times 10^{-3}\sqrt{k/\phi}; \tag{12/3.1b}$$

$$k_{ro,iw} = 0.70 + 4 \times 10^{-3}\sqrt{k/\phi}; \tag{12/3.1c}$$

$$k_{rw,or} = 0.30 + 2 \times 10^{-3}\sqrt{k/\phi}; \tag{12/3.1d}$$

with k in md and ϕ as a decimal fraction.

The reservoir is producing under bottom water drive, and the reservoir pressure remains above the bubble point at all times. *As a consequence, a gradual rise in the water/oil contact is expected.*

Calculate the relative permeability pseudo-curves for oil and water *for flow parallel to the layering* (approximately horizontal), ignoring any capillary pressure effects ($P_{c,ow} = 0$).

Repeat these calculations with the layers in inverted order (i.e. layer 5 on top, layer 1 at the bottom).

Solution

Firstly, we will calculate S_{iw}, S_{or}, $k_{ro,iw}$ and $k_{rw,or}$ using Eqs. (12/3.1), and tabulate (Table E12/3.1) the results along with the other data that are available:

We now consider a block of reservoir – a parallelepiped with a base 1×1 m and a height equal to the thickness of the reservoir ($h_t = 15$ m). The volume of this block is therefore

$$V_t = 15 \text{ m}^3.$$

The pore volume in this block is:

$$V_p = \sum_{j\,1}^{5} h_j \phi_j = 4.07, \text{ m}^3$$

Table E12/3.1

Layer No.	h (m)	ϕ (frac)	S_{iw}	S_{or}	k (md)	$k_{ro,iw}$	$k_{rw,or}$
1	1	0.40	0.15	0.30	1000	0.90	0.40
2	3	0.30	0.20	0.37	200	0.80	0.35
3	5	0.25	0.21	0.39	100	0.78	0.34
4	4	0.28	0.20	0.38	150	0.79	0.35
5	2	0.20	0.22	0.40	50	0.76	0.33

$h_t = 15$

so its average porosity must be:

$$\bar{\varphi} = \frac{V_p}{V_t} = 0.271\bar{3}.$$

The average permeability of the block, for flow parallel to the layering, is:

$$\bar{k} = \frac{\sum_{1}^{5} {}_j h_j k_j}{h_t} = \frac{2800}{15} = 186.\bar{6} \text{ md}.$$

To calculate the relative permeability curves as a function of the height of the water/oil contact we should construct a table in which, for each of the layers, and for the two limiting situations where $S_w = S_{iw}$ and $S_w = 1 - S_{or}$, we list (Table E12/3.2):

- the pore volume $V_{p,j}$ in the jth block;
- the volume of water $V_{w,j}$ in the jth block;
- the conductivity $(hkk_{ro,iw})_j$ to oil in the jth block;
- the conductivity $(hkk_{rw,or})_j$ to water in the jth block.

Since $\sum_{1}^{5} {}_j h_j k_j = 2800$ md$\times$ m, we can immediately calculate:

a) *water/oil contact at the base of layer 5*
   ($S_w = S_{iw}$ in all 5 layers):

$$\bar{S}_{iw} = \frac{0.8145}{4.07} = 0.2001$$

$$\bar{k}_{ro,iw} = \frac{2320}{2800} = 0.8286$$

$$k_{rw} = 0.00$$

b) *water/oil contact at the top of layer 1*
   ($S_w = 1 - S_{or}$) in all 5 layers:

$$\bar{S}_w = \frac{2.5439}{4.07} = 0.6250 \quad [\bar{S}_{or} = 0.3750]$$

$$\bar{k}_{rw,or} = \frac{1023}{2800} = 0.3654$$

$$k_{ro} = 0.00$$

The results for positions of the water/oil contact between these two extremes will be found in Table E12/3.3.

Table E12/3.2

Layer	Pore volume	$S_o = 1 - S_{iw}$		$S_o = S_{or}$	
	$V_{p,j}$	$V_{w,j}$	$(hkk_{ro,iw})_j$	$V_{w,j}$	$(hkk_{rw,or})_j$
N	m^3	m^3	md $\times$ m	m^3	md $\times$ m
1	0.40	0.0600	900	0.2800	400
2	0.90	0.1800	480	0.5670	210
3	1.25	0.2625	390	0.7625	170
4	1.12	0.2240	474	0.6944	210
5	0.40	0.0880	76	0.2400	33
Totals	4.07	0.8145	2320	2.5439	1023

Table E12/3.3

Position of water/oil contact	Layers invaded by water					$\Sigma_j V_w$	$\Sigma_j hkk_{ro}$	$\Sigma_j hkk_{rw}$	$\overline{\overline{S}}_w$	$\bar{k}_{ro}$	$\bar{k}_{rw}$
	1	2	3	4	5	(m^3)	(md × m)	(md× m)	dimension-less	dimension-less	dimension-less
Base layer 5						0.8145	2320	zero	0.2001	0.8286	zero
Top layer 5					×	0.9665	2244	33	0.2375	0.8014	0.0118
Top layer 4				×	×	1.4369	1770	243	0.3530	0.6321	0.0868
Top layer 3			×	×	×	1.9369	1380	413	0.4759	0.4929	0.1475
Top layer 2		×	×	×	×	2.3239	900	623	0.5710	0.3214	0.2225
Top layer 1	×	×	×	×	×	2.5439	zero	1023	0.6250	zero	0.3654

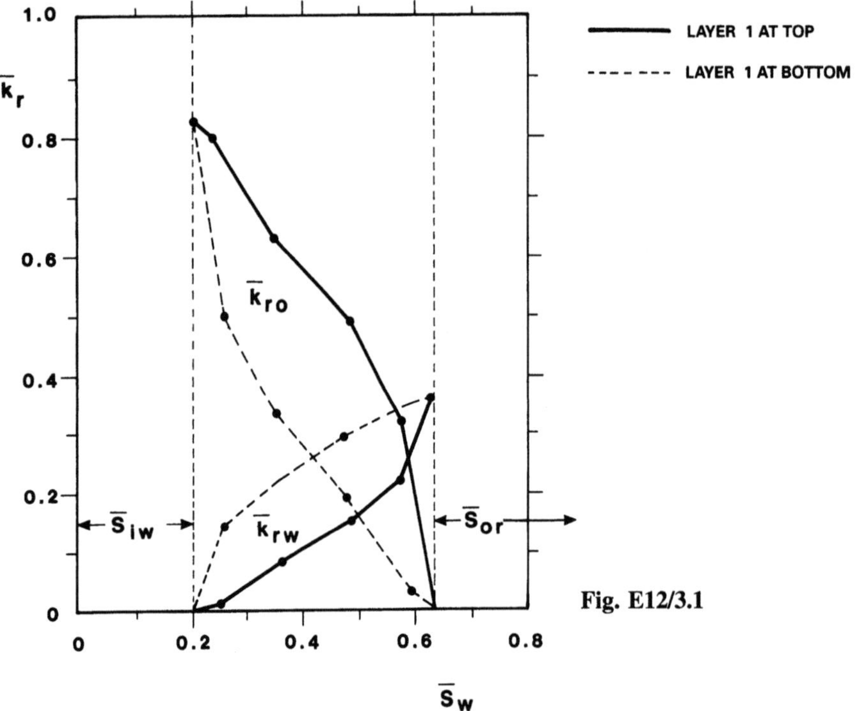

Fig. E12/3.1

The values of $\Sigma_j V_w$, $\Sigma_j hkk_{ro}$ and $\Sigma_j hkk_{rw}$ have been calculated simply by summing the appropriate values for each layer (as listed in Table E12/3.2), taking into account the relevant saturation conditions existing at each time step ($S_o = 1 - S_{iw}$ or $S_o = S_{or}$ in each layer).

If, as we assume, capillary pressure is negligible, we have piston-like displacement, meaning that S_o can only be either $(1 - S_{iw})$ or S_{or}.

In Table E12/3.4, the calculations have been repeated for the case where the reservoir has been overturned, so that layer 1 is now at the bottom.

The results for both sets of calculations are presented in Fig. E12/3.1: there is a noticeable difference between the two scenarios.

Table E12/3.4

Position of water/oil contact	Layers invaded by water					$\Sigma_j V_w$	$\Sigma_j hkk_{ro}$	$\Sigma_j hkk_{rw}$	$\overline{\overline{S}}_w$	$\bar{k}_{ro}$	$\bar{k}_{rw}$
	1	2	3	4	5	(m³)	(md × m)	(md × m)	dimension-less	dimension-less	dimension-less
Base layer 1						0.8145	2320	zero	0.2001	0.8286	zero
Top layer 1	×					1.0345	1420	400	0.2542	0.5071	0.1429
Top layer 2	×	×				1.4215	940	610	0.3493	0.3357	0.2179
Top layer 3	×	×	×			1.9215	550	780	0.4721	0.1964	0.2786
Top layer 4	×	×	×	×		2.3919	76	990	0.5877	0.0271	0.3536
Top layer 5	×	×	×	×	×	2.5439	zero	1023	0.6250	zero	0.3654

We can conclude from this that an accurate zonation of the reservoir is essential.

$$\Diamond \; \Diamond \; \Diamond$$

Exercise 12.4

This exercise, furnished by Prof. Guido Gottardi of the Faculty of Engineering at the University of Bologna, provides a computer program for the calculation of relative permeability and capillary pressure pseudo-curves for oil/water systems in multilayered reservoirs with vertical communication, for the two flow regimes: segregated and dispersed (Sect. 12.5.2.2).

A test data set is included at the end so that the program can be checked. The relative permeability and capillary pressure pseudo-curves calculated using the test data set are presented in Figs. E12/4.1 and 4.2 (this page) and in Figs. 12/4.3 and 4.4 (page 113).

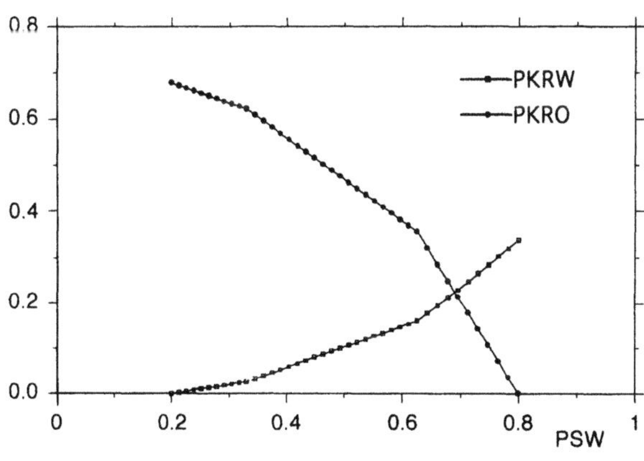

Fig. E12/4.1. *Relative permeability pseudo-curves for water* (PKRW) *and oil* (PKRO) *in segregated flow*

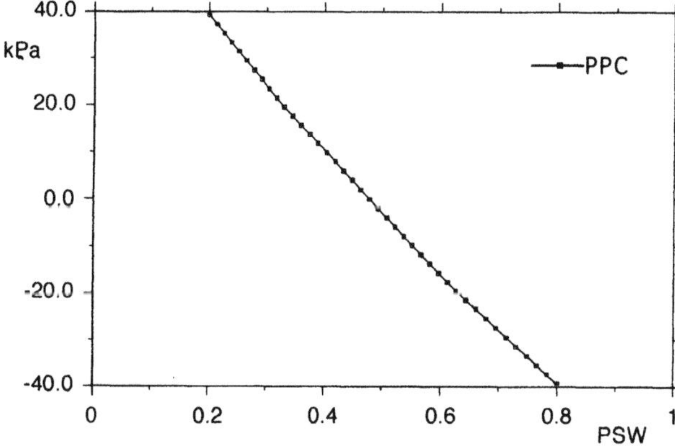

Fig. E12/4.2. *Water-oil capillary pressure pseudo-curve in segregated flow*

```
      PROGRAM MLAY_VE
C---------------------------------------------------------------------
C
C
C         Calculates the relative permeability and capillary pressure
C         pseudo-curves from rock relative permeability and capillary
C         pressure curves, assuming vertical equilibrium (VE), in the
C         dispersed or segregated flow regime, for a reservoir consist-
C         ing of n layers of different properties.
C         Input data:
C
C         NLAY          = number of layers into which the reservoir is
C                         subdivided
C         NI            = number of integration intervals for each layer
C         NP            = number of positions of the water-oil table
C         KODE          = flag          = 0 for dispersed flow
C                                        = 1 for segregated flow
C
C         DENW          = water density [kg/m^3]
C         DENO          = oil density   [kg/m^3]
C         DIPANG        = angle of dip of the layers [degrees]
C
C         H(I)          = thickness of I-th layer [m]
C         PERM(I)       = permeability of I-th layer [md]
C         PHI(I)        = porosity of I-th layer [fraction]
C         SWC(I)        = irreducible water saturation of I-th layer
C                         [fraction]
C         SOR(I)        = residual oil saturation of I-th layer [fraction]
C         KROSWC(I)     = relative permeability to oil at irriducible water
C                         saturation in the I-th layer
C         KRWSOR(I)     = relative permeability to water at residual oil
C                         saturation in the I-th layer
C
C         NPPL(L)       = number of values in the capillary pressure and
C                         relative permeability tables for the L-th layer,
C                         where: L=1, NLAY
C
C         PC(I,L)       = capillary pressure table for L-th layer [kPa]
C         SW(I,L)       = water saturation table for L-th layer [fraction]
C         KRW(I,L)      = water relative permeability table for L-th layer
C         KRO(I,L)      = oil relative permeability table for L-th layer,
C                         where: I=1,NPPL(L); L=1,NLAY
C---------------------------------------------------------------------
      PARAMETER (LP1=30,LP2=20,LP3=200,LP4=100)
C     LP1=maximum number of lines in rock property tables
C     LP2=maximum number of layers
C     LP3=maximum number of numerical integration points for
C         each layer
C     LP4=maximum number of pseudo values for each pseudo
C         fuction
C
C---------------------------------------------------------------------
      REAL KRW,KRO,INTSW,INTKRW,INTKRO,KRWT,KROT,KROSWC,KRWSOR
C
      CHARACTER*15 F14,F16,ANS*1
C
C
      DIMENSION Z(LP3,LP2),PC(LP1,LP2),SW(LP1,LP2),KRW(LP1,LP2),
     &          KRO(LP1,LP2)
      DIMENSION ZT(LP3,LP2),PCT(LP3,LP2),SWT(LP3,LP2),KRWT(LP3,LP2),
     &          KROT(LP3,LP2)
      DIMENSION H(LP2),PERM(LP2),PHI(LP2),SWC(LP2),SOR(LP2),
     &          KROSWC(LP2),KRWSOR(LP2)
      DIMENSION INTSW(LP3,LP2),INTKRW(LP3,LP2),INTKRO(LP3,LP2),
     &          NPPL(LP2)
      DIMENSION PSW(LP4),PKRW(LP4),PKRO(LP4),PPC(LP4),ZWOC(LP4)
      DIMENSION ZTOP(LP2),ZBOT(LP2),SSW(LP2),SKRW(LP2),SKRO(LP2)
C
C-------READING OF FILE NAMES
C
```

```fortran
 8400     WRITE(*,8020)
          READ(*,8010) F14
          WRITE(*,8030)
          READ(*,8010) F16
          WRITE(*,8050)
          READ(*,8010) ANS
          IF(ANS.EQ.'N') GOTO 8400
          IF(ANS.EQ.'E') STOP
          IF(ANS.NE.'Y') GOTO 8400
C
 8010     FORMAT(A)
 8020     FORMAT(/,1X,' INPUT DATA FILE NAME  > ',$)
 8030     FORMAT(/,1X,' OUTPUT FILE NAME      > ',$)
 8045     FORMAT(/,1X,' PLOTTER FILE NAME     > ',$)
 8050     FORMAT(/,1X,' OK ?   (Y/N/E)        > ',$)
C
C-----OPENING OF FILES
C
          OPEN(14,FILE=F14,STATUS='OLD',FORM='FORMATTED')
          OPEN(16,FILE=F16,STATUS='NEW',FORM='FORMATTED')
          OPEN(15,STATUS='SCRETCH',FORM='FORMATTED')
C
C
          WRITE(*,*) 'Please wait program MLAY_VE is running'
C
C-----COPY INPUT FILE (14) INTO FILE (15) ELIMINATING
C-----COMMENT AND EMPTY LINES
C
          CALL CMNT(14,15)
C
          READ(15,*) NLAY,NI,NP,KODE,DENW,DENO,DIPANG
          WRITE(16,200) NLAY,NI,NP,KODE,DENW,DENO,DIPANG
  200     FORMAT(5X,'NUMBER OF LAYERS (NLAY).................. =',I5,/,
         &       5X,'NUMBERS OF INTERVALS FOR INTEGRATION..... =',I5,/,
         &       5X,'NUMBER OF VALUES FOR EACH PSEUDO FUNCTION =',I5,/,
         &       5X,'DISPERSED (0) OR SEGREGATED (1) FLOW KODE.=',I5,/,
         &       5X,'DENSITY OF WATER [KG/M^3]................ =',F8.2,/,
         &       5X,'DENSITY OF OIL   [KG/M^3]................ =',F8.2/
         &       5X,'RESERVOIR DIP ANGLE(DEGREES)............. =',F8.2,//)
C
C-----CHECK ON DIMENSIONING
C
          IF(NLAY.GT.LP2.OR.NI+1.GT.LP3.OR.NP.GT.LP4) THEN
            WRITE(*,*) 'ARRAY BOUNDARY ERROR!!! CHECK DIMENSIONING'
            PAUSE
            STOP
          ENDIF
C
          WRITE(16,250)
  250     FORMAT(//,T16,'THICKNESS, PERMEABILITY AND POROSITY TABLE:',/,
         &          T16,'------------------------------------------',//,
         & T10,'I',T16,'H',T22,'PERM',T32,'PHI',T39,'SWC',T47,'SOR',
         & T55,'KROSWC',T63,'KRWSOR'/)
          DO I=1,NLAY
                READ(15,*) H(I),PERM(I),PHI(I),SWC(I),SOR(I),
         &          KROSWC(I),KRWSOR(I)
                WRITE(16,260) I,H(I),PERM(I),PHI(I),SWC(I),
         &          SOR(I),KROSWC(I),KRWSOR(I)
          ENDDO
  260     FORMAT(5X,I5,7F8.2)
          HT=0.0
          DENP=0.0
          DENK=0.0
C
          DO I=1,NLAY
             DENP=DENP+PHI(I)*H(I)
             DENK=DENK+PERM(I)*H(I)
             HT=HT+H(I)
          ENDDO
```

```
C
            DIPANG=DIPANG*3.14159265/180.
            DGAM=0.00981*(DENW-DENO)*SIN(DIPANG)
            READ(15,*) (NPPL(I),I=1,NLAY)
            WRITE(16,210) (NPPL(I),I=1,NLAY)
   210      FORMAT(//,T5,'NUMBER OF LINES OF DATA FOR EACH LAYER :',//,
        &   T5,(10I5))
C
C           READ IN THE PERMEABILITY AND CAPILLARY PRESSURE TABLES
C           FOR DISPERSED CASE ONLY
C
            IF(KODE.EQ.0) THEN
            WRITE(16,220)
   220      FORMAT(//,T16,'ROCK CAPILLARY AND PERMEABILITY TABLES:',/,
        &            T16,'-----------------------------------------',//)
            DO L=1,NLAY
                    WRITE(16,230) L
                    WRITE(16,235)
   230      FORMAT(/,T5,'LAYER NUMBER:',I5,//)
   235      FORMAT(T10,'N ',T20,' PC',T32,'SW',T43,'KRW',T55,'KRO'/)
              NPL=NPPL(L)
                DO 15 I=1,NPL
                  READ(15,*) PC(I,L),SW(I,L),KRW(I,L),KRO(I,L)
                  WRITE(16,240) I,PC(I,L),SW(I,L),KRW(I,L),KRO(I,L)
   240      FORMAT(5X,I5,4F12.3)
                ENDDO
            ENDDO
            ENDIF
C
            IF(KODE.EQ.0) THEN
                PCMAX=0.
                DO L=1,NLAY
                  IF(PC(1,L).GT.PCMAX) PCMAX=PC(1,L)
                ENDDO

                ZWOC(1)=-PCMAX/DGAM
                DZWC=(HT-ZWOC(1))/(NP-1)
                DO I=2,NP
                  ZWOC(I)=ZWOC(I-1)+DZWC
                ENDDO
            ELSE IF(KODE.EQ.1) THEN
                ZWOC(1)=0.0
                DZWC=HT/(NP-1)
                DO 33 I=2,NP
                  ZWOC(I)=ZWOC(I-1)+DZWC
                ENDDO
C           ENDIF
            WRITE(16,400)
   400      FORMAT(//,T16,'BOTTOM AND TOP COORDINATES OF EACH LAYER',/,
        &            T16,'----------------------------------------',//,
        &          T10,'L',T20,'ZBOT(L)',T35,'ZTOP(L)',/)
            ZTOP(1)=H(1)
            ZBOT(1)=0.0
            DO L=2,NLAY
                ZTOP(L)=ZTOP(L-1)+H(L)
                ZBOT(L)=ZBOT(L-1)+H(L-1)
            ENDDO
            DO L=1,NLAY
              WRITE(16,405) L,ZBOT(L),ZTOP(L)
            ENDDO
   405      FORMAT(T6,I5,2F15.2)
C
C
C-------CALCULATE THE MEAN WATER PSEUDO-SATURATION PSW(I), THE
C       PSEUDO-RELATIVE PERMEABILITIES TO WATER AND OIL PKRW(I)
C       AND PKRO(I), AND THE PSEUDO-CAPILLARY PRESSURE PPC(I) AS
C       A FUNCTION OF THE HEIGHT OF THE WATER/OIL CONTACT ZWOC
C       (ABOVE THE BASE OF THE BOTTOM LAYER)
C
```

```fortran
            WRITE(16,425)
  425       FORMAT(//,T26,'COMPUTED PSEUDO FUCTION TABLE:',/,
     &             T26,'---------------------------------',/
     &      /T10,'I ',T17,'ZWOC',T30,'PPC',T42,'PSW',
     &       T54,'PKRW',T66,'PKRO'/)
C
C-------MAIN LOOP ON WATER TABLE POSITIONS
C
            NI1=NI+1
            DO 40 IP=1,NP
              SUMW=0.
              SUMKW=0.
              SUMKO=0.
X             WRITE(16,*) 'ZWOC:',ZWOC(IP)
C
C------LOOP ON LAYERS
C
            IF(KODE.EQ.0) THEN

            DO 45 L=1,NLAY
X                 WRITE(16,*) 'LAYER:',L
X                 WRITE(16,1000)
X1000       FORMAT(T2,'I',T7,'Z',T15,'CP',T22,'SWT',T29,'KRWT',T37,'KROT')
            DZ=(ZTOP(L)-ZBOT(L))/NI
            Z(1,L)=ZBOT(L)

            DO I=2,NI1
              Z(I,L)=Z(I-1,L)+DZ
            ENDDO
C
            DO 55 I=1,NI1
              CP=DGAM*(Z(I,L)-ZWOC(IP))
              IF(Z(I,L).LE.ZWOC(IP)) THEN
                SWT(I,L)=1.-SOR(L)
                KRWT(I,L)=KRWSOR(L)
                KROT(I,L)=0.0
              ELSE IF(Z(I,L).GT.ZWOC(IP)) THEN
                CALL INTERP(PC,SW,LP1,LP2,CP,SWT(I,L),NPPL,L)
                CALL INTERP(SW,KRW,LP1,LP2,SWT(I,L),KRWT(I,L),NPPL,L)
                CALL INTERP(SW,KRO,LP1,LP2,SWT(I,L),KROT(I,L),NPPL,L)
              ENDIF
X             WRITE(16,*) I,Z(I,L),CP,SWT(I,L),KRWT(I,L),KROT(I,L)
   55       CONTINUE

            CALL QTFG(Z,SWT,INTSW,LP3,LP2,NI1,L)
            SSW(L)=INTSW(NI1,L)
            CALL QTFG(Z,KRWT,INTKRW,LP3,LP2,NI1,L)
            SKRW(L)=INTKRW(NI1,L)
            CALL QTFG(Z,KROT,INTKRO,LP3,LP2,NI1,L)
            SKRO(L)=INTKRO(NI1,L)
            SUMW=SUMW+PHI(L)*SSW(L)
            SUMKW=SUMKW+PERM(L)*SKRW(L)
            SUMKO=SUMKO+PERM(L)*SKRO(L)
X           WRITE(16,1100)
 1100       FORMAT(/,T4,'SSW',T11,'SKRW',T21,'SKRO',T32,'SUMW',T40,
     &              'SUMKW',T48,'SUMKO'/)
X           WRITE(16,*) SSW(L),SKRW(L),SKRO(L),SUMW,SUMKW,SUMKO
   45       CONTINUE

            ELSE IF(KODE.EQ.1) THEN

              DO L=1,NLAY
                IF(ZWOC(IP).LE.ZBOT(L)) THEN
                  SUMW=SUMW+PHI(L)*H(L)*SWC(L)
                  SUMKO=SUMKO+PERM(L)*H(L)*KROSWC(L)
                ELSE IF(ZWOC(IP).GT.ZBOT(L).AND.ZWOC(IP).LT.ZTOP(L)) THEN
                  SUMW=SUMW+PHI(L)*(ZWOC(IP)-ZBOT(L))*(1.-SOR(L))
                  SUMKW=SUMKW+PERM(L)*(ZWOC(IP)-ZBOT(L))*KRWSOR(L)
                  SUMW=SUMW+PHI(L)*(ZTOP(L)-ZWOC(IP))*SWC(L)
```

```fortran
                        SUMKO=SUMKO+PERM(L)*(ZTOP(L)-ZWOC(IP))*KROSWC(L)
                    ELSE IF(ZWOC(IP).GE.ZTOP(L)) THEN
                        SUMW=SUMW+PHI(L)*H(L)*(1.-SOR(L))
                        SUMKW=SUMKW+PERM(L)*H(L)*KRWSOR(L)
                  ENDIF
                ENDDO
            ENDIF

                PSW(IP)=SUMW/DENP
                PKRW(IP)=SUMKW/DENK
                PKRO(IP)=SUMKO/DENK
                PPC(IP)=DGAM*(0.5*HT-ZWOC(IP))
C
            WRITE(16,430) IP,ZWOC(IP),PPC(IP),PSW(IP),PKRW(IP),PKRO(IP)
C
   40     CONTINUE
  430     FORMAT(5X,I5,5F12.4)
          CLOSE(15)
              CLOSE(16)
          STOP
          END
C         ----------------------------------------
          SUBROUTINE INTERP(X,Y,K,M,XX,YY,N,J)
C         ----------------------------------------
C         PERFORM LINEAR INTERPOLATION
C           J = material index, M,K = dimensions of X,Y array
C           N(J)= maximum number of elements for material J.
C
          DIMENSION X(K,M),Y(K,M),N(M)
C
          JF=0
          IF(X(1,J).LT.X(2,J)) JF=1
C
          IF(JF.EQ.0.AND.XX.GE.X(1,J)) THEN
           YY=Y(1,J)
           RETURN
          ELSE IF(JF.EQ.0.AND.XX.LE.X(N(J),J)) THEN
           YY=Y(N(J),J)
           RETURN
          ELSE IF(JF.EQ.1.AND.XX.LE.X(1,J)) THEN
           YY=Y(1,J)
           RETURN
          ELSE IF(JF.EQ.1.AND.XX.GE.X(N(J),J)) THEN
           YY=Y(N(J),J)
           RETURN
          ENDIF
C
          DO 10 I=2,N(J)
            IF(JF.EQ.0.AND.XX.LE.X(I,J)) GO TO 10
            IF(JF.EQ.1.AND.XX.GE.X(I,J)) GO TO 10
            YY=Y(I-1,J)+(XX-X(I-1,J))*(Y(I,J)-Y(I-1,J))/(X(I,J)-X(I-1,J))
            RETURN
   10     CONTINUE
          END
C         ----------------------------------------
          SUBROUTINE QTFG(X,Y,Z,NR,NC,NDIM,L)
C         ----------------------------------------
C         PERFORM INTEGRATION
C
          DIMENSION X(NR,NC),Y(NR,NC),Z(NR,NC)
C
          SUM2=0.
          IF(NDIM-1)4,3,1
C
C         INTEGRATION LOOP
    1     DO 2 I=2,NDIM
          SUM1=SUM2
          SUM2=SUM2+.5*(X(I,L)-X(I-1,L))*(Y(I,L)+Y(I-1,L))
    2     Z(I-1,L)=SUM1
```

```
       3 Z(NDIM,L)=SUM2
       4 RETURN
         END
C        ------------------------------
         SUBROUTINE CMNT(FILEIN,FILEOU)
C        ------------------------------
C
C        COPY FILEIN INTO FILEOU REMOVING COMMENTS
C        LINES (LINES WITH A PERIOD) AND BLANK LINES
C
         INTEGER FILEIN,FILEOU
         CHARACTER*132 LINEA
       1 READ(FILEIN,1000,END=2) LINEA
         N=1
         DO WHILE(LINEA(N:N).NE.';'.AND.N.LT.LEN(LINEA))
           N=N+1
         ENDDO
         M=N-1
         DO WHILE(LINEA(M:M).EQ.' '.AND.M.GT.0)
           M=M-1
         ENDDO
         IF(M.GT.0) WRITE(FILEOU,1000) LINEA(1:M)
         GOTO 1
       2 CLOSE(FILEIN)
         REWIND(FILEOU)
         RETURN
    1000 FORMAT(A)
         END
```

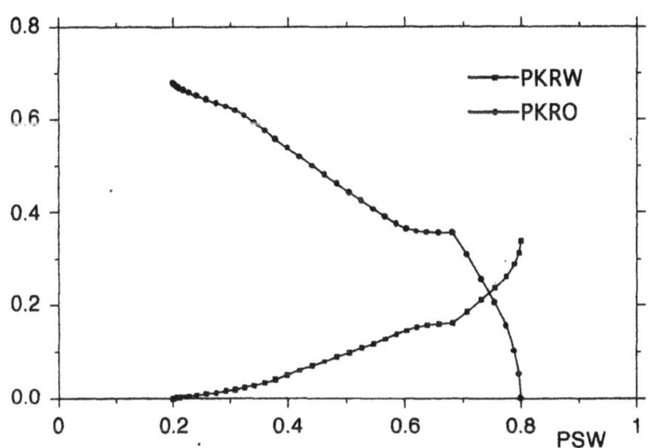

Fig. E12/4.3. *Relative permeability pseudo-curves for water* (PKRW) *and oil* (PKRO) *in dispersed flow*

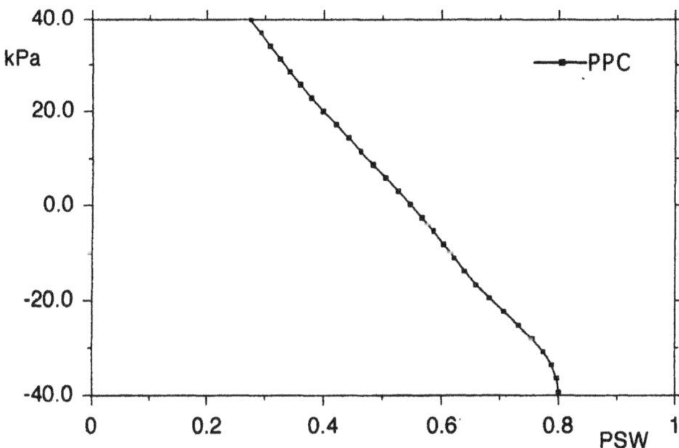

Fig. E12/4.4. *Water-oil capillary pressure pseudo-curve in dispersed flow*

TEST CASE

```
; NLAY      NI   NP  KODE     DENW     DENO    DIPANG
;                             [kg/m^3] [kg/m^3] [degree]
;---------------------------------------------------------
    3       100   41   0      1040.    840.      0.0
;
;  H    PERM      PHI    SWC    SOR    KROSWC   KRWSOR
; [m]   [md]
;---------------------------------------------------------
   10.    50.     0.15   0.20   0.20    0.50     0.24
   20.   100.     0.17   0.20   0.20    0.60     0.30
   10.   200.     0.20   0.20   0.20    0.80     0.40


   NPPL(NLAY)
    6   7   6
;---------------------------------------------------------
; CAPILLARY PRESSURE AND RELPERM TABLE   OF EACH LAYER
;---------------------------------------------------------
;layer n.1
;  PC       SW            KRW             KRO
; [kPa]
;---------------------------------------------------------
  34.47     .20          0.00            0.50
  20.68     .25          0.03            0.40
  13.79     .40          0.05            0.21
   6.89     .69          0.18            0.02
   3.45     .78          0.23            0.002
   0.0      .80          0.24            0.00
;---------------------------------------------------------
;
;layer n.2
;  PC       SW            KRW             KRO
; [kPa]
;---------------------------------------------------------
  27.58     .20          0.0             0.60
  20.68     .22          0.001           0.55
  17.24     .24          0.003           0.50
  13.79     .29          0.02            0.40
  10.34     .45          0.07            0.18
   6.89     .63          0.17            0.05
   0.0      .80          0.30            0.00
;---------------------------------------------------------
;
;layer n. 3
;  PC       SW            KRW             KRO
; [kPa]
;---------------------------------------------------------
  27.58     .20          0.00            0.80
  24.13     .20          0.00            0.80
  20.68     .20          0.00            0.80
  13.79     .20          0.00            0.80
   8.27     .40          0.008           0.30
   0.00     .80          0.40            0.00
```

Exercise 12.5

Suppose the 5 layers in Exercise 12.3 are *not* in vertical communication, but are separated by thin shale intercalations.

The reservoir is being produced by waterflood. The injected water viscosity is 0.5 cP under in situ conditions, and the reservoir oil has a viscosity of 5 cP.

Using the data provided in Ex. 12.3, calculate the curve for the vertical invasion efficiency $E_I(f_w)$ as a function of the fraction of water in the production stream.

Table E12/5.1

Layer	h (m)	ϕ	S_{iw}	S_{or}	k (md)	$k_{ro,iw}$	$k_{rw,or}$
1	1	0.40	0.15	0.30	1000	0.90	0.40
2	3	0.30	0.20	0.37	200	0.80	0.35
3	5	0.25	0.21	0.39	100	0.78	0.34
4	4	0.28	0.20	0.38	150	0.79	0.35
5	2	0.20	0.22	0.40	50	0.76	0.33

Solution

The data from Ex. 12.3 that are needed to solve this problem are presented in Table E12/5.1.

For the physical model to be used for the calculation, we consider a block of reservoir rock with the following dimensions:

length: $L = 100$ m
width: $W = 100$ m
thickness: $h = 15$ m (= reservoir thickness)

For each of the 5 layers we calculate

$$M_{wo} = \frac{k_{rw,or}}{k_{ro,iw}} \frac{\mu_o}{\mu_w}$$

$$N_{p,u} = LWh\phi(1 - S_{iw} - S_{or}) \tag{12.17b'}$$

$$q(t = 0) = \frac{khWk_{ro,iw}}{\mu_o} \frac{\Delta p}{L} \tag{12.12}$$

$$t_{BT} = \frac{1 + M_{wo}}{M_{wo}} \frac{N_{p,u}}{2q(t = 0)} \tag{12.19b}$$

Since the choice of Δp is completely arbitrary, we can choose a value which conveniently makes t_{BT} maximum (in the least permeable layer) equal to 100 days.

In layer 5, where $M_{wo} = 4.34$ and $N_{p,u} = 1520$ m^3, this requires a value for $q(t = 0)$ given by:

$$100 = \frac{5.34}{4.34} \frac{1520}{2q_5(t = 0)}$$

from which:

$$q_5(t = 0) = 9.351 \text{ m}^3/\text{day}.$$

Therefore, from Eq. (12.12), we get:

$$9.351 = 50 \times 2 \times 100 \times 0.76 \frac{\Delta p}{\mu_o L}$$

so that

$$\frac{\Delta p}{\mu_o L} = 1.23041 \times 10^{-3} \text{ (arbitrary units)}.$$

We can now determine $q(t = 0)$ and t_{BT} for the other 4 layers (Table E12/5.2).

When calculating $E_I(t)$ and $f_w(t)$ – from which we will get $E_I(f_w)$ – if we use the following time sequence:

12.171   20.000   30.000   40.243   49.431   57.000   64.048   76.000   88.000   100.000

we will include all the t_{BT} values appearing in Table E12/5.2.

Table E12/5.3 lists $x(t)/L$ calculated for each layer [Eq. (12.18b)], and in Table E12/5.4 you will find $q_o(t)$ and $q_w(t)$ [Eq. (12.15)], calculated under reservoir conditions. (The * denotes that water breakthrough has occurred.)

Finally, Table E12/5.5 contains the calculated flow rates for oil, $Q_o(t)$; water, $Q_w(t)$; and the cumulative production $N_p(t)$, all under reservoir conditions, for the whole block model. In this table,

Table E12/5.2

Layer	M_{wo}	$N_{p,\mu}$ (m^3)	$q(t=0)$ (m^3/day)	t_{BT} (days)
1	4.44	2 200	110.737	12.171
2	4.38	3 870	59.060	40.243
3	4.36	5 000	47.986	64.048
4	4.43	4 704	58.322	49.431
5	4.34	1 520	9.351	100.000

Table E12/5.3

t (days)	x/L Layer No.				
	1	2	3	4	5
12.171	1.0	0.20139	0.12260	0.16093	0.07716
20.0	1.0	0.35341	0.20873	0.27786	0.12949
30.0	1.0	0.59394	0.32984	0.45053	0.19994
40.243	1.0	1.0	0.47210	0.67570	0.27713
49.431	1.0	1.0	0.62484	1.0	0.35169
57.0	1.0	1.0	0.78370	1.0	0.41785
64.048	1.0	1.0	1.0	1.0	0.48426
76.0	1.0	1.0	1.0	1.0	0.61138
88.0	1.0	1.0	1.0	1.0	0.76882
100.0	1.0	1.0	1.0	1.0	1.0

Table E12/5.4

t (days)	Liquid production rate (m^3/day) at reservoir p and T Layer No.				
	1	2	3	4	5
12.171	491.67	69.93	52.99	66.62	9.94
20.0	491.67*	81.21	57.18	74.31	10.39
30.0	491.67*	109.04	64.34	89.57	11.05
40.243	491.67*	258.68	75.43	122.31	11.89
49.431	491.67*	258.68*	92.55	258.37	12.82
57.0	491.67*	258.68*	121.16	258.37*	13.78
64.048	491.67*	258.68*	209.22	258.37*	14.91
76.0	491.67*	258.68*	209.22*	258.37*	17.66
88.0	491.67*	258.68*	209.22*	258.37*	22.90
100.0	491.67*	258.68*	209.22*	258.37*	40.58

Note: Before breakthrough, each layer produces only oil; after breakthrough, only water. Water production is indicated by an *

the status just prior to water breakthrough in any of the layers is denoted by a superscript ($-$) on the time, while that just after breakthrough is denoted by a ($+$). Each layer produces only oil before breakthrough, and only water after.

From these results, we can calculate $f_w(t)$ under reservoir conditions, and $E_I(t)$. These values are included in Table E12/5.5, and $E_I(f_w)$ is plotted against f_w in Fig. E12/5.1.

There are a few important points to make about this diagram.

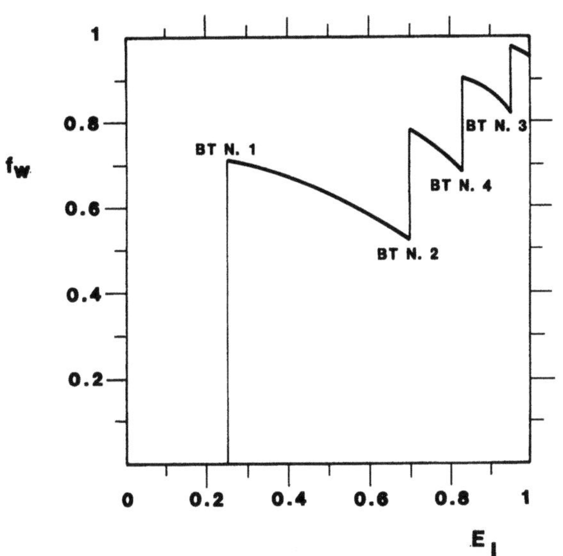

Table E12/5.5

t (days)	Q_o (m³/d)	Q_w (m³/d)	N_p (m³)	f_w	E_1
12.171(−)	691.2	zero	4 467	zero	0.2582
12.171(+)	199.5	491.7	4 467	0.7114	0.2582
20.0	223.1	491.7	6 115	0.6879	0.3536
30.0	274.0	491.7	8 571	0.6422	0.4956
40.243(−)	468.3	491.7	12 030	0.5122	0.6956
40.243(+)	209.6	750.4	12 030	0.7816	0.6956
49.431(−)	363.7	750.4	14 433	0.6735	0.8346
49.431(+)	105.4	1,008.7	14 433	0.9054	0.8346
57.0	135.0	1,008.7	15 328	0.8820	0.8863
64.048(−)	224.1	1,008.7	16 510	0.8182	0.9547
64.048(+)	14.9	1,217.9	16 510	0.9879	0.9547
76.0	17.7	1,217.9	16 703	0.9857	0.9658
88.0	22.9	1,217.9	16 943	0.9815	0.9797
100(−)	40.6	1,217.9	17 294	0.9678	1.0000
100(+)	zero	1,258.5	17 294	1.0000	1.0000

Note: The situation immediately before water breakthrough is indicated by $t^{(-)}$, and that immediately after by $t^{(+)}$. Q_o, Q_w, N_p and f_w are all expressed under reservoir conditions

As would be expected, there is a sharp increase in f_w when water breaks through in each layer. It may not be quite so obvious why f_w decreases between successive breakthroughs.

This can be explained by noting that, while those layers in which water has already broken through will produce water at a constant rate (according to our basic assumptions), the oil production from the remaining layers will *increase* because the advance of the more mobile water reduces the hydraulic impedance in these layers.

Obviously, if $M_{wo} < 1$, the opposite behaviour would be observed.

In multilayered reservoirs with no vertical communication between layers and $M_{wo} > 1$, the decline in water cut f_w after the breakthrough of water in one of the layers has often been observed, but rarely explained.

◊ ◊ ◊

Exercise 12.6

An oil reservoir has been developed with a well spacing of 80 acres. At one of the wells, the permeable rock has a total thickness of 50 m.

The upper 30 m are oil-bearing, and the underlying 20 m of water constitutes the local aquifer. The rock properties are:

– porosity:	ϕ	= 0.25
– horizontal permeability:	k_r	= 500 md
– relative permeability to oil at S_{iw}:	$k_{ro,iw}$	= 0.9
– relative permeability to water at S_{or}:	$k_{rw,or}$	= 0.4

The reservoir fluid properties are:

– oil volume factor:	B_{of}	= 1.25 m^3/m^3
– oil density:	ρ_o	= 780 kg/m^3
– oil viscosity:	μ_o	= 2 cP
– water density:	ρ_w	= 1030 kg/m^3
– water viscosity:	μ_w	= 0.6 cP

Calculate the critical flow rate for water coning if the interval open for production has a length of:

– case a: 6 m
– case b: 18 m

measured from the top of the formation.

Consider the two assumptions:

1) the rock is isotropic ($k_v = k_r$)
2) the rock is anisotropic, with $k_v = 0.1 k_r$.

Solution

Since 1 acre = 4046.856 m^2, the drainage area of the well in question is

$$A = 80 \times 4046.856 = 323\,748 \text{ m}^2$$

Assuming the drainage area to be circular, the equivalent radius is

$$r_e = \sqrt{\frac{A}{\pi}} = 321 \text{ m}$$

The penetration ratios for the two cases specified are:

– case a: $b_D = 6/30 = 0.2$
– case b: $b_D = 18/30 = 0.6$

For these two values of b_D we calculate [Eqs. (12.48b, c and d)]:

$$A(0.2) = 0.995091$$

$$B(0.2) = 3.597640$$

$$C(0.2) = 1.566011$$

$$A(0.6) = 1.000041$$

$$B(0.6) = 6.700616$$

$$C(0.6) = 2.864839$$

From Eq. (12.49) we now calculate the term:

$$Q = 5.256 \times 10^{-6} h_o^2 (\rho_w - \rho_o) \frac{k_r k_{ro,iw}}{B_{of}\mu_o} =$$

$$= 5.256 \times 10^{-6} \times 30^2 \times (1030 - 780) \frac{500 \times 0.9}{1.25 \times 2} = 212.868$$

which is common to all the cases under consideration.

Isotropic reservoir

If $k_v = k_r$, Eq. (12.45c) gives us

$$r_{De} = \frac{321}{30}\sqrt{1} = 10.70$$

With this value of r_{De}, Eq. (12.48a) becomes

$$\Omega(10.7; 0.2) = \frac{0.995091}{3.597640 + 1.566011 \ln 10.7} = 0.13614$$

$$\Omega(10.7; 0.6) = \frac{1.000041}{6.700616 + 2.864839 \ln 10.7} = 0.07413$$

Therefore,

$$q_{o,crit}(b = 6 \text{ m}; k_v = k_r) = 212.868 \times 0.13614 = 29.0 \text{ m}^3/\text{day}$$

$$q_{o,crit}(b = 18 \text{ m}; k_v = k_r) = 212.868 \times 0.07413 = 15.8 \text{ m}^3/\text{day}$$

Anisotropic reservoir

If $k_v = 0.1k_r$, Eq. (12.45c) gives us

$$r_{De} = \frac{321}{30}\sqrt{0.1} = 3.384$$

Therefore,

$$\Omega(3.384; 0.2) = \frac{0.995091}{3.597640 + 1.566011 \ln 3.384} = 0.18071$$

$$\Omega(3.384; 0.6) = \frac{1.000041}{6.700616 + 2.864839 \ln 3.384} = 0.09811$$

and

$$q_{o,crit}(b = 6 \text{ m}; k_v = 0.1k_r) = 212.868 \times 0.18071 = 38.5 \text{ m}^3/\text{day}$$

$$q_{o,crit}(b = 18 \text{ m}; k_v = 0.1k_r) = 212.868 \times 0.09811 = 20.9 \text{ m}^3/\text{day}$$

Note how low the oil production rates have to be in all four cases to avoid water coning.

It would probably be necessary to accept some degree of water production in order to achieve a viable production of oil.

$$\Diamond \; \Diamond \; \Diamond$$

Exercise 12.7

For the well in Ex. 12.6, calculate the time to breakthrough by coning of water, and the subsequent variation in the water cut, when the well is produced at a constant rate of 100 m³/day of liquid under surface conditions. The well penetrates only 6 m of the reservoir ($b_D = 0.2$).

Solution

We shall use the equations:

$$(H_{D,v})_{BT} = 5.256 \times 10^{-6} \frac{h_o^2(\rho_w - \rho_o)k_r}{q_o B_{of}\mu_o}(1 - b_D) \qquad (12.55b)$$

and

$$(t_D)_{BT} = 4.181 \times 10^{-7} \frac{(\rho_w - \rho_o)k_v(1 + M_{wo}^\alpha)}{\mu_o \phi h_o} t_{BT} \qquad (12.55a)$$

with

$$\alpha = 0.5 \quad \text{for } M_{\text{wo}} \leq 1$$
$$\alpha = 0.6 \quad \text{for } 1 < M_{\text{wo}} \leq 10.$$

All parameters are expressed in practical metric units.

If we limit ourselves to the case where $k_{\text{v}} = k_{\text{r}}$, we have in this example:

$$\phi \qquad = 0.25$$
$$h_{\text{o}} \qquad = 30 \text{ m}$$
$$b_{\text{D}} \qquad = 0.2$$
$$k_{\text{r}} = k_{\text{v}} = 500 \text{ md}$$
$$k_{\text{ro,iw}} \quad = 0.9$$
$$k_{\text{rw,or}} \quad = 0.4$$
$$B_{\text{of}} \quad = 1.25$$
$$\rho_{\text{o}} \qquad = 780 \text{ kg/m}^3$$
$$\mu_{\text{o}} \qquad = 2 \text{ cP}$$
$$\rho_{\text{w}} \qquad = 1030 \text{ kg/m}^3$$
$$\mu_{\text{w}} \qquad = 0.6 \text{ cP.}$$

We can assume that $B_{\text{w}} = 1.00$.

We have:

$$M_{\text{wo}} = \frac{k_{\text{rw,or}}}{k_{\text{ro,iw}}} \frac{\mu_{\text{o}}}{\mu_{\text{w}}} = \frac{0.4}{0.9} \frac{2}{0.6} = 1.48$$

and:

$$M_{\text{wo}}^{\alpha} \qquad = M_{\text{wo}}^{0.6} = 1.266$$

$$(H_{\text{D,v}})_{\text{BT}} = 5.256 \times 10^{-6} \frac{30^2 \times (1030 - 780) \times 500}{100 \times 1.25 \times 2}(1 - 0.2)$$

$$= 1.89216$$

$$(t_{\text{D}})_{\text{BT}} \quad = 4.181 \times 10^{-7} \frac{(1030 - 780) \times 500 \times (1 + 1.266)}{2 \times 0.25 \times 30} t_{\text{BT}}$$

$$= 7.895 \times 10^{-3} t_{\text{BT}} \text{ (days)}$$

from which we derive

$$t_{\text{BT}} = 126.66(t_{\text{D}})_{\text{BT}} \text{ (days)}$$

Using the equation proposed by Sobocinski and Cornelius[23]:

$$(t_{\text{D}})_{\text{BT}}^{\text{SOBO}} = \frac{(H_{\text{D,v}})_{\text{BT}}}{4} \frac{16 + 7(H_{\text{D,v}})_{\text{BT}} - (H_{\text{D,v}})_{\text{BT}}^2}{7 - 2(H_{\text{D,v}})_{\text{BT}}}, \tag{12.54a}$$

we calculate:

$$(t_{\text{D}})_{\text{BT}}^{\text{SOBO}} = 2.7221$$

Using the equation proposed by Bournazel and Jeanson[23]:

$$(t_{\text{D}})_{\text{BT}}^{\text{BOUR}} = \frac{(H_{\text{D,v}})_{\text{BT}}}{3 - 0.7(H_{\text{D,v}})_{\text{BT}}}, \tag{12.54b}$$

we get:

$$(t_{\text{D}})_{\text{BT}}^{\text{BOUR}} = 1.1293.$$

Therefore,

$$t_{\text{BT}}^{\text{SOBO}} = 126.66 \times 2.7221 = 345 \text{ days}$$
$$t_{\text{BT}}^{\text{BOUR}} = 126.66 \times 1.1293 = 143 \text{ days.}$$

The ratio of these two estimates is

$$\frac{t_{\text{BT}}^{\text{SOBO}}}{t_{\text{BT}}^{\text{BOUR}}} = 2.4$$

This confirms the observation made in Sect. 12.6.2 that the correlation of Sobocinski and Cornelius gives an appreciably higher result for $(t_{\text{D}})_{\text{BT}} = f(H_{\text{D,V}})_{\text{BT}}$ than that of Bournazel and Jeanson.

To calculate f_{w} at any time after water breakthrough, we will use Kuo and Des Brisay's correlation[23]:

$$f_{\text{w}} = 0 \quad \text{for} \quad \left(\frac{t}{t_{\text{BT}}}\right) < 0.56$$

$$\frac{f_{\text{w}}}{(f_{\text{w}})_{\text{lim}}} = 0.94 \ \log\left(\frac{t}{t_{\text{BT}}}\right) + 0.24$$

$$\text{for } 0.56 \le (t/t_{\text{BT}}) \le 5.7$$

$$f_{\text{w}} = (f_{\text{w}})_{\text{lim}} \text{ for } \left(\frac{t}{t_{\text{BT}}}\right) > 5.7$$

with

$$(f_{\text{w}})_{\text{lim}} = \frac{h_{\text{w}} M_{\text{wo}}}{h_{\text{w}} M_{\text{wo}} + \dfrac{B_{\text{w}}}{B_{\text{of}}} h_{\text{o}}}. \tag{12.58}$$

Since:

$$h_{\text{w}} = 20 \text{ m},$$

we calculate

$$(f_{\text{w}})_{\text{lim}} = \frac{1.48 \times 20}{1.48 \times 20 + \dfrac{30}{1.25}} = 0.552$$

If we take the more cautious estimate of 143 days for t_{BT} provided by the Bournazel and Jeanson correlation, we obtain the values for $f_{\text{w}}(t)$ listed in Table E12/7.1 and presented graphically in Fig. E12/7.1.

Note that these calculations do not take into account any possible global rise in the level of *the average water/oil contact* in the period under consideration. If this were to occur, the increase in h_{w} and decrease in h_{o} would result in an increase in $(f_{\text{w}})_{\text{lim}}$.

Referring to the results in Table E12/7.1, we deduce that after 815 days, the well has produced:

$$N_{\text{p}} = 54\,300 \text{ m}^3 \text{ stock tank oil}$$

$$W_{\text{p}} = 27\,200 \text{ m}^3 \text{ water}$$

Table E12/7.1

t/t_{BT}	t (days)	f_{w}
0.56	80	0.000
0.60	86	0.017
0.80	114	0.082
1.00	143	0.132
1.25	179	0.183
1.50	215	0.224
1.75	250	0.259
2.00	286	0.289
2.50	358	0.339
3.00	429	0.380
4.00	572	0.445
5.00	715	0.495
5.70	815	0.525

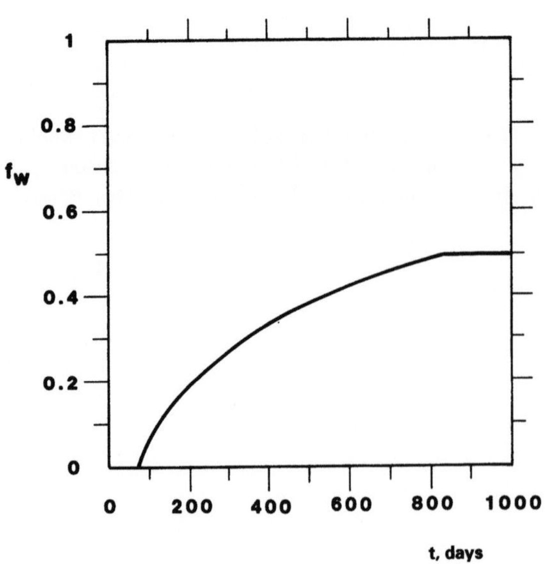

with an average water cut of 0.33 under surface conditions.

Continued production at a total liquid rate of 100 m^3/day beyond the 815th day will yield 47.5 m^3/day of stock tank oil.

For comparison, if the production rate is held at the critical rate for coning (29 m^3/day stock tank oil, no water), only 23 600 m^3 of oil will have been recovered in the first 815 days. The same rate can of course be maintained subsequently.

In this particular case, it is preferable to produce above the critical rate for coning.

◊ ◊ ◊

13 The Simulation of Reservoir Behaviour Using Numerical Modelling

13.1 Introduction

As mentioned in Chap. 10, it was not until the late 1950s – a little more than 30 years ago – that the means became available for reservoir engineers to make detailed simulations of the "internal" behaviour of a reservoir. They could now study and predict the variation with time of pressure and fluid saturations at any point in the reservoir, as well as the oil, gas and water production rates (or injection rates in the case of gas or water) in each well. This was made possible through *numerical modelling*.

Numerical models are by no means the only ones in use for the simulation of reservoir behaviour – there are also analytical and physical models. Analytical models are used, for example, to interpret well tests (Chaps. 6 and 7) and to predict water influx into the reservoir (Chap. 9). An example of the physical type is the "scaled model"[6,43]. An artificial porous medium is constructed, replicating, to scale, the geometrical and physical properties of the reservoir. Applying certain well-defined scaling laws[12], it is possible to reproduce the conditions of flow and saturation (but almost never pressure) that would exist in reality.

Physical models are considerably more accurate than numerical models because they do not require discretisation in space (subdivision of the porous medium into blocks) and time (subdivision into time steps). They are, however, extremely costly and slow. In addition, they are not suited to modelling processes involving the exchange of mass between phases such as when gas evolves from, or redissolves back into, oil. They have been more or less superseded by the numerical type of model, although over the last few years some of the more advanced oil and gas companies have started to use physical models again, to study sections of reservoirs where the production processes are too complex to be handled numerically.

13.2 The Philosophy and Methodology Behind Numerical Modelling

In numerical modelling, the flow of each fluid in the reservoir is described by the generalised Darcy equation (Sect. 3.5.2.2), while the continuity equation takes care of the conservation of mass. The mass flow rates of fluids produced from or injected into wells must of course be included. The volumetric and phase behaviour of the reservoir fluids are modelled through the relevant PVT parameters, or equations of state.

The initial conditions of pressure and saturation must then be defined for each fluid at every point in the reservoir. The conditions at the outer boundary of the reservoir volume to be simulated can be: specified pressure (constant or varying with time) – the *Dirichlet* condition; or specified flow rate (constant or varying with time) – the *Neumann* condition.

The incorporation of the generalised Darcy equation into the continuity equation, taking into account the equation of state of the fluids, results in an equation which completely describes the reservoir behaviour.

As we shall see in later sections, this is a second-order partial differential equation, with coefficients which are functions of the variables (pressure and saturation). Numerical solution of this equation is achieved by "relaxation" (or "discretisation"), which transforms it into a finite difference equation. This involves discretising – or subdividing – the reservoir volume (and surrounding aquifer, if any) into contiguous blocks conforming to a predetermined grid pattern. The time is also discretised into a series of time steps.

A typical procedure can be summarised as follows:

1. Definition of a *geological model* for the reservoir and, if present, the aquifer, in terms of geometry, zonation, spatial distribution of rock properties, initial saturations and pressures.
2. Specification of the thermodynamic properties of the reservoir fluids, including possible variations both areally and vertically. Definition and normalisation of the relative permeability and capillary pressure curves or pseudo-curves.
3. Configuration of the most appropriate gridding for subdivision of the reservoir into blocks, and calculation of the interblock transmissivities.
4. "Initialisation" of the model: this assigns local initial values of the petrophysical, thermodynamic and dynamic parameters of the rock and fluids to each grid block, as well as the initial saturations. Verification of the initial equilibrium conditions.
5. "History matching" or replication of the production history of the reservoir (if it has one). The model is run, each well having been assigned its actual record of production or injection rates, and the pressures, water/oil ratios (WOR) and gas/oil ratios (GOR) computed at each well are compared with measured data where available. The reservoir description and model parameters can be modified *within reasonable limits* and the model rerun as necessary until a satisfactory match is obtained.

   At the successful conclusion of the history matching phase, the final version of the model is said to have been "validated".
6. Prediction of the reservoir behaviour under any future development program that the engineer may wish to specify (schedules of oil production, or water or gas injection, in each well, addition of or shutting in of wells). This is using the model in "forecast mode".

Numerical models are therefore very powerful tools for evaluating reservoir behaviour under a wide variety of producing conditions, enabling the profitability of different development and production programs to be compared. It goes without saying that the reliability of the results of a simulation is only as good as the data upon which the model was based. The well-known saying "garbage in, garbage out" holds true even in the world of numerical modelling!

Unfortunately, it is not uncommon for reservoir engineers to be so impressed by the power and performance of numerical models (sometimes using several hours of CPU time) that they overlook this simple truth. They then assign to results (which they pass on to the technical and economic decision-makers in the company) a degree of confidence that far exceeds the reliability of their input data.

The mystique that surrounded numerical modelling for the first 20 or so years has by now been largely dispelled, and a competent engineer fully appreciates the overwhelming importance of focusing his efforts on determining the most probable *geological model* for the reservoir, and its internal structure – leaving the task of data processing and numerical modelling to the experts.

For this reason, the present chapter will only touch on those technical aspects of numerical modelling that might improve the reservoir engineer's ability to converse with the modelling expert. A number of publications are available for a deeper study of the topic: the classical work of Peaceman[28], Crichlow[11], Aziz and Settari[1], the excellent book by Thomas[39], etc. This chapter will, however, deal in some depth with the methods used to process the basic data, and the interpretation of the results of the model.

A very important fact to bear in mind at all times is that the "validated model" resulting from history matching (see 5 above) *is not necessarily unique*. In other words, there can be more than one model (each consisting of a different spatial distribution of parameters) which will reproduce the production history of the reservoir to within the error associated with the input data. Only one of them will be correct.

A good history match, therefore, by no means confirms that the model represents the real reservoir. So, while the validation phase is a *mandatory* part of the study of a reservoir by numerical modelling (whenever the reservoir has some production data against which the model can be matched), *this validation does not guarantee that the model will be capable of correctly forecasting future reservoir behaviour.*

The basic input parameters to the model (such as: the vertical permeability, the relative permeability and capillary pressure curves, the extent and transmissibility of the aquifer, the presence and location of faults and permeability barriers) are subject to various degrees of uncertainty. It is therefore necessary to evaluate the effects that different assumptions about these parameters will have on the results (the predicted oil, gas and water production, and pressure throughout the reservoir), by running *a series* of simulations with the model. These *sensitivity studies* – of the effects of parameter uncertainty – define a range of possible reservoir behaviours, from which the engineer can choose what he judges to be the most probable (or, perhaps, the most cautious) estimate. In the author's experience, sensitivity studies have proven to be the most realistic application of numerical modelling in the forecasting of reservoir behaviour.

13.3 Types of Numerical Model

13.3.1 Classification Based on the Way in Which the Flow Equations Are Discretised

There are two principal ways to "discretise" (Sect. 13.2) the partial differential equations which describe the flow of fluids in the reservoir:

- the finite difference method;
- the finite element method.

Rather than provide a comprehensive review of their characteristics and differences, this chapter will deal only with what is of direct interest to a reservoir engineer.

The finite difference method is limited to a parallelepiped grid-block geometry (orthogonal or non-orthogonal in the vertical direction, with square, rectangular,

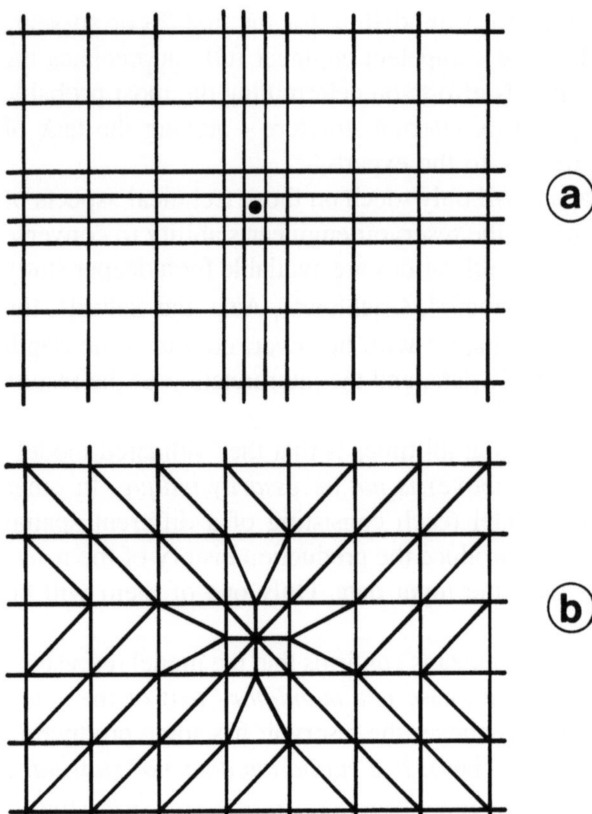

Fig. 13.1. *Configuring a finer mesh in the region of a well:* **a** *finite difference model;* **b** *finite element model*

rhombic or rhomboidal bases). The finite element method, on the other hand, allows for a large degree of flexibility, including a choice of polygonal shapes and prisms, provided, of course, that they are able to encompass the entire reservoir volume.

In the horizontal two-dimensional geometries that we shall consider here, it is much easier to design gridding in parts of the reservoir where a more precise description of flow is required with finite element models. For example, in the region around a well, a finer mesh will model the locally higher pressure and saturation gradients more accurately than a coarser one.

With the finite difference method, it is impossible to restrict the fine gridding solely to the near-well region where it is required (Fig. 13.1a). With the finite element method, however, it can be localised by using, for instance, triangular blocks (Fig. 13.1b). Despite this useful advantage, the finite element method [19,50] has difficulty handling the flow of highly compressible fluids (gas). For this reason, it has not yet found universal use by reservoir engineers.

In the 1980s, the techniques of "zooming"[45] and discretisation of the grid block containing the well [5,31] (assuming steady-state flow within the block) were developed, making it possible to model flow behaviour in the presence of discontinuities (such as faults) and the steep pressure and saturation gradients around wells quite accurately with the finite difference method.

13.3.2 Classification Based on Reservoir Geometry

Numerical models classified according to the geometry of the reservoir fall into three main categories:

- one-dimensional (1D);
- two-dimensional (2D);
- three-dimensional (3D).

One-dimensional models are, in turn, subdivided into (Fig. 13.2):

- horizontal
- inclined
- vertical

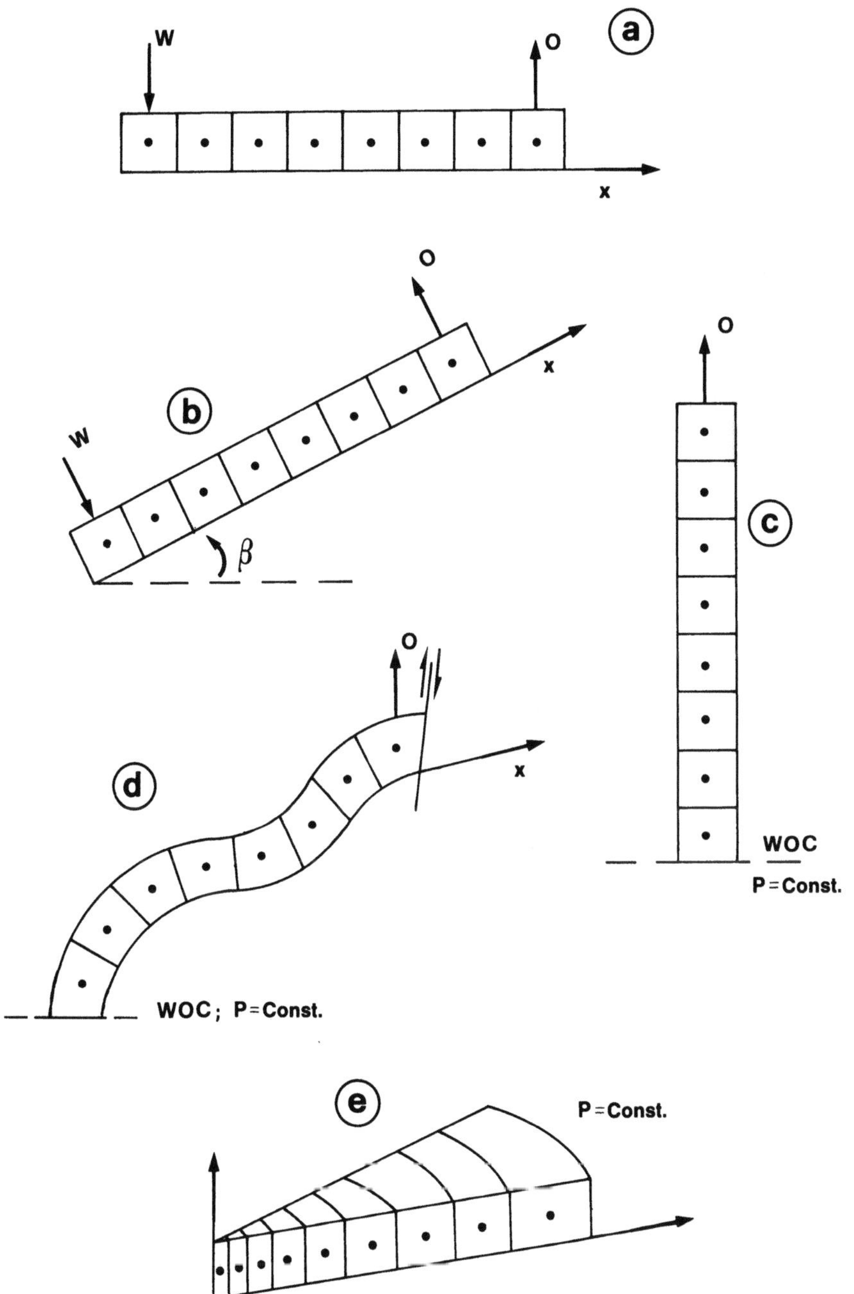

Fig. 13.2. *One-dimensional models:* **a** *horizontal;* **b** *inclined;* **c** *vertical;* **d** *curvilinear coordinates;* **e** *radial*

- curvilinear coordinates
- radial.

Two-dimensional models are subdivided into (Fig. 13.3):

- horizontal
- vertical (in other words, a cross section of the reservoir)
- radial

In two-dimensional reservoir cross sections, a curvilinear coordinate system is often used so that the shape of the reservoir top and base can be followed closely. There is, of course, only one type of three-dimensional model (Fig. 13.4). Here, too, curvilinear coordinates are frequently used.

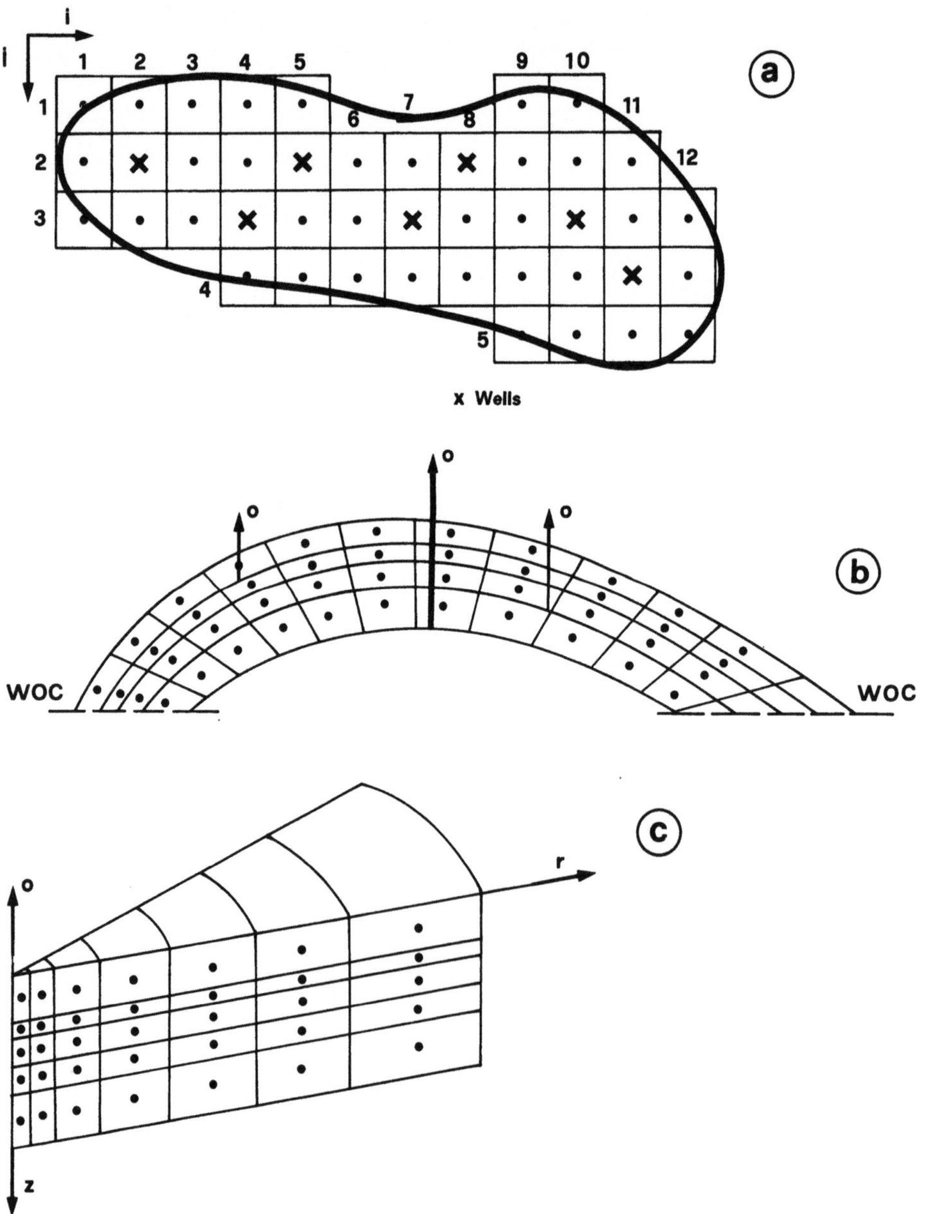

Fig. 13.3. *Two-dimensional models:* **a** *horizontal;* **b** *reservoir cross section;* **c** *radial*

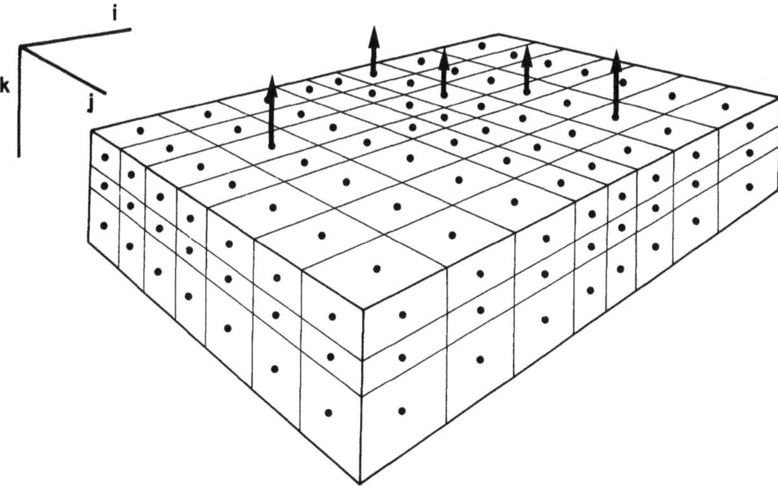

Fig. 13.4. *Three-dimensional model*

13.3.3 Classification Based on the Number and Nature of the Mobile Phases

Classification according to the number of mobile fluid phases in the reservoir is as follows:

- monophasic or single-phase (1P);
- biphasic or two-phase (2P);
- triphasic or three-phase (3P).

In a monophasic model, the moving fluid can be:

- oil above its bubble point, in the presence of S_{iw};
- gas above its dew point, in the presence of S_{iw};
- water ($S_w = 1.00$) in hydrological models.

In a biphasic model, the possibilities are:

- oil above its bubble point + water;
- oil (or condensate) below saturation pressure + gas in equilibrium, in the presence of S_{iw};
- gas above its dew point + water at $S_w > S_{iw}$.

In a triphasic model, there is one case:

- oil (or condensate) below saturation pressure + gas in equilibrium + water at $S_w > S_{iw}$.

In the two- and three-phase models, where oil (or condensate) is present below its saturation pressure together with gas in equilibrium, the exchange of mass between gas and liquid phases can be treated in two ways:

1. *Gas composition does not vary with pressure* over the entire life of the reservoir, and we can describe the effects of the liberation and redissolution of gas by means of the usual PVT parameters B_{od}, B_{of}, R_{sd}, R_{sf}, μ_o: the reservoir hydrocarbon system is simulated by a two-component model consisting of oil taken to $p = 0.1013$ MPa and 288 K by "flash" liberation (Sects. 2.3.2.1 and 10.5.1), and gas of constant composition that was dissolved in it.

This simplistic representation of a reservoir hydrocarbon system is called a *"black oil"* model. It describes the thermodynamic behaviour of medium-heavy oils with reasonable accuracy.

2. *The exchange of mass between gas and liquid phases (including water) is described for each hydrocarbon component present in the reservoir system*: in this case, the composition of the gaseous phase can vary with time (and, consequently, so can that of the two liquid phases, oil and water).

This type of model, much more complex and more demanding in terms of computing time, is called the *"compositional"* model. It describes, among other things, how the compositions of the three phases – liquid hydrocarbons/gaseous hydrocarbons/water + dissolved hydrocarbons – change with time.

From a practical viewpoint, we do not know the equilibrium constants (K-values) describing the gas/liquid phase partitions for *all* the individual components present in the reservoir system. For this reason, and to reduce the computational complexity, it is common practice to group components together into perhaps 12 to 20 *"pseudo-components"*. The behaviour of the system is then modelled in terms of these. This is the *"pseudo-compositional"* model.

Pseudo-compositional models are particularly well adapted to the simulation of volatile oil reservoirs below the bubble point, and condensate gas reservoirs below the dew point, where the equilibrium composition of the hydrocarbon phases varies during the course of production. They are also suited to the modelling of oil reservoirs into which gas is being injected, where there is a high proportion of intermediate (C_2–C_6) and heavy (C_{7+}) hydrocarbons.

13.4 The Continuity Equation

We have already encountered the continuity equation when it was used to derive the diffusivity equation for a single phase fluid (oil in Sect. 5.2; gas in Sect. 7.3; and water in Sect. 9.4) for one-dimensional horizontal radial flow. We will now derive these equations in their most general form for three-dimensional geometry and polyphasic flow.

We will use the same Cartesian system of coordinates as before: (x, y, z) with z positive downwards. $\boldsymbol{u}_i$ is the Darcy velocity of fluid i flowing in a porous medium, which it partially or totally saturates.

If u_{ix}, u_{iy} and u_{iz} are the components of $\boldsymbol{u}_i$ in the directions of the axes:

$$\boldsymbol{u}_i = u_{ix}\,\boldsymbol{i} + u_{iy}\,\boldsymbol{j} + u_{iz}\,\boldsymbol{k}, \tag{13.1}$$

where $\boldsymbol{i}, \boldsymbol{j}, \boldsymbol{k}$ are unit vectors in the x-, y- and z-axis directions respectively.

OABCDEFG in Fig. 13.5 is an elementary parallelepiped of porous medium with sides $\mathrm{d}x$, $\mathrm{d}y$, $\mathrm{d}z$.

If u_{ix} is the value of the component of $\boldsymbol{u}_i$ perpendicular to the face OABC, and ρ_i is the density of fluid i (at local p and T) flowing into the face, then:

$$V_{\mathrm{m},1} = \rho_i u_{ix} \tag{13.2}$$

is the mass velocity (= the mass flow rate per unit area) of fluid i entering the parallelepiped through face OABC.

The corresponding mass velocity leaving the opposite face DEFG is:

$$V_{\mathrm{m},2} = \rho_i u_{ix} + \frac{\partial}{\partial x}(\rho_i u_{ix})\,\mathrm{d}x. \tag{13.3}$$

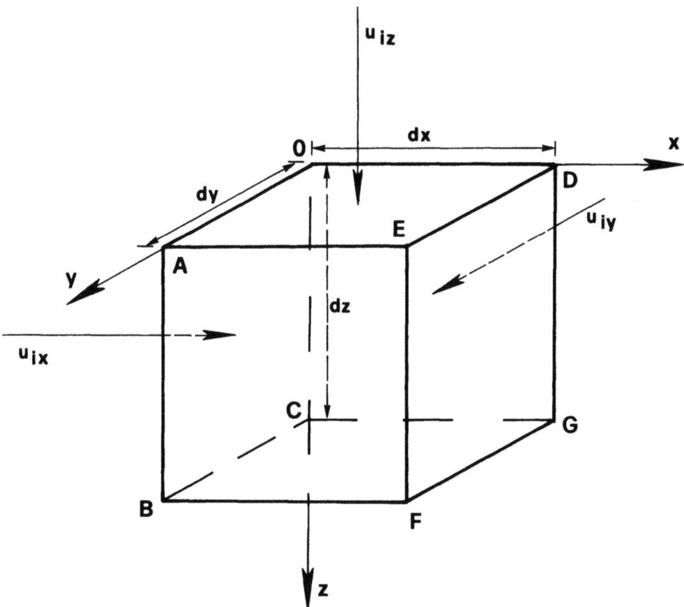

Fig. 13.5. *Schematic of an elementary parallelepiped*

The area of each of these faces is $dy \times dz$, and in time dt the change in mass of fluid i inside the elementary parallelepiped due to component of flow in the x-direction will be:

$$dm_x = (V_{m,1} - V_{m,2})\, dy\, dz\, dt$$

$$= -\frac{\partial(\rho_i u_{ix})}{\partial x}\, dx\, dy\, dz\, dt. \tag{13.4}$$

If we sum the components along the three axes we have:

$$dm_i = dm_x + dm_y + dm_z$$

$$= -\left[\frac{\partial(\rho_i u_{ix})}{\partial x} + \frac{\partial(\rho_i u_{iy})}{\partial y} + \frac{\partial(\rho_i u_{iz})}{\partial z}\right] dx\, dy\, dz\, dt, \tag{13.5}$$

where dm_i is the total change in the mass of fluid i in time dt.

If S_i is the local saturation of fluid i in the porous medium, and $\phi\, dx\, dy\, dz$ the pore volume in the parallelepiped, the mass of fluid i it contains is:

$$m_i = \phi S_i \rho_i\, dx\, dy\, dz, \tag{13.6a}$$

so that its rate of change with time will be:

$$dm_i = \frac{\partial}{\partial t}(\phi S_i \rho_i)\, dx\, dy\, dz\, dt. \tag{13.6b}$$

To satisfy the conservation of mass, the values of dm_i expressed in Eq. (13.5) and Eq. (13.6b) must be equal. Therefore:

$$\frac{\partial(\rho_i u_{ix})}{\partial x} + \frac{\partial(\rho_i u_{iy})}{\partial y} + \frac{\partial(\rho_i u_{iz})}{\partial z} = -\frac{\partial}{\partial t}(\phi S_i \rho_i). \tag{13.7}$$

Equation (13.7) is *the continuity equation for fluid i in cartesian coordinates.*

In vectorial algebra, the divergence of a vector v is defined as:

$$\text{div } V = \nabla \cdot V = \left(\frac{\partial}{\partial x}i + \frac{\partial}{\partial y}j + \frac{\partial}{\partial z}k \right)(V_x i + V_y j + V_z k)$$

$$= \frac{\partial V_x}{\partial x} + \frac{\partial V_y}{\partial y} + \frac{\partial V_z}{\partial z}, \tag{13.8}$$

where $\nabla \cdot$ (the "del" or "nabla" operator) is an alternative notation more commonly used in English language texts.

Since ρ is a scalar quantity, Eq. (13.7) can be written in the form:

$$\nabla \cdot (\rho u) = -\frac{\partial}{\partial t}(\phi \rho S). \tag{13.9}$$

The term on the left in this equation is the *"transport term"*, and that on the right the *"accumulation term"*. The names refer to the way in which the terms are derived.

In two-dimensional radial coordinates (Fig. 13.3c), the continuity equation is of the form:

$$\frac{\partial}{\partial r}(\rho_i u_{ir}) + \frac{\partial}{\partial z}(\rho_i u_{iz}) = -\frac{\partial}{\partial t}(\phi S_i \rho_i). \tag{13.10}$$

This can be stated in terms of vectorial algebra as:

$$\nabla \cdot (\rho u) = -\frac{\partial}{\partial t}(\phi \rho S),$$

which is the same as Eq. (13.9). Here, we have applied the vector algebra relationship:

$$\text{div } V = \nabla \cdot V = \left(\frac{\partial}{\partial r}l + \frac{\partial}{\partial z}k \right)(V_r l + V_z k)$$

$$= \frac{\partial V_r}{\partial r} + \frac{\partial V_z}{\partial z} \tag{13.11}$$

l and k being unit vectors in the direction of the r- and z-axes respectively. You will find the derivation of Eq. (13.10) in Exercise 13.1.

13.5 The Flow Equation

Firstly, we should assemble the following important equations:

Continuity: $\qquad \nabla \cdot (\rho u) = -\dfrac{\partial}{\partial t}(\phi \rho S),$ (13.9)

Potential: $\qquad \Phi = \displaystyle\int_o^p \frac{dp}{\rho} - gD,$ (3.28a)

Generalised Darcy: $u = -\dfrac{[k]k_r \rho}{\mu}\nabla\Phi,$ (3.49)

Fluid properties: $\quad \rho = \rho(p, T),$ (13.12a)

$\qquad\qquad\qquad \mu = \mu(p, T).$ (13.12b)

In Eq. (3.28a), D is the depth, positive downwards.

In Eq. (3.49), the relative permeability k_r has been introduced for the case where $S < 1$, and for consistency in this chapter, $\nabla\Phi$ replaces the grad Φ notation used in Chap. 3.

Now, by definition:

$$\nabla = \text{grad} = \frac{\partial}{\partial x}\boldsymbol{i} + \frac{\partial}{\partial y}\boldsymbol{j} + \frac{\partial}{\partial z}\boldsymbol{k} \tag{13.13a}$$

and, since Φ is a scalar quantity:

$$\nabla\Phi = \text{grad}\,\Phi = \frac{\partial\Phi}{\partial x}\boldsymbol{i} + \frac{\partial\Phi}{\partial y}\boldsymbol{j} + \frac{\partial\Phi}{\partial z}\boldsymbol{k}. \tag{13.13b}$$

Still referring to Eq. (3.49), $[k]$ is a symmetrical tensor of the second order:

$$[k] = \begin{vmatrix} k_{11} & k_{12} & k_{13} \\ k_{21} & k_{22} & k_{23} \\ k_{31} & k_{32} & k_{33} \end{vmatrix} \quad \text{with } k_{ij} = k_{ji}. \tag{13.14a}$$

If the (x, y, z) axes of our Cartesian system coincide with the principal axes of the tensor, we have:

$$[k] = \begin{vmatrix} k_x & 0 & 0 \\ 0 & k_y & 0 \\ 0 & 0 & k_z \end{vmatrix}. \tag{13.14b}$$

Combining Eqs. (3.28a) and (3.49), and ignoring the gradient of ρ in (x, y, z) space (this being a second-order term), we get:

$$\boldsymbol{u} = -\frac{[k]k_r}{\mu}\nabla(p - \rho g D)$$

$$= -\frac{[k]k_r}{\mu}(\nabla p - \rho g \nabla D). \tag{13.15}$$

If $\tilde{q}_m$ is the mass flow rate of fluid extracted from the system per unit volume of rock through the producing wells, Eqs. (13.9) and (13.15) give us:

$$\nabla \cdot \left[\frac{[k]k_r\rho}{\mu}(\nabla p - \rho g \nabla D) \right] - \tilde{q}_m = \frac{\partial}{\partial t}(\phi\rho S). \tag{13.16a}$$

This is the *generalised equation of flow for a single fluid* in a mono- or polyphasic system.

In a monophasic system:

$p = p(x, y, z, t) = $ pressure at any point in the reservoir,

$S = 1$,

$k_r = 1$.

In a polyphasic system:

$p = p(x, y, z, t) = $ phase pressure at any point in the reservoir

(measured in a particular phase)

$0 < S < 1$,

$k_r = k_r(S)$.

In both cases:

$$\rho = \rho(p, T), \tag{13.12a}$$

$$\mu = \mu(p, T). \tag{13.12b}$$

Exercise 13.2 will show how Eq. (13.16a) can be expressed in a more general notation as:

$$\nabla \cdot \left[\frac{\alpha[k]k_r\rho}{\mu}(\nabla p - \rho g \nabla D) \right] - \alpha \tilde{q}_m = \alpha \frac{\partial(\phi \rho S)}{\partial t}, \tag{13.16b}$$

where:

In a one-dimensional geometry:

$$\alpha(x) \equiv A(x) = \text{cross-sectional area of the porous medium}$$
$$\text{(a function of } x\text{),}$$

In a two-dimensional geometry:

$$\alpha(x, y) \equiv h(x, y) = \text{thickness of the porous medium [a function of } (x, y)\text{],}$$

In a three-dimensional geometry:

$$\alpha(x, y, z) = 1.$$

13.6 Single Phase Flow of a Slightly Compressible Fluid

In the case of single phase (monophasic) flow, Eq. (13.16b) becomes:

$$\nabla \cdot \left[\frac{\alpha[k]\rho}{\mu}(\nabla p - \rho g \nabla D) \right] - \alpha \tilde{q}_m = \alpha \frac{\partial(\phi \rho)}{\partial t}. \tag{13.17}$$

For simplicity, we will consider a *horizontal* two-dimensional porous medium (D = const.) *of constant thickness h*, in which a single phase fluid is flowing.

Referring to Eqs. (13.8) and (13.13), we can write Eq. (13.17) in algebraic notation as:

$$\frac{\partial}{\partial x}\left(\frac{k_x \rho}{\mu}\frac{\partial p}{\partial x}\right) + \frac{\partial}{\partial y}\left(\frac{k_y \rho}{\mu}\frac{\partial p}{\partial y}\right) - \tilde{q}_m = \frac{\partial(\phi \rho)}{\partial t}. \tag{13.18}$$

We now introduce the following compressibility terms:

$$c = \frac{1}{\rho}\left(\frac{\partial \rho}{\partial p}\right)_T = \text{fluid compressibility,} \tag{2.31}$$

$$c_f = \frac{1}{\phi}\left(\frac{\partial \phi}{\partial p}\right)_{\bar{\sigma}} = \text{pore compressibility,} \tag{3.9d}$$

$$c_t = c_f + c = \text{total compressibility,} \tag{5.5b}$$

so that, after expanding the derivatives in Eq. (13.18), we get:

$$\frac{\partial}{\partial x}\left(\frac{k_x}{\mu}\frac{\partial p}{\partial x}\right) + \frac{\partial}{\partial y}\left(\frac{k_y}{\mu}\frac{\partial p}{\partial y}\right) + c\left[\frac{k_x}{\mu}\left(\frac{\partial p}{\partial x}\right)^2 + \frac{k_y}{\mu}\left(\frac{\partial p}{\partial y}\right)^2\right]$$

$$-\frac{\tilde{q}_m}{\rho} = \phi c_t \frac{\partial p}{\partial t}. \tag{13.19}$$

If the compressibility c of the fluid is small (oil above its bubble point, or water), the third term in Eq. (13.19) is negligibly small and we are left with:

$$\frac{\partial}{\partial x}\left(\frac{k_x}{\mu}\frac{\partial p}{\partial x}\right) + \frac{\partial}{\partial y}\left(\frac{k_y}{\mu}\frac{\partial p}{\partial y}\right) - \tilde{q}_v = \phi c_t \frac{\partial p}{\partial t}, \tag{13.20}$$

where $\tilde{q}_v$ is the volumetric flow rate of fluid per unit volume of rock.

Equation (13.20) is the two-dimensional flow equation for a single phase fluid in a porous medium.

It can be written in a more general form as:

$$\nabla \cdot \left[\frac{\alpha[k]}{\mu}\left(\nabla p - \rho g \nabla D\right)\right] - \alpha \tilde{q}_v = \alpha \phi c_t \frac{\partial p}{\partial t}, \tag{13.21}$$

where α means the same as in Eq. (13.16b).

In the special case where $k_x = k_y = k = $ const. and $h = $ const., if we assume that μ is independent of pressure p, Eq. (13.20) simplifies to:

$$\frac{\partial^2 p}{\partial x^2} + \frac{\partial^2 p}{\partial y^2} = \frac{\mu}{k}\left(\phi c_t \frac{\partial p}{\partial t} + \tilde{q}_v\right). \tag{13.22a}$$

We now introduce the Laplace operator "del-squared":

$$\nabla^2 = \frac{\partial^2}{\partial x^2} + \frac{\partial^2}{\partial y^2}, \tag{13.23}$$

so that:

$$\nabla^2 p = \frac{\mu}{k}\left(\phi c_t \frac{\partial p}{\partial t} + \tilde{q}_v\right). \tag{13.22b}$$

In pseudo-steady state flow (see Chap. 5), $\partial p/\partial t = f(x, y)$ is independent of t, and Eq. (13.22) reduces further, to *Poisson's equation*:

$$\nabla^2 p = f(x, y). \tag{13.24}$$

In steady-state flow ($\partial p/\partial t = 0$, $\Sigma \tilde{q}_v = 0$), on the other hand, Eq. (13.22) becomes what is known as the *Laplace equation*:

$$\nabla^2 p = 0. \tag{13.25}$$

At this point, we should recall that a second-order partial differential equation in n independent variables is defined as either elliptical, parabolic or hyperbolic depending on which of the following forms it can be transformed to:

elliptical equation:

$$\sum_1^n {}_k \frac{\partial^2 u}{\partial x_k^2} + \cdots = 0, \tag{13.26a}$$

parabolic equation:

$$\sum_1^n {}_k \frac{\partial^2 u}{\partial x_k^2} + \alpha_1 \frac{\partial u}{\partial x_1} + \cdots = 0, \tag{13.26b}$$

hyperbolic equation:

$$\sum_1^{n-1} {}_k \frac{\partial^2 u}{\partial x_k^2} - \frac{\partial^2 u}{\partial x_n^2} + \cdots = 0. \tag{13.26c}$$

The first partial derivative has only been presented in Eq. (13.26b) because in this case α_1 *must be non-zero*.

We have already encountered two typical examples of *elliptical* equations: Poisson's equation [Eq. (13.24)] and the Laplace equation [Eq. (13.25)].

The *parabolic* form includes the Fourier equation, or heat transfer equation:

$$\nabla^2 u - \frac{1}{K}\frac{\partial u}{\partial t} = 0,$$ (13.27a)

where:

K = thermal transmissibility

and Eq. (13.20) for single phase flow, when k and μ are constants:

$$\nabla^2 p - \frac{\phi \mu c_t}{k}\frac{\partial p}{\partial t} - \frac{\mu}{k}\tilde{q}_v = 0.$$ (13.27b)

The classification "parabolic" is also extended to include the flow equation when the derivative coefficients are functions of the variables [p in single phase flow, p and S (saturation) in polyphasic flow].

The wave equation:

$$\nabla^2 u - \frac{1}{v^2}\frac{\partial^2 u}{\partial t^2} = 0,$$ (13.28)

where v is the propagation velocity, is an example of a *hyperbolic* equation.

13.7 Single Phase Flow of Gas

From a reservoir engineering viewpoint, ideal gases are of little practical interest. In this section we shall look at the single phase flow ($S_w = S_{iw}$) of a real gas (see Sect. 2.3.1).

Since ρ_g is usually very small, we can ignore the effects of gravity, and Eq. (13.17) simplifies to:

$$\nabla \cdot \left[\frac{\alpha [k]\rho_g}{\mu_g}\nabla p\right] - \alpha \tilde{q}_{m,g} = \alpha \frac{\partial (\phi \rho_g)}{\partial t}.$$ (13.29)

From Eq. (2.14) we know that:

$$\rho_g = 120.32\frac{Mp}{zT}.$$ (2.14)

In Eq. (13.29), if we ignore the pore compressibility (much smaller than that of the gas), and insert Eq. (2.14), we obtain:

$$\nabla \cdot \left[\frac{\alpha [k]p}{z\mu_g}\nabla p\right] - \beta \tilde{q}_{m,g} = \alpha \phi \frac{\partial}{\partial t}\left(\frac{p}{z}\right),$$ (13.30)

where: $\beta = \dfrac{\alpha T}{120.32M}$.

The *real gas pseudo-pressure* is defined as (Sect. 7.2):

$$m(p) = 2\int_{p_o}^{p} \frac{p}{z\mu_g}\, dp,$$ (7.2)

from which:

$$\nabla m = \left(\frac{\partial m}{\partial p}\right)_T \nabla p = \frac{2p}{z\mu_g}\nabla p$$ (13.31a)

and:

$$\frac{\partial m}{\partial t} = \left(\frac{\partial m}{\partial p}\right)_{\mathrm{T}} \frac{\partial p}{\partial t} = \frac{2p}{z\mu_g}\frac{\partial p}{\partial t}. \qquad (13.31b)$$

The isothermal gas compressibility is defined as:

$$c_g = \frac{1}{\rho_g}\left(\frac{\partial \rho_g}{\partial p}\right)_{\mathrm{T}}.$$

Substituting for ρ_g from Eq. (2.14), this becomes:

$$c_g = \frac{z}{p}\frac{\partial}{\partial p}\left(\frac{p}{z}\right)_{\mathrm{T}}. \qquad (13.32)$$

Therefore:

$$\frac{\partial}{\partial t}\left(\frac{p}{z}\right) = \frac{\partial}{\partial p}\left(\frac{p}{z}\right)_{\mathrm{T}}\frac{\partial p}{\partial t} = \mu_g\left[\frac{z}{p}\frac{\partial}{\partial p}\left(\frac{p}{z}\right)_{\mathrm{T}}\right]\left[\frac{p}{z\mu_g}\frac{\partial p}{\partial t}\right]$$

$$= \frac{1}{2}\mu_g c_g\frac{\partial m}{\partial t}. \qquad (13.33)$$

With Eqs. (13.31) and (13.33), we can express Eq. (13.30) as:

$$\nabla\cdot(\alpha[k]\nabla m) - 2\beta\tilde{q}_m = \alpha\phi\mu_g c_g\frac{\partial m}{\partial t}. \qquad (13.34)$$

This is *the flow equation for the single phase flow of a real gas*.

In practice, the product $\mu_g c_g$ is usually assumed to be independent of the pressure, even though, as we saw in Sect. 7.3, this is often not the case, especially when large decreases in pressure occur in the reservoir.

13.8 Multiphase Flow

We will consider the completely general case of a *three-dimensional* porous medium in which three phases – oil, gas ($S_g > S_{gc}$) and water ($S_w > S_{iw}$) – are flowing at the same time.

According to Eq. (13.16a), we can write, for each phase:

oil phase:

$$\nabla\cdot\left[\frac{[k]k_{ro}\rho_o}{\mu_o}\left(\nabla p_o - \rho_o g\nabla D\right)\right] - \tilde{q}_{m,o} = \frac{\partial}{\partial t}(\phi\rho_o S_o), \qquad (13.35a)$$

gas phase:

$$\nabla\cdot\left[\frac{[k]k_{rg}\rho_g}{\mu_g}\left(\nabla p_g - \rho_g g\nabla D\right)\right] - \tilde{q}_{m,g} = \frac{\partial}{\partial t}(\phi\rho_g S_g), \qquad (13.35b)$$

water phase:

$$\nabla\cdot\left[\frac{[k]k_{rw}\rho_w}{\mu_w}\left(\nabla p_w - \rho_w g\nabla D\right)\right] - \tilde{q}_{m,w} = \frac{\partial}{\partial t}(\phi\rho_w S_w). \qquad (13.35c)$$

We assume the oil to be medium-heavy. According the the "black oil" model of Sect. 13.3.3, this means it can be considered as consisting of two components: oil of density $\rho_{o,sc}$ under standard conditions, with no dissolved gas; and gas of density $\rho_{g,sc}$ under standard conditions ($p = 0.1013$ MPa and $T = 288.2$ K).

The formation volume factor of the oil, B_o, and the gas solubility in the oil, R_s, are obtained by "composite liberation". At each step during the differential liberation of the gas at reservoir temperature (Sect. 2.3.2.1), a sample of reservoir oil is taken from reservoir to standard conditions (0.1013 MPa, 288.2 K) through a series of separations which replicate, in the laboratory, the stages, pressures and temperatures of the field separator system. The parameters derived in this way are B_o and R_s.

For heavy oils, B_o and R_s values can be calculated from the corresponding "differential liberation" values B_{od} and R_{sd} (Sect. 2.3.2.1) using the procedure described in Sect. 2.5 of L.P. Dake's book[53].
Then:

$$\rho_o = \frac{\rho_{o,sc} + \rho_{g,sc} R_s}{B_o}. \tag{13.36a}$$

Likewise:

$$\rho_w = \frac{\rho_{w,sc} + \rho_{g,sc} R_{sw}}{B_w}. \tag{13.36b}$$

Furthermore [Eq. (2.8c)]:

$$\rho_g = \frac{\rho_{g,sc}}{B_g}. \tag{13.36c}$$

Using the relationships in Eqs. (13.36), Eqs. (13.35) become:

$$\rho_{o,sc} \, \nabla \cdot \left[\frac{[k]k_{ro}}{\mu_o B_o} (\nabla p_o - \rho_o g \nabla D) \right] + \rho_{g,sc} \nabla \cdot \left[\frac{[k]k_{ro} R_s}{\mu_o B_o} (\nabla p_o - \rho_o g \nabla D) \right]$$
$$- \tilde{q}_{m,o} = \rho_{o,sc} \frac{\partial}{\partial t} \left(\frac{\phi S_o}{B_o} \right) + \rho_{g,sc} \frac{\partial}{\partial t} \left(\frac{\phi R_s S_o}{B_o} \right), \tag{13.37a}$$

$$\rho_{g,sc} \, \nabla \cdot \left[\frac{[k]k_{rg}}{\mu_g B_g} (\nabla p_g - \rho_g g \nabla D) \right] - \tilde{q}_{m,g} = \rho_{g,sc} \frac{\partial}{\partial t} \left(\frac{\phi S_g}{B_g} \right), \tag{13.37b}$$

$$\rho_{w,sc} \, \nabla \cdot \left[\frac{[k]k_{rw}}{\mu_w B_w} (\nabla p_w - \rho_w g \nabla D) \right] + \rho_{g,sc} \nabla \cdot \left[\frac{[k]k_{rw} R_{sw}}{\mu_w B_w} (\nabla p_w - \rho_w g \nabla D) \right]$$
$$- \tilde{q}_{m,w} = \rho_{w,sc} \frac{\partial}{\partial t} \left(\frac{\phi S_w}{B_w} \right) + \rho_{g,sc} \frac{\partial}{\partial t} \left(\frac{\phi R_{sw} S_w}{B_w} \right). \tag{13.37c}$$

Note that, in Eq. (13.37a), the terms containing $\rho_{g,sc}$ express the mass balance for the gas *dissolved* in the oil, while those containing $\rho_{o,sc}$ express the mass balance of *degasified* oil. Similar comments apply to the terms containing $\rho_{g,sc}$ and $\rho_{w,sc}$ in Eq. (13.37c), this time for gas dissolved in water, and degasified water. Equation (13.37b) describes the mass balance for *free* gas in the reservoir, at a saturation S_g.

By adding together Eqs. (13.37a, b and c) and grouping the terms in $\rho_{o,sc}$, $\rho_{g,sc}$ and $\rho_{w,sc}$, we obtain, for each of the three *components* of the system:
Degasified oil under standard conditions:

$$\nabla \cdot \left[\frac{[k]k_{ro}}{\mu_o B_o} (\nabla p_o - \rho_o g \nabla D) \right] - \tilde{q}_{v,o} = \frac{\partial}{\partial t} \left(\frac{\phi S_o}{B_o} \right). \tag{13.38a}$$

Gas at standard conditions:

$$\nabla \cdot \left[\frac{[k]k_{\mathrm{rg}}}{\mu_g B_g} (\nabla p_g - \rho_g g \, \nabla D) \right] + \nabla \cdot \left[\frac{[k]k_{\mathrm{ro}} R_s}{\mu_o B_o} (\nabla p_o - \rho_o g \, \nabla D) \right]$$

$$+ \nabla \cdot \left[\frac{[k]k_{\mathrm{rw}} R_{\mathrm{sw}}}{\mu_w B_w} (\nabla p_w - \rho_w g \, \nabla D) \right] - (\tilde{q}_{v,g})_{\mathrm{tot}}$$

$$= \frac{\partial}{\partial t} \left[\phi \left(\frac{S_g}{B_g} + \frac{R_s S_o}{B_o} + \frac{R_{\mathrm{sw}} S_w}{B_w} \right) \right]. \tag{13.38b}$$

Degasified water under standard conditions:

$$\nabla \cdot \left[\frac{[k]k_{\mathrm{rw}}}{\mu_w B_w} (\nabla p_w - \rho_w g \, \nabla D) \right] - \tilde{q}_{v,w} = \frac{\partial}{\partial t} \left(\frac{\phi S_w}{B_w} \right), \tag{13.38c}$$

where:

$$\tilde{q}_{v,o} = \frac{\tilde{q}_{m,o}}{\rho_{o,\mathrm{sc}} + R_s \rho_{g,\mathrm{sc}}}, \tag{13.39a}$$

$$\left(\tilde{q}_{v,g} \right)_{\mathrm{tot}} = \frac{\tilde{q}_{m,g}}{\rho_{g,\mathrm{sc}}} + q_{o,\mathrm{sc}} R_s + q_{w,\mathrm{sc}} R_{\mathrm{sw}}, \tag{13.39b}$$

$$\tilde{q}_{v,w} = \frac{\tilde{q}_{m,w}}{\rho_{w,\mathrm{sc}} + R_{\mathrm{sw}} \rho_{g,\mathrm{sc}}} \tag{13.39c}$$

are the *volumetric* flow rates of oil, gas and water produced from the reservoir, under standard conditions (0.1013 MPa and 288.2 K).

If water or gas are being injected, their flow rates are given negative signs.

Equations (13.38) are the *flow equations for a medium-heavy oil (black oil) below its bubble point* in a triphasic system oil/water/gas. Note that these are equations *for each component* of the system, *not for each phase*.

The pair of equations for three-dimensional biphasic flow can be derived from the polyphasic equations:

Gas-oil in the presence of S_{iw}:

Eq. (13.38a) + Eq. (13.38b) with $R_{\mathrm{sw}} = 0$

Undersaturated oil-water:

Eq. (13.38a) + Eq. (13.38c)

Gas-water:

Eq. (13.38b) with $R_s = 0$ + Eq. (13.38c).

If we introduce the term α, as defined in Eq. (13.16b), into Eqs. (13.38), they can also be used in one- and two-dimensional geometries.

13.9 Generalised Compositional Equation for Multiphase Flow

In the case of a very light oil (a "volatile oil"), or a condensate gas, we have to use a compositional description (Sect. 13.3.3) of the phases present in the reservoir. Their compositions, of course, vary over the producing life of the reservoir.

$C_{i,o}$, $C_{i,g}$ and $C_{i,w}$ are the *mass fractions* of component i in the "liquid hydrocarbon", "gaseous hydrocarbon" and "water" phases respectively:

$$C_{i,o} = \frac{m_i}{m_o}; \quad C_{i,g} = \frac{m_i}{m_g}; \quad C_{i,w} = \frac{m_i}{m_w}, \tag{13.40}$$

where:

$$m_i = \text{mass of the } i\text{th component}$$
$$m_o, m_g, m_w = \text{masses of the liquid hydrocarbon, gaseous hydrocarbon}$$
$$\text{and water phases.}$$

The equation for the conservation of mass for component i is derived by summing Eqs. (13.35) for the oil, gas and water phases, and writing each in terms of the mass fraction of component i:

$$\nabla \cdot \left[\frac{C_{i,o}\rho_o[k]k_{ro}}{\mu_o} \left(\nabla p_o - \rho_o g \, \nabla D \right) \right] + \nabla \cdot \left[\frac{C_{i,g}\rho_g[k]k_{rg}}{\mu_g} \left(\nabla p_g - \rho_g g \, \nabla D \right) \right]$$

$$+ \nabla \cdot \left[\frac{C_{i,w}\rho_w[k]k_{rw}}{\mu_w} \left(\nabla p_w - \rho_w g \, \nabla D \right) \right] - C_{i,o}\tilde{q}_{m,o} - C_{i,g}\tilde{q}_{m,g} - C_{i,w}\tilde{q}_{m,w}$$

$$= \frac{\partial}{\partial t} \left[\phi \left(C_{i,o}\rho_o S_o + C_{i,g}\rho_g S_g + C_{i,w}\rho_w S_w \right) \right], \tag{13.41}$$

where the $\tilde{q}_m$ terms are the mass flow rates measured under reservoir conditions.

For N components, there will be N mass balance equations of the form of Eq. (13.41).

The unknowns in this system of N equations are:

Unknowns	Number
$C_{i,o}$; $C_{i,g}$; $C_{i,w}$	3N
ρ_o; ρ_g; ρ_w	3
k_{ro}; k_{rg}; k_{rw}	3
μ_o; μ_g; μ_w	3
p_o; p_g; p_w	3
S_o; S_g; S_w	3
	Total: 3N+15

The partitioning of each of the N components between the gas and oil, and gas and water* phases is defined by means of the *equilibrium constants* or *K-values*:

$$K_{i,go} = \frac{C_{i,g}}{C_{i,o}} = C_{i,go} \left(p_g, p_o, T, \text{composition of the gas and oil phases} \right),$$

$$\tag{13.42a}$$

$$K_{i,gw} = \frac{C_{i,g}}{C_{i,w}} = C_{i,gw} \left(p_g, p_w, T, \text{composition of the gas and water phases} \right).$$

$$\tag{13.42b}$$

---

* We ignore the solubility of oil in water, which is extremely small.

The density of each phase, and the equilibrium constants of each component, are described by the respective equations of state. The numbers of equations available are:

Equations			Number
N mass balance equations [Eq. (13.41)]			N
$\sum_{1}^{N} {}_i C_{i,\mathrm{o}} = 1;$	$\sum_{1}^{N} {}_i C_{i,\mathrm{g}} = 1;$	$\sum_{1}^{N} {}_i C_{i,\mathrm{w}} = 1;$	3
$\rho_\mathrm{o} = \rho_\mathrm{o}(p, T);$	$\rho_\mathrm{g} = \rho_\mathrm{g}(p, T);$	$\rho_\mathrm{w} = \rho_\mathrm{w}(p, T);$	3
$\mu_\mathrm{o} = \mu_\mathrm{o}(p, T);$	$\mu_\mathrm{g} = \mu_\mathrm{g}(p, T);$	$\mu_\mathrm{w} = \mu_\mathrm{w}(p, T);$	3
$K_{i,\mathrm{go}}, \quad i = 1, \ldots, N$			N
$K_{i,\mathrm{gw}}, \quad i = 1, \ldots, N$			N
$k_\mathrm{rg} = f(S_\mathrm{g});$	$k_\mathrm{ro} = f(S_\mathrm{g}, S_\mathrm{o});$	$k_\mathrm{rw} = f(S_\mathrm{w})$	3
$P_{\mathrm{c,ow}} = p_\mathrm{o} - p_\mathrm{w};$	$P_{\mathrm{c,go}} = p_\mathrm{g} - p_\mathrm{o}$		2
$S_\mathrm{o} + S_\mathrm{g} + S_w = 1$			1
		Total:	3N+15

Since we have as many equations as unknowns, the system is fully defined and can be solved. The 15 equations listed above, together with a set of initial and boundary conditions, make up the most general formulation of the compositional equation for polyphasic flow.

As we shall see in Exercise 13.3, by applying appropriate simplifying assumptions, it is possible to derive the equation of flow used in *moderately volatile oil models* as well as the special case of black oils [Eq. (13.38)].

13.10 Discussion of the Flow Equations

13.10.1 Monophasic Flow

In the case of single phase flow, described by Eq. (13.21), there is only one unknown – $p(x, y, z, t)$ – to determine. All the coefficients appearing in Eq. (13.21) (except α and g) are functions of p: their values will therefore be affected by variations of p in space and time.

13.10.2 Biphasic Flow

As an example, consider a two-phase system consisting of undersaturated oil and water, described by Eqs. (13.38a) and (13.38c). At first glance, there would appear to be four unknowns in these equations: p_o, p_w, S_o and S_w, all functions of (x, y, z, t).

In fact, the relationships:

$$p_\mathrm{o} = p_\mathrm{w} + P_{\mathrm{c,ow}}(S_\mathrm{w}) \tag{13.43a}$$

and

$$S_\mathrm{o} = 1 - S_\mathrm{w}, \tag{13.43b}$$

where $P_{\mathrm{c,ow}}$ is the water/oil capillary pressure in the *imbibition* cycle (Sect. 3.4.4.4), reduce the number of unknowns to two, and when we have two equations (13.38a + 13.38b) and two unknowns, we can find a unique solution.

The system can be solved for the two pressures p_o and p_w, or one of the pressures and a saturation.

If we wish to solve for the pressures, we will have:

$$S_w = P_{c,ow}^{-1}(p_o - p_w), \tag{13.43c}$$

where $P_{c,ow}^{-1}$ is the inverse function of $P_{c,ow}$ (not the reciprocal!).

Note that while all the other coefficients in Eqs. (13.38a) and (13.38b) are functions only of the pressure (except α and g which are constants), the transmissibility:

$$T = \frac{k k_r}{\mu B} \tag{13.44}$$

is also a function of the saturation through the relative permeability term k_r.

Therefore:

$$dT = k_r \frac{d}{dp}\left(\frac{k}{\mu B}\right) dp + \frac{k}{\mu B}\frac{dk_r}{dS} dS. \tag{13.45}$$

This is the *"fully implicit"* method and will be referred to again in Sect. 13.14.3.4 on the solution of the polyphasic flow equation.

13.10.3 Triphasic Flow

In a three-phase system of oil, gas and water, flow is described by Eqs. (13.38a,b and c). The six apparent unknowns here – three pressures: p_o, p_g, p_w; and three saturationss: S_o, S_g and S_w – are in fact only three, by virtue of the following relationships:

$$p_g = p_o + P_{c,go}(S_g), \tag{13.46a}$$

$$p_w = p_o - P_{c,ow}(S_w) \tag{13.46b}$$

and

$$S_g = 1 - S_o - S_w, \tag{13.46c}$$

where $P_{c,go}$ is the gas/oil capillary pressure in the *drainage* cycle, and $P_{c,ow}$ is the oil/water capillary pressure in the *imbibition* cycle. Since we have three equations, we can find unique solutions to the three unknowns.

The equations are usually arranged so that the unknowns are p_o, S_o, and S_w. Values are then calculated for S_g, p_g and p_w using Eqs. (13.46). The comments made in the preceding section about the coefficients appearing in the equations, especially the transmissibility T, are also applicable to the triphasic case.

13.11 Basic Principles of the Finite Difference Method

13.11.1 Discretisation

We have already seen in the preceding chapters that the flow equations cannot be solved analytically. We therefore have to resort to numerical methods. The approach most commonly used in the simulation of reservoir behaviour is the method of finite differences, introduced in Sect. 13.3.

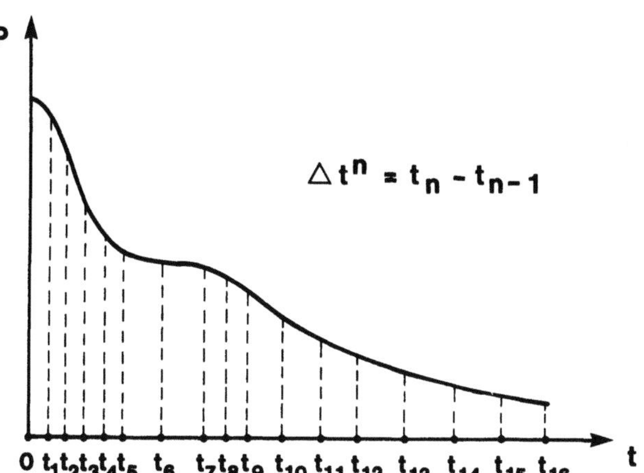

Fig. 13.6. *Time discretisation*

The reservoir volume is *discretised* by means of a *grid* (or *mesh*) appropriate to the system being studied (Sect. 13.3.2), and the time is discretised into time steps Δt_i, of equal or different durations (Fig. 13.6).

We have now gone from a physical model with *continuously variable parameters* (k, ϕ, p_o, p_g, p_w, S_o, S_w, S_g, etc.) to a numerical model with *localised parameters*, in which a different set of parameter values (constants and dependent variables) can be assigned at a point in *each* elementary volume (*grid block*) making up the discretised reservoir.

13.11.2 Types of Grid (or Mesh)

Parameter values can be specified (in the case of the input constants) or calculated (in the case of the dependent variables) either at the centre points of the grid blocks (*centre-block* system; Fig. 13.7a) or at their corners (*corner-point* or *lattice* system, Fig. 13.7b). The centre-block type of grid is particularly suited to models where there is predetermined flow across the external boundaries of the reservoir (the Neumann condition). If a zero flow is required across the boundary into a certain block, an "imaginary" block is joined to its outer face with the phase pressures held at the same values as those in the real block: this ensures a local flow rate of zero across the boundary. The corner-point grid system is best suited to modelling constant pressure external boundaries (the Dirichlet condition). In this case, the pressure $p(x, y, z)$ can be specified at the outer corners of each grid block. Because the large majority of reservoirs have fixed rates (most often zero) across their outer boundaries, the grid type most commonly used in numerical modelling is generally centre-block. The convention is to identify the blocks by means of the *indices i, j* and k in the x-, y- and z-directions respectively (Fig. 13.8). Time steps are identified by *superscripts* (not to be confused with exponents!).

For example:

$$p^n_{w,2,4,3}$$

is the pressure in the water phase, at time t_n, in the block at $x = 2$, $y = 4$, $z = 3$.

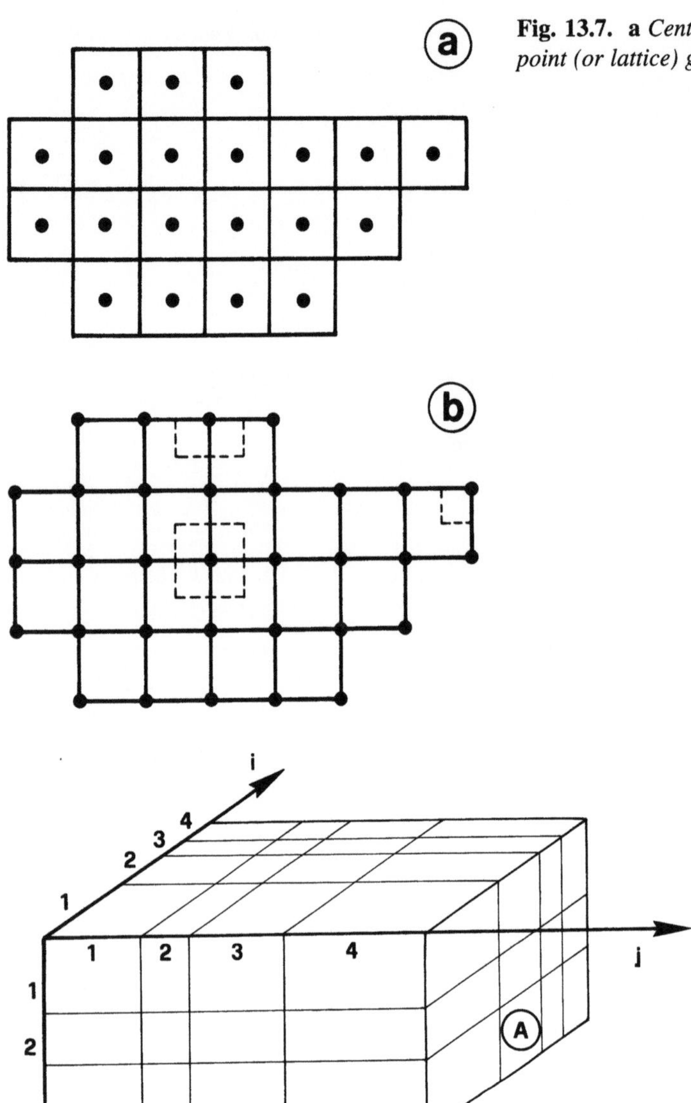

Fig. 13.7. **a** *Centre-block grid* and **b** *corner-point (or lattice) grid*

Fig. 13.8. *Block numbering in a 3D model. Block A is at (2,4,3)*

13.11.3 Discretisation of Derivatives into Finite Differences

Suppose a function $f(x)$, of class C^k, is continuous in the interval (a, b), with continuous derivatives down to the kth. We can develop Taylor's series as follows [using the notation $\Delta x^l = (\Delta x)^l$]:

$$f(x_0 + \Delta x) = f(x_0) + \frac{\Delta x}{1!}\left(\frac{\mathrm{d}f}{\mathrm{d}x}\right)_{x_0} + \frac{\Delta x^2}{2!}\left(\frac{\mathrm{d}^2 f}{\mathrm{d}x^2}\right)_{x_0}$$

$$+ \frac{\Delta x^3}{3!}\left(\frac{\mathrm{d}^3 f}{\mathrm{d}x^3}\right)_{x_0} + \cdots + \frac{\Delta x^k}{k!}\left(\frac{\mathrm{d}^k f}{\mathrm{d}x^k}\right)_{x_0} \tag{13.47a}$$

and

$$f(x_0 - \Delta x) = f(x_0) - \frac{\Delta x}{1!}\left(\frac{\mathrm{d}f}{\mathrm{d}x}\right)_{x_0} + \frac{\Delta x^2}{2!}\left(\frac{\mathrm{d}^2 f}{\mathrm{d}x^2}\right)_{x_0}$$
$$- \frac{\Delta x^3}{3!}\left(\frac{\mathrm{d}^3 f}{\mathrm{d}x^3}\right)_{x_0} + \cdots + (-1)^k \frac{\Delta x^k}{k!}\left(\frac{\mathrm{d}^k f}{\mathrm{d}x^k}\right)_{x_0}. \tag{13.47b}$$

From Eq. (13.47a) we have:

Forward-difference derivative:

$$\left(\frac{\mathrm{d}f}{\mathrm{d}x}\right)_{x_0} = \frac{f(x_0 + \Delta x) - f(x_0)}{\Delta x} + \mathrm{O}(\Delta x) \tag{13.48a}$$

and from Eq. (13.47b):

Backward-difference derivative:

$$\left(\frac{\mathrm{d}f}{\mathrm{d}x}\right)_{x_0} = \frac{f(x_0) - f(x_0 - \Delta x)}{\Delta x} + \mathrm{O}(\Delta x). \tag{13.48b}$$

If we subtract Eq. (13.47b) from Eq. (13.47a) we get:

Central-difference derivative:

$$\left(\frac{\mathrm{d}f}{\mathrm{d}x}\right)_{x_0} = \frac{f(x_0 + \Delta x) - f(x_0 - \Delta x)}{2\Delta x} + \mathrm{O}(\Delta x^2). \tag{13.48c}$$

Equations (13.48a, b and c) are the *finite difference approximations to the first derivative of the function f(x)*.

Note that the central-difference formulation is a higher-order approximation than the forward- or backward-difference [$\mathrm{O}(\Delta x^2)$ as opposed to $\mathrm{O}(\Delta x)$].

If we add Eqs. (13.47a) and (13.47b) we get:

$$\left(\frac{\mathrm{d}^2 f}{\mathrm{d}x^2}\right)_{x_0} = \frac{f(x_0 + \Delta x) - 2f(x_0) + f(x_0 - \Delta x)}{\Delta x^2} + \mathrm{O}(\Delta x^2). \tag{13.49}$$

This is the *finite difference approximation to the second derivative of the function f(x)*.

We will look first at a one-dimensional discretisation. Fig. 13.9 represents any three adjacent blocks $i - 1, i$ and $i + 1$, of equal length Δx, where:

$$\left(\frac{\mathrm{d}f}{\mathrm{d}x}\right)_i = \frac{f_{i+1} - f_i}{\Delta x} \qquad \text{forward-differencing}$$

$$= \frac{f_i - f_{i-1}}{\Delta x} \qquad \text{backward-differencing}$$

$$= \frac{f_{i+1} - f_{i-1}}{2\Delta x} \qquad \text{central-differencing} \tag{13.50}$$

$$\left(\frac{\mathrm{d}^2 f}{\mathrm{d}x^2}\right)_i = \frac{f_{i+1} - 2f_i + f_{i-1}}{\Delta x^2}. \tag{13.51}$$

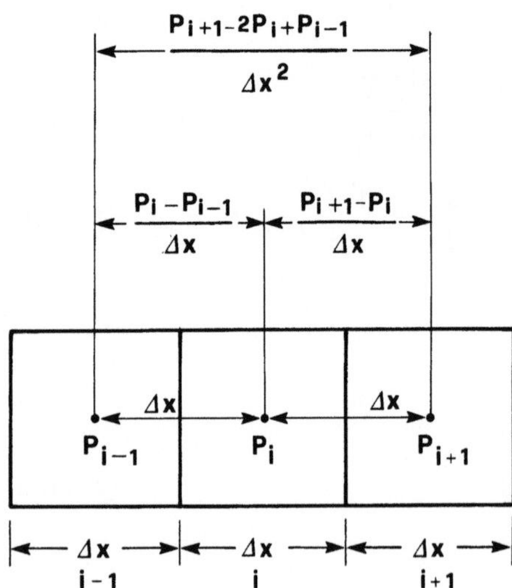

Fig. 13.9. *Schematic of the first and second derivatives by finite differences*

Analogously, the derivative of $f(x, t)$ with respect to time at the ith block will be:

$$\left(\frac{\partial f}{\partial t}\right)_i = \frac{f_i^{n+1} - f_i^n}{\Delta t},\tag{13.52}$$

where $f_i^{n+1} = f(x_i, t_{n+1}); \quad f_i^n = f(x_i, t_n).$

13.11.4   Choice of Time for the Spatial Derivative, and the Conditions for Stability

The Fourier equation for a simple one-dimensional system is (Sect. 13.6):

$$\frac{\partial^2 u}{\partial x^2} = \frac{1}{K}\frac{\partial u}{\partial t}.\tag{13.53}$$

We know the values of u^n at time t^n in each block, and we want to calculate each u^{n+1} at time t^{n+1}.

The following simplified notations will be used:

$$\Delta^2 u = u_{i-1} - 2u_i + u_{i+1};\tag{13.54a}$$

$$\Delta_t u = u_i^{n+1} - u_i^n;\tag{13.54b}$$

$$\gamma \quad = \frac{\Delta t}{\Delta x^2}.\tag{13.54c}$$

There are three methods for solving Eq. (13.53) by finite differences:

1. Explicit method

In order to calculate $\partial^2 u/\partial x^2$ using Eq. (13.51), we use the values of u at time t_n (Fig. 13.10):

$$\frac{u_{i-1}^n - 2u_i^n + u_{i+1}^n}{\Delta x^2} = \frac{1}{K}\frac{u_i^{n+1} - u_i^n}{\Delta t}.\tag{13.55a}$$

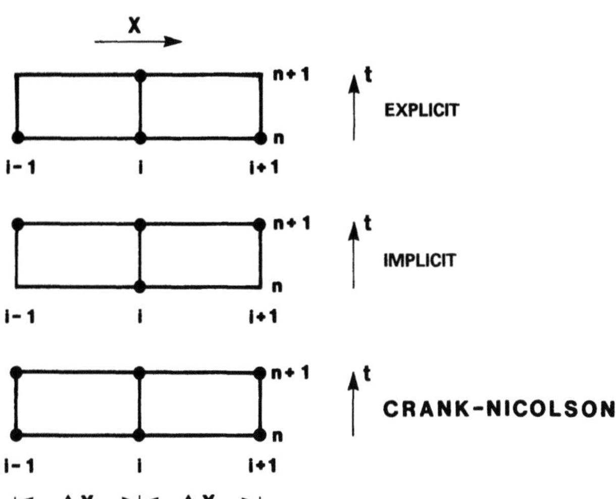

Fig. 13.10. *Schematic representations of the three methods for approximating the second derivative in space*

In other words, using the notation defined in Eqs. (13.54c):

$$\gamma \Delta^2 u^n = \frac{1}{K}\Delta_t u. \tag{13.55b}$$

In this case, the only unknown is u_i^{n+1}.

2. Implicit method

All values of u in the spatial derivative are taken at time t_{n+1} (Fig. 13.10):

$$\gamma \Delta^2 u^{n+1} = \frac{1}{K}\Delta_t u. \tag{13.56}$$

There are therefore three unknowns (u_{i-1}^{n+1}, u_i^{n+1}, u_{i+1}^{n+1}) each time Eq. (13.56) is written.

3. The Crank-Nicolson method

Average values of u are assumed at times t_n and t_{n+1} for the evaluation of the spatial derivative (Fig. 13.10):

$$\gamma \frac{\Delta^2 u^{n+1} + \Delta^2 u^n}{2} = \frac{1}{K}\Delta_t u. \tag{13.57}$$

The three unknowns that appeared in the implicit method also appear in this case.

An analysis of stability[28] reveals that the explicit method, which is the simplest approach as far as numerical calculation is concerned, is only stable when:

$$\gamma = \frac{\Delta t}{\Delta x^2} < \frac{1}{2}. \tag{13.58}$$

This generally implies very small time steps (with $\Delta x = 100$ m, $\Delta t \approx 1.5$ h).

The solution becomes unstable quite rapidly when Δt exceeds the limit expressed in Eq. (13.58) (Fig. 13.11).

The implicit and Crank-Nicolson methods, on the other hand, are *unconditionally stable* for any value of γ. Consequently, only these two are used for the finite difference approximation of the spatial derivative of the function $f(x, y, z, t)$. The Crank-Nicolson method is the more widely used of the two.

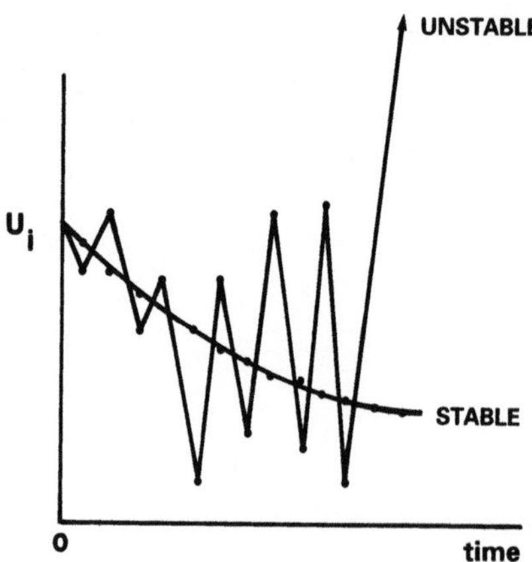

Fig. 13.11. *Stable and unstable solutions to Eq. (13.53)*

It would be wrong to assume that the presence of three unknowns in a single equation (u_{i-1}^{n+1}, u_i^{n+1}, u_{i+1}^{n+1}) renders the problem indeterminate.

Let us consider the Fourier equation [Eq. (13.53)], with our one-dimensional system subdivided into N blocks. The initial conditions $u_i^0 (1 \le i \le N)$ are specified in each block, with a boundary value $u_N(t)$ in block N.

Using the implicit method (for example), we have:

$$\frac{u_{i-1}^{n+1} - 2u_i^{n+1} + u_{i+1}^{n+1}}{\Delta x^2} = \frac{1}{K}\frac{u_i^{n+1} - u_i^n}{\Delta t}, \tag{13.59}$$

where, if $\gamma = \Delta t / \Delta x^2$ as in Eq. (13.54c):

$$f_i(u) = -\gamma K u_{i-1}^{n+1} + (2\gamma K + 1)u_i^{n+1} - \gamma K u_{i+1}^{n+1} - u_i^n = 0.$$

We define the following terms:

$$a_i = -\gamma K = \frac{\partial f_i}{\partial u_{i-1}^{n+1}}, \tag{13.60a}$$

$$b_i = 2\gamma K + 1 = \frac{\partial f_i}{\partial u_i^{n+1}}, \tag{13.60b}$$

$$c_i = -\gamma K = \frac{\partial f_i}{\partial u_{i+1}^{n+1}}, \tag{13.60c}$$

$$d_i = u_i^n \text{(known)}, \tag{13.60d}$$

so that Eq. (13.59) becomes:

$$a_i u_{i-1}^{n+1} + b_i u_i^{n+1} + c_i u_{i+1}^{n+1} = d_i. \tag{13.61}$$

Equation (13.61) is written for each group of three adjacent blocks ($i - 1, i, i + 1$), stepping along one block at a time. $u_N(t)$ is fixed by the outer boundary condition. This produces the following set of equations [the superscripts ($n + 1$) have been omitted to simplify the presentation]:

$$\left\{\begin{aligned}
b_1 u_1 + c_1 u_2 &&&= d_1 \\
a_2 u_1 + b_2 u_2 + c_2 u_3 &&&= d_2 \\
a_3 u_2 + b_3 u_3 + c_3 u_4 &&&= d_3 \\
&\vdots && \vdots \\
a_{N-2} u_{N-3} + b_{N-2} u_{N-2} + c_{N-2} u_{N-1} &&&= d_{N-2} \\
a_{N-1} u_{N-2} + b_{N-1} u_{N-1} &&&= d_{N-1}.
\end{aligned}\right. \tag{13.62}$$

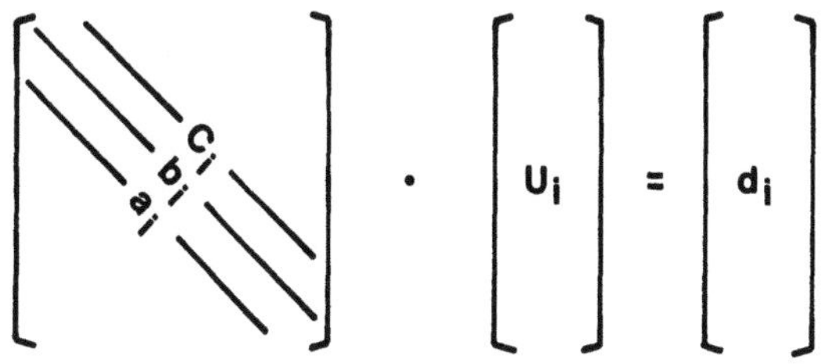

Fig. 13.12. *Schematic of Eq. (13.63)*

This system of $(N-1)$ equations in $(N-1)$ unknowns is fully defined and can be solved numerically for all values of u_i^{n+1} for $1 \le i \le (N-1)$.

Equation (13.62) is written in matrix form as:

$$Au = d, \tag{13.63}$$

where A is the matrix of coefficients, u is the vector of unknowns, and d is the vector of known terms.

A is a *tridiagonal matrix*. All its non-zero terms appear on three diagonals – the principal diagonal and the one on either side of it (Fig. 13.12). The terms appearing in A consist of the derivatives of the function $f(u)$ with respect to each variable [Eqs. (13.60a–13.60c)]. Therefore, *the matrix A is the Jacobian of $f(u)$*:

$$A = J(f(u)). \tag{13.64}$$

Multiplying both terms in Eq. (13.63) by the inverse matrix A^{-1}, we get:

$$A^{-1}Au = Iu = A^{-1}d. \tag{13.65a}$$

Therefore:

$$u = A^{-1}d. \tag{13.65b}$$

This *inversion* of the coefficient matrix provides a means of solving Eq. (13.62). However, it requires considerable computing time, and is only used when the number of blocks (N) is small.

13.11.5 Rounding and Truncation Errors

Computers perform numerical calculations in *scientific notation* ($3.569812E12 = 3.569812 \times 10^{12}$), and most of them use a *fixed word length*, meaning that the numbers ("words") are made up of a fixed number of "characters": 7 to 12 in the case of "single precision", and 14 to 24 in "double precision". If a number is longer than the allowed word length, the extra characters are lost, leading to a *rounding error*. In an algebraic summation of two numbers of widely different orders of magnitude, this can result in the loss of significant figures in the smaller term.

This so-called *cancellation error* can be reduced by working in double precision and using techniques which prevent operations occurring between numbers of very different orders of magnitude.

Truncation or *discretisation errors*, on the other hand, cannot be eliminated. These arise from discrepancies between the true value of the derivative and its finite difference approximation [Eqs. (13.48) and (13.49)]. This error tends to zero with the discretisation step size Δx [$0(\Delta x)$] or its square [$0(\Delta x^2)$], so an improvement can be obtained by using a finer mesh. There is, however, a practical limit to how far this can be taken, because the number of blocks will increase in inverse proportion to their individual volumes, and the computing time is proportional to a power of the number of blocks. Therefore, a certain degree of discretisation error has to be accepted. Unfortunately, the magnitude of the error itself cannot always be evaluated precisely.

A third class of error, which becomes very significant in the simulation of polyphasic flow (to be discussed in depth in Sect. 13.14), arises from the phenomenon of *numerical dispersion*.

13.12 Numerical Simulation of Single Phase Flow

13.12.1 The Finite Difference Equation for Single Phase Flow

Equation (13.21) for monophasic flow is *parabolic* (Sect. 13.6), becoming *elliptical* when $\partial p/\partial t = $ const. (pseudo-steady state flow) or zero (steady state flow). In orthogonal cartesian coordinates (x, y, z), with z increasing downwards, it takes the form:

$$\frac{\partial}{\partial x}\left(\frac{k_x}{\mu}\frac{\partial p}{\partial x}\right) + \frac{\partial}{\partial y}\left(\frac{k_y}{\mu}\frac{\partial p}{\partial y}\right) + \frac{\partial}{\partial z}\left[\frac{k_z}{\mu}\left(\frac{\partial p}{\partial z} - \rho g\right)\right]$$
$$- \tilde{q}_v = \phi c_t \frac{\partial p}{\partial t}. \tag{13.66}$$

We will now see how the finite difference approximation to a partial differential flow equation [Eq. (13.66) in particular] can be derived directly as an alternative to discretising the derivative (Sect. 13.11.3). For simplicity, we will consider a *horizontal* two-dimensional reservoir, *of constant thickness h*. A part of the mesh is shown in Fig. 13.13, with grid-block (i, j) and the four adjacent blocks $(i-1, j)$, $(i+1, j)$, $(i, j-1)$ and $(i, j+1)$, not necessarily of the same dimensions.

To derive the flow equation by finite differences, we write the mass balance equation for block (i, j) using a method which is analogous to that used in Sect. 13.5 to derive the flow equation in partial derivative form. We will start in the x-direction. The parameters are defined at the centre of each block. Firstly, we calculate the volumetric flow-rate between the centres of blocks $(i-1, j)$ and (i, j) in what amounts to a linear system (Fig. 13.14). The average permeability in the x-direction between the two block centres (the *interblock* or *linking permeability*) is the *harmonic mean* of $(k_x)_{i-1,j}$ and $(k_x)_{i,j}$. It is assigned to the mid-point between the two centres and is designated $(k_x)_{i-1/2,j}$:

$$(k_x)_{i-1/2,j} = \frac{(k_x)_{i-1,j}(k_x)_{i,j}}{(k_x)_{i-1,j}\Delta x_i + (k_x)_{i,j}\Delta x_{i-1}}(\Delta x_{i-1} + \Delta x_i). \tag{13.67a}$$

If $\Delta x_{i-1} = \Delta x_i$, we have:

$$(k_x)_{i-1/2,j} = \frac{2(k_x)_{i-1,j}(k_x)_{i,j}}{(k_x)_{i-1,j} + (k_x)_{i,j}}. \tag{13.67b}$$

The interblock permeability is sometimes calculated as the *arithmetic mean*:

$$(k_x)_{i-1/2,j} = \frac{(k_x)_{i-1,j}\Delta x_{i-1} + (k_x)_{i,j}\Delta x_i}{\Delta x_{i-1} + \Delta x_i}, \tag{13.68a}$$

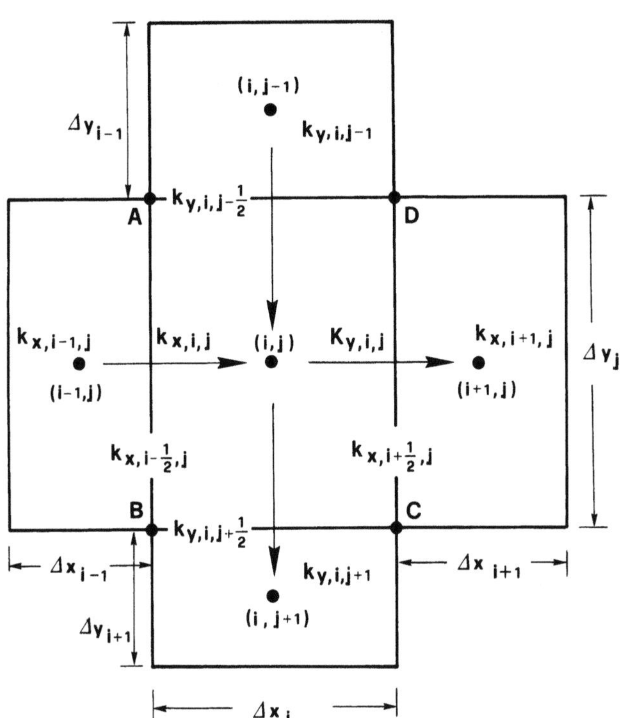

Fig. 13.13. *Configuration of the block dimensions and permeabilities used to derive the single phase flow equation by the finite difference method*

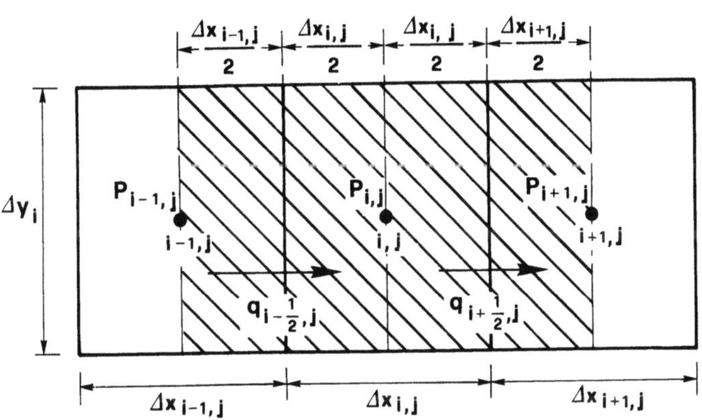

Fig. 13.14. *Flow calculation in the x-direction*

and, if $\Delta x_{i-1} = \Delta x_i$:

$$(k_x)_{i-1/2,j} = \frac{(k_x)_{i-1,j} + (k_x)_{i,j}}{2}. \tag{13.68b}$$

The Darcy equation for one-dimensional horizontal flow is:

$$q = -\frac{kA}{\mu L}\Delta p,$$

so the volume of fluid passing from block $(i-1, j)$ to block (i, j) in time Δt (Fig. 13.14) can be expressed as:

$$q_{i-1/2,j} = \frac{(k_x)_{i-1/2,j}\,h\,\Delta y_i}{\mu\dfrac{\Delta x_{i-1} + \Delta x_i}{2}}\left(p_{i-1,j} - p_{i,j}\right)\Delta t. \tag{13.69a}$$

Here, h is the size of the block in the z-direction. Next, defining the *interblock transmissibility* from block $(i - 1, j)$ to block (i, j) as:

$$(T_x)_{i-1/2,j} = \frac{2(k_x)_{i-1/2,j} h \Delta y_i}{\mu(\Delta x_{i-1} + \Delta x_i)}, \tag{13.70}$$

we have:

$$q_{i-1/2,j} = (T_x)_{i-1/2,j}(p_{i-1,j} - p_{i,j})\Delta t. \tag{13.69b}$$

Analogously, the flow rate leaving block (i, j) into block $(i + 1, j)$ will be:

$$q_{i+1/2,j} = (T_x)_{i+1/2,j}(p_{i,j} - p_{i+1,j})\Delta t. \tag{13.69c}$$

If we ignore any changes in ρ in the pressure range $p_{i+1,j} \leq p \leq p_{i-1,j}$, the mass of fluid in block (i, j) will change by the following amount in time Δt as a result of the flow in the x-direction:

$$\begin{aligned}(\Delta q_m)_x = \rho[&(T_x)_{i-1/2,j}(p_{i-1,j} - p_{i,j}) \\ &- (T_x)_{i+1/2,j}(p_{i,j} - p_{i+1,j})]\Delta t. \end{aligned} \tag{13.71}$$

Using the notation:

$$\begin{aligned}\Delta_x T_x \Delta_x p = &(T_x)_{i+1/2,j}(p_{i+1,j} - p_{i,j}) \\ &- (T_x)_{i-1/2,j}(p_{i,j} - p_{i-1,j}), \end{aligned} \tag{13.72a}$$

Eq. (13.71) becomes:

$$(\Delta q_m)_x = \rho \, \Delta_x T_x \Delta_x p \, \Delta t. \tag{13.73a}$$

We can repeat the above procedure in the y-direction to get:

$$(\Delta q_m)_y = \rho \, \Delta_y T_y \Delta_y p \, \Delta t \tag{13.73b}$$

with:

$$\begin{aligned}\Delta_y T_y \Delta_y p = &(T_y)_{i,j+1/2}(p_{i,j+1} - p_{i,j}) \\ &- (T_y)_{i,j-1/2}(p_{i,j} - p_{i,j-1}). \end{aligned} \tag{13.72b}$$

If $p_{i,j}^{n+1}$ and $p_{i,j}^n$ are the pressures in block (i, j) at times t_{n+1} and t_n respectively ($\Delta t = t_{n+1} - t_n$), the volume of fluid in the block will change in time Δt by an amount:

$$\Delta V_{i,j} = \phi c_t h \Delta x_i \Delta y_i (p_{i,j}^{n+1} - p_{i,j}^n). \tag{13.74a}$$

Ignoring any variation in ρ with pressure, we can write:

$$(\Delta q_m)_{\text{acc}} = \rho \phi c_t h \Delta x_i \Delta y_i (p_{i,j}^{n+1} - p_{i,j}^n). \tag{13.74b}$$

We define two new terms:

$$\tilde{q}_v = \frac{\tilde{q}_m}{\rho} = \text{volumetric flow rate, per unit pay thickness, of fluid extracted}$$
$$\text{from blocks containing wells.}$$

$$\alpha_{i,j} \quad = c_t \phi h \Delta x_i \Delta y_j. \tag{13.75a}$$

The mass balance equation for block (i, j)

$$(\Delta q_m)_x + (\Delta q_m)_y - \tilde{q}_m \Delta t = (\Delta q_m)_{\text{acc}}$$

can be expressed as follows:

$$\Delta_x T_x \Delta_x p + \Delta_y T_y \Delta_y p - \tilde{q}_v = \frac{\alpha_{i,j}}{\Delta t}(p_{i,j}^{n+1} - p_{i,j}^n). \tag{13.76}$$

This is the *finite difference equation for two-dimensional horizontal flow* in a layer of constant thickness.

In three-dimensional systems the flow in the z-direction must be taken into consideration. If gravitational effects are included, the flow term $\Delta_z T_z \Delta_z p$ takes on a form somewhat different from Eq. (13.72):

$$\Delta_z T_z \Delta_z p = (T_z)_{i,j,k+1/2}\left(p_{i,j,k+1} - p_{i,j,k} - \rho g \frac{\Delta z_{i,j,k+1} + \Delta z_{i,j,k}}{2}\right)$$

$$- (T_z)_{i,j,k-1/2}\left(p_{i,j,k} - p_{i,j,k-1} - \rho g \frac{\Delta z_{i,j,k} + \Delta z_{i,j,k-1}}{2}\right). \quad (13.77)$$

This equation is derived in Exercise 13.4.

In three dimensions, we have:

$$\alpha_{i,j,k} = c_t\,\phi\,\Delta x_i\,\Delta y_j\,\Delta z_k. \qquad (13.75b)$$

In compact notation, Eq. (13.76) is:

$$\sum_s \Delta_s T_s \Delta_s p - \tilde{q}_v = \frac{\alpha}{\Delta t}\left(p_{i,j,k}^{n+1} - p_{i,j,k}^n\right)$$

$$(s = x, y, z), \qquad (13.78)$$

with Eqs. (13.77) and (13.75b) taking care of the z-direction.

From Eqs. (13.71) and (13.66), incorporating the relationships expressed in Eqs. (13.69) and (13.70), we get:

$$\frac{\partial}{\partial x}\left(\frac{k_x}{\mu}\frac{\partial p}{\partial x}\right) = \frac{1}{\overline{\Delta x}}\left[\frac{2\left(\dfrac{k_x}{\mu}\right)_{i+1/2,j}\left(p_{i+1,j} - p_{i,j}\right)}{\Delta x_{i+1,j} + \Delta x_{i,j}}\right.$$

$$\left. - \frac{2\left(\dfrac{k_x}{\mu}\right)_{i-1/2,j}\left(p_{i,j} - p_{i-1,j}\right)}{\Delta x_{i,j} + \Delta x_{i-1,j}}\right], \qquad (13.79)$$

where $\overline{\Delta x} = (\Delta x_{i+1,j} + 2\Delta x_{i,j} + \Delta x_{i-1,j})/2$. An equivalent expression can be derived for the y-axis.

Equation 13.79 has second-order precision $[0(\Delta x^2)]$ and is the finite difference expression of an important component of the flow equation. This will be dealt with in sect. 13.14 on polyphasic flow.

13.12.2 Matrix Form of the Finite Difference Equation for Single Phase Flow

In solving Eq. (13.78), all the values of p appearing in the flow term are taken at time t_{n+1}, and are therefore unknown, or *implicit*:

$$\sum_s \Delta_s T_s \Delta_s p^{n+1} - \tilde{q}_v = \frac{\alpha_{i,j,k}}{\Delta t}\left(p_{i,j,k}^{n+1} - p_{i,j,k}^n\right), \qquad (13.80)$$

while the transmissibility T_x [Eq. (13.70) and its equivalents for T_y and T_z] are taken at time t_n, and are known, or *explicit*.

It should be pointed out that, in monophasic flow, the *fully implicit* method would rarely be used to solve Eq. (13.78) (except in the case of very low permeability reservoirs or high pressure gradients). As long as T (which contains k and μ) does not vary significantly over the pressure range p^n to p^{n+1}, T^{n+1} can be assumed to be the same as T^n, which is known.

Referring to Fig. 13.15, if N_x, N_y and N_z are the numbers of grid blocks in the three directions, then for each block Eq. (13.80) will contain three pressures in a one-dimensional system, five in a two-dimensional system, and seven in a three-dimensional system.

We now have a system of $N_x N_y N_z$ linear algebraic equations in as many unknowns $p_{i,j,k}$, which must be solved at each time step. To get an idea of the corresponding coefficient matrices A for this system:

$$\mathbf{Ap} = \mathbf{d}, \tag{13.63}$$

we will consider a simple two-dimensional model consisting of three blocks in the x-direction and four blocks in the y-direction ($N_x = 3$, $N_y = 4$).

Since each block (i, j) is only in communication with the four adjacent blocks $(i - 1, j)$, $(i + 1, j)$, $(i, j - 1)$ and $(i, j + 1)$, the non-zero terms in the matrix will be located as shown by "x" in Fig 13.16. This is a *banded matrix*, with its non-zero

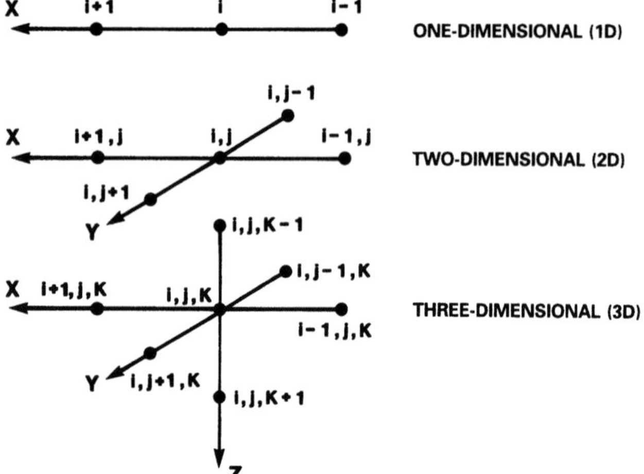

Fig. 13.15. *Connections between neighbouring blocks in 1D, 2D and 3D*

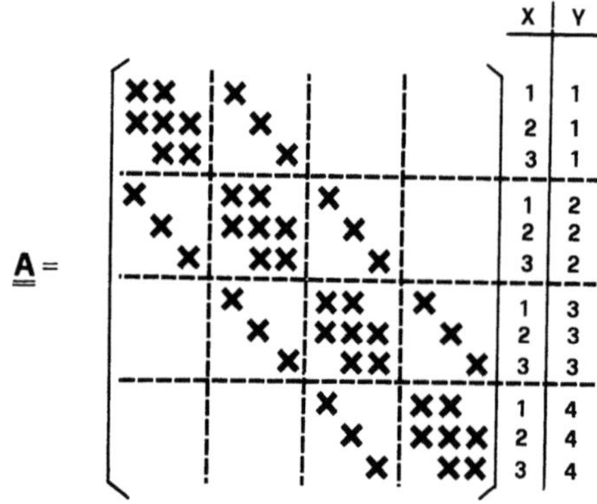

Fig. 13.16. *Location of the non-zero terms in the coefficient matrix for $N_x = 3$ and $N_y = 4$. (From Ref. 39, 1982, Prentice Hall. Reprinted with permission of Prentice Hall)*

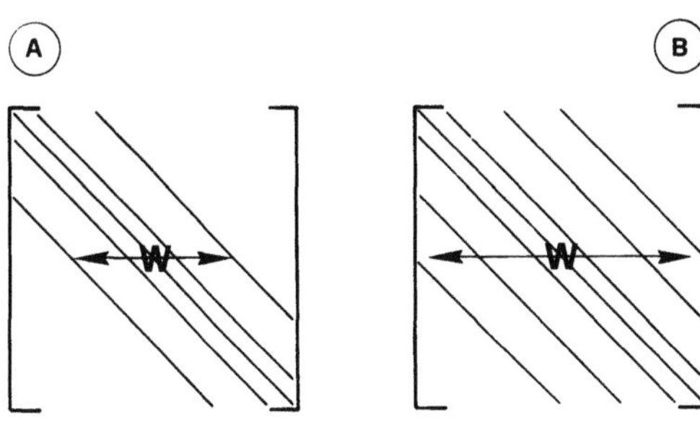

Fig. 13.17. A *Pentadiagonal matrix for a two-dimensional geometry;* **B** *heptadiagonal matrix for a three-dimensional geometry. (From Ref. 39, 1982, Prentice Hall. Reprinted with permission of Prentice Hall)*

terms along five diagonals, the middle one of which is the principal diagonal. In other words, this is a *pentadiagonal matrix*. It will be apparent from Fig. 13.16 that matrices of this type can also be regarded as consisting of submatrices which are *tridiagonal blocks* (3 × 3). In the submatrices on the principal diagonal, only the terms (1,3) and (3,1) are zero. In the submatrices on the two adjacent diagonals, all terms are zero except those on their principal diagonals.

In one-dimensional models, the coefficient matrices are *tridiagonal* [Eq. (13.62) and Fig. 13.12], while in three dimensions we have *heptadiagonal* banded matrices (Fig. 13.17B) where the non-zero terms lie on seven diagonals, the central one of which is the principal diagonal. The width W of the band is $(2N_x + 1)$ in two-dimensional models, and $(2N_x N_y + 1)$ in three-dimensional models. When N_x and N_y are large, the matrix A, although remaining essentially banded, tends to become *sparse*, given the considerable size of the bands.

At each time step, the system [Eq. (13.63)] of $N_x N_y N_z$ algebraic equations with as many unknowns $p_{i,j,k}$ can be solved by *direct methods* (subject to possible reordering of the matrix) or by *iterative methods*.

13.13 Solution of the Linear Algebraic Equations Derived from the Discretisation of the Flow Equation

13.13.1 Introduction

In this chapter we will just look at some of the main features of the methods used to solve the linear algebraic equations derived by discretising the flow equation. For a more detailed study the reader is referred to the list of published material[33,36,44] at the end of the chapter.

Strictly speaking, this topic falls in the domain of the expert concerned with the application of numerical modelling to reservoir simulation. However, the reservoir engineer should have at least a basic grasp of the concepts and methodology, if only to be able to share a common terminology with the experts.

13.13.2 Direct Methods[16]

13.13.2.1 The Gaussian Elimination Method

Consider a simple system of four equations with four unknowns describing the single phase flow of a fluid in a horizontal linear reservoir consisting of four blocks (Fig. 13.18).

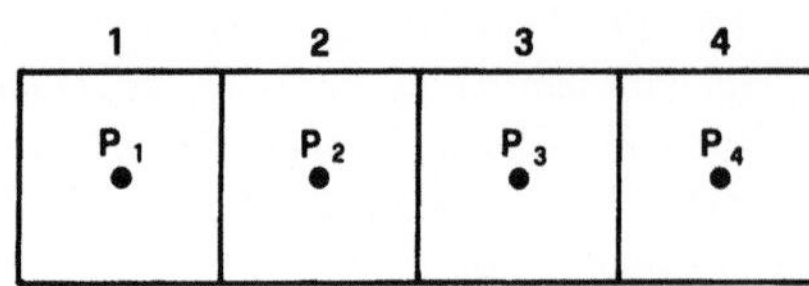

Fig. 13.18. *The four-block system represented by Eq. (13.81)*

Equation (13.62) is rewritten below in a simplified notation omitting the indices $n+1$ which denote the pressures p_i^{n+1}:

$$\begin{cases} a_{11}p_1 + a_{12}p_2 & = d_1 \\ a_{21}p_1 + a_{22}p_2 + a_{23}p_3 & = d_2 \\ a_{32}p_2 + a_{33}p_3 + a_{34}p_4 = d_3 \\ a_{43}p_3 + a_{44}p_4 = d_4. \end{cases} \tag{13.81}$$

Block 1 is in contact with block 2 only, so only pressures p_1 and p_2 will appear in its particular Eq. 13.81. The same goes for block 4, which will only have the terms p_3 and p_4 appearing.

Gauss's method of elimination, when applied to a system of n equations with n unknowns, involves triangularising the coefficient matrix by operating on the rows and columns and solving the resulting system of equations, starting from the one containing a single unknown.

Systems of equations with tridiagonal matrices, frequently encountered in reservoir modelling, can only be solved directly by the *Thomas algorithm*[33], a variant of Gauss's method. The procedure is as follows:

For the first block ($i = 1$), Eq. (13.81) gives:

$$p_1 = \frac{d_1}{a_{11}} - \frac{a_{12}}{a_{11}}p_2 = c_{11} - c_{12}p_2. \tag{13.82a}$$

Substituting this into Eq. (13.81) for the second block ($i = 2$) we have:

$$a_{21}(c_{11} - c_{12}p_2) + a_{22}p_2 + a_{23}p_3 = d_2, \tag{13.82b}$$

from which:

$$p_2 = \frac{d_2 - a_{21}c_{11} - a_{23}p_3}{a_{22} - a_{21}c_{12}} = c_{22} - c_{23}p_3. \tag{13.82c}$$

This in turn is substituted into the equation for the third block:

$$a_{32}(c_{22} - c_{23}p_3) + a_{33}p_3 + a_{34}p_4 = d_3, \tag{13.82d}$$

so that:

$$p_3 = \frac{d_3 - a_{32}c_{22} - a_{34}p_4}{a_{33} - a_{32}c_{23}} = c_{33} - c_{34}p_4. \tag{13.82e}$$

Finally, Eq. (13.82e) is substituted into Eq. (13.81) for the fourth block:

$$a_{43}(c_{33} - c_{34}p_4) + a_{44}p_4 = d_4 \tag{13.82f}$$

and

$$p_4 = \frac{d_4 - a_{43}c_{33}}{a_{44} - a_{43}c_{34}}. \tag{13.82g}$$

p_4 can now be calculated. It is then an easy matter to calculate, in order, p_3, p_2 and p_1.

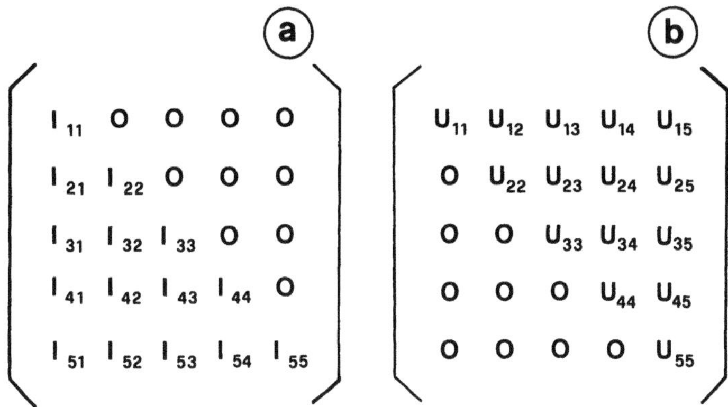

Fig. 13.19. a *lower triangular matrix of order 5;* **b** *upper triangular matrix of order 5*

13.13.2.2 Triangular Matrices

A *lower triangular matrix*, L, is a square matrix (same number of rows and columns) in which $a_{ij} = 0$ for $i < j$ (a_{ij} being the element in the ith row, jth column). Similarly, an *upper triangular matrix*, U, is a square matrix in which $a_{ij} = 0$ for $i > j$. Figure 13.19 illustrates lower and upper triangular matrices of order 5.

A system of linear algebraic equations with triangular coefficient matrices can be solved immediately by the method of substitution, because one of the unknowns is, in fact, known. For example, in a system with a coefficient matrix like Fig. 13.19a, we have:

$$l_{11}x_1 = d_1, \tag{13.83a}$$

so that:

$$x_1 = \frac{d_1}{l_{11}}. \tag{13.83b}$$

Consequently, in the second equation we will have:

$$l_{21}x_1 + l_{22}x_2 = d_2, \tag{13.83c}$$

$$x_2 = \frac{1}{l_{22}}\left(d_2 - \frac{l_{21}}{l_{11}}d_1\right), \tag{13.83d}$$

etc.

As you can see, *a system whose coefficient matrix is triangular can be solved directly, and quite simply.* This point is very relevant to the next section.

13.13.2.3 Factorisation of the Coefficient Matrix

Any square matrix A of the Nth order can be expressed as the product of two matrices L and U, also of order N:

$$A = LU, \tag{13.84a}$$

where L is a lower triangular matrix, and U an upper, *in which all the elements on the principal diagonal have unit value*:

$$\begin{pmatrix} a_{11} & a_{12} & \cdots & a_{1N} \\ a_{21} & a_{22} & \cdots & a_{2N} \\ & \vdots & & \\ a_{N1} & a_{N2} & \cdots & a_{NN} \end{pmatrix} = \begin{pmatrix} l_{11} & & & \\ l_{21} & l_{22} & & \\ & \vdots & & \\ l_{N1} & l_{N2} & \cdots & l_{NN} \end{pmatrix}$$

$$\times \begin{pmatrix} 1 & u_{12} & u_{13} & \cdots & u_{1N} \\ & 1 & u_{23} & \cdots & u_{2N} \\ & & \vdots & & \\ & & & & 1 \end{pmatrix} . \qquad (13.84\text{b})$$

In Eqs. (13.84), the matrix A has been *factorised*.

When dealing with banded matrices such as those that result from discretising the flow equation, the elements l_{ij} of L can be calculated using Crout's algorithm[13] if the elements u_{ij} of U are predetermined, or vice versa.

For an Nth order banded matrix A with elements a_{ij} and *odd* bandwidth W where:

$$L_1(i) = \max\left(i, \frac{2i - W + 1}{2}\right), \qquad (13.85\text{a})$$

$$L_2(j) = \max\left(j, \frac{2j - W + 1}{2}\right), \qquad (13.85\text{b})$$

$$L_3(i) = \max\left(1, \frac{2i - W - 1}{2}\right), \qquad (13.85\text{c})$$

we have:

$$l_{ij} = a_{ij} - \sum_{\substack{k \\ L_1}}^{j-1} l_{ik} u_{kj} \quad [j = L_1(i), L_1 + 1, \ldots, i], \qquad (13.85\text{d})$$

$$u_{ij} = \frac{1}{l_{ij}} \left(a_{ij} - \sum_{\substack{k \\ L_2}}^{i-1} l_{ik} u_{kj}\right) \quad [j = i + 1, \ldots, L_3(i)]. \qquad (13.85\text{e})$$

Equations 13.85 allow us to calculate the elements of L when the values of all elements of U are predetermined (in particular $u_{ii} = 1$), as is most commonly the case; or to calculate the elements of U when the values in L are predetermined.

We now consider a system of algebraic equations:

$$Ax = b. \qquad (13.63)$$

This can always be transformed by factorisation into:

$$LUx = b. \qquad (13.86\text{a})$$

We now introduce the auxiliary vector y, of order N, such that:

$$Ux = y, \qquad (13.86\text{b})$$

so that Eq. (13.86a) becomes:

$$Ly = b. \qquad (13.86\text{c})$$

Since L is triangular, Eq. (13.86c) can be solved very quickly by the method of substitution (Sect. 13.13.2.2) to obtain the N components of **y**. When input to Eq. (13.86b), these allow an equally rapid calculation of the N components $(x_1, x_2, \ldots x_n)$ of x to be performed, again by substitution.

The procedure consists, then, of three steps:

1. Factorisation of the coefficient matrix A into two triangular matrices L and U.
2. Calculation of the components $(y_1, y_2, \ldots, y_n)$ of the auxiliary vector y through Eq. (13.86c) (forward calculation).
3. Calculation of the unknowns $(x_1, x_2, \ldots, x_n)$ through Eq. (13.86b) (back calculation).

13.13.2.4 *The Ordering of Sparse Matrices*

The method of factorisation, although very quick, does involve some time wastage through the execution of a number of unnecessary operations, such as the summation and multiplication of any zero elements lying within the band containing the non-zero elements of the matrix. In addition, factorisation can "create" non-zero elements in the triangular matrices L and U which will increase computing time for the subsequent solution by substitution. There is a technique which, if correctly applied, can optimise the calculation procedure. It is called the *ordering of sparse matrices*, and it transforms a sparse matrix into a matrix (or the product of two matrices) structured in such a way that both the computing time and memory requirement are minimised. This is too specialised a topic to describe in any detail here. The point to remember is that this is the most efficient *direct* method for the solution of systems of equations obtained from the discretisation of the flow equation. For further reading, please refer to the work of Woo et al.[49], Price and Coats[32] and George and Liu[18].

13.13.3 **Iterative Procedures**[44]

13.13.3.1 *Introduction*

Iterative procedures involve the technique of successive approximations. Starting from estimated (or guessed) values for each unknown, the first iteration calculates a *first approximation* to each of them, using one of a number of algorithms that are available for this purpose.

These first approximations are then input as the starting values for the second iteration, and the algorithm produces a set of *second approximations* to each unknown. The iterations are repeated until the approximation to each unknown has converged to a stable value. The *iteration number* is indicated by a superscript (not to be confused with an exponent!): thus, x^1 is the value of the first approximation (i.e. from iteration number 1), x^2 is the value of the second approximation (iteration number 2), etc.

The index normally used for a general iteration number is k, with $(k + 1)$ for the next, and so on. The more efficient an iteration procedure is, the smaller will be the number of iterations required for convergence to a stable solution. With the exception of Thomas's method for tridiagonal matrices, iterative methods usually require less computer memory than direct methods. In addition, where large systems of equations with banded coefficient matrices are concerned, iterative methods can be faster than direct methods (including the method of matrix ordering mentioned in Sect. 13.13.2.4). Consequently, the iterative approach is often used when the number of grid-blocks, N, is very large. In this case, the initial values (at $t = 0$) are used as the first estimates of the unknowns to start the iterations in the first time step t_1. For every subsequent time t_{n+1}, the values computed in the previous time step t_n are used as starting values.

Referring to the generic system:

$$\begin{cases} a_{11}x_1 + a_{12}x_2 + \cdots + a_{1N}x_N &= b_1 \\ a_{21}x_1 + a_{22}x_2 + \cdots + a_{2N}x_N &= b_2 \\ &\cdots \\ a_{N1}x_1 + a_{N2}x_2 + \cdots + a_{NN}x_N &= b_N, \end{cases}$$

(13.87a)

the iterative process begins by rearranging each of the N equations so that the equation in row "i" of the matrix defines the ith unknown x_i explicitly:

$$\begin{cases} x_1 &= \dfrac{1}{a_{11}}\left[b_1 - (a_{12}x_2 + a_{13}x_3 + \cdots + a_{1N}x_N)\right] \\ x_2 &= \dfrac{1}{a_{22}}\left[b_2 - (a_{21}x_1 + a_{23}x_3 + \cdots + a_{2N}x_N)\right] \\ &\cdots \\ x_N &= \dfrac{1}{a_{NN}}\left[b_N - \left(a_{N1}x_1 + a_{N2}x_2 + \cdots + a_{N,N-1}x_{N-1}\right)\right]. \end{cases}$$

(13.87b)

We can write this in a compact form as:

$$x_i = \frac{1}{a_{ii}}\left[b_i - \sum_{1}^{i-1}{}_j a_{ij}x_j - \sum_{i+1}^{N}{}_j a_{ij}x_j\right].$$

(13.88)

The main methods used to solve Eq. (13.87) will be described next.

13.13.3.2   The Point Jacobi Method (Simultaneous Displacement)

In the point Jacobi method, the $(k + 1)$th iteration on x_i, written as x_i^{k+1}, is calculated by introducing into Eq. (13.88) the values of x_j^k obtained from the preceding iteration k. We get:

$$x_i^{k+1} = \frac{1}{a_{ii}}\left[b_i - \sum_{1}^{i-1}{}_j a_{ij}x_j^k - \sum_{i+1}^{N}{}_j a_{ij}x_j^k\right].$$

(13.89)

For each iteration we calculate the absolute of the mean of the residuals:

$$R^{k+1} = \frac{1}{N}\sum_{1}^{N}{}_i |R_i^{k+1}|,$$

(13.90a)

where:

$$R_i^{k+1} = a_{i1}x_1^{k+1} + a_{i2}x_2^{k+1} + \cdots + a_{iN}x_N^{k+1} - b_i.$$

(13.90b)

R_{k+1} should approach zero, without oscillation, as we converge on the exact value of x_i. The point Jacobi method tends to converge too slowly to be suited to numerical reservoir modelling.

13.13.3.3 *The Point Gauss–Seidel Method (Successive Displacement)*

Returning to Eq. (13.89), you will notice that when we calculate x_i^{k+1}, we already know all the values of x_j^{k+1} for $j < i$.

The point Gauss–Seidel method uses the values of x_j^{k+1} for $j < i$, and x_j^k for $j > i$, to calculate x_i^{k+1}.

We have:

$$x_i^{k+1} = \frac{1}{a_{ii}} \left[b_i - \sum_{1}^{i-1} {}_j a_{ij} x_j^{k+1} - \sum_{i+1}^{N} {}_j a_{ij} x_j^k \right] \tag{13.91}$$

with the convergence criterion that

$$R^{k+1} < \epsilon. \tag{13.90c}$$

This method is about twice as fast as the point Jacobi, because it needs roughly half the number of iterations. However, it is still too slow to be useful in the numerical modelling of reservoir behaviour.

13.13.3.4 *Point Successive Over-Relaxation (SOR)*

If we subtract x_i^k from both sides of Eq. (13.91) (x_i^k was calculated in the previous iteration), we have:

$$x_i^{k+1} - x_i^k = \frac{1}{a_{ii}} \left[b_i - \sum_{1}^{i-1} {}_j a_{ij} x_j^{k+1} - \sum_{i}^{N} {}_j a_{ij} x_j^k \right], \tag{13.92a}$$

where $(x_i^{k+1} - x_i^k)$ is the difference between the value of x_i in current iteration and in the previous one.

In the method of point successive over-relaxation (SOR), the term on the right of Eq. (13.92a) is multiplied by a coefficient $\omega > 1$ so as to speed up the convergence. ω is known as the *relaxation* or *iteration parameter*. We now have:

$$x_i^{k+1} = x_i^k + \omega \left\{ \frac{1}{a_{ii}} \left[b_i - \sum_{1}^{i-1} {}_j a_{ij} x_j^{k+1} - \sum_{i}^{N} {}_j a_{ij} x_j^k \right] \right\}. \tag{13.92b}$$

It can be demonstrated that the following must be true:

$$1 < \omega < 2. \tag{13.92c}$$

The point Gauss–Seidel method corresponds to the case where $\omega = 1$ (no over-relaxation). The optimum value of the iteration parameter, ω_{opt}, is the one which minimises the number of iterations required to achieve convergence. Optimising ω is critical to the efficiency of the iteration process. When $\omega < \omega_{\mathrm{opt}}$, the convergence is slower, while for $\omega > \omega_{\mathrm{opt}}$ the iterations become unstable, with x_i oscillating about the exact solution.

Equations exist[20,34] to calculate ω_{opt} for the simpler cases such as monophasic flow in a homogeneous, isotropic reservoir. For the more complex cases, it becomes a trial and error process, or one based on experience.

Equation (13.92b) is the *point SOR* algorithm, so called because the procedure is applied to each variable in turn. The method can, however, be applied simultaneously to all the variables in any line in the system of equations (line SOR or LSOR), or any two adjacent lines (2LSOR), or to a block (block SOR). You should refer to Varga's classic text[44] for a description of these methods.

Another method of over-relaxation, very widely used in reservoir modelling, is the *strongly implicit procedure* (SIP). This was proposed by Stone[37] for two-dimensional single phase systems, and extended by Weinstein, Stone and Kwan[47,48] to three-dimensional polyphasic systems.

13.13.3.5 A Simple Example of the Use of Iterative Methods

We can appreciate – albeit in a very rudimentary way – the differences in behaviour of these different iterative methods, and their respective calculation procedures, by considering a simple system of two equations in two unknowns:

$$\begin{cases} 2x_1 + x_2 = 1 \\ x_1 + 2x_2 = 5. \end{cases} \tag{13.93}$$

The solution can be obtained by any of the elementary direct methods of algebra (substitution, Cramer): $x_1 = -1$; $x_2 = 3$.

In Table 13.1 you will find the algorithms and results from successive iterations for the Jacobi and Gauss–Seidel methods. Table 13.2 illustrates the algorithm and results from the point SOR method, with different values of relaxation parameter between 1.05 and 1.2.

Table 13.1. *The solution of Eq. (13.93) by iterative methods not using over-relaxation*

Method	Jacobi			Gauss–Seidel		
Solution algorithm	$x_1^{k+1} = -0.5x_2^k + 0.5$ $x_2^{k+1} = -0.5x_1^k + 2.5$			$x_1^{k+1} = -0.5x_2^k + 0.5$ $x_2^{k+1} = -0.5x_1^{k+1} + 2.5$		
	k	x_1	x_2	k	x_1	x_2
	0	1.00000	0.00000	0	1.00000	0.00000
	1	0.50000	2.00000	1	0.50000	2.25000
	2	−0.50000	2.25000	2	−0.62500	2.81250
	3	−0.62500	2.75000	3	−0.90625	2.95313
	4	−0.87500	2.81250	4	−0.97656	2.98828
	5	−0.90625	2.93750	5	−0.99414	2.99707
	6	−0.96875	2.95313	6	−0.99854	2.99927
	7	−0.97656	2.98438	$\boxed{7}$	−0.99963	2.99982
	8	−0.99219	2.98828			
	9	−0.99414	2.99609			
	10	−0.99805	2.99707	$\square$ $\Delta < 0.1\%$ for both x_1 and x_2		
	11	−0.99854	2.99902			
	$\boxed{12}$	−0.99951	2.99927			

Table 13.2. *Solution of Eq. (13.93) by the point SOR method*

Solution algorithm

$$x_1^{k+1} = x_1^k - \omega(x_1^k + 0.5x_2^k - 0.5)$$
$$x_2^{k+1} = x_2^k - \omega(0.5x_1^{k+1} + x_2^k - 2.5)$$

$\omega = 1.05$			$\omega = 1.08$			$\omega = 1.10$			$\omega = 1.15$			$\omega = 1.20$		
k	x_1	x_2	k	x_1	x_2	k	x_1	x_2	k	x_1	x_2	k	x_1	x_2
0	1.00000	0.00000	0	1.00000	0.00000	0	1.00000	0.00000	0	1.00000	0.00000	0	1.00000	0.00000
1	0.47500	2.37563	1	0.46000	2.45160	1	0.45000	2.50250	1	0.42500	2.63063	1	0.40000	2.76000
2	−0.74596	2.89785	2	−0.82066	2.94703	2	−0.87138	2.97900	2	−1.00136	3.05619	2	−1.13600	3.12960
3	−0.95907	2.98362	3	−0.98574	2.99654	3	−1.00131	3.00282	3	−1.03210	3.01003	3	−1.05056	3.00442
4	−0.99345	2.99738	☐4	−0.99927	2.99988	4	−1.00142	3.00050	☐4	−1.00095	2.99904	4	−0.99254	2.99464
5	−0.99895	2.99958	5	−1.00000	3.00000	☐5	−1.00013	3.00002	5	−0.99931	2.99974	5	−0.99828	3.00004
☐6	−0.99983	2.99993				6	−1.00000	3.00000	6	−0.99996	3.00001	☐6	−1.00037	3.00021
7	−0.99997	2.99999							7	−1.00001	3.00000	7	−1.00005	2.99999
8	−1.00000	3.00000										8	−0.99998	2.99999
												9	−1.00000	3.00000

☐ $\Delta < 0.1\%$ for both x_1 and x_2

If we fix a convergence criterion of less than 0.1% from the exact solution, the final results that will be obtained by the three methods can be summarised as follows:

Method	Number of iterations to achieve $\Delta < 0.1\%$	Comments
Jacobi	12	
Gauss–Seidel	7	
SOR with:		
$\omega = 1.05$	6	
$\omega = 1.08$	4	
$\omega = 1.10$	5	Oscillates
$\omega = 1.15$	4	Oscillates
$\omega = 1.20$	6	Oscillates

As you will see, the most efficient method as far as computing time is concerned is the SOR with $\omega = 1.08$. If $\omega > \omega_{\mathrm{opt}}$, more iterations are in general required to satisfy the convergence criterion *for the first time*, after which the solution oscillates, only settling on the correct value after a certain number of extra iterations.

If the exact value of ω_{opt} cannot be calculated, it is better to use a value slightly higher than that estimated. This will avoid what may be a very slow asymptotic convergence on the solution, and will still satisfy the convergence criterion (perhaps by "overshooting"), normally defined as in Eqs. (13.90).

13.13.4 Methods of Alternating Directions

13.13.4.1 Introduction

The various methods of alternating directions are among the most widely used, owing to their speed and low memory requirement. The alternating direction implicit (ADI) method, a method of *direct* solution, is only sufficiently stable in bidimensional systems with a low degree of heterogeneity. The iterative alternating direction implicit (IADI) method is an *iterative* SOR approach, suitable for three-dimensional, or strongly heterogeneous two-dimensional, systems. Both ADI and IADI achieve stable solutions with reasonably sized time steps Δt.

Where heterogeneity is very pronounced, where local flow rates are high, or where very small blocks have to be used (for example, a two-dimensional radial well model), the direct methods such as the ordering of sparse matrices (Sect. 13.13.2.4), block SOR or SIP will be necessary if reasonable time steps are to be used. These are computationally very heavy in terms of time and memory requirements.

13.13.4.2 The ADI method

The flow equation for a two-dimensional system is:

$$\Delta_x T_x \Delta_x p + \Delta_y T_y \Delta_y p - \tilde{q}_v = \frac{\alpha_{i,j}}{\Delta t} \left(p_{i,j}^{n+1} - p_{i,j}^{n} \right). \tag{13.76}$$

In the ADI[30] method, each time step Δt is divided into two equal parts, $\Delta t/2$, and the values of p at time $(t_n + 1/2\Delta t)$ are denoted by $p^{n+1/2}$. In the first step (the "x-sweep"), the values of p in the terms $\Delta_y T_y \Delta_y p$ in Eq. (13.76) are kept constant at time t_n, and values of $p^{n+1/2}$ are calculated with the equation:

$$\Delta_x T_x \Delta_x p^{n+1/2} + \Delta_y T_y \Delta_y p^n - \tilde{q}_v = \frac{2\alpha_{i,j}}{\Delta t} \left(p_{i,j}^{n+1/2} - p_{i,j}^n \right). \qquad (13.94a)$$

p^n are known, having been calculated in the preceding time step at t^n, and the coefficient $2\alpha_{i,j}/\Delta t$ arises from the fact that we are using a time step $\Delta t/2$.

In the second step (the "y-sweep"), the values of p in the terms $\Delta_x T_x \Delta_x p$ in Eq. (13.76) are kept constant at time $(t_n + 1/2)$, and values of p^{n+1} are calculated with the equation:

$$\Delta_x T_x \Delta_x p^{n+1/2} + \Delta_y T_y \Delta_y p^{n+1} - \tilde{q}_v = \frac{2\alpha_{i,j}}{\Delta t} \left(p_{i,j}^{n+1} - p_{i,j}^{n+1/2} \right). \qquad (13.94b)$$

This completes the ADI procedure.

Only three unknowns appear in Eq. (13.94a): $p_{i-1,j}^{n+1/2}$, $p_{i,j}^{n+1/2}$ and $p_{i+1,j}^{n+1/2}$; and three in Eq. (13.94b): $p_{i,j-1}^{n+1}$, $p_{i,j}^{n+1}$ and $p_{i,j+1}^{n+1}$. The systems of algebraic equations corresponding to Eqs. (13.94) will therefore be tridiagonal, instead of pentadiagonal – as would be the case if Eq. (13.76) had been solved in a single step from t^n to t^{n+1}. Consequently, they can be solved with Thomas's algorithm (Sect. 13.13.2.1), which is very fast and requires little memory.

Peaceman[28] showed that in 2D the ADI method is unconditionally stable for any value of Δt, although this is not the case in 3D. Nevertheless, when significant heterogeneities are present, or T_x and T_y are very different, or production rates are high, it may prove necessary to use very small time steps to avoid instability in the form of oscillations in the pressure p (Fig. 13.11).

In such cases, and in 3D, the IADI method must be used.

13.13.4.3 The IADI Method

The IADI method[14] uses successive iterations to derive p^{n+1} from p^n computed in the preceding time step at t_n. Each *iteration* is subdivided into three successive subiterations, corresponding to "sweeps" in the x-, y- and z-directions respectively (see Sect. 13.13.4.2).

For the three sweeps, $k + 1/3$, $k + 2/3$ and $k + 1$, we proceed as follows:

- in the first subiteration, the pressure p in the terms $\Delta_y T_y \Delta_y p$ and $\Delta_z T_z \Delta_z p$ in Eq. (13.78) is held constant at p^k, which is known, and $p^{k+1/3}$ is calculated from the term $\Delta_x T_x \Delta_x p^{k+1/3}$ (the x-sweep),
- in the second subiteration, the pressure p in the term $\Delta_z T_z \Delta_z p$ is held constant at p^k; the $p^{k+1/3}$ just calculated is used for the term $\Delta_x T_x \Delta_x p^{k+1/3}$, and $p^{k+2/3}$ is calculated from the term $\Delta_y T_y \Delta_y p^{k+2/3}$ (the y-sweep),
- in the third and final subiteration, $p^{k+1/3}$ (known) is used for the term $\Delta_x T_x \Delta_x p^{k+1/3}$, $p^{k+2/3}$ (just calculated) is used for the term $\Delta_y T_y \Delta_y p^{k+2/3}$, and p^{k+1} is calculated from the term $\Delta_z T_z \Delta_z p^{k+1}$ (the z-sweep).

To speed up the convergence, an over-relaxation term is introduced into each subiteration. This performs the same function as ω in the SOR method, but is formulated differently. To illustrate this, we will look at the first subiteration, for which Eq. (13.78) becomes:

$$\Delta_x T_x \Delta_x p^{k+1/3} + \Delta_y T_y \Delta_y p^k + \Delta_z T_z \Delta_z p^k - \tilde{q}_v =$$

$$= \frac{\alpha_{i,j,k}}{\Delta t} \left(p_{i,j,k}^{k+1/3} - p_{i,j,k}^n \right). \tag{13.95a}$$

This can also be expressed as:

$$\Delta_x T_x \Delta_x p^{k+1/3} + \Delta_y T_y \Delta_y p^k + \Delta_z T_z \Delta_z p^k - \frac{\alpha_{i,j,k}}{\Delta t} p_{i,j,k}^{k+1/3}$$

$$= -\frac{\alpha_{i,j,k}}{\Delta t} \left(p_{i,j,k}^n - \tilde{q}_v \right) = -C, \tag{13.95b}$$

where:

$$C = \frac{\alpha_{i,j,k}}{\Delta t} \left(p_{i,j,k}^n - \tilde{q}_v \right). \tag{13.95c}$$

Here, C is known because $p_{i,j,k}^n$ has been calculated in the previous time step and $\tilde{q}_v$ is an external input (the well production).

In order to speed up the convergence process, we add to the right-hand side of Eq. (13.95b) a term proportional to the difference between the p calculated in the current subiteration and the p calculated at the end of the preceding *complete* iteration k for the block i,j,k.

Equation (13.95b) then becomes:

$$\Delta_x T_x \Delta_x p^{k+1/3} + \Delta_y T_y \Delta_y p^k + \Delta_z T_z \Delta_z p^k - \frac{\alpha_{i,j,k}}{\Delta t} p_{i,j,k}^{k+1/3} =$$

$$= H_k(p_{i,j,k}^{k+1/3} - p_{i,j,k}^k) - C, \tag{13.96a}$$

$$\text{where } H_k = \sigma_k(\Sigma T), \text{ with} \tag{13.96b}$$

$$\sigma_k \qquad = \text{the relaxation parameter, and}$$

$$\begin{aligned}
\Sigma T \quad &= (T_x)_{i+1/2,j,k} + (T_x)_{i-1/2,j,k} \\
&\quad + (T_y)_{i,j+1/2,k} + (T_y)_{i,j-1/2,k} \\
&\quad + (T_z)_{i,j,k+1/2} + (T_z)_{i,j,k-1/2}.
\end{aligned}$$

The relaxation parameter σ_k is varied from one complete iteration to the next in such a way as to accelerate the convergence of p^k towards the stable solution p^{n+1}.

If we rearrange the equation so that all the known terms are on the right-hand side, and the terms to be calculated are on the left, the equations describing the three subiterations (or sweeps) of the IADI method will be:

x-sweep

$$\Delta_x T_x \Delta_x p^{k+1/3} - \left(\frac{\alpha_{i,j,k}}{\Delta t} + H_k \right) p_{i,j,k}^{k+1/3}$$

$$= - \left\{ \Delta_y T_y \Delta_y p^k + \Delta_z T_z \Delta_z p^k + H_k p_{i,j,k}^k + C \right\}. \tag{13.97a}$$

y-sweep

$$\Delta_y T_y \Delta_y p^{k+2/3} - \left(\frac{\alpha_{i,j,k}}{\Delta t} + H_k \right) p_{i,j,k}^{k+2/3}$$

$$= - \left\{ \Delta_x T_x \Delta_x p^{k+1/3} + \Delta_z T_z \Delta_z p^k + H_k p_{i,j,k}^k + C \right\}. \tag{13.97b}$$

z-sweep

$$\Delta_z T_z \Delta_z p^{k+1} - \left(\frac{\alpha_{i,j,k}}{\Delta t} + H_k\right) p_{i,j,k}^{k+1}$$

$$= -\left\{\Delta_x T_x \Delta_x p^{k+1/3} + \Delta_y T_y \Delta_y p^{k+2/3} + H_k p_{i,j,k}^k + C\right\}. \tag{13.97c}$$

Only three unknown values of p appear in each of these three equations. For example, in Eq. (13.97a) we have $p_{i-1,j,k}^{k+1/3}$, $p_{i,j,k}^{k+1/3}$ and $p_{i+1,j,k}^{k+1/3}$.

The systems of algebraic equations represented by Eqs. (13.97a–c) are therefore tridiagonal, and can be solved easily using Thomas's algorithm (Sect. 13.13.2.1). Upon completion of the iterations, the correctness of the convergence of p^k to the correct value p^{n+1} can be verified by calculating the discrepancy R between the flow terms and the accumulation term at time t_{n+1}. This is done for each block:

$$R_{i,j,k} = \sum_s \Delta_s T_s \Delta_s p_{i,j,k}^{n+1} - (\tilde{q}_v^{n+1})_{i,j,k}$$

$$- \frac{\alpha_{i,j,k}}{\Delta t} \left(p_{i,j,k}^{n+1} - p_{i,j,k}^n\right) \quad (s = x, y, z), \tag{13.98a}$$

following which the absolute value of the residual for all $N_x N_y N_z$ blocks in the model can be totalled:

$$\epsilon = \sum_{i}^{N_x} \sum_{j}^{N_y} \sum_{k}^{N_z} |R_{i,j,k}|. \tag{13.98b}$$

If the convergence is satisfactory, ϵ should be very small (typically less than 0.1% of the total fluid volume of the reservoir).

13.14 Numerical Simulation of Multiphase Flow

13.14.1 The Finite Difference Equation for Multiphase Flow

The finite difference equations approximating the multiphase flow of medium-heavy oil (black oil) are derived by the same procedure as used in Sect. 13.12.1 for single phase flow. The derivation is illustrated in Exercise 13.5.

$$\sum_s \Delta_s T_{o,s}(\Delta_s p_o - \rho_o g \Delta_s D) - q_{o,sc} = \frac{V_b}{\Delta t}\Delta_t \left(\frac{\phi S_o}{B_o}\right), \tag{13.99a}$$

$$\sum_s \Delta_s T_{g,s}(\Delta_s p_g - \rho_g g \Delta_s D) + \sum_s \Delta_s R_s T_{o,s}(\Delta_s p_o - \rho_o g \Delta_s D)$$

$$- (q_{g,sc})_{tot} = \frac{V_b}{\Delta t}\Delta_t \left(\frac{\phi S_g}{B_g} + \frac{R_s \phi S_o}{B_o}\right), \tag{13.99b}$$

$$\sum_s \Delta_s T_{w,s}(\Delta_s p_w - \rho_w g \Delta_s D) - q_{w,sc} = \frac{V_b}{\Delta t}\Delta_t \left(\frac{\phi S_w}{B_w}\right), \tag{13.99c}$$

where $s = x, y, z$. $R_s =$ the solubility of gas in oil (we have ignored $R_{s,w}$, which is negligible). B_o, B_g and B_w are the formation volume factors. $q_{g,sc}$ is the total produced gas rate under standard conditions (the sum of the associated gas and free gas).

The following auxiliary equations also apply:

$$S_o + S_g + S_w = 1, \tag{13.99d}$$

$$P_{c,ow} = p_o - p_w, \tag{13.99e}$$

$$P_{c,go} = p_g - p_o. \tag{13.99f}$$

In addition: $V_b = \Delta x_i \Delta y_j \Delta z_k$

$\quad q_{l,sc} \quad$ = the rate at which fluid ($l = o, g, w$) is extracted from the ith block, measured under standard conditions,

$$\Delta_t \left(\frac{\phi S_l}{B_l} \right) = \left(\frac{\phi S_l}{B_l} \right)^{n+1} - \left(\frac{\phi S_l}{B_l} \right)^{n} \quad (l = o, g, w).$$

The notation $\Delta_s T_{l,s} \Delta_s p$ has the same significance as in Eq. (13.72), with

$$T_{l,s} = \frac{\alpha_s k_s k_{rl}}{\mu_l B_l \overline{\Delta s}} \quad (l = o, g, w), \tag{13.100}$$

where: α_s = cross-sectional area of the block (i, j, k) in the direction perpendicular to flow,

$\quad k_{rl}$ = relative permeability to the fluid l in the region between two adjacent block centres (the interblock or linking relative permeability).

There are two important points regarding the interblock relative permeability. We consider two blocks B_1 and B_2 which are adjacent in the s-direction ($s = x, y, z$), in which the fluid $l(l = o, g, w)$ is flowing from block B_1 (the upstream block) to block B_2 (the downstream block). If S_1 is the saturation of fluid l in B_1, and S_2 is its saturation in B_2, *the interblock relative permeability is taken to be at saturation S_1 (the "upstream weighted relative permeability").*

$$k_{rl} = k_{rl}(S_{\text{upstream}}). \tag{13.101a}$$

The use of any weighted average of the form:

$$k_{rl} = w k_{rl}(S_1) + (1 - w) k_{rl}(S_2) \tag{13.101b}$$

$$(0 \leq w \leq 1)$$

will always introduce errors into the results.

To illustrate this, take the simple example of a one-dimensional biphasic system (oil and water) in which the flow is from block B_1 to B_2 ($p_{B1} > p_{B2}$), with:

$\quad$ in $\quad B_1 : \quad S_o = S_{or} \quad k_{ro} = 0,$

$\quad$ in $\quad B_2 : \quad S_o > S_{or} \quad k_{ro} > 0.$

For any value of $w < 1$, Eq. (13.101b) will always give an interblock relative permeability $k_{ro} > 0$, which means that there would be a movement of *residual* oil from B_1 to B_2!

Referring to Eqs. (13.70) and (13.100), for a block (i,j,k) of dimensions Δx_i, Δy_j, Δz_k in which oil is flowing in the x-direction (horizontally) from block $(i + 1, j, k)$ to block $(i - 1, j, k)$, we have:

$$\Delta_x T_{ox} \Delta_x p = \frac{2(k_{ro})_{i+1,j,k}(k_x)_{i+1/2,j,k} \Delta y_j \Delta z_k}{\mu_o B_o (\Delta x_i + \Delta x_{i+1})} \left(p_{i+1,j,k} - p_{i,j,k} \right)$$
$$- \frac{2(k_{ro})_{i,j,k}(k_x)_{i-1/2,j,k} \Delta y_j \Delta z_k}{\mu_o B_o (\Delta x_{i-1} + \Delta x_i)} \left(p_{i,j,k} - p_{i-1,j,k} \right). \tag{13.102}$$

Todd[42] proposed an alternative solution to the problem. The interblock relative permeability is calculated by extrapolating the relative permeability gradient obtained from the values of k_{rl} existing in the two blocks immediately upstream of the interface through which the fluid l is flowing.

With reference to Fig. (13.20a), suppose there is flow from block $(i-1)$ to block $(i+1)$, because $p_{l,i-1} > p_{l,i+1}$. Then:

$$(k_{rl})_{i+1/2} = (k_{rl})_i + \left(\frac{\partial k_{rl}}{\partial x}\right)_i \frac{\Delta x_i}{2}$$

$$= (k_{rl})_i + \frac{(k_{rl})_i - (k_{rl})_{i-1}}{\Delta x_i + \Delta x_{i-1}} \Delta x_i. \tag{13.103a}$$

Todd's method achieves a significantly greater accuracy than the preceding approach which assumes that the interblock relative permeability is the same as that for the block immediately upstream. This also allows larger blocks to be used, which reduces computing time. However, the method requires a number of precautionary measures.

Referring to Figs. 13.20a, b, we consider the case of the displacement of oil by water in a one-dimensional system, with the direction of flow from left to right.

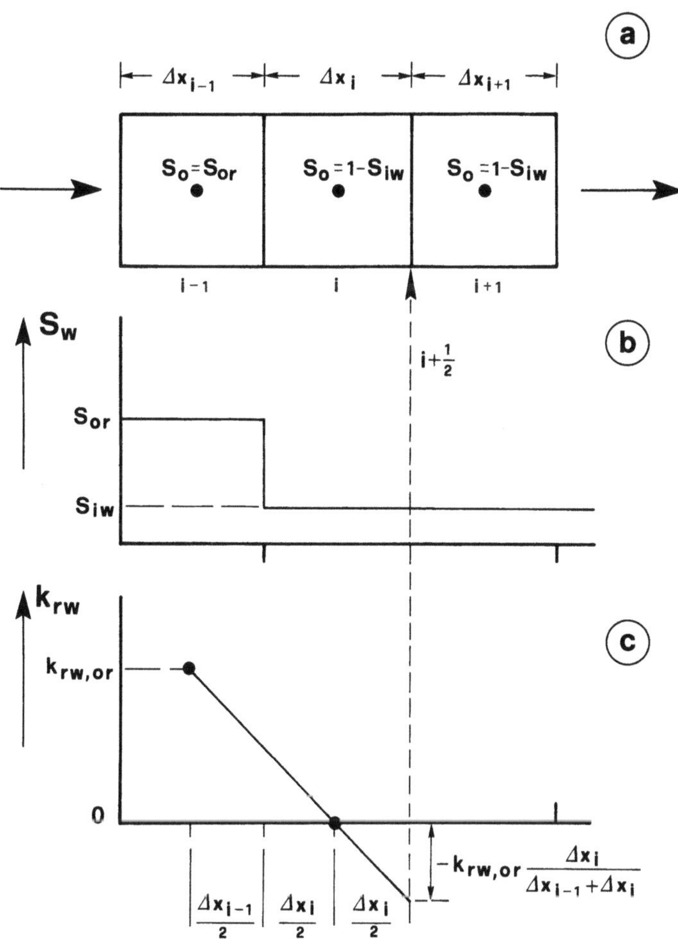

Fig. 13.20a–c. *Overshoot of the value of $(k_{rw})_{i+1/2}$ using Todd's method [Eq. (13.103a)]. (From Ref. 53. Reprinted with permission of Elsevier Science Publishers and Prof. L.P. Dake*

In block $(i-1)$, $S_o = S_{or}$, so $(k_{rw})_{i-1} = k_{rw,or}$; in block i, $S_o = 1 - S_{wi}$ and $(k_{rw})_i = 0$.

Applying Eq. (13.103a), we would calculate:

$$(k_{rw})_{i+1/2} = -k_{rw,or}\frac{\Delta x_i}{\Delta x_i + \Delta x_{i-1}}, \tag{13.103b}$$

which would be consistent with the flow of water *from block (i+1) to block i*. We can eliminate similar impossibilities by imposing the following conditions:

$(k_{rl})_{\text{interblock}} \geq 0$,

$(k_{rl})_{\text{interblock}} \leq$ the larger of the permeabilities k_{rl} of the two blocks immediately upstream and downstream of the interface.

Before going on to describe the numerical solution of the equation for polyphasic flow, we will look at the phenomenon of *numerical dispersion* (Sect. 13.11.5) and how it can be minimised by functioning the relative permeability.

13.14.2 Numerical Dispersion and Its Treatment in Multiphase Flow

A simple 1D 2P vertical oil/water system (see Sects. 13.3.2 and 13.3.3) is illustrated in Fig. 13.21. ϕ, k_z, $k_{ro}(S_o)$, $k_{rw}(S_w)$, Δz and the cross-sectional area A are the same in all blocks. If we assume the capillary transition zone to be of negligible thickness, the initial conditions $(t_o = 0)$ are: $S_w = 1$ in blocks 1 and 2, and $S_o = 1 - S_{iw}$ in blocks 3, 4, 5 and 6.

The boundary conditions are:

- water injection rate $q_{w,in}B_w = $ constant at the lower face of block 1;
- fluid production rate $q_oB_o + q_{w,out}B_w = q_{w,in}B_w$ from the upper face of block 6.

The flow is steady state, and the flow equation is therefore elliptical. By time $t_1 = t_0 + \Delta t$, a certain quantity of water will have entered block 3, and its water saturation will have changed from S_{iw} to $S_{w,3}^1 > S_{iw}$. Reading from the *rock relative permeability curves* (labelled R in Fig. 13.22), the relative permeability to water in block 3 will have increased to $k_{rw,3}^1 > 0$.

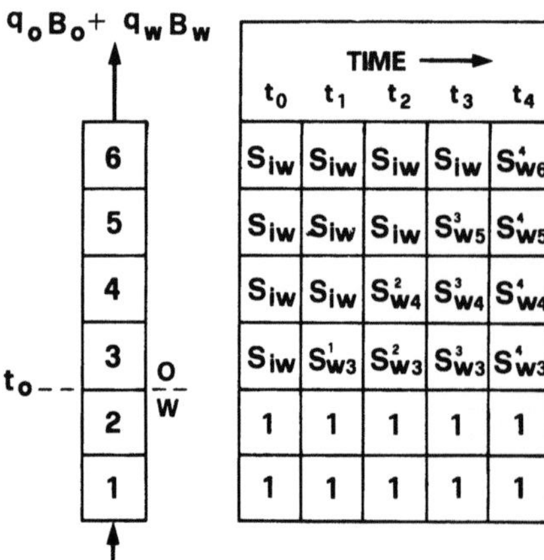

Fig. 13.21. *Variation in the water saturation with time for the model in Section 13.14.2*

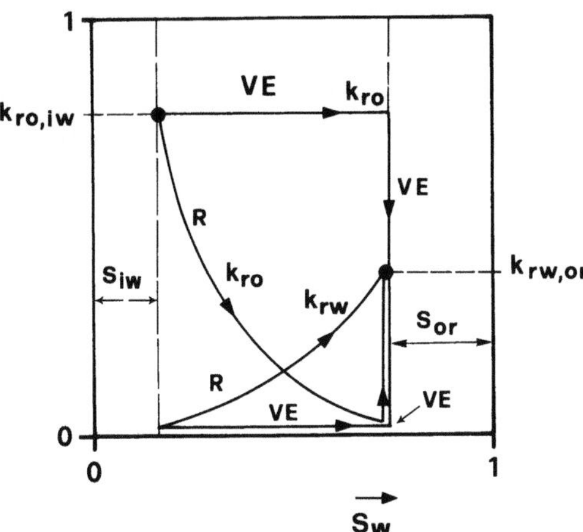

Fig. 13.22. *Oil/water relative permeability curves for dispersed flow (R) and segregated flow in the vertical direction (VE)*

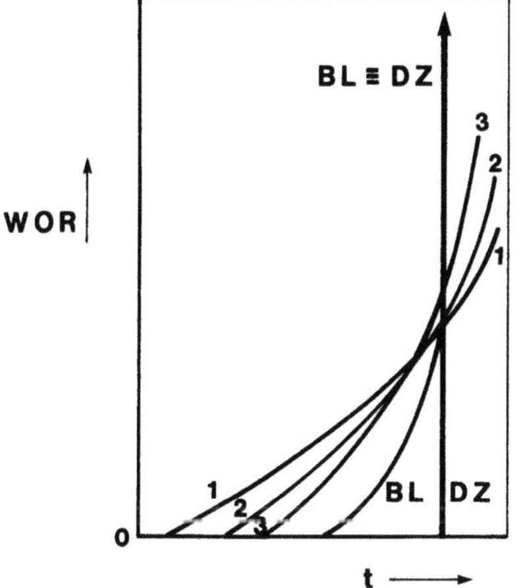

Fig. 13.23. *Variation in the WOR with time predicted by Buckley-Leverett (BL), Dietz (DZ) and some other dispersed flow hypotheses. Note that all of these cases result in the same final volume of produced oil (WOR = ∞)*

Since we decided in Sect. 13.14.1 that we would use the *upstream* values of the relative permeabilities, we use $k_{rw,3}^{1}$ for the movement of water from block 3 to block 4. At time $t_2 = t_0 + 2\Delta t$ a certain quantity of water will have flowed from block 3 to 4, and its water saturation will have increased from S_{iw} to $S_{w,4}^{2} > S_{iw}$. In addition, $k_{rw,4}^{2} > 0$.

Proceeding in this way, we can deduce that at time $t_4 = t_0 + 4\Delta t$ water will appear in block 6, and will start to be produced. From this moment, WOR $= (q_w/q_o) > 0$.

As time progresses, S_w increases in all blocks – particularly in block 6. Consequently, the WOR will increase with time in the manner of curve 1 in Fig. 13.23, until S_w reaches $1 - S_{or}$ in block 6, at which point no more oil will be produced ($S_o = S_{or}$, and WOR $= \infty$).

The way in which the numerical model predicts the behaviour of the WOR and, in fact, all the phenomena associated with displacement, is in complete contrast with the Buckley-Leverett theory (Sect. 11.3.2). However, this theory has been amply validated by experimental work on porous media.

It has been well established that the displacement of oil by water takes place with the formation of a shock front in the form of a discontinuity in S_w. [This does not include the case of the downwardly concave fractional flow curve $f_w = f_w(S_w)$

encountered when the oil is very viscous.] Water production from the exit face of block 6 only begins when the water front has completely traversed the porous medium.

Subsequently, according to the Buckley-Leverett model, the WOR increases progressively towards an infinite value. The actual curve WOR $= f(t)$ therefore looks like curve BL in Fig. 13.23.

If the process occurs in vertical equilibrium (VE) – segregated flow (piston-like displacement) (Dietz, Sect. 11.4), $S_o = S_{or}$ behind the front and $S_w = S_{iw}$ ahead of it – oil production ceases as soon as the front reaches the exit face of the porous medium, and the WOR goes to infinity (curve DZ in Fig. 13.23).

As the number of blocks in the model is increased, WOR $= f(t)$ will approximate more and more closely the real curve (curves 1 to 2 to 3 in Fig. 13.23), and will coincide exactly when $N_z = \infty$.

The discrepancy between the real curve and the one derived from a numerical model is referred to as *numerical dispersion*. This is because discretisation produces the same kind of error as would be observed if the rock characteristics (in particular, heterogeneity) were such that the water and oil became mixed (*dispersed*) instead of remaining separate (*segregated*) across the displacement front.

The physical phenomenon of dispersion is common when the two fluids are *miscible*. It was described in Sect. 12.7.2 in the context of the movement of tracers in injection water. Numerical dispersion errors can be reduced by making the grid finer. However, this results in a corresponding increase in computing time, and an alternative method, involving modification of the relative permeability curves, has been devised for vertical displacement.

If the displacement takes place in VE (Dietz's piston-like displacement), for flow in the vertical direction, z, we use the following values for the relative permeability of the *upstream block*:

$$\text{for } (1 - S_{iw}) \geq S_o > S_{or} \quad \begin{aligned} k_{ro} &= k_{ro,iw}, \\ k_{rw} &= 0. \end{aligned}$$
$$\text{for } S_o = S_{or} \quad\quad\quad\quad \begin{aligned} k_{ro} &= 0, \\ k_{rw} &= k_{rw,or}. \end{aligned}$$

This amounts to using the curves labelled "VE" in Fig. 13.22.

In this way, no water movement can occur from one block to the next until $S_w < (1 - S_{or})$, and the calculated WOR follows the curve DZ in Fig. 13.23.

The relative permeability curves "VE" in Fig. 13.22 represent the *relative permeability pseudo-curves* (Sect. 11.5) *for vertical flow* assuming a *fully segregated* flow regime (VE).

Flow parallel to the bedding plane (x- and y-directions) can be treated in a similar manner. The calculation of the pseudo-curves of relative permeability and capillary pressure was described in Sects. 11.5, 12.5.2.1 and 12.5.2.2.

In *dispersed flow*, which occurs, for example, when low permeability rocks with an extensive capillary transition zone are produced at high rates, the rock relative permeability curves themselves are used (curve R in Fig. 13.22). In this case, numerical dispersion replicates to some extent the physical dispersion caused by the presence of the transition zone. The criteria by which we can determine whether the flow is segregated or dispersed can be found in Sects. 11.4.1 and 11.4.2.

Problems arise when the flow regime is intermediate between segregated and dispersed. Common practice here is to conduct a *sensitivity study* (Sect. 13.2), to model the behaviour of the reservoir under each of the two limiting flow regimes. This provides an appreciation of the possible error associated with any prediction that might be made subsequently.

In the special case of the vertical migration ("*percolation*")[10] of gas liberated from oil below its bubble point, the segregated flow model is highly effective. The gas starts to migrate upwards ($k_{rg} > 0$) as soon as S_g exceeds the critical value S_{gc} (Sect. 3.5.2.5). It follows that, for two blocks (i, j, k) and $(i, j, k + 1)$ adjacent in the vertical direction, the vertical transmissivity $(T_{g,z})_{i,j,k+1/2}$ becomes non-zero as soon as $S_g > S_{gc}$ in block $(i, j, k + 1)$.

If the time step Δt is large, it may happen that the numerical model will calculate a volume of gas transferred from block $(i, j, k+1)$ to the overlying block (i, j, k) which is larger than the actual volume of mobile gas (or even the liberated gas) in block $(i, j, k + 1)$.

In this case:

$$\left(T_{g,z}\right)_{i,j,k+1/2}\left(p_{i,j,k+1}^{n+1} - p_{i,j,k}^{n+1} - \rho_g g\frac{\Delta z_{k+1} + \Delta z_k}{2}\right)\Delta t$$

$$> \left[\frac{V_b\phi}{B_g}(S_g - S_{gc})\right]_{i,j,k+1}^{n}$$

$$\text{or even:} \quad > \left[\frac{V_b\phi}{B_g}S_g\right]_{i,j,k+1}^{n}. \tag{13.104}$$

If Δt is large enough, the value of S_g calculated in the lower block may even be *negative*. In any case, the phenomenon accelerates rapidly so that all the liberated gas migrates to the top of the reservoir – in the model, that is, but not in reality!

The obvious cure for this is to reduce the size of the time step Δt, but this of course will increase computing time. An alternative approach[10] is to reduce the interblock transmissivity $T_{g,z}$ by means of a scaling coefficient $\beta < 1$. This is computed automatically block by block so as to honour the stability condition expressed by the equality between the two components of Eq. (13.104).

13.14.3 Solution of the Finite Difference Equation for Multiphase Flow

13.14.3.1 Choice of Solution Method

There are three types of dependent variable in the set of polyphasic flow Eqs. (13.99a–f): pressure, p; saturation, S; transmissibility, T. These parameters are all functions of time as well as position x, y, z. More specifically, the local values of p and S are directly dependent on time, while T, being a function of S (via k_{rs}) and of p [via k, μ and B, Eq. (13.45)] is indirectly dependent on time.

The solution methods for multiphase flow are classified according to the number of dependent variable types that are calculated implicitly at each time step.

First-order implicit (leap frog or IMPES) methods: only the pressures are treated implicitly and calculated at each time step t_{n+1}. Saturations and transmissibilities are frozen at time t_n, and calculated explicitly at t_{n+1} once the p^{n+1} are known.

Second-order implicit (simultaneous solution) methods: pressures and saturations are computed implicitly at t_{n+1}. Transmissibilities are frozen at time t_n, and calculated explicitly at t_{n+1} once the p^{n+1} and S^{n+1} are known.

Third-order implicit (fully implicit formulation) methods: all three dependent variable types are computed implicitly at t_{n+1}. The solution algorithm incorporates the relationships linking T to S and p.

The higher the order of the method, the more stable it tends to be, so that longer time steps Δt can be used without impeding convergence to the real solution. However, this is traded off against an increase in computing time and memory requirement. To counter this problem, iterative procedures have been devised to solve for S and T which, when applied to first- or second-order methods, effectively upgrade their performance to that of the second or third order respectively.

13.14.3.2 *The IMPES (IMplicit Pressure/Explicit Saturation) Method*

In the IMPES methods, the flow Eqs. (13.99a–f) are first rearranged. The derivative of the saturation with respect to time is eliminated using Eq. (13.99d), and two of the pressures p_o, p_g, p_w are eliminated using Eqs. (13.99e and f).

Suppose that p_w is the pressure remaining. The resulting equation is solved for p_w^{n+1} *with the coefficients from time* t_n. The relative permeabilities and capillary pressures which, with other parameters, appear in the coefficients, are computed *explicitly* using the saturations at t_n. Having solved for p_w^{n+1}, we then calculate the saturations, the capillary pressures, and p_o^{n+1} and p_g^{n+1} *explicitly* at t_{n+1}.

The first implicit-explicit method was the "leap-frog"[15]. This did not achieve widespread usage because it was superseded a year later by the IMPES method proposed in 1960 by Sheldon, Harris and Bavly[35]. This in turn was improved by Stone and Garder[38] in 1961.

These are all *first-order implicit* methods, because only one dependent variable type – pressure – is solved implicitly while the other two [the saturations and the coefficients (which are functions of p and S)] are held at their values at t_n. They are then solved explicitly. To illustrate the algebraic reduction of the set of polyphasic flow equations to a single equation in p we will look at the simple case of a two-phase oil/water system.

In Eq. (13.38a), which is a differential equation describing the flow of oil, the right-hand term can be expanded as:

$$\frac{\partial}{\partial t}\left(\phi\frac{S_o}{B_o}\right) = \frac{\phi}{B_o}\frac{\partial S_o}{\partial t} + \frac{S_o}{B_o}\frac{\mathrm{d}\phi}{\mathrm{d}p_o}\frac{\partial p_o}{\partial t} - \frac{\phi S_o}{B_o^2}\frac{\mathrm{d}B_o}{\mathrm{d}p_o}\frac{\partial p_o}{\partial t}$$

$$= \frac{\phi}{B_o}\left[\frac{\partial S_o}{\partial t} + S_o\left(\frac{1}{\phi}\frac{\mathrm{d}\phi}{\mathrm{d}p_o}\right)\frac{\partial p_o}{\partial t} + S_o\left(-\frac{1}{B_o}\frac{\mathrm{d}B_o}{\mathrm{d}p_o}\right)\frac{\partial p_o}{\partial t}\right]. \quad (13.105)$$

Substituting from Eqs. (2.15), (3.8c) and (13.99e):

$$B_o\frac{\partial}{\partial t}\left(\phi\frac{S_o}{B_o}\right) = \phi\left[\frac{\partial S_o}{\partial t} + S_o\left(c_f + c_o\right)\frac{\partial}{\partial t}\left(p_w + P_{c,ow}\right)\right]. \quad (13.106a)$$

Similarly, the right-hand term in Eq. (13.38c) can be written as:

$$B_w\frac{\partial}{\partial t}\left(\phi\frac{S_w}{B_w}\right) = \phi\left[\frac{\partial S_w}{\partial t} + S_w\left(c_f + c_w\right)\frac{\partial p_w}{\partial t}\right]. \quad (13.106b)$$

Summing the terms in Eq. (106a and b), and ignoring $\partial P_{c,ow}/\partial t$, which is usually negligibly small unless there are rapid changes of saturation with time, we have:

$$B_o \frac{\partial}{\partial t}\left(\phi \frac{S_o}{B_o}\right) + B_w \frac{\partial}{\partial t}\left(\phi \frac{S_w}{B_w}\right)$$

$$= \phi\,(c_f + c_o S_o + c_w S_w)\,\frac{\partial p_w}{\partial t}, \qquad (13.107)$$

since:

$$\frac{\partial}{\partial t}(S_o + S_w) = 0.$$

We now multiply both sides of Eq. (13.99a) by B_o, and (13.99c) by B_w. Adding the equations together term by term we obtain, with Eq. (13.107):

$$B_o \sum_s \Delta_s T^n_{o,s}(\Delta_s p^{n+1}_w + \Delta_s P^n_{c,ow} - \rho^n_o g \Delta_s D)$$

$$+ B_w \sum_s \Delta_s T^n_{w,s}(\Delta_s p^{n+1}_w - \rho^n_w g \Delta_s D) - q_{o,sc} B_o - q_{w,sc} B_w$$

$$= \frac{V_b \phi}{\Delta t}\,[c_f + S_o c_o + S_w c_w]^n\left(p^{n+1}_w - p^n_w\right). \qquad (13.108)$$

All coefficients are evaluated at t_n, where their values are known.

Equation (13.108) is solved in the same way as the single phase flow equation described in Sect. 13.13. Once values of p^{n+1}_w have been computed, the *explicit* calculation of S^{n+1}_w and S^{n+1}_o (and hence p^{n+1}_o) follows, using Eqs. (13.99c, d and e) respectively.

The IMPES method becomes unstable when there are rapid changes of saturation with time in several parts of the reservoir (around the wells, for example).

Consider the simple case of two contiguous blocks, i and $i + 1$, in a model where water is being injected so as to flow from (i) to $(i + 1)$ (Fig. 13.24).

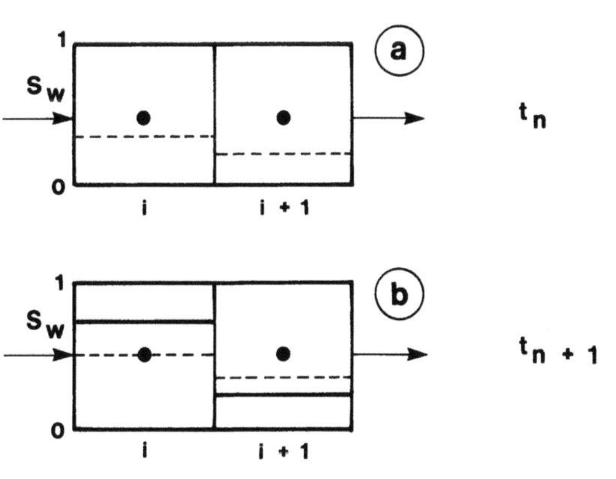

Fig. 13.24a–c. *Illustration of the instability in the computed water saturations caused by the use of transmissibilities calculated explicitly. In each case, the dashed line is the correct solution, and the solid line the calculated value. (From Ref. 53. Reprinted with permission of Elsevier Science Publishers and of Prof. L.P. Dake)*

The true values of S_w are indicated by the dashed lines in Fig. 13.24, and the solid lines represent the calculated values.

At time t_{n+1}, since the value we use for $S_{w,i}$ is $S_{w,i}^n < S_{w,i}^{n+1}$, the corresponding transmissibility $T_{w,i+1/2}^n$ will be less than the real value. Too little water will therefore flow from (i) to $(i + 1)$ in the time interval $(t_{n+1} - t_n)$, so that $S_{w,i}^{n+1}$ will be too high in block (i), as illustrated in Fig. 13.24b. Consequently, $T_{w,i+1/2}^{n+1}$ will be too large.

In the next time step $\Delta t = t_{n+2} - t_{n+1}$, therefore, too great a volume of water will be calculated as flowing from (i) to $(i + 1)$, leading to an $S_{w,i}^{n+2}$ which is too small, as in Fig. 13.24c. This results in an oscillation in the local value of S_w (and S_o). The effects of this may be transmitted to neighbouring blocks, making the whole model unstable.

Two very different techniques are used to reduce this instability:

 – the iterative IMPES method (IMPESIT),
 – semi-implicit treatment of the transmissibility.

The iterative IMPES method employs an iterative procedure (Sect. 13.13.3) to solve the system of $N_x N_y N_z$ linear equations represented by Eq. (13.108). S_o^k, S_g^k and S_w^k are calculated explicitly in each iteration (k). From them, k_{ro}^k, k_{rg}^k and k_{rw}^k, and hence $T_{s,o}^k$, $T_{s,g}^k$ and $T_{s,w}^k$, are determined for use in the next iteration $(k + 1)$. Consequently, when convergence on p_w^{n+1} is attained, the computed transmissibilities correspond to the values of S_o, S_g and S_w at the last-but-one iteration, which differ very little from S_o^{n+1}, S_g^{n+1} and S_w^{n+1}. The capillary pressures $P_{c,wo}$ and $P_{c,go}$ can be treated in the same way as the relative permeabilities. This method, the *Newtonian iteration method*, is used in many other areas for the numerical solution of pseudo-linear differential equations (i.e. with non-linear coefficients).

For the semi-implicit treatment of the transmissibility[25-27], we refer to Eq. (13.100):

$$T_{l,s} = \frac{\alpha_s}{\Delta S} \left(\frac{k_s}{\mu_l B_l} \right)_p (k_{rl})_{S_l} . \tag{13.109}$$

This can be expanded as a Taylor series, as far as the first term:

$$T_{l,s}^{k+1} = T_{l,s}^k + \frac{\alpha_s}{\Delta s} \left(\frac{k_s}{\mu_l B_l} \right)_{p^k} \left(\frac{dk_{rl}}{dS_l} \right)_{S_l^k} (S_l^{k+1} - S_l^k)$$

$$+ \frac{\alpha_s}{\Delta s} (k_{rl})_{S_l^k} \left[\frac{d}{dp} \left(\frac{k_s}{\mu_l B_l} \right) \right]_{p^k} (p^{k+1} - p^k) + O(\Delta S_l^2, \Delta p^2). \tag{13.110a}$$

Since $(k_s / \mu_l B_l)$ hardly varies with pressure, the last term can be omitted. Therefore:

$$T_{l,s}^{k+1} = T_{l,s}^k + \frac{\alpha_s}{\Delta s} \left(\frac{k_s}{\mu_l B_l} \right)_{p^k} \left(\frac{dk_{rl}}{dS_l} \right)_{S_l^k} \delta S_l, \tag{13.110b}$$

where the derivative with respect to S_l is evaluated numerically, and δS_l is chosen to be slightly larger than the maximum expected value of $(S_l^{k+1} - S_l^k)$.

This method, too, produces transmissivities that are only very slightly behind the true value at t_{n+1}. Although both the IMPESIT and semi-implicit methods achieve stable solutions, even when there are strong heterogeneities and high flow rates, they are not unconditionally stable. The semi-implicit method converges with a relatively small number of iterations, but computation time is longer per iteration owing to the complexity of the determination of the coefficients.

Whichever solution method is used, each phase should be checked for mass balance (oil, gas and water, under standard conditions) at the end of each time step. As an example, we can examine the mass balance of the oil phase under standard conditions.

Equation (13.99a), which describes the conservation of mass of degassed oil (Sect. 13.8) for a general block (i, j, k), can be written as follows:

$$R_{i,j,k} = \left[\frac{\frac{V_b}{\Delta t} \Delta_t \left(\frac{\phi S_o}{B_o} \right)}{\sum_s \Delta_s T_{o,s}(\Delta_s p_o - \rho_o g \Delta_s D) - q_{o,sc}} \right]_{i,j,k} - 1 = 0.$$

(13.111a)

The criterion for mass balance in (i, j, k) will be that:

$$|R_{i,j,k}| < \epsilon_1,$$

(13.111b)

where ϵ_1 is a very small, arbitrary, value.

Summing over all $N_x N_y N_z$ blocks in the model, and employing the notation:

$$\sum_N = \sum_i^{N_x} \sum_j^{N_y} \sum_k^{N_z}$$

we have:

$$R_T = \frac{\sum_N \left[\frac{V_b}{\Delta t} \Delta_t \left(\frac{\phi S_o}{B_o} \right) \right]}{\sum_N \left[\sum_s \Delta_s T_{o,s} (\Delta_s p_o - \rho_o g \Delta_s D) \right] - \sum_N q_{o,sc}} - 1 = 0.$$

(13.112a)

But, by definition:

$$\sum_N \left[\sum_s \Delta_s T_{o,s} (\Delta_s p_o - \rho_o g \Delta_s D) \right] = 0,$$

(13.112b)

since, for each block, the mass flow rate entering through one face is equal and of opposite sign to the flow rate leaving an adjacent block *through the same face*.

Therefore:

$$|R_T| = \left| \frac{\sum_N \left[\frac{V_b}{\Delta t} \Delta_t \left(\frac{\phi S_o}{B_o} \right) \right] - \sum_N q_{o,sc}}{\sum_N q_{o,sc}} \right|.$$

(13.113)

If global mass balance has been satisfied for degasified oil throughout the whole reservoir, then, the following condition must be true:

$$|R_T| < \epsilon_2,$$

(13.114)

where ϵ_2 is a very small, arbitrary, value (typically < 0.001).

Equations analogous to Eqs. (13.111a) and (13.113) can be derived for gas and for water. The mass balance criteria expressed by Eqs. (13.111b) and (13.114) also apply for these phases. Time steps and areas of the model where instability is developing can be identified by displaying iso-ϵ_1 contours and the curve $\epsilon_2(t)$ on the computer screen or hard copy. It is also useful to output iso-S_o, iso-S_g and iso-S_w maps at selected times for the same purpose.

13.14.3.3   The Method of Simultaneous Solution

In the method of simultaneous solution[15], the three Eqs. (13.99) (for a three component system), or the appropriate pair of equations for a two component system, are solved simultaneously, using the values of the *coefficients* determined in the preceding time step. This makes the method *second-order implicit*: the saturations and pressures are in fact calculated implicitly, while the coefficients are kept explicit.

In a three-component system there are six unknowns: p_o, p_g, p_w, and S_o, S_g and S_w. Three of these can be eliminated by substituting from the auxiliary Eqs. (13.99d, e and f), as explained in Sect. 13.10.

The system of $3N_x N_y N_z$ linear algebraic equations is usually solved for:

- p_o, p_g, p_w: this requires reliable capillary pressure curves for $P_{c,ow}$ and $P_{c,go}$ so that good estimates of S_g and S_o [and S_w via Eq. (13.99d)] can be obtained.
- one pressure (usually p_o) and two saturations (usually S_g and S_w): the explicit calculation of the other two pressures is of no particular practical interest, so accurate capillary pressure data is not necessary.

The four unknowns in a two-component system can be reduced to 2 using the auxiliary Eqs. (13.99d and e) (or f).

Again, there is a choice of unknowns to solve for:

- the two pressures: for example, in an oil/water system, we have:

$$S_w = P_{c,ow}^{-1}(p_o - p_w) \tag{13.43c}$$

and a reliable capillary pressure curve $P_{c,ow} = f(S_w)$ is essential.
- one pressure and one saturation: the other saturation can be calculated from Eq. (13.99d). An accurate capillary pressure curve is therefore not necessary.

The resulting system of algebraic equations has a coefficient matrix of the tridiagonal block type. Solution by direct methods (factorisation, matrix ordering; Sect. 13.13.2) or iteration (IADI, block SOR, SIP; Sect. 13.13.3) becomes very heavy when there are more than a few thousand blocks in the model, even using vector or parallel processing methods.

As for the IMPES method, every time step should be checked for mass balance in each block and in the entire model [Eqs. (13.111b) and (13.114)].

Although more stable than IMPES, the method of simultaneous solution is not unconditionally stable and may require very small time steps to converge if there are strong heterogeneities or sudden changes in flow rate (typically, the start-up of production from the reservoir or from a group of wells).

To enhance stability, an iterative procedure with recalculation of the coefficients at each iteration, or the semi-implicit method for the calculation of the transmissivities, can be incorporated in the simultaneous solution method. These have been described in Sect. 13.14.3.2 on IMPESIT. This makes it slightly less than *third-order implicit*, the transmissibilities being updated to the last-but-one iteration.

13.14.3.4   The Fully Implicit Method

In the fully implicit formulation[2], the pressure and saturation variables, and the coefficients – transmissibility and capillary pressure – that depend on them, are all solved implicitly. They are computed simultaneously at time t_{n+1} from the values that were determined at t_n.

This is therefore a *third order implicit* method.

T^{n+1} and P_c^{n+1} can be expressed by expanding T and P_c as Taylor's series as far as the first term:

$$T^{n+1} = T^n + \Delta T = T^n + \left(\frac{\partial T}{\partial S}\right)^n \Delta_t S + \mathrm{O}(\Delta S^2), \tag{13.115a}$$

$$P_c^{n+1} = P_c^n + \Delta P_c = P_c^n + \left(\frac{dP_c}{dS}\right)^n \Delta_t S + \mathrm{O}(\Delta S^2). \tag{13.115b}$$

In Eq. (11.115a) the dependence of T on pressure has been ignored, in view of the negligible variation of k, μ and B between p^n and p^{n+1}.

When Eqs. (115a and b) are substituted into the flow Eqs. (13.99), they become non-linear in the unknowns, because of the $\Delta_s p \times \Delta_t S$ term.

This non-linearity disappears if we assume second-order terms to be negligibly small (see Exercise 13.6), and the flow Eqs. (13.99) become *fully linear* because the coefficients now no longer depend on the unknowns.

The number of variables can be reduced to 3 by means of the auxiliary Eqs. (13.99d, e and f) (Sect. 13.14.1): the set of dependent variables chosen is generally (p_o, p_g, p_w) or (p_o, S_w, S_g).

The resulting system of $3N$ equations in $3N$ unknowns (N being the number of blocks $N_x N_y N_z$) can be solved by direct methods (matrix ordering and factorisation) or by iteration (block SOR or SIP).

Each time step takes about five to ten times as long as the IMPES method, but because of its higher stability, the fully implicit method can be run with longer, and therefore fewer, time steps.

It is particularly useful for modelling wells [a two-dimensional (r, z) coning study, for instance], where correct simulation of the flow behaviour requires the mesh close to the well to be fine. The use of very small blocks renders the first and second-order implicit methods unstable, even with small time steps.

Generally speaking, in a typical numerical reservoir model certain areas of the reservoir, where there are strong heterogeneities or high flow rates, will require a fully implicit solution, while other "quieter" areas could be handled satisfactorily using IMPES or IMPESIT methods (Sect. 13.14.3.2). The use of a fully implicit solver for the entire model would, in this case, be a clear example of "overkill".

The adaptive implicit method (AIM)[40] treats the reservoir on an area by area basis, and automatically selects the lowest order implicit method that can solve the local situation in each area. This is done for each time step. The fully implicit method is only used when necessary. AIM is the most efficient method available today for the numerical simulation of reservoir behaviour. Among its subroutines are all the implicit solution methods for polyphasic flow, as well as the most powerful direct and iterative procedures for the solution of systems of algebraic equations.

13.15 Introduction to the Study of Reservoir Behaviour Through Numerical Modelling

To illustrate the methodology behind the study of reservoir behaviour by numerical modelling, we will take the case of a black oil model (Sect. 13.3.3). This uses most of the techniques employed in numerical modelling.

We require a detailed *geological model* which provides the overall reservoir geometry, plus accurate definition of the *internal structure of the reservoir* (zonation, areal permeability and porosity variations, faults, fluid contacts, etc.) and the extent and characteristics of the underlying or flanking aquifer if present.

The construction of the geological model itself will not be described. This is based on *well data* (logs, cores), structural and stratigraphic interpretations of *3D seismic data*, the study of *reservoir rock outcrops* (if any), and the analysis of all available *paleogeographic, sedimentological, tectonic* and *diagenetic* data from the basin and the reservoir region. Of course, the properties of the rock between wells cannot be measured directly from logs or cores. They must be inferred from indirect sources such as stratigraphic interpretations of 3D seismic surveys, or analysis of the lithology and petrophysical characteristics of surface outcrops of the same formation.

There is, consequently, a high degree of uncertainty about the interwell zones (continuity of different layers, spatial distribution of porosity, permeability, initial water saturation, etc). The description of the internal structure of the reservoir requires a probabilistic treatment.

The reservoir rock contains a well-defined and unique distribution of lithological, petrophysical and fluid dynamic properties, and the reservoir model should consist of a deterministic description of these terms. However, because knowledge of the rock characteristics between wells is poor, the description must be probabilistic.

Several schools of thought exist, and development of suitable techniques is being actively pursued. Currently, the most widely used approach is the so-called *"two-stage stochastic modelling of reservoir heterogeneities"*[54,59,63].

In stage 1 of the process, the *large-scale* heterogeneities associated with different geological bodies (facies) are modelled either *deterministically* [if the density and quality of the known ("hard") data is sufficient to define the spatial continuity and extent of the different sedimentary bodies or flow units], or by a *discrete stochastic model.*

Marked point processes[55], Markov fields[56], truncated random functions[58] and two-point histograms[57] are the most common methods for generating the discrete stochastic model, appropriately conditioned so as to honour any hard information from wells, outcrops and 3D seismics. Simple layer-cake architecture can usually be handled deterministically, while jigsaw and labyrinth[65] reservoir types can only be modelled stochastically.

In stage 2, the spatial variations of the petrophysical properties *within* each sedimentary body or flow unit are modelled using a stochastic model of continuous type. Indicator kriging[61], Gaussian random fields[62] and fractals[60] are the most common methods used for estimating parameter distributions in continuous models. Recent examples of the application of two-stage stochastic modelling have been presented by Alabert and Massonat[51], Damsleth et al.[54] and Rudkiewicz et al.[64]

When a stochastic approach is used in building the geological model of a reservoir, each "realisation" (i.e. each distribution of reservoir parameters obtained by repeatedly applying the stochastic model to the simulation of reservoir architecture) has, by definition, an equal probability of reproducing the actual internal structure of the reservoir.

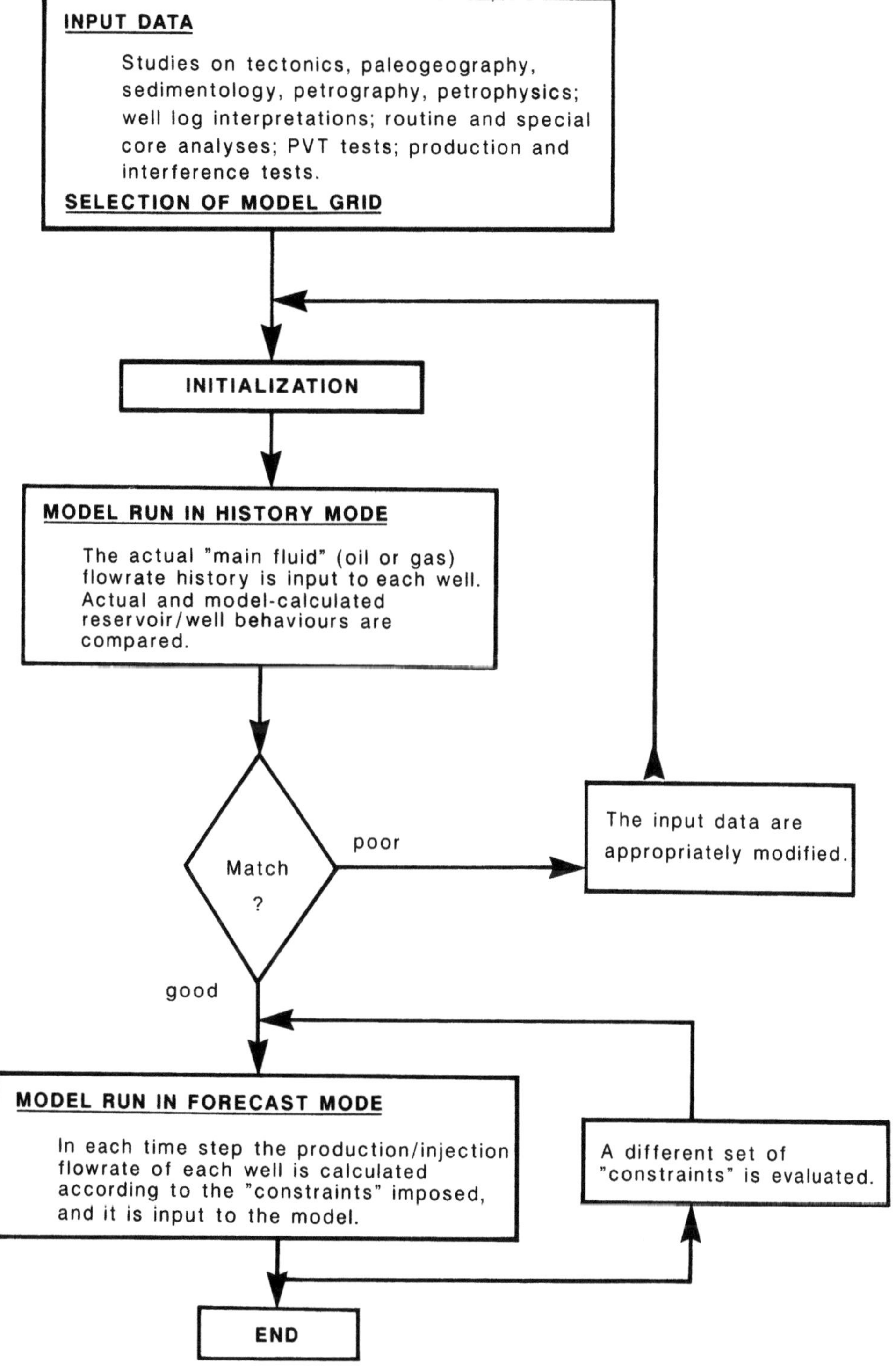

Fig. 13.25. *Flow chart for a numerical modelling study*

Unfortunately, when such "equiprobable" geological models are used in forecasting the production performance of the reservoir (via numerical simulation models), differing results are usually obtained. Although this provides an assessment of the indeterminacy of the predicted reservoir performance, the problem still remains of how to select the most probable prediction – the one on which investment and general economic decisions on field development and exploitation will be based.

This is why a maximum of hard information (well, outcrop and seismic) must be gathered during the life of the field.[52] Data acquisition should be continued until the amount and quality of information available are sufficient to at least make stage 1 of the two-stage process completely deterministic.

The numerical modelling study is performed in four stages, as illustrated in Fig. 13.25:

- gathering, editing and preprocessing of basic data,
- initialisation of the model,
- validation of the model by history matching the past performance of the reservoir (if already in production),
- forecasting the future behaviour of the reservoir, with sensitivity studies on critical parameters, over their ranges of uncertainty.

Each stage will now be described in some detail.

13.16 Gathering, Editing and Preprocessing the Basic Data

13.16.1 Choice of Gridding for Reservoir Discretisation

Part of the preprocessing phase involves assigning a value for each of the basic parameters to each individual block. A decision must therefore be made during the data gathering phase about the gridding to be used in the model (Sect. 13.3.1).

Discretisation in the (x, y) plane should incorporate a finer mesh in regions where the fluid flow needs to be modelled with a greater degree of detail, such as where there is a high concentration of wells, or a boundary. Coarse gridding can be used within the aquifer. This is illustrated in Fig. 13.26a.

Once the gridding has been configured, the grid-block positions are defined by the coordinate pair (i, j) in the x- and y-axis directions. The $N_x N_y$ blocks have dimensions $\Delta x_1, \Delta x_2, \ldots, \Delta x_{Nx}$ and $\Delta y_1, \Delta y_2, \ldots, \Delta y_{Ny}$ in the (x, y) plane.

In the z-direction, each zone (sedimentary unit) is subdivided vertically into a certain number of blocks, depending on its thickness and the amount of detail required. The vertical position of a block [or plane of blocks (x,y)] is defined as k, starting with $k = 1$ at the top of the reservoir.

By using a curvilinear (x, y) coordinate system it is possible to make the successive surfaces $k = 1, 2, \ldots, N_z$ conform to the upper and lower boundaries of each zone, as shown in Fig. 13.26b.

"Dead" blocks – parts of the mesh which lie outside the physical reservoir and aquifer – are flagged with a suitable indicator. Some situations may require grid blocks to be assigned a number of special indicators so that they may be distinguished from others or grouped together for special purposes. Examples of this are a reservoir divided into concessions or leases belonging to different operators, or where there are several gathering centres on the same reservoir: the fluid produced per lease or gathering centre can then be computed automatically.

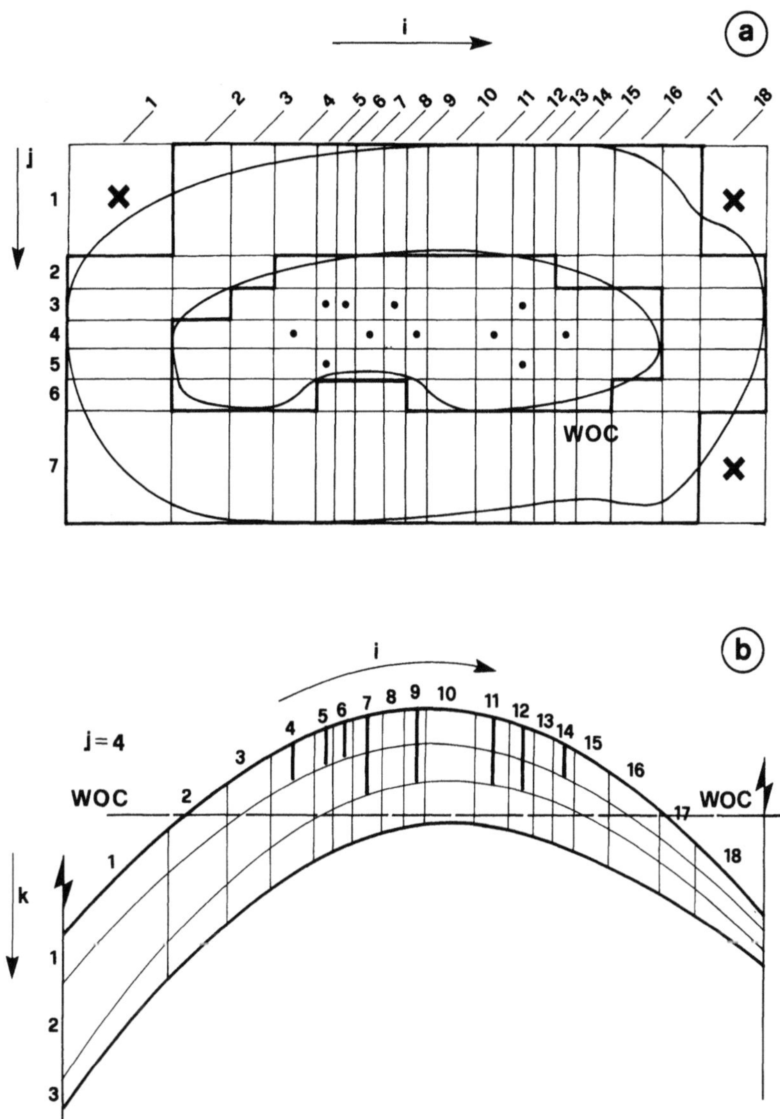

Fig. 13.26 a,b. *Schematic of the three-dimensional discretisation of a reservoir and aquifer*

13.16.2 Gross and Net Pay Thickness for Each Layer

The geological model provides the isobath map of the top of each layer. The map of the bottom of a layer, of course, corresponds to the top of the layer below. If not already generated by computer, these maps can be digitised using an optical scanner. The data preprocessor will then automatically compute the iso-thickness maps for each layer and, interpolating between contours, assign the dimension Δz to each grid block. At this stage, we are dealing with the *gross* pay thickness (Sect. 4.3.3). In order to calculate the pore volumes and transmissibilities, we also need the *net* pay thickness.

The geological model will provide iso-value maps of the net/gross ratio of each layer, based on core and log analysis. These maps are digitised if necessary, and the preprocessor assigns a net pay value to each grid block (gross thickness $\Delta z \times$ local net/gross ratio).

13.16.3 Porosity and Pore Compressibility

The analysis of cores provides values of porosity under laboratory conditions. These have to be corrected to initial reservoir pressure using the pore compressibility c_f, also measured from cores (Sect. 3.4.3). Generally speaking, relatively few wells in the field are cored, so it is essential to integrate core porosity data with log-derived porosities (Sect. 3.3).

Porosity logs such as the CNL, FDC and BHC (Sects. 3.2 and 8.5) are calibrated against core porosities where available, so that estimates of porosity are available in all wells. The arithmetic mean porosity *with respect to the net pay thickness* is then calculated for each layer in each well. Iso-porosity curves are then constructed for each layer over the entire reservoir + aquifer and digitised if necessary. The preprocessor can then assign a mean porosity value to each grid block. The pore compressibility c_f in each layer is the arithmetic mean derived from core analysis.

13.16.4 Horizontal Permeability

Information about the horizontal permeability (or, more specifically, the permeability parallel to the bedding plane) can only be obtained from cores (Sect. 3.5.1) and from flowing and buildup well tests (Chaps. 6 and 7). Core analysis yields the *absolute* permeability (Sect. 3.5.1.2) of a plug, and is a very localised measurement. A well test provides a value of the *effective* permeability (Sect. 3.5.2.2) averaged over the full local pay thickness and a relatively large area.

To fill in the missing information in wells that were not cored, a correlation is sought between well logs and all available core data in a given layer (Sect. 3.5.1.8) – porosity from logs $= f(\log k_h)$. This will also show whether or not the layers as defined are statistically homogeneous (Sect. 3.6.1).

Once established, these permeability-porosity correlations can be used to derive the point k_h values from porosity logs in the uncored wells. Next, the mean permeability $\bar{k}_h$ is computed in each layer in each well, as the geometric mean of the point k_h values. These mean values must then be compared with the effective permeabilities obtained from well tests. If there are significant differences (as there might be in a microfractured formation, for example), correction factors are applied to $\bar{k}_h$ so as to match the local well-test values. The degree of correction required can, of course, vary in different areas of the reservoir.

We have so far assumed that the reservoir rock is isotropic in the plane of sedimentation – i.e. $k_x = k_y = k_h$. If interwell interference tests (Sect. 6.11) or tracer tests (Sect. 12.7.2) provide clear quantitative evidence of the existence of anisotropy (K_{ani}) in the permeability, the local values of $\bar{k}_x$ and $\bar{k}_y$ can be evaluated in the following manner.

By definition:

$$K_{ani} = \frac{\bar{k}_y}{\bar{k}_x}. \tag{13.116a}$$

Therefore:

$$\bar{k}_x = \frac{2\bar{k}_h}{1 + K_{ani}} \tag{13.116b}$$

and

$$\bar{k}_y = \frac{2K_{\text{ani}}\bar{k}_{\text{h}}}{1 + K_{\text{ani}}}. \tag{13.116c}$$

An iso-perm map can now be drawn for each layer, using $\bar{k}_{\text{h}}$, or $\bar{k}_x$ and $\bar{k}_y$ separately. Finally, after digitising and passing through the preprocessor, values of k_{h}, or k_x and k_y, are assigned to the grid blocks.

13.16.5 Vertical Permeability

In a layer which is statistically homogeneous, the vertical permeability k_v for each block is derived from core data processed in the same way as k_{h} in the previous section. If a layer consists of several strata, each with a thickness h_i and vertical permeability $k_{v,i}$ (*all non-zero*), the block average vertical permeability k_v is given by the *harmonic mean* of the set of $k_{v,i}$. This was demonstrated in Exercise 3.5.

$$\bar{k}_v = \frac{\sum\limits_{1}^{n} {}_i h_i}{\sum\limits_{1}^{n} {}_i \dfrac{h_i}{k_{v,i}}}. \tag{13.117}$$

The situation where there are statistically distributed *impermeable* intercalations[9] ($k_v = 0$) in a layer requires special consideration. This is typical of clastic formations consisting of alternating sands and shales. The statistical distribution of the lateral extent of a shale stratum depends on the energy of deposition, as illustrated in the classic diagram[46] in Fig. 13.27.

The presence of non-continuous impermeable beds, distributed statistically in terms of size and position, will obviously impede vertical fluid movement by imposing a tortuous flow path. Consequently, k_v is reduced (Fig. 13.28).

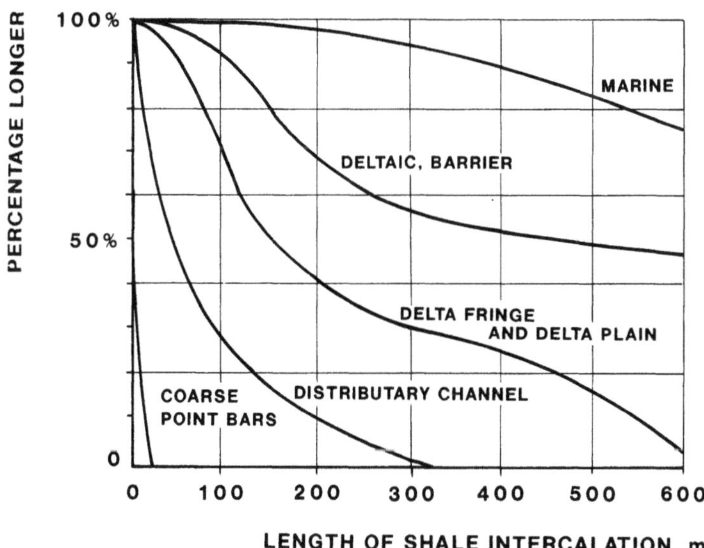

Fig. 13.27. *Probabilistic distribution of the continuity of shale and silt intercalations, as a function of the depositional environment. (From Ref. 46, 1982, Society of Petroleum Engineers of AIME. Reprinted with permission of the SPE)*

Fig. 13.28. *The influence of shale lenses on the flow path. (From Ref. 4, 1987, Society of Petroleum Engineers of AIME. Reprinted with permission of the SPE)*

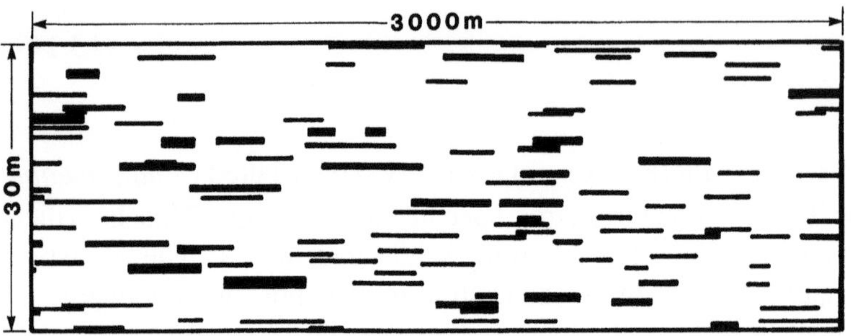

Fig. 13.29. *Distribution of shale lenses in a reservoir rock, generated by the stochastic method of Haldorsen[21]*

It is extremely important to use the correct value of k_v in the numerical model in order to simulate correctly vertical displacement (bottom water drive, gas cap expansion), coning and the vertical migration of gas liberated from oil below the bubble point.

Haldorsen[21,22] has developed a method for generating the statistical distribution of impermeable lenses (Fig. 13.29) using the evidence of their vertical distribution available from cores and well logs. The method generates the dimensions and positions of the lenses, and honours the constraints at the well bore by reproducing the observed distributions. The complete code is presented in Haldorsen's Ph.D. thesis[21] submitted at the University of Texas in Austin.

Using a three-dimensional single phase numerical mini-model[7] which incorporates the statistical distribution of the shale beds, the three components of the fluid velocity u can be calculated[3,4] each corresponding to the same potential gradient along each axis.

$$\frac{\partial \Phi}{\partial x} = \frac{\partial \Phi}{\partial y} = \frac{\partial \Phi}{\partial z} = \frac{\partial \Phi}{\partial s}. \tag{13.118a}$$

From Darcy's Eq. (3.31), with:

$$k_h = \frac{k_x + k_y}{2}, \tag{13.118b}$$

we have:

$$\frac{k_{\mathrm{v}}}{k_{\mathrm{h}}} = \frac{2u_z}{u_x + u_y} = K_{\mathrm{aniv}}. \tag{13.119}$$

Mini-models are used to determine the vertical anisotropy K_{aniv} for a number of representative cases in each layer of the reservoir. The local values of the coefficient then permit k_{v} to be derived from k_{h}.

To a first approximation, K_{aniv} can be estimated simply from a knowledge of the dimensions and frequency of the impermeable lenses.

We define:

F_{s}: fraction of the rock volume occupied by impermeable lenses,
$\bar{l}$: average length of the lenses (m),
$\bar{w}$: average width of the lenses (m),
$\bar{f}$: average number of lenses per metre of pay thickness (m^{-1}).

Based on the results of a large number of simulations using mini-models, Begg[4] proposed the relationship:

$$K_{\mathrm{aniv}} = \frac{\bar{k}_{\mathrm{v}}}{\bar{k}_{\mathrm{h}}} = \frac{1 - F_{\mathrm{s}}}{(1 + \bar{f}\bar{d})^2}, \tag{13.120a}$$

where

$$\bar{d} = \frac{\bar{w}}{6\bar{l}}(3\bar{l} - \bar{w}). \tag{13.120b}$$

As an example, fine clay laminations ($F_{\mathrm{s}} = 0.01$), 5 m long and 1 m wide, occurring at a frequency of three per metre will produce an anisotropy of [Eqs. (13.120)]:

$$K_{\mathrm{aniv}} = 0.172.$$

The vertical permeability has been reduced to about one-sixth of its horizontal value by the presence of the thin shale laminae.

13.16.6 The Correlation Parameter

In the equations for the normalisation of capillary pressure curves (Sect. 3.4.4.7) and those for relative permeability (Sect. 3.5.4), a correlation parameter appears $\sqrt{(k/\phi)}$ which is essentially an indicator of the tortuosity of the porous medium on a microscopic scale. The local value of this parameter figures in the calculation of the capillary pressure and relative permeability curves for each block (Sects. 13.16.7 and 13.16.8), so a value is assigned to each block by the preprocessor, based on the local $\bar{k}_{\mathrm{h}}$ and $\bar{\phi}$.

13.16.7 Capillary Pressure Curves

Water/oil capillary pressure curves $P_{\mathrm{c,ow}} = f(S_{\mathrm{w}})$ for the imbibition process, and gas/oil curves $P_{\mathrm{c,go}} = f(S_{\mathrm{g}})$ for the drainage process are obtained from cores in each layer. They are normalised individually (Sect. 3.4.4.7) to determine the Leverett J-functions $J_{\mathrm{wo}} = f(S_{\mathrm{w}})$ and $J_{\mathrm{go}} = f(S_{\mathrm{g}})$ for each layer.

The relationship between P_c and S can then be derived for each block:

$$P_{c,wo}(S_w) = J_{wo}(S_w)\sigma_{wo} \cos \Theta_{wo}\sqrt{\frac{\phi}{k}}, \tag{13.121a}$$

$$S_w = P_{c,wo}^{-1}(S_w) \tag{13.121b}$$

and

$$P_{c,go}(S_g) = J_{go}(S_g)\sigma_{go} \cos \Theta_{go}\sqrt{\frac{\phi}{k}}, \tag{13.122a}$$

$$S_g = P_{c,go}^{-1}(S_g), \tag{13.122b}$$

where the term $P_c^{-1}(S)$ represents the inverse function of $P_c(S)$.

Equations (13.121) and (13.122) are sufficient to calculate values of S_w and S_g corresponding to known values of $P_{c,ow}$ and $P_{c,go}$, and vice versa, during the initialisation of the model and subsequent simulation. The use of Eqs. (13.121) and (13.122) presupposes that σ_{wo}, $\cos \Theta_{wo}$, σ_{go} and $\cos \Theta_{go}$ have been evaluated either experimentally or through correlations (Sects. 2.3.2.4 and 3.4.4.1).

13.16.8 Relative Permeability Curves

The procedure for the relative permeability curves is similar to that for capillary pressure. All the imbibition curves $k_{ro}(S_w)$ and $k_{rw}(S_w)$ measured from cores in the same layer are normalised by the method explained in Sect. 3.5.4. We will then have the following relationships for each layer:

$$k_{ro}^* = \frac{k_{ro}}{k_{ro,iw}} = f(S_w^*), \tag{13.123a}$$

$$k_{rw}^* = \frac{k_{rw}}{k_{rw,or}} = f(S_w^*), \tag{13.123b}$$

$$S_w^* = \frac{S_w - S_{iw}}{1 - S_{iw} - S_{or}}, \tag{13.123c}$$

$$k_{ro,iw} = f_1\left(\sqrt{\frac{k}{\phi}}\right), \tag{13.123d}$$

$$k_{rw,or} = f_2\left(\sqrt{\frac{k}{\phi}}\right), \tag{13.123e}$$

$$S_{iw} = f_3\left(\sqrt{\frac{k}{\phi}}\right), \tag{13.123f}$$

$$1 - S_{iw} - S_{or} = f_4\left(\sqrt{\frac{k}{\phi}}\right). \tag{13.123g}$$

These relationships are implemented in the model either analytically or as tables. They are used, together with the correlation parameter $\sqrt{k/\phi}$, to calculate the *rock* values of $k_{ro}(S_w)$ and $k_{rw}(S_w)$ for each block.

An equivalent set of equations exists for biphasic oil/gas systems:

$$k_{ro}^* = \frac{k_{ro}}{k_{ro,gc}} = f(S_g^*), \tag{13.124a}$$

$$k_{rg}^* = \frac{k_{rg}}{k_{rg,or}} = f(S_g^*), \tag{13.124b}$$

$$S_g^* = \frac{S_g - S_{gc}}{1 - S_{gc} - S_{or} - S_{iw}}, \tag{13.124c}$$

$$k_{ro,gc} = f_5\left(\sqrt{\frac{k}{\phi}}\right), \tag{13.124d}$$

$$k_{rg,or} = f_6\left(\sqrt{\frac{k}{\phi}}\right), \tag{13.124e}$$

$$S_{gc} = f_7\left(\sqrt{\frac{k}{\phi}}\right), \tag{13.124f}$$

$$1 - S_{gc} - S_{or} - S_{iw} = f_8\left(\sqrt{\frac{k}{\phi}}\right). \tag{13.124g}$$

These are treated in the same way as Eqs. (13.123a–g).

These two sets of equations deal the *two-phase flow* of oil/water and oil/gas systems.

In the *three-phase flow* of gas/oil/water, k_{rw} is a function only of S_w, and can be calculated using Eqs. (13.123b, c, e, f, g); k_{rg} is a function only of S_g and can be calculated using Eqs. (13.124b, c, e, f, g). k_{ro}, on the other hand, *is a function of both S_w and S_g.*

The relative permeability to oil in three-phase flow can be calculated from Stone's correlation[17], which, in its simplified form, is:

$$k_{ro}(S_g, S_w) = k_{ro,iw}\left[\frac{k_{ro}(S_w)}{k_{ro,iw}} + k_{rw}(S_w)\right]\left[\frac{k_{ro}(S_g)}{k_{ro,iw}} + k_{rg}(S_g)\right]$$
$$- [k_{rw}(S_w) + k_{rg}(S_g)], \tag{13.125}$$

where $k_{ro}(S_w)$, $k_{rw}(S_w)$ and $k_{ro,iw}$ are calculated through Eqs. (13.123), and $k_{ro}(S_g)$, $k_{rg}(S_g)$ and $k_{ro,gc}$ through Eqs. (13.124).

13.16.9 PVT Properties

The treatment of PVT properties is very straightforward if PVT analysis shows that the oil has the same bubble point p_b over the entire reservoir and at any depth. The following relationships are implemented in the model in analytical or tabular form:

$$\textit{For } p > p_b: \quad \begin{aligned} c_o &= f(p - p_b), \\ c_w &= f(p - p_b) \approx \text{const}, \\ \mu_o &= f(p - p_b), \\ \mu_w &= f(p - p_b) \approx \text{const.} \end{aligned} \tag{13.126a}$$

For $p \leq p_b$: $B_o(p)$ $B_w(p)$, $z(p)$ of the gas liberated

$R_s(p)$ $R_{sw}(P)$ from the oil,

$\mu_o(p)$ $\mu_w(p) \approx$ const.   (13.126b)

Also required are $\rho_{g,sc}$ of the liberated gas (assumed independent of pressure), $\rho_{o,sc}$ (or API gravity), and $\rho_{w,sc}$.

The formation volume factor of the oil, $B_o(p)$, and the solubility $R_s(p)$ of the gas in the oil, are obtained by "composite liberation", as described in Sect. 13.8.

In many oil reservoirs, especially where the pay is very thick, the bubble point p_b is observed to vary (usually decreasing) with increasing depth D. In some cases, there is also a lateral variation in the type of oil (characterised by a change in the oil gravity), so that the relationship $p_b = f(D)$ differs in different areas of the reservoir.

Let us assume that, in a certain reservoir, we have a single type of oil and therefore a single relationship of the form:

$$p_b = p_b(D). \tag{13.127}$$

In this case, the set of equations in Eq. (13.126b) can be derived from experimental PVT data up to a pressure equal to the maximum bubble point $p_{b,max}$ encountered in the reservoir. For values of $p_b(D)$ less than $p_{b,max}$, Eqs. (13.126b) are used up to $p = p_b(D)$, above which Eq. (13.126a) is used.

If several types of oil are present in the reservoir [and there are, consequently, several relationships of the form Eq. (13.127)], the above procedure must be repeated for each type, using the appropriate relationships defined by Eqs. (13.126) and (13.127).

When running the model, however, the movement of oil through the reservoir results in the mixing of oils of different types and bubble points in the grid blocks. This complication[41] necessitates a periodic recomputation of the PVT properties of the oil in each block.

13.16.10   Initial Gas/Oil and Water/Oil Contacts

The analysis of well logs (Sects. 3.2, 8.6 and 8.7), formation tester data (Sect. 8.2), and the fluids produced during drill-stem testing, usually provide accurate enough information in the drilling phase to determine the average depths of the initial water/oil and gas/oil contacts (WOC and GOC respectively). These contact depths are used by the model to compute the initial vertical distribution of S_w and S_g (and S_o) throughout the reservoir ("model initialisation").

13.16.11   Initial Pressure and Temperature at the Reference Datum, and the Temperature Gradient

Conventionally, the reference datum for pressures is taken as the water/oil contact or, if one does not exist, the maximum depth at which mobile oil is observed. The static pressure and temperature measured at, or extrapolated to, the datum, are input to the model, together with the vertical temperature gradient of the reservoir.

13.17 Initialisation of the Model

13.17.1 Calculation of Geometrical Parameters

When the model is initialised, the pore volume is first calculated for each block (i, j, k):

$$\Delta V_{p,i,j,k} = \phi_{i,j,k} \Delta x_i \Delta y_j \left[\frac{\text{net pay}}{\text{gross pay}} \Delta z_k \right]. \tag{13.128}$$

These values are stored in an array.

The interblock permeabilities [Eq. (13.67)] and single phase transmissibilities T [$k_r = 1$, Eq. (13.70)] are also calculated and stored.

13.17.2 Calculation of Phase Pressures

Starting from the pressure p_{WOC} at the water/oil contact at depth z_{WOC}, the initial pressures in the oil and water phases at depth $z_{i,j,k}$ are computed at the centre (or node) of each block, as illustrated in Fig. 13.30.

Since the capillary pressure $P_{c,\text{ow}}$ is zero at the WOC, we have:

$$p_w(z_{i,j,k}) = p_{\text{WOC}} - g\rho_w(z_{\text{WOC}} - z_{i,j,k}), \tag{13.129a}$$

$$p_o(z_{i,j,k}) = p_{\text{WOC}} - g\rho_o(z_{\text{WOC}} - z_{i,j,k}), \tag{13.129b}$$

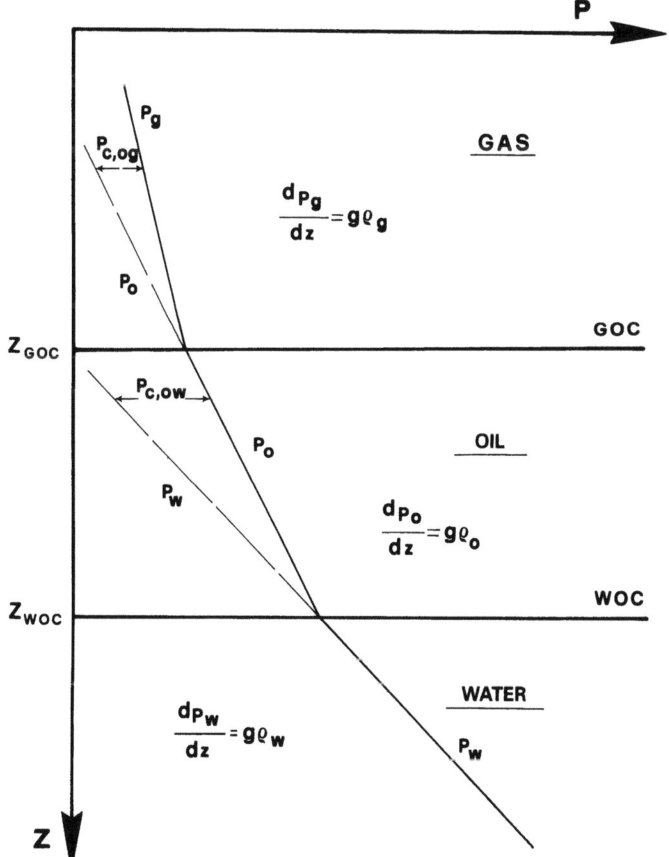

Fig. 13.30. *Vertical variation of the pressure in the water, oil and gas phases, and of the capillary pressures $P_{c,\text{ow}}$ and $P_{c,\text{go}}$ in a virgin reservoir*

where ρ_o is calculated from $R_s(p)$ and $B_o(p)$ using Eq. (13.36a), and ρ_w from R_{sw} and B_w with Eq. (13.36b).

$p_w(z)$ and $p_o(z)$ are calculated from the WOC up to the centre of block $k = 1$.

In the same manner, the initial pressure in the gas phase is computed starting from the gas/oil contact at depth z_{GOC} (Fig. 13.30).

Since $P_{c,go}$ at the GOC is zero:

$$p_g(z_{i,j,k}) = p_o(z_{GOC}) - g\rho_g(z_{GOC} - z_{i,j,k}), \tag{13.129c}$$

where ρ_g is the gas density calculated from $z(p)$, as defined in Eq. (2.8c).

13.17.3 Calculation of Capillary Pressures and Saturations

The capillary pressures at each node (block centre) are obtained directly from:

$$P_{c,ow}(z) = p_o(z) - p_w(z), \tag{13.130a}$$

$$P_{c,go}(z) = p_g(z) - p_o(z). \tag{13.130b}$$

The initialisation procedure calculates the values of S_w, S_o and, above the GOC, S_g, for each block, and stores them in three arrays. Capillary pressure curves, derived from the normalised curves which have already been introduced into the model (Sect. 13.16.7), are used for this purpose.

At each well axis, these saturations can be compared with the values obtained from the analysis of well logs (Sects. 3.2 and 8.5–8.6). If significant discrepancies are observed, either the capillary pressure curves $P_{c,ow}$ and $P_{c,go}$ or the log interpretations will have to be revised until agreement is achieved.

13.17.4 Calculation of the Initial Volume of Fluids in the Reservoir

In the initialisation phase, the program calculates a number of useful parameters for each layer, or for each lease area, and for the entire reservoir. In the following list, Σ_T represents the sum over all blocks in the layer, lease or reservoir, as appropriate, and ΔV_p is the pore volume in a single block:

$$V_p = \text{pore volume} = \Sigma_T \, \Delta V_p, \tag{13.131a}$$

$$N B_{oi} = \text{volume of oil in the reservoir} = \Sigma_T S_o \, \Delta V_p, \tag{13.131b}$$

$$W B_{wi} = \text{volume of water in the oil zone} = (\Sigma_T S_w \, \Delta V_p)_{oil}, \tag{13.131c}$$

$$\bar{S}_w = \text{average water saturation in the oil zone}$$

$$= \left(\frac{W B_{wi}}{N B_{oi} + W B_{wi}} \right)_{oil}, \tag{13.131d}$$

$$G B_{gi} = \text{volume of free gas in the reservoir}$$

$$= \Sigma_T S_g \Delta V_p, \tag{13.131e}$$

$$m = \frac{\text{volume of gas cap}}{\text{volume of oil}} = \frac{G B_{gi}}{N B_{oi}}, \tag{13.131f}$$

$$N = \text{oil in place (standard conditions)}$$

$$= \Sigma_T \left(\frac{S_o \Delta V_p}{B_{oi}} \right), \tag{13.131g}$$

$$G = \text{gas in place (standard conditions)}$$

$$= \left[\Sigma_T \frac{S_g \Delta V_p}{B_{gi}} \right]_{gas} + \left[\Sigma_T \frac{R_{si} S_o \Delta V_p}{B_{oi}} \right]_{oil}. \tag{13.131h}$$

These values of V_p, NB_{oi}, GB_{gi}, N and G are compared with the values calculated by the volumetric method (Chap. 4) and any differences are reconciled by adjusting the input parameters.

13.17.5 Plotting the Iso-Value Maps

Next, the automatic digitisation of the iso-value maps (iso-pay, iso-net/gross ratio, iso-porosity, iso-k_h, iso-k_v, etc.) performed by optical scanner in the preprocessing phase, is checked for errors. The input data array to the numerical model (one value of each parameter for each block) is passed through a post-processor which replots the iso-value maps. By comparison with the original field maps, any errors in digitisation can be identified and corrected.

This same post-processor will be used later to map the results computed by the model. It is also used to calculate and map the equivalent hydrocarbon column (EHC) thickness:

$$\text{EHC} = h_n \bar{\phi}(1 - \bar{S}_w) \tag{4.11}$$

over the plane of the reservoir.

Figure 13.31 is an example of a map produced in this way. (Alternative three-dimensional presentations can be obtained using suitable graphics software.) These maps are of particular interest in offshore reservoirs: to optimise the economics of field development, the production platforms must be centred over the regions with the highest reserve concentrations. Wells drilled from these platforms will exhibit the highest productivity, because of the greater pay thickness, and will achieve the most effective reservoir drainage.

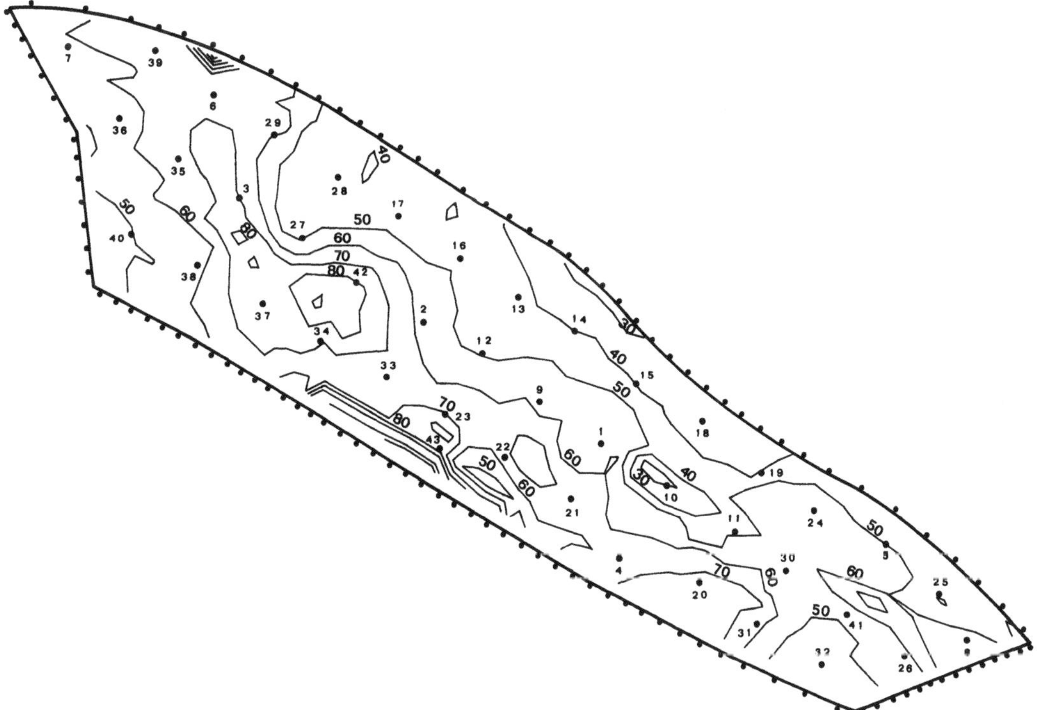

Fig. 13.31. *Iso-thickness map of the equivalent hydrocarbon column (EHC)*

13.18 Choice of Model and Method of Simulation

13.18.1 Choice of Model

With an AIM type of numerical simulator (Sect. 13.14.3.4), no choice of solution method needs to be made. This is selected automatically (from implicit-explicit, through simultaneous solution, to fully implicit) region by region throughout the reservoir, with redefinition of the regions in each time step. AIM simulators are always equipped with a number of procedures to solve the systems of algebraic equations which are derived from the discretisation of the flow equations (Sect. 13.13), and the appropriate one will be applied for the solution method adopted.

For the other types of numerical simulator, the operator must use his own experience to choose the optimum solution method (Sect. 13.14.3.1). The IMPES method or its iterative variant IMPESIT (Sect. 13.14.3.2) are usually adequate for the simulation of most reservoirs and development scenarios. Higher order solution methods should only be necessary when there is strong heterogeneity or faulting (or other discontinuities).

The choice of solution procedure is also a matter for the operator. It is advisable to try several alternatives over a few time steps in order to identify the most stable one. A simulator written for, or compatible with, a parallel or a vector computer, should always be preferred if available. Computation time is considerably shorter than with the serial computers, and costs are lower. This enables more cases to be investigated in the sensitivity study phase (Sect. 13.2) for the same computing time.

As a general rule, there is no advantage in using a solution method of higher order than the one required (i.e. avoid overkill). As for the reliability of the data used in the model: *the quality of the results can never be superior to the quality of the data input to the model, regardless of the amount of overkill applied to the solution.*

13.18.2 Convergence and Stability Criteria

When solving algebraic equations iteratively (Sects. 13.13.3 and 13.13.4), convergence is considered to have been achieved when the criterion expressed in Eq. (13.90) is satisfied. The convergence term ϵ must therefore be specified, along with a maximum number of iterations above which the process will be terminated if convergence has not been reached. An appropriate message would be displayed in this case.

Likewise, the stability of the solution method is judged by the mass balance relationship expressed in Eq. (13.114). Again, the convergence term ϵ and the maximum number of iterations must be specified whenever iterative solution methods are employed.

A restart procedure is also required. This transfers the current results arrays and other useful data from memory to hard disk at predetermined time steps. This will allow the simulation to be restarted from the last time step should the computation be interrupted – either intentionally or by a system crash.

13.18.3 Simulation of the Displacement Process: Choice of Relative Permeability and Capillary Pressure Pseudo-Curves

Section 11.4.1 discussed the simulation of the displacement of oil by water in a two-dimensional (x, y) system by means of an equivalent one-dimensional system using relative permeability and capillary pressure "pseudo-curves". The same principle can of course be applied to the displacement of oil by gas.

In the case of a three-dimensional model, every layer (k = const.) of blocks is in fact a three-dimensional system whose thickness Δz_k varies with the position (i, j) of the blocks within the layer. We will now see how, in three-dimensional systems (x, y, z) of this kind, the displacement process *in the direction parallel to the bedding plane* can be simulated with an equivalent two-dimensional (x, y) model. Furthermore, if we only consider a vertical section i = const., or j = const., with the width W of a single block, the problem reduces to that of simulating displacement in a two-dimensional system by means of an equivalent one-dimensional model (Sect. 11.4). It was explained in Sect. 11.4 that displacement could take place in a *dispersed* regime, in *vertical equilibrium (VE)* (also referred to as a *segregated* or *level* regime), or in the intermediate *partially segregated* regime.

It is worthwhile to recap briefly the physical significance of the terms "dispersed" and "segregated". In a *dispersed regime* the vertical permeability of the rock and/or the ratio of the block thickness to the height of the capillary transition zone are so small that gravitational equilibrium – the separation of gas + S_{or}/oil/water + S_{or} – is not attained during the time it takes the displacing fluid to traverse the block. In this case the actual *rock relative permeability curves* obtained from laboratory measurements on core plugs, and already initialised in the model (Sect. 13.16.8), can be used directly in the simulation of the displacement process.

If, on the other hand, the vertical permeability is high and the displacement velocity is low enough, the gas, oil and water phases may be able to separate out (or *segregate*) into gas + S_{or}/oil/water + S_{or} while the displacing fluid traverses the block. In this equilibrium condition, the displacement can be visualised as a gradual rise (in the case of water/oil) or fall (in the case of gas/oil) of the capillary transition zone, while the zone itself retains the same saturation profile.

Displacement *parallel to the bedding plane* is modelled using *relative permeability pseudo-curves* (described in Sect. 11.5). In Sect. 11.5.3 we saw that, for a water/oil system, if the block height Δz_k was much larger than the thickness h_c of the capillary transition zone, the k_r pseudo-curves were linear (Fig. 11.32). In a numerical model they are computed from the values of S_{iw}, $k_{ro,iw}$ and $1 - S_{or}$, and $k_{rw,or}$ which were input in the initialisation phase [Eqs. (13.123d–g)].

The water/oil capillary pressure pseudo-curves are derived by displacing the rock capillary pressure curves progressively towards the top of the layer, and calculating the mean values of S_w and $P_{c,ow}$ for each block at each displacement. The displacement of oil by gas is treated in a similar manner. Where the block height Δz_k is comparable to the capillary transition zone thickness h_c, the pseudo-curves of k_r and P_c are calculated from the rock curves by the method described in Sect. 11.5.2.

The procedure to determine whether displacement occurs in a segregated or dispersed regime is described in Sect. 11.4.1. It involves choosing a number of characteristic vertical sections *in each layer* and studying each of these sections using two different mini-models[7] (Fig. 11.24).

In the first of these mini-models, which we will call A, we retain the discretisation already configured for the reservoir model, and the relative permeability and capillary pressure pseudo-curves calculated as described above. In the second, B, *the mesh is made much finer in the z-axis direction*, and the true rock curves are used for k_r and P_c. The displacement is simulated with both models, with the highest fluid velocity that can be realistically expected in the reservoir.

If the results obtained from mini-model A are acceptably close to those from B, the displacement is in *segregated* flow. The pseudo-curves of k_r and P_c derived for VE conditions should therefore be used in the model. If the two sets of results differ, A is rerun using the rock curves for k_r and P_c. If the results from A and B now agree, the displacement is in *fully dispersed* flow. In this case, the rock curves should be used in the reservoir model. If the results still do not agree, the displacement is in *partially segregated* flow.

Jacks et al.[23] and, using a different formulation, Kyte and Berry[24], proposed that *dynamic pseudo-curves* should be used in this situation. The basic steps in the calculation of the dynamic pseudo-curves for the displacement of oil by water are as follows. (Oil displacement by gas involves a similar procedure.)

In the simulation using mini-model B (finer grid and rock k_r and P_c curves), successive groups of columns are isolated at the end of the displacement (Fig. 13.32). *For each time step*, and for each group of blocks – for instance, the four blocks 1-2-3-4 of model B corresponding to the single block 1 of model A – an auxiliary program is used to calculate the following values from the results held in memory during the simulation:

- $\bar{S}_w$: average water saturation in the group of blocks;
- $\bar{\Phi}_{o,in}$, $\bar{\Phi}_{w,in}$, $\bar{p}_{o,in}$, $\bar{p}_{w,in}$: average potential and pressures in the oil and water phases at the inlet face of the group of blocks;
- $\bar{\Phi}_{o,out}$, $\bar{\Phi}_{w,out}$, $\bar{p}_{o,out}$, $\bar{p}_{w,out}$: average potential and pressures in the oil and water phases at the outlet face of the group of blocks;
- oil flow rate;
- water flow rate.

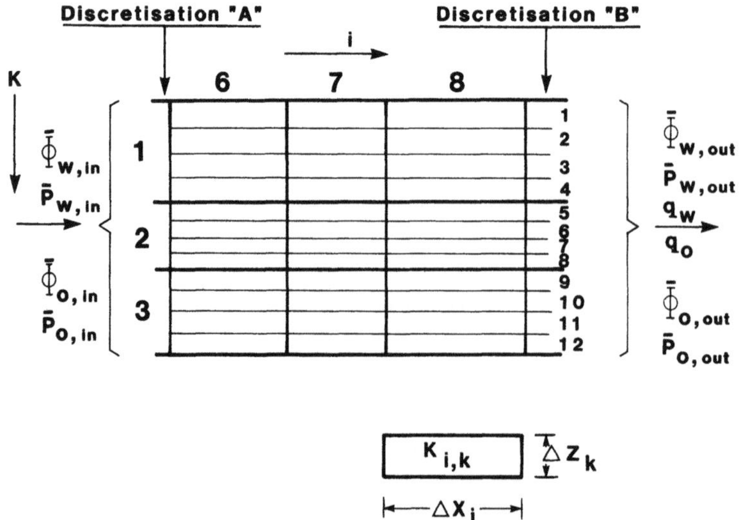

Fig. 13.32. *Schematic of a group of blocks, showing the mesh configuration in the two mini-models A and B, for the calculation of the dynamic relative permeability and capillary pressure pseudo-curves*

The average permeability of the group of blocks is calculated with the series/parallel equation:

$$\bar{k} = \frac{n}{m} \cdot \frac{\sum\limits_{1}^{m} {}_k \sum\limits_{1}^{n} {}_i \Delta x_{i,k}}{\sum\limits_{1}^{m} {}_k \sum\limits_{1}^{n} {}_i \Delta z_{i,k}} \sum\limits_{1}^{m} {}_k \left(\sum\limits_{1}^{n} {}_i \frac{\Delta x_{i,k}}{k_{i,k} \Delta z_{i,k}} \right)^{-1}, \qquad (13.132)$$

where, in a mini-model consisting of m rows and n columns of blocks, $\Delta x_{i,k}$ .and $\Delta z_{i,k}$ are the width and height of the block (i, k) – the ith block in the kth row – and $k_{i,k}$ is the corresponding rock permeability.

At each time step we calculate the average values $\bar{k}_{ro}$, $\bar{k}_{rw}$ and $\bar{S}_w$ for the group of blocks in question, using the generalised Darcy equation. We then construct the dynamic pseudo-curves for the relative permeabilities $\bar{k}_{ro}(\bar{S}_w)$ and $\bar{k}_{rw}(\bar{S}_w)$, which will be used to simulate two-phase flow in the corresponding block of model A (block 1 in this case).

We also calculate, at each time step, the average value $\bar{P}_{c,ow} = \bar{p}_o - \bar{p}_w$, from which we can derive the dynamic pseudo-curve for the capillary pressure for each group of blocks. This procedure possesses the useful characteristic of *"remembering" the refined grid with which the calculations were made*[24]. This means that using the dynamic pseudo-curves of k_r and P_c obtained in this way in a model with coarse gridding (model A) predicts the same $\bar{S}_w$, q_o and q_w behaviour as model B with the finer gridding.

The displacement flow regime changes progressively from segregated to dispersed as the displacement velocity increases. The dynamic pseudo-curves must therefore be calculated for a range of fluid velocities, and introduced in the model as functions of velocity.

In heterogeneous formations, the procedure outlined above can be used to calculate the water/oil dispersion coefficient as a dynamic term. Displacement is represented as a dispersive phenomenon[7], where the rock characteristics are defined by means of average end values $(\bar{S}_{iw}, \bar{k}_{ro,iw})$, $(\bar{S}_{or}, \bar{k}_{rw,or})$ and the dispersion coefficient is a velocity-dependent variable.

Reservoir engineers must always bear in mind that *the results of a simulation are strongly influenced by the manner in which the displacement has been modelled* (segregated, partially segregated or dispersed regime). It is therefore *essential* to run a number of mini-model studies before embarking on a full-blown simulation of reservoir behaviour, so as to choose the curves, pseudocurves or dynamic pseudocurves that most aptly describe the displacement process. Since the layer thickness, vertical permeability and displacement velocity may vary from place to place, it may also be necessary to use a different approach in different layers or parts of layers.

The preceding discussion describes the case where *the displacement is along the bedding plane* (layer of blocks k = const. in the model). For displacement in the z-direction, the notes in Sect. 13.14.2 on numerical dispersion can be applied to segregated flow. For dispersed flow, the rock relative permeability curves are used (just as they were for horizontal displacement).

A useful insight into the true nature of the displacement flow regime can often be obtained when validating the model by history matching (Sect. 13.19), when the simulated and actual well behaviour are compared. The treatment of the k_r and

P_c curves should be carefully modified until reasonable agreement is achieved. If doubts remain about the flow regime (observed and computed behaviour cannot be reconciled), two separate runs of the simulator may have to be performed – one assuming dispersed flow, and one assuming segregated. Generally, the results obtained will differ significantly, and the choice of flow regime is a critical factor in the success of a reservoir simulation.

13.19 Validation of the Numerical Model by Matching the Reservoir History

13.19.1 Scope and Methodology of the Validation Process

If the reservoir has been on production, the numerical model set up and initialised as described in the previous sections should be "validated" by reproducing the past production and "history matching" the calculated behaviour against field measurements. In the case of an oil reservoir, the simulation is run by imposing the recorded oil production rates $q_{o,sc}(t)$ at each well, and the model calculates the corresponding gas and water rates $q_{g,sc}(t)$ and $q_{w,sc}(t)$, and the grid-block pressures $p_o(t)$.

In a gas reservoir, the recorded gas production rates $q_{g,sc}(t)$ are imposed at each well, and the water rate $q_{w,sc}(t)$ and grid-block pressures $p_g(t)$ are calculated. The model is considered to have been "validated" if the computed flow rates and static pressures at the wells agree with the values measured during the producing history of the reservoir, to within experimental error.

In order to achieve an acceptable history match, some of the parameters input to the initialisation phase will almost certainly need adjusting. This is one of the most delicate stages of the study, often more of an art than a science, whose success is highly dependent on the experience and judgement of the analysts. The parameters most frequently in need of adjustment are the horizontal and vertical permeabilities (to match the pressures), and the gas/oil and water oil relative permeabilities (to match the GORs and WORs). Note that these are also the parameters with the highest associated uncertainties in the initialisation phase of the model.

Another important rule to remember is that *the model that has been validated by history matching may not be unique*[8]. *Alternative models* (distributions of the physical parameters) *may exist which are capable of reproducing the same reservoir history*. Obviously, only the one which corresponds to the real reservoir parameters will be able to predict future behaviour effectively. This explains why the adjustment of the reservoir parameters is so critical. In order to obtain a model that matches the true reservoir behaviour as closely as possible, every parameter change must be reviewed in the context of the geological model by the interdisciplinary group conducting the study.

13.19.2 Data Needed for History Matching

Model validation requires data on the production history of all the wells in the reservoir, with a reasonable estimate of their accuracy.

The following well data must be provided:

- completion history, including:
  - record of each completion

- intervals open to production
- well stimulations and their results
- squeezes and other cement jobs
- production records:
  - oil, $[q_{o,sc}(t)]$
  - gas, $[q_{g,sc}(t)]$
  - water, $[q_{w,sc}(t)]$
- fluid properties at different times:
  - API oil gravity
  - gas density (γ_g)
  - water salinity
- records of all static bottom-hole pressure measurements $[p_o(t)]$
- production test reports, including the pressure buildups (Chaps. 6 and 7), productivity indices (J_o), and skin factors (S).
- production logs (PLT; Sect. 8.3), to define the relative contributions from the open intervals in each well
- cased hole monitoring logs to track the movement of the gas/oil contact [CNL (perhaps with FDC), TDT; Sect. 8.6] and oil/water contact (TDT, GST, chlorine log; Sect. 8.7) behind casing (i.e. in intervals not open to the well bore).

It is important to realise that there can be quite considerable errors associated with the measurement of gas and water flow rates by standard methods in oil wells, and in water flow rates in gas producers. These errors are carried through into the GOR and WOR figures. The reservoir engineer is therefore advised to check these data very carefully with the field personnel to assess the reliability of the figures they have provided. Assigning a greater degree of accuracy to field data than they actually have will lead to unnecessary adjustments being made to model parameters in an attempt to achieve a satisfactory match with the actual past history. Such adjustments have no physical significance, and represent wasted effort. In addition to the well data listed above, records of the average temperatures and pressures of the separators during the producing history will be required for the calculation of the equilibrium constants, from which the field GOR will be computed.

13.19.3 Use of the Model in History Matching

In the history matching phase, the measured production of the "principal" fluid (oil in an oil reservoir, gas in a gas reservoir) is specified at each well, and the model calculates the "induced" flow rates (gas and water in an oil reservoir, water and condensates in a gas reservoir).

In reservoirs where wells may be producing from several layers at the same time, there is the additional problem of apportioning the flow rates of the "principal" fluid among the layers. The most commonly used approach is to pro-rate the total oil flow rate of each well according to the product hkk_{ro}/μ_o of each of the blocks that is open to flow. A similar approach is used for gas wells.

Let us take an oil well producing from n layers as an example. For the ith layer open to the well bore, we will have, from Eq. (3.49):

$$q_{o,sc,i} = \alpha \left(\frac{hkk_{ro}\rho_o}{\mu_o B_o} \right)_i (\Phi_{o,i} - \Phi_{wf}), \qquad (13.133a)$$

where $q_{o,sc,i}$ = oil flow rate from layer i, under standard conditions,

α = geometrical constant for the system,

$\Phi_{o,i}$ = average potential of the block containing the well in the ith layer,

Φ_{wf} = bottom-hole flowing potential (the same for all layers),

k_{ro} = a function of the mean oil saturation in the block containing the well in the ith layer,

B_o = composite oil formation volume factor for gas liberation under surface separator conditions,

$\rho_o, \mu_o = f(p_{o,i})$ in the block containing the well in the ith layer.

Peaceman[29] showed that the average potential calculated for a block of area (Δx^2) containing a well is equal to the flowing potential at a distance $0.2\Delta x$ from the axis of the well. Therefore, from Eq. (5.46):

$$\alpha = \frac{2\pi}{\ln\dfrac{0.2\Delta x}{r_w} - \dfrac{3}{4} + S},$$

(13.134)

assuming pseudo-steady state flow in the area under consideration.

Equation (13.133a) can now be written as:

$$q_{o,sc} = \sum_{1}^{n}{}_i q_{o,sc,i}$$

$$= \alpha \sum_{1}^{n}{}_i \left[\left(\frac{hkk_{ro}\rho_o}{\mu_o B_o} \right)_i (\Phi_{o,i} - \Phi_{wf}) \right]$$

$$= J_o(\bar{\Phi}_o - \Phi_{wf}),$$

(13.135)

where $\bar{\Phi}_o$ is the average potential in the oil phase in the grid mesh penetrated by the well, and J_o is the average productivity index.

The value of α derived from Eq. (13.135) is used in Eq. (13.133a) to calculate the flow rate $q_{o,sc,i}$ from each layer. If we assume there is no coning (Sect. 12.6), the associated water and gas flow rates are calculated as follows:

For gas:

$$q_{g,sc,i} = \alpha \left(\frac{hkk_{rg}\rho_g}{\mu_g B_g} \right)_i (\Phi_{g,i} - \Phi_{wf}) + R_{s,i} q_{o,sc,i}.$$

(13.133b)

By making the simplification that $\Phi_{g,i} = \Phi_{o,i}$, Eqs. (13.133a and b) can be combined to give:

$$q_{g,sc,i} = \left[\frac{k_{rg}\rho_g}{\mu_g B_g} \frac{\mu_o B_o}{k_{ro}\rho_o} + R_s \right]_i q_{o,sc,i}.$$

(13.136a)

In the same way, for water we will obtain:

$$q_{w,sc,i} = \left[\frac{k_{rw}\rho_w}{\mu_w B_w} \frac{\mu_o B_o}{k_{ro}\rho_o} \right]_i q_{o,sc,i}.$$

(13.136b)

Then:

$$q_{g,sc} = \sum_{1}^{n}{}_i q_{g,sc,i},$$

(13.137a)

$$q_{w,sc} = \sum_{1}^{n} {}_i q_{w,sc,i},$$ (13.137b)

$$GOR = q_{g,sc}/q_{o,sc},$$ (13.138a)

$$WOR = q_{w,sc}/q_{o,sc}.$$ (13.138b)

If the gas/oil or water/oil contacts are close to the well, coning is a possibility and correlations must be introduced into the model for the GOR and WOR:

$GOR = f(q_{o,sc}$, distance from the top of the perforated interval to the GOC);
$WOR = f(q_{o,sc}$, distance from the bottom of the perforated interval to the WOC).

These relationships can be calculated in the manner described in Sect. 12.6, or derived[26] through a numerical well model (r, z). Of course, if PLT measurements of the actual principal fluid flow rates from each layer are available in some or all of the wells, they can be input directly to the model. However, the "induced" flow rates of the associated fluids (gas and/or water) are never used as input data, for two reasons:

- the gas and water flow rates measured during the reservoir's production history might not be consistent with the distribution of S_g and S_w calculated by the model. A typical example of this would be the imposition of water production from a well for which the simulation has not even indicated the arrival of water, nor any coning).
- it leaves the possibility of validating the model by history matching computed GORs and WORs.

13.19.4 Checking the History Match

On completion of each history match run, a post-processor displays the results of the model as follows:

For the reservoir:

- iso-pressure maps at specified times, on which are also presented any well static pressures that were measured at the same times;
- iso-saturation maps $(S_o, S_g$ and $S_w)$ at specified times.

For each well:

- graphs of the variation of the following calculated parameters with time:
  - bottom-hole flowing pressure (p_{wf})
  - static reservoir pressure (p_{ws}) (the pressure of the block containing the well)
  - gas flow rate $(q_{g,sc})$
  - water flow rate $(q_{w,sc})$
  - gas/oil ratio (GOR)
  - water/oil ratio (WOR)
  - produced oil API gravity (when the oil properties vary over the reservoir)
  - produced water salinity (when different layers in the reservoir are in communication with aquifers of different salinities).

The corresponding measured production data are presented on each well plot for comparison. In the case where a buildup test was not of sufficient duration for the pressure to reach the static reservoir pressure, the method described by Peaceman[29] should be used to facilitate comparison with the computed static pressure.

According to Peaceman, the pressure calculated by the model for the block containing the well at a given time should be equal to the pressure observed on the buildup plot in the same well at an elapsed time Δt_s after shut-in, where:

$$\Delta t_s = C \frac{\phi \mu_o c_t A}{k}. \tag{13.139}$$

In this equation $C = 0.0178$ in SI units
$= 3065$ in practical metric units (Table 6.1)
$= 2.941 \times 10^6$ in oil-field units (Table 6.1)
and $A =$ area of the mesh containing the well.

Adjustment of the model to obtain a good history match is made much easier by the ability to compare graphically the variations between the computed parameters and the measured production data. The gas and water iso-saturation maps computed by the model must also be compared with any data available from cased hole logs (Sects. 8.6 and 8.7), and the simulated and measured iso-pressure maps must be reconciled. The model validation phase ends when a reasonable history match has been obtained. Just what is meant by "reasonable" is open to discussion.

13.20   Forecasting Reservoir Behaviour

13.20.1   Introduction

Once the model has been initialised and validated through history matching against past production data (if available), it can be used to predict the reservoir behaviour that would result from different development scenarios (different production rates, drilling of infill wells, depletion drive versus water or gas injection, condensate gas cycling, etc.).

Forecasting the economic viability of reservoir development programs is the most useful and effective application of numerical models. The results are quantified in terms of a discounted cash flow, where past and future prices are converted to "dollars-of-the-day" using a specified discount rate. In this way, the investment in wells, surface facilities, transport, etc., and production costs, can be compared with the commercial value of the oil and gas extracted over the producing life of the reservoir. The programme which offers the best return on investment or, alternatively, the healthiest discounted cash flow, is obviously the one that is the most economically attractive.

13.20.2   Production Constraints

13.20.2.1   Well Constraints

For each of the wells currently in the model, and at each location where a future well might be drilled in order to maintain field production at the desired level, all intervals which are possible candidates for completion are listed in order of priority. In this way, the model will be able to "open" additional intervals when a completion no longer satisfies certain criteria. These criteria refer to the bottom-hole and wellhead flowing pressures p_{wf} and p_{tf}, the oil production rate and the GOR and WOR, and the gas production rate and WGR in the case of gas fields.

The *pressure* constraints are:

- the maximum pressure drawdown ($p_{ws} - p_{wf}$) that can be sustained without inducing mechanical problems such as sanding;
- the lowest bottom-hole or wellhead flowing pressure that will still allow the well to produce naturally (without requiring artificial lift such as gas-lift, centrifugal or sucker rod pumping);
- vertical lift performance curves for a range of possible well completion designs. From these, the optimum completion design can be selected to ensure effective production to surface. If the reservoir pressure declines, it may in fact be necessary to change the completion design.

The *flow rate* constraints are:

- the minimum oil or gas production rate at which the well can be economically operated with the current completion. When the computed flow rate falls below this threshold, the model automatically recompletes the well by opening the next interval in the priority list
- maximum GOR and WOR that can be tolerated without posing problems with lifting the fluid to surface and the subsequent disposal of gas and water.
  Each well in the model has a set of correlations for coning. When a GOR or WOR limit is exceeded, the well is automatically recompleted according to the interval priority list.

Once all the productive layers listed for a well have been completed, as soon as the oil or gas flow rate falls below the economic limit or the GOR or WOR exceed their maximum values, the well is shut down.

In water injection wells, upper limits are placed on ($p_{wf} - p_{ws}$) so that the formation will not be fractured, and a minimum injection rate i_w is imposed to keep the well operational. A priority list of intervals for possible completion can also be assigned to injection wells. If, when all these intervals have been opened for injection, it eventually proves impossible to achieve the minimum injection rate even with the maximum allowed value of ($p_{wf} - p_{ws}$), the well will be shut down.

All of these constraints are applied via subroutines which are called regularly by the model to calculate the rate at which each well can in fact be produced. The average percentage downtime of all the wells in the reservoir (when they will be off production for maintenance or other work on the downhole or surface equipment) must also be specified. The average rates and cumulative volumes of produced oil and injected water computed by the model are then reduced by the percentage downtime.

13.20.2.2 *Surface Facility Constraints*

The produced well fluids are transported through flow lines to the gathering and treatment centres, of which there may be several per reservoir. Here, they must be treated so that they conform to the standards laid out in the sales contract (water content, H_2S, solids, salt, vapour pressure in the case of crude oil, dew point for water and hydrocarbons in the case of gas). Highly viscous or waxy crudes require special processing in order to be transported by pipeline or tanker.

The disposal of water separated from the crude oil or gas must also be taken into consideration. Some or all of the separator gas, too, will have to be flared off or compressed and reinjected into the reservoir if it cannot be sold.

The construction of the surface facilities is designed in the project planning phase, and it will not be possible to exceed the design specifications of the equipment once the field is operational – for example:

- maximum oil, water and gas production rates
- maximum GOR and WOR
- maximum water salinity.

Should the value of one of the parameters be exceeded, some form of intervention will be necessary at well level – for instance, choking back selected wells, or shutting in wells producing with a high water or gas rate. The optimum well intervention strategy and choice of production rates to adopt can be determined with a subroutine which is invoked when one of the operating limits of the treatment units or separators is exceeded.

13.20.2.3   Reservoir Constraints

The principal reservoir constraints are the quantities of oil and gas to be produced as a function of time. These will be clearly stated in the terms of the sales contracts and, in some cases, they will figure in the licence for the concession, or will be imposed by the local regulatory body.

If the total rate of production that could be realised from all of the wells (subject to the constraints outlined in Sect. 13.20.2.1) is greater than the contractual requirement from the reservoir, it will be necessary to plan a strategy for the cutting back of some well rates, or the temporary closure of others. If, on the other hand, the existing wells cannot meet the specified total productivity, more wells will obviously have to be drilled, in the locations determined at the beginning of the study (Sect. 13.20.2.1). The siting and timing of these new wells, too, will be done in a carefully planned manner.

A discounted cash flow study will determine whether it is economically viable to drill these extra wells – an alternative course of action would be to reduce the commercial production quotas for oil and gas, if existing contractual obligations permitted. A subroutine will be called automatically by the numerical model as and when necessary to determine the best strategy for cut backs or in-fill drilling, bearing in mind that the flow rates must satisfy the operating constraints associated with the wells and surface treatment facilities and separators.

13.20.3   Calculation of Well Flow Rates

13.20.3.1   Introduction

For each hypothetical development program simulated by the model, the most important result is its prediction of the way the flow rates and cumulative production of oil, gas and water will vary with time, given the constraints associated with the wells, surface equipment and reservoir (Sect. 13.20.2). In addition, the iso-pressure and water, gas and oil iso-saturation maps produced by the post-processor (Sect. 13.19.4) provide important information that may suggest other possible development strategies, such as the drilling of in-fill wells in poorly drained areas, the implementation of water or gas injection, or the location of additional injection wells for pressure maintenance or to improve the volumetric sweep efficiency (Sect. 12.5).

Since the total reservoir production is the sum of the production from individual wells, it is worthwhile considering carefully the methodology to be used for the calculation of well flow rates and bottom-hole flowing pressures. Usually, the constraints applied to the wells during the initial phase of exploitation are the flow rates required to achieve the desired total production from the reservoir (Sect. 13.20.2.3). When the bottom-hole flowing pressure in any well reaches the specified minimum, the well is allowed to continue producing at this pressure. The oil flow rate will then begin to decrease as the reservoir pressure declines further (Fig. 13.33).

The well is now said to be producing at its "deliverability", instead of at a "fixed rate".

We will now look at how to calculate the production parameters of a well – bottom-hole flowing pressure (p_{wf}), and the flow rates of oil ($q_{o,sc}$), gas ($q_{g,sc}$) and water ($q_{w,sc}$) – in the most general case of a well producing from several layers in the same reservoir, each of which is represented by a single grid block in the z-direction (Fig. 13.34). U represents the uppermost layer, L the lowermost, and K any layer.

Assuming hydrostatic equilibrium exists in the well bore, the pressure potential must be the same at any point in the well:

$$\Phi_U = \Phi_K = \Phi_L. \tag{13.140a}$$

If we ignore the dynamic pressure losses when the fluid is in motion, Eq. (13.140a) can be written as:

$$p_{wf,K} = p_{wf,U} + g\bar{\rho}\Delta z_K, \tag{13.140b}$$

where: $p_{wf,U}$ = well-bore pressure opposite layer U,
$\quad\ p_{wf,K}$ = well-bore pressure opposite layer K,
$\quad\ \bar{\rho}$ = average density of the well fluids,
$\quad\ \Delta z_K$ = $z_K - z_U$.

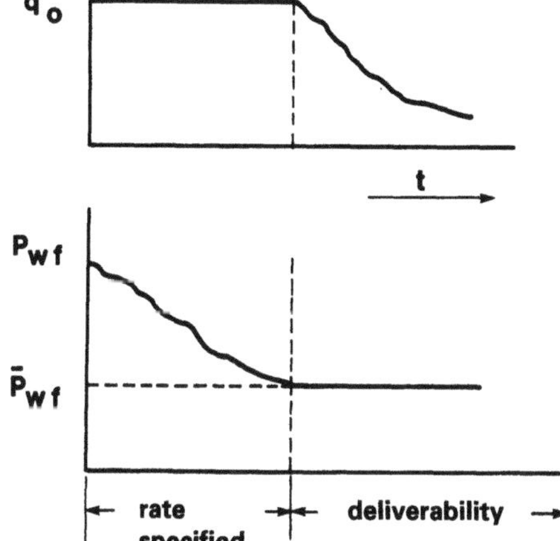

Fig. 13.33. *Production rate and bottom-hole pressure variations in a well producing initially at a fixed rate, and then at deliverability. (From Ref. 39, 1982, Prentice Hall. Reprinted with permission of Prentice Hall)*

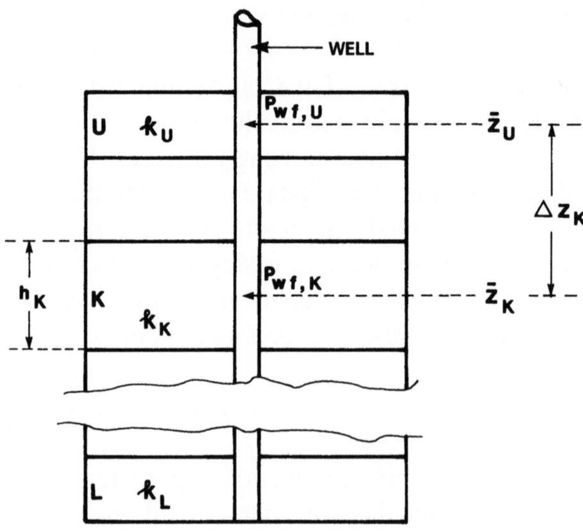

Fig. 13.34. *Schematic of a well producing from several layers in the same reservoir. (From Ref. 39, 1982, Prentice Hall. Reprinted with permission of Prentice Hall)*

13.20.3.2   Wells Producing at Fixed Rates

For a layer K open to the well bore, and a generic fluid type l ($l = o, g, w$), starting from the definition of the productivity index J in Eq. (5.56a) we can write:

$$q_{l,\mathrm{sc},K} = J_{l,K}(\bar{p}_K - p_{\mathrm{wf},K}), \tag{13.141a}$$

where: $q_{l,\mathrm{sc},K}$: flow rate of fluid l produced from layer K, measured under surface conditions,

$\qquad\quad J_{l,K}$: productivity index of layer K for fluid l,

$\qquad\quad \bar{p}_K$: average pressure in the block in layer K containing the well,

$\qquad\quad p_{\mathrm{wf},K}$: well-bore flowing pressure opposite layer K.

In pseudo-steady state flowing conditions, Eqs. (13.133a) and (13.134) give:

$$J_{l,K} = \frac{2\pi h_K k_K k_{rl}(S_{l,K})}{\mu_l B_l} \; \frac{1}{\ln \dfrac{0.2\Delta x}{r_{\mathrm{w}}} - \dfrac{3}{4} + S}. \tag{13.142}$$

As seen in Sect. 13.19.3 the flowing potential at a distance $0.2\,\Delta x$ from a well is equal to the average potential calculated by the model for the block, of area $(\Delta x)^2$, containing the well (Peaceman[29]).

From Eqs. (13.140b) and (13.141a) we have:

$$q_{l,\mathrm{sc},K} = J_{l,K}(\bar{p}_K - p_{\mathrm{wf},U} - g\bar{\rho}\Delta z_K). \tag{13.141b}$$

When the well produces at a fixed rate $\bar{q}_{\mathrm{o,sc}}$:

$$\bar{q}_{\mathrm{o,sc}} = \sum_{L}^{U} {}_K q_{\mathrm{o,sc},K}, \tag{13.143}$$

which, taking Eq. (13.141b) into account, gives:

$$p_{\mathrm{wf},U} = \frac{\displaystyle\sum_{L}^{U} {}_K [J_{\mathrm{o},K}(\bar{p}_K - g\bar{\rho}\Delta z_K)] - \bar{q}_{\mathrm{o,sc}}}{\displaystyle\sum_{L}^{U} {}_K J_{\mathrm{o},K}}. \tag{13.144}$$

Equation (13.144) allows us to calculate the bottom-hole flowing pressure at the top of the reservoir (layer U) for a given fixed oil flow rate $\bar{q}_{o,sc}$. The simplest method for subdividing $\bar{q}_{o,sc}$ into the individual flow rates $q_{o,K}$ from each layer is to use Eq. (13.141b), with values of $J_{o,K}$ derived from Eq. (13.142). For this, k_{ro} is evaluated at the saturations S_g and S_w computed by the model for block K at the preceding time step t_{n-1}.

This *explicit* method may sometimes produce unstable solutions for $S_{o,K}$ and $\bar{p}_k$, especially if the volume of oil produced from block K during time interval $(t_n - t_{n-1})$ represents a significant fraction of the volume of oil in the block. In fact, the use of $S_{g,K}^{n-1}$ and $S_{w,K}^{n-1}$ leads to an overestimation of $k_{ro,K}^n$ and, consequently, of $q_{o,K}^n$ – the flow rate of oil from block K at time t_n, *measured under reservoir conditions* ($q_{o,K}^n = B_o q_{o,sc,K}^n$). The model will then overestimate $S_{g,K}^n$ and $S_{w,K}^n$ and underestimate $\bar{p}_K^n$.

It follows that the model will underestimate k_{ro}^{n+1} and, therefore, $q_{o,k}^{n+1}$ at the next time step t_{n+1}. These oscillations in block pressures and saturations may, in extreme cases, lead to instability (Fig. 13.35).

The calculation of $q_{o,K}$ by an implicit method avoids this problem, and produces a very stable model – however, a large number of calculations are required. The method used is based on the fact that at time t_n the oil flow rate $q_{o,K}^n$ from layer K is a function of the values of $\bar{p}_K$, $p_{wf,K}$ and (through k_{ro}) S_g and S_w, in block K at time t_n:

$$q_{o,K} = f_K(\bar{p}_K, p_{wf,K}, S_{g,K}, S_{w,K}). \tag{13.145a}$$

Therefore:

$$dq_{o,K} = \frac{\partial q_{o,K}}{\partial \bar{p}_K} d\bar{p}_K + \frac{\partial q_{o,K}}{\partial p_{wf,K}} dp_{wf,K}$$
$$+ \frac{\partial q_{o,K}}{\partial S_{g,K}} dS_{g,K} + \frac{\partial q_{o,K}}{\partial S_{w,K}} dS_{w,K}. \tag{13.145b}$$

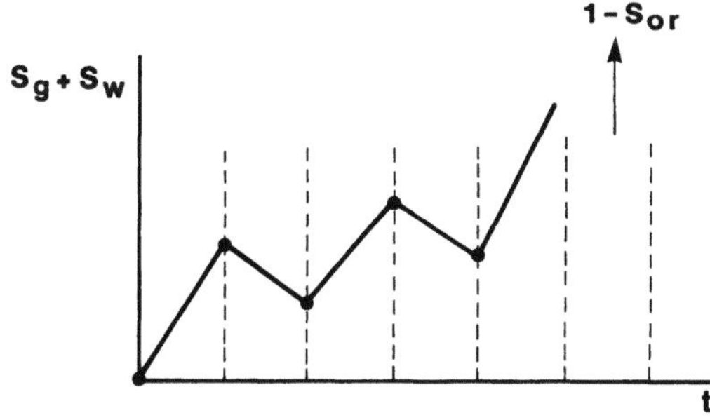

Fig. 13.35. *Oscillating well flow rates and saturations in the block containing the well, leading to instability. This can occur when the oil flow rate is solved explicitly*

If we define B_o as the formation volume factor of the oil for "composite liberation" at surface separator conditions, from Eqs. (13.141b) and (13.142) we can derive:

$$\frac{\partial q_{o,K}}{\partial \bar{p}_K} = J_{o,K} B_o, \tag{13.146a}$$

$$\frac{\partial q_{o,K}}{\partial p_{wf,K}} = -J_{o,K} B_o, \tag{13.146b}$$

$$\frac{\partial q_{o,K}}{\partial S_{g,K}} = (\bar{p}_K - p_{wf,K}) \frac{2\pi h_K k_K}{\mu_o} \frac{1}{\ln \dfrac{0.2\Delta x}{r_w} - \dfrac{3}{4} + S} \frac{\partial k_{ro}}{\partial S_{g,K}}, \tag{13.146c}$$

$$\frac{\partial q_{o,K}}{\partial S_{w,K}} = (\bar{p}_K - p_{wf,K}) \frac{2\pi h_K k_K}{\mu_o} \frac{1}{\ln \dfrac{0.2\Delta x}{r_w} - \dfrac{3}{4} + S} \frac{\partial k_{ro}}{\partial S_{w,K}}. \tag{13.146d}$$

Since $\bar{q}_{o,sc} = \text{constant}$, $d\bar{q}_{o,sc} = 0$. Therefore, ignoring the variation of $\bar{p}$ between t_{n-1} and t_n and remembering that $\bar{q}_{o,sc} = \bar{q}_o/B_o$, we can combine Eqs. (13.143) and (13.140b) to produce:

$$\sum_{L}^{U} {}_K \left[J_{o,K}^{n-1} \left(d\bar{p}_K - dp_{wf,U} \right) \right.$$

$$+ \frac{2\pi h_K k_K}{\mu_o B_o} \frac{1}{\ln \dfrac{0.2\Delta x}{r_w} - \dfrac{3}{4} + S}$$

$$\left. \times \left(\frac{\partial k_{ro}}{\partial S_{g,K}} dS_{g,K} + \frac{\partial k_{ro}}{\partial S_{w,K}} dS_{w,K} \right) \left(\bar{p}_K^n - p_{wf,U}^n - g\bar{\rho}\Delta z_K \right) \right]$$

$$= 0, \tag{13.147}$$

where, by finite differences:

$$d\bar{p}_K = p_K^n - p_K^{n-1} \qquad dS_{g,K} = S_{g,K}^n - S_{g,K}^{n-1},$$

$$dp_{wf,U} = p_{wf,U}^n - p_{wf,U}^{n-1} \qquad dS_{w,K} = S_{w,K}^n - S_{w,K}^{n-1}.$$

and the partial derivatives of k_{ro} with respect to S_g and S_w are evaluated at time t_{n-1}.

Equation (13.147) has four variables: $p_{wf,U}$, $\bar{p}_K$, $S_{g,K}$ and $S_{w,K}$. Three of these appear in Eqs. (13.99) describing the reservoir behaviour – in particular, the behaviour of the series of overlying blocks from which the well is producing.

Now that we have Eq. (13.147), which is limited to these same blocks, we can also calculate $p_{wf,U}^n$ implicitly. With Eq. (13.140b) we can then obtain $p_{wf,K}^n$. Once we have p_K^n, $p_{wf,K}^n$ and (through $S_{g,K}^n$ and $S_{w,K}^n$) k_{ro}^n, $J_{o,K}^n$ follows, and $q_{o,K}^n$ can then be calculated from Eq. (13.141a).

The calculations of the "induced flow rates" of water and gas are then derived from the obvious relationships:

$$q_{w,sc,K}^n = q_{o,sc,K}^n \frac{J_{w,K}^n}{J_{o,K}^n}, \tag{13.148a}$$

$$q_{g,sc,K}^n = q_{o,sc,K}^n \left(R_s^n + \frac{J_{g,K}^n}{J_{o,K}^n} \right), \tag{13.148b}$$

where the k_{rw} and k_{rg} appearing in $J^n_{\text{w},K}$ and $J^n_{\text{g},K}$ are calculated using $S^n_{\text{w},K}$ and $S^n_{\text{g},K}$ respectively.

By solving implicitly for $p_{\text{wf},K}$, $q_{\text{o,sc},K}$, $q_{\text{g,sc},K}$ and $q_{\text{w,sc},K}$ for all layers in the manner described above, all risk of flow rate instability is eliminated.

13.20.3.3 Wells Producing at Deliverability

This case was introduced in Sect. 13.20.3.1: when the bottom-hole flowing pressure in a well reaches the minimum value for which the produced fluids can still be lifted to the surface, the flow rate is reduced progressively in such a way as to prevent the flowing pressure from falling below the limit. The well is then said to be producing at its "deliverability", or at "constant bottom-hole pressure" $p_{\text{wf,U}}$. In this case, $dp_{\text{wf},K} = 0$ for any K, and Eq. (13.147) will not be required to calculate $dp_{\text{wf,U}}$, because it is always zero.

Equation (13.144), on the other hand, still describes the relationship between the well and the column of blocks K from which it is producing.

There is an alternative *semi-implicit* method of evaluating the flow rate. When Eq. (13.142) is expanded as a series as far as the first term, we get (ignoring the effect of $\bar{p}_K$ variations on μ_{o} and B_{o}):

$$J^n_{\text{o},K} = J^{n-1}_{\text{o},K} + \left(\frac{\partial J_{\text{o},K}}{\partial S_{\text{g}}}\right)^{n-1} dS_{\text{g}} + \left(\frac{\partial J_{\text{o}K}}{\partial S_{\text{w}}}\right)^{n-1} dS_{\text{w}}, \tag{13.149}$$

which, substituting for $J^{n-1}_{\text{o},K}$, becomes:

$$J^n_{\text{o},K} = \frac{2\pi h_K k_K}{\mu_{\text{o}} B_{\text{o}}} \frac{1}{\ln\dfrac{0.2\Delta x}{r_{\text{w}}} - \dfrac{3}{4} + S} \left[k^{n-1}_{\text{ro}} + \left(\frac{\partial k_{\text{ro}}}{\partial S_{\text{g}}}\right)^{n-1} dS_{\text{g}} \right.$$
$$\left. + \left(\frac{\partial k_{\text{ro}}}{\partial S_{\text{w}}}\right)^{n-1} dS_{\text{w}} \right]. \tag{13.150}$$

k^{n-1}_{ro} and its derivatives with respect to S_{g} and S_{w} are known terms, having been calculated at time t_{n-1}; and $dS_{\text{g}} = (S^n_{\text{g}} - S^{n-1}_{\text{g}})$, $dS_{\text{w}} = (S^n_{\text{w}} - S^{n-1}_{\text{w}})$.

We also know that:

$$q^n_{\text{o,sc},K} = J^n_{\text{o},K}(\bar{p}^n_k - p_{\text{wf,U}} - g\bar{\rho}\Delta z_K). \tag{13.141c}$$

$q^n_{\text{o,sc},K}$ is then calculated semi-implicitly by solving Eqs. (13.150) and (13.141c) simultaneously with Eqs. (13.99) for the blocks containing the well. The "induced" flow rates of water and gas can then be derived by the method described in Sect. 13.20.3.2 for constant rate production, using Eqs. (13.148).

13.21 Summary

Numerical models have been used by engineers to forecast reservoir behaviour for a good 30 years now. These models, and the numerical techniques they use, are physically and mathematically highly complex. In addition, powerful computers, with a large memory capacity and a high calculation speed, are required to run them. Perhaps for these reasons, there is often a tendency for the engineer to have more confidence in the results of a simulation than is actually justified.

It should never be forgotten that these results can, at the very best, only be as good as the basic data input to the model: the computer is not yet an "intelligent" entity, just a "number cruncher". It cannot increase the validity of the data with which it is supplied.

It has been pointed out a number of times in this chapter that it is absolutely essential for the reservoir engineer to devote sufficient time to the critical initial phase of constructing a reliable geological model, before proceeding to the simulation phase.

It is a fallacy to believe that the use of a more sophisticated model (for example, three dimensional, triphasic, fully compositional with EOS) will necessarily produce more reliable results than a simpler model (for example two dimensional, triphasic, black oil). The only certainty is that the more sophisticated model will need more computing time!

Every so often a new numerical model appears, but the engineer should try to avoid being seduced by the latest "fashion". He should use his experience and good sense to choose a model which is *adequate* for the problem on hand, given the constraints imposed by the quantity and quality of data available.

One of the most fruitful applications of numerical modelling is the *qualitative* (or, at most, semi-quantitative) comparison of the reservoir behaviour predicted for a number of different development scenarios (number and location of wells, the addition of in-fill wells during the production life cycle, well stimulation, recompletions, the implementation of water or gas injection at a specified time, etc).

Even though the results predicted for the different scenarios may not be *quantitatively* valid, provided the geological model upon which the simulations have been based is reasonably correct, they will be extremely useful for *comparative* purposes. It is only by making successive refinements through repeated model validation as more production data become available over the life of the reservoir (Sect. 13.19) that a model capable of truly *quantitative* forecasting will be obtained. As a rule of thumb, a reliable forecast can only be made as far into the future as the length of the production history used to validate the model.

References

1. Aziz K, Settari A (1979) Petroleum reservoir simulation. Applied Science Publishers, Barking, UK
2. Bansái PP, Harper JL, Mac Donald AE, Moreland EE, Odeh AS, Trimble RH (1979) A strongly coupled, fully implicit, three dimensional, three phase reservoir simulator. Paper SPE 8329 presented at the 54th SPE Annu Fall Meet, Las Vegas, Sept 23–26
3. Begg SH, Chang DM, Haldorsen HH (1985) A simple statistical method for calculating the effective vertical permeability of a reservoir containing discontinuous shales. Paper SPE 14271 presented at the 60th SPE Annu Fall Meet, Las Vegas, Sept 22–25
4. Begg SH, Carter RR, Dranfield P (1987) Assigning effective values to simulator grid-block parameters in heterogeneous reservoirs. Paper SPE 16754 presented at the 62nd SPE Annu Fall Meet, Dallas, Sept 27–30
5. Chappelaar JE, Williamson AE (1981) Representing wells in numerical reservoir simulation. Part I. Theory; Part II. Implementation. Soc Pet Eng J June: 323–344: Trans, AIME 271
6. Chierici GL (1976) Modelli per la simulazione del comportamento dei giacimenti petroliferi. Una revisione critica. Boll Assoc Min Subalpina 13(3): 237–251

7. Chierici GL (1985) Petroleum reservoir engineering in the year 2000. Energy Exploration Exploitation 3: 173–193

8. Chierici GL, Pizzi G, Ciucci GM (1967) Water drive gas reservoirs: uncertainty in reserves evaluation from past history. J Pet Technol Feb: 237–244: Discussion in J Pet Technol July: 965–967; Trans, AIME 240

9. Chierici GL, Ciucci GM, Sclocchi G, Terzi L (1978) Water drive from interbedded shale compaction in superpressured gas reservoirs – a model study. J Pet Technol June: 937–946

10. Coats KH (1968) A treatment of the gas percolation problem in simulation of three-dimensional, three-phase flow in reservoirs. Trans, AIME 243: 413–419 (Part II)

11. Crichlow HB (1977) Modern reservoir engineering – a simulation approach. Prentice-Hall, Englewood Cliffs

12. Croes GA, Schwarz N (1955) Dimensionally scaled experiments and theories on the water drive process. Trans, AIME 204: 35–42

13. Crout PD (1941) A short method for evaluating determinants and solving systems of linear equations with real and complex coefficients. Trans, AIEE 60: 1235–45

14. Douglas J, Rachford HH (1956) On the numerical solution of heat conduction problems in two or three space variables. Trans Am Math Soc 62: 421–439

15. Douglas J, Peaceman DW, Rachford HH (1959) A method for calculating multi-dimensional immiscible displacement. Trans, AIME 216: 297–308

16. Faddeeva VN (1959) The computational methods of linear algebra. Dover Publications, New York

17. Fayers FJ (1987) An extension of Stone's method I and conditions for real characteristics in three-phase flow. Paper SPE 16965 presented at the 62nd SPE Annu Fall Meet, Dallas, Sept 27–30

18. George A, Liu JWH (1981) Computer solution of large sparse positive definite systems. Prentice-Hall, Englewood Cliffs

19. Gottardi G (1986) Modello numerico per i giacimenti di gas naturale col metodo di Galerkin. Boll Assoc Min Subalpina 23 (2–3): 171–191

20. Hageman LA, Kellog RB (1968) Estimating optimum overrelaxation parameters. Math Comp 22: 60–68

21. Haldorsen HH (1983) Reservoir characterization procedure for numerical simulation. PhD Thesis, University of Texas at Austin

22. Haldorsen HH, Lake LW (1984) A new approach to shale management in field-scale models. Soc Pet Eng J Aug: 447–457; Trans, AIME 277

23. Jacks HH, Smith OJ, Mattax CC (1973) The modelling of a three-dimensional reservoir with a two-dimensional reservoir simulator – the use of dynamic pseudo-functions. Soc Pet Eng J June: 175–185

24. Kyte JR, Berry DW (1975) New pseudo functions to control numerical dispersion. Soc Pet Eng J Aug: 269–276

25. Letkeman JP, Ridings RL (1970) A numerical coning model. Soc Pet Eng J Dec: 418–424; Trans, AIME 249

26. Mac Donald RC, Coats KH (1970) Methods for numerical simulation of water and gas coning. Soc Pet Eng J Dec: 425–436; Trans, AIME 249

27. Nolen JS, Berry DW (1972) Tests of stability and time – step sensitivity of semi – implicit reservoir simulation techniques. Soc Pet Eng J June: 253–266; Trans, AIME 253

28. Peaceman DW (1977) Fundamentals of numerical reservoir simulation. Elsevier, Amsterdam

29. Peaceman DW (1978) Interpretation of well-block pressures in numerical reservoir simulation. Soc Pet Eng J June: 183–194; Trans, AIME 265

30. Peaceman DW, Rachford HH (1955) The numerical solution of parabolic and elliptic differential equations. SIAM J 3 (1): 28–41

31. Pedrosa OA Jr, Aziz K (1986) Use of a hybrid grid in reservoir simulation. SPE Reservoir Eng Nov: 611–621; Trans, AIME 281

32. Price HS, Coats KH (1974) Direct methods in reservoir simulation. Soc Pet Eng J June: 295–308; Trans, AIME 257

33. Richtmeyer RD, Morton KW (1967) Difference methods for initial – value problems, 2nd edn. Interscience New York

34. Rigler AK (1965) Estimation of the successive over-relaxation factor. Math Comp 19 (90): 302–307

35. Sheldon JW, Harris CD, Bavly D (1960) A method for general reservoir behavior simulation on digital computers. Paper 1521-G presented at the 35th SPE Annu Fall Meet, Denver. Oct 2–5

36. Smith GD (1965) Numerical solution of partial differential equations. Oxford University Press, London

37. Stone HL (1968) Iterative solutions of implicit approximations of multi-dimensional partial differential equations. SIAM J Sept: 530–538

38. Stone HL, Garder AO Jr (1961) Analysis of gas cap or dissolved gas drive reservoirs. Trans, AIME 222: 92–104 (Part II)

39. Thomas GW (1982) Principles of hydrocarbon reservoir simulation, 2nd edn. Prentice Hall, Englewood Cliffs

40. Thomas GW, Thurnau DH (1983) Reservoir simulation using an adaptive implicit method. Soc Pet Eng J Oct: 759–768; Trans, AIME 275

41. Thomas LK, Lumpkin WB, Reheis GM (1976) Reservoir simulation of variable bubble-point problems. Trans, AIME 261: 10–16

42. Todd MR, O'Dell PM, Hirasaki GJ (1972) Methods for increased accuracy in numerical reservoir simulators. Soc Pet Eng J Dec: 515–530; Trans, AIME 253

43. van Meurs P (1957) The use of transparent three-dimensional models for studying the mechanism of flow processes in oil reservoirs. Trans, AIME 210: 295–301

44. Varga RS (1962) Matrix iterative analysis. Prentice-Hall, Englewood Cliffs

45. von Rosenberg DU (1982) Local mesh refinement for finite difference methods. Paper SPE 10979 presented at the 57th SPE Fall Meet, New Orleans, Sept 26–29

46. Weber KJ (1982) Influence of common sedimentary structures on fluid flow in reservoir models. J Pet Technol March: 665–672; Trans, AIME 273

47. Weinstein HG, Stone HL, Kwan TV (1969) Iterative procedure for solution of systems of parabolic and elliptic equations in three dimensions. Ind Eng Chem Fund 8: 281–287

48. Weinstein HG, Stone HL, Kwan TV (1970) Simultaneous solution of multiphase flow problems. Soc Pet Eng J June: 99–110

49. Woo PT, Roberts SJ, Gustavson FG (1973) Application of sparse matrix techniques in reservoir simulation. Paper SPE 4544 presented at the 48th SPE Annu Fall Meet, Las Vegas, Sept 30–Oct 3

50. Young LC (1981) A finite-element method for reservoir simulation. Soc Pet Eng J Feb: 115–128; Trans, AIME 271

51. Alabert FG, Massonat GJ (1990) Heterogeneity in a complex turbiditic reservoir: stochastic modelling of facies and petrophysical variability. Paper SPE 20604 presented at the 65th Annu Conf and Exhibition of SPE, New Orleans, Sept 23–26

52. Chierici GL (1992) Economically improving oil recovery by advanced reservoir management. J Pet Sci Eng 8 (3): 205–219

53. Dake LP (1978) Fundamentals of reservoir engineering. Elsevier, Amsterdam

54. Damsleth E, Tjølsen CB, Omre KH, Haldorsen HH (1990) A two-stage stochastic model applied to a North Sea reservoir. Paper SPE 20605 presented at the 65th Annu Conf and Exhibition of SPE, New Orleans, Sept 23–26

55. Delhomme AEK, Giannesini JF (1979) New reservoir description techniques improve simulation results in Hassi-Messaoud Field, Algeria. Paper SPE 8435 presented at the 54th Annu Conf and Exhibition of SPE, Las Vegas, Sept 23–26

56. Farmer CL (1988) The generation of stochastic fields of reservoir parameters with specified geostatistical distributions. Mathematics of Oil Production. Clarendon Press, Oxford pp 235–252

57. Farmer CL (1989) The mathematical generation of reservoir geology. Paper presented at the IMA Eur Conf on the Mathematics of oil recovery, Cambridge, UK, July 25–27

58. Guérillot D, Rudkiewicz JL, Ravenne Ch, Renard G (1989) An integrated model for computer-aided reservoir description: from outcrop study to fluid flow simulations. Proc 5th Eur Symp on Improved oil recovery, Budapest, April 25–27, pp 651–660

59. Haldorsen HH, Damsleth E (1990) Stochastic modelling. J Pet Technol April: 404–412; Discussion in J Pet Technol July: 929–930

60. Hewett TA (1986) Fractal distributions of reservoir heterogeneity and their influence on fluid transport. Paper SPE 15386 presented at the 61st Annu Conf and Exhibition of SPE, New Orleans, Oct 5–8

61. Journel AG, Alabert FG (1990) New method for reservoir mapping. J Pet Technol Feb: 212–218

62. Matheron G (1973) The intrinsic random functions and their applications. Adv Appl Probability 5: 439–468
63. Omre H, Jorde K, Gulbrandsen B, Nystuen JP (1992) Heterogeneous models – field examples, Chap 6. SPOR monograph – Recent advances in improved oil recovery methods. Norwegian Petroleum Directorate, Stavanger, Norway
64. Rudkiewicz JL, Guérillot D, Galli A, HERESIM Group (1990) An integrated software for stochastic modelling of reservoir lithology with an example from the Yorkshire Middle Jurassic. In: Buller AT (ed) North Sea oil and gas reservoirs II. Graham & Trotman, London, pp 399–406
65. Weber KJ, van Geuns LC (1990) Framework for constructing clastic reservoir simulation models. J Pet Technol Oct: 1248–1297; Trans, AIME 289

EXERCISES

Exercise 13.1

Derive the continuity equation in radial coordinates [Eq. (13.10) in the main text].

Solution

In a system of radial coordinates with the z-axis vertical, we consider an elementary volume (Fig. E13/1.1) consisting of an annular ring with a square cross-section contained between two concentric cylinders of radius r and $(r + dr)$, centred on the z-axis, and two horizontal planes at depths z and $(z + dz)$. ABCD is the section of this ring on the (r, z) plane.

We will next consider, in turn, the two components of the Darcy velocity vector u in the r- and z-directions.

ρ_i is the density of fluid i at in situ p and T. The mass velocity of this fluid entering the elemental volume in the radial direction (across AB) is therefore:

$$V_{r,1} = \rho_i u_{ir}.$$

The mass velocity of fluid leaving in the radial direction (across CD) is:

$$V_{r,2} = \rho_i u_{ir} + \frac{\partial}{\partial r}(\rho_i u_{ir})\,dr.$$

Therefore:

$$dV_r = V_{r,1} - V_{r,2} = -\frac{\partial}{\partial r}(\rho_i u_{ir})\,dr.$$

Since the cross-sectional area to radial flow is:

$$A_r = 2\pi r\,dz,$$

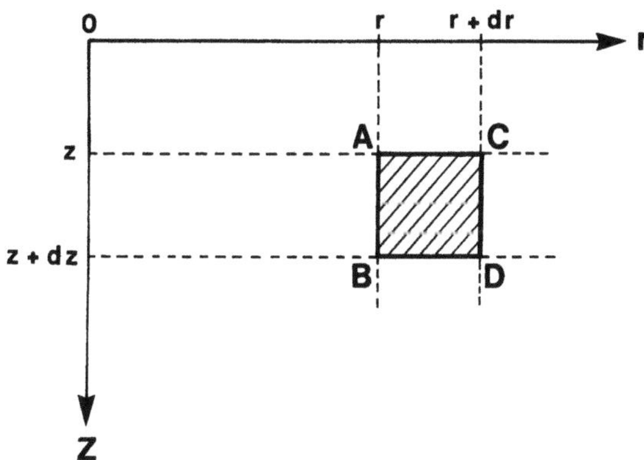

Fig. E13/1.1

the change in the mass of fluid contained in the volume in time dt will be:

$$dm_r = -2\pi r \frac{\partial}{\partial r}(\rho_i u_{ir})\, dr\, dz\, dt.$$

Looking now at flow in the vertical direction, the mass velocity of fluid entering the volume (across AC) is:

$$V_{z,1} = \rho_i u_{iz},$$

and leaving the volume (across BD):

$$V_{z,2} = \rho_i u_{iz} + \frac{\partial}{\partial z}(\rho_i u_{iz})\, dz.$$

Therefore:

$$dV_z = V_{z,1} - V_{z,2} = -\frac{\partial}{\partial z}(\rho_i u_{iz})\, dz.$$

Since the cross-sectional area to vertical flow is:

$$A_z = 2\pi r\, dr,$$

the change in the mass of fluid contained in the volume in time dt will be:

$$dm_z = -2\pi r \frac{\partial}{\partial z}(\rho_i u_{iz})\, dr\, dz\, dt.$$

The total change in mass in time dt as a result of fluid flow will therefore be:

$$dm = dm_r + dm_z = -2\pi r \left[\frac{\partial}{\partial r}(\rho_i u_{ir}) + \frac{\partial}{\partial z}(\rho_i u_{iz}) \right] dr\, dz\, dt. \qquad (13/1.1)$$

This is the *transport term.*
The volume of the annular ring is:

$$V = 2\pi r\, dr\, dz.$$

If ϕ is the porosity of the medium, and S_i the saturation of fluid i, then the mass of this fluid contained in the elemental volume is:

$$m = 2\pi r(\phi \rho_i S_i)\, dr\, dz.$$

It follows that, in time dt, the change in the mass of fluid i in the elemental volume will be:

$$dm = \frac{\partial m}{\partial t}dt = 2\pi r \frac{\partial}{\partial t}(\phi \rho_i S_i)\, dr\, dz\, dt. \qquad (13/1.2)$$

This is the *accumulation term.*
Equating Eqs. (13/1.1) and (13/1.2) and simplifying, we get:

$$\frac{\partial}{\partial r}(\rho_i u_{ir}) + \frac{\partial}{\partial z}(\rho_i u_{iz}) = -\frac{\partial}{\partial t}(\phi \rho_i S_i).$$

This is the continuity equation in radial coordinates.

Exercise 13.2

Derive the expression for the coefficient α in Eq. (13.16b) for one- and two-dimensional geometries.

Solution

One-Dimensional Geometry

Consider a system with the geometry illustrated in Fig. E13/2.1. Flow is essentially one-dimensional, but the cross-sectional area $A(x)$ varies with x.

The mass flow rate across any section $A(x)$ is:

$$q_{m,1} = \rho u A(x),$$

where ρ is the density of the fluid, and u is its Darcy velocity.

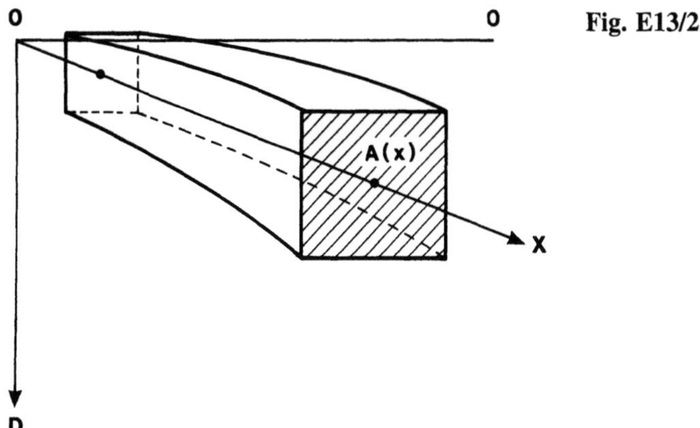

Fig. E13/2.1

Across a nearby section a distance dx away, the mass flow rate will be:

$$q_{m,2} = q_{m,1} + dq = \rho u A(x) + \frac{\partial}{\partial x}[\rho u A(x)]\,dx.$$

Therefore, the change in mass in time dt in the elemental volume contained between the two surfaces is (the *transport term*):

$$dm_t = -\frac{\partial}{\partial x}[\rho u A(x)]\,dx\,dt.$$

If ϕ is the porosity of the medium, and S the saturation of fluid in question, then the change in mass of fluid in the elemental volume in time dt can also be expressed as (the *accumulation term*):

$$dm_a = \frac{\partial}{\partial t}[A(x)dx\rho\phi S]\,dt = A(x)\frac{\partial}{\partial t}(\rho\phi S)\,dx\,dt.$$

Now if fluid is extracted from the rock at a mass flow rate *per unit volume of rock* of $\tilde{q}_m$, the mass flow rate leaving the elemental volume in time dt is (the *source term*):

$$dm_s = -[A(x)dx]\tilde{q}_m\,dt.$$

Conservation of mass requires that:

$$dm_t + dm_s = dm_a,$$

so that:

$$-\frac{\partial}{\partial x}[\rho u A(x)] - A(x)\tilde{q}_m = A(x)\frac{\partial}{\partial t}(\phi\rho S). \tag{13/2.1}$$

The generalised Darcy equation states that:

$$u = -\frac{kk_r\rho}{\mu}\frac{\partial\Phi}{\partial x}.$$

Since

$$\Phi = \int_0^P \frac{dp}{\rho} - gD,$$

where D = depth of point measured from an arbitrary reference datum, this becomes:

$$u = -\frac{kk_r}{\mu}\left(\frac{\partial p}{\partial x} - \rho g\frac{dD}{dx}\right). \tag{13/2.2}$$

Substituting for u in Eq. (13/2.1) we get:

$$\frac{\partial}{\partial x}\left[\frac{A(x)kk_r\rho}{\mu}\left(\frac{\partial p}{\partial x} - \rho g\frac{dD}{dx}\right)\right] - A(x)\tilde{q}_m$$

$$= A(x)\frac{\partial}{\partial t}(\phi\rho S).$$

This is the same as Eq. (13.16b) for a one-dimensional system, with $\alpha = A(x)$.

Two-Dimensional Geometry

Consider a two-dimensional system in a plane which may or may not be horizontal. The thickness of the porous medium is a function $h(x, y)$ of the local coordinates.

An elemental volume of this porous medium consists of a rectangular parallelepiped whose base is dx · dy and height $h(x, y)$.

Following a similar reasoning to that for the one-dimensional geometry, we have:

a) *transport term*

x-direction:

$$-\frac{\partial}{\partial x}\left[\rho u_x h(x, y)\mathrm{d}y\right]\mathrm{d}x\ \mathrm{d}t.$$

y-direction:

$$-\frac{\partial}{\partial y}\left[\rho u_y h(x, y)\mathrm{d}x\right]\mathrm{d}y\ \mathrm{d}t.$$

total:

$$\mathrm{d}m_t = -\left\{\frac{\partial}{\partial x}\left[\rho u_x h(x, y)\right] + \frac{\partial}{\partial y}\left[\rho u_y h(x, y)\right]\right\}\mathrm{d}x\ \mathrm{d}y\ \mathrm{d}t.$$

b) *accumulation term*

$$\mathrm{d}m_a = \frac{\partial}{\partial t}\left[h(x, y)\mathrm{d}x\ \mathrm{d}y\ \phi\rho S\right]\mathrm{d}t$$

$$= h(x, y)\frac{\partial}{\partial t}(\phi\rho S)\,\mathrm{d}x\ \mathrm{d}y\ \mathrm{d}t.$$

c) *source term*

$$\mathrm{d}m_s = -[h(x, y)\mathrm{d}x\ \mathrm{d}y]\tilde{q}_m\mathrm{d}t$$

$$= -h(x, y)\tilde{q}_m\mathrm{d}x\ \mathrm{d}y\ \mathrm{d}t.$$

For mass conservation we have:

$$\mathrm{d}m_t + \mathrm{d}m_s = \mathrm{d}m_a,$$

so that:

$$-\frac{\partial}{\partial x}\left[\rho u_x h(x, y)\right] - \frac{\partial}{\partial y}\left[\rho u_y h(x, y)\right] - h(x, y)\tilde{q}_m$$

$$= h(x, y)\frac{\partial}{\partial t}(\phi\rho S). \tag{13/2.3}$$

From Eq. (13/2.2) we can write that:

$$u_x = -\frac{k_x k_r}{\mu}\left(\frac{\partial p}{\partial x} - \rho g\frac{\partial D}{\partial x}\right),$$

$$u_y = -\frac{k_y k_r}{\mu}\left(\frac{\partial p}{\partial y} - \rho g\frac{\partial D}{\partial y}\right).$$

Substituting for u_x and u_y into Eq. (13/2.3), we obtain:

$$\frac{\partial}{\partial x}\left[\frac{h(x, y)k_x k_r\rho}{\mu}\left(\frac{\partial p}{\partial x} - \rho g\frac{\partial D}{\partial x}\right)\right] + \frac{\partial}{\partial y}\left[\frac{h(x, y)k_y k_r\rho}{\mu}\left(\frac{\partial p}{\partial y} - \rho g\frac{\partial D}{\partial y}\right)\right] - h(x, y)\tilde{q}_m$$

$$= h(x, y)\frac{\partial}{\partial t}(\phi\rho S),$$

or, using symbolic notation:

$$\nabla\cdot\left[\frac{h(x, y)[k]k_r\rho}{\mu}(\nabla p - \rho g\ \nabla D)\right] - h(x, y)\tilde{q}_m = h(x, y)\frac{\partial}{\partial t}(\phi\rho S).$$

This is the same as Eq. (13.16b) for the two-dimensional case, with h as a function of position.

◇ ◇ ◇

Exercise 13.3

From the generalised compositional Eq. (13.41) for polyphasic flow, commonly used in the simulation of the behaviour of volatile oil and condensate reservoirs, derive the appropriate equations for the following two cases:

a) a moderately volatile oil reservoir,
b) a medium to heavy oil (black oil).

Solution

Firstly, to recap Eq. (13.41):

$$
\nabla \cdot \left[\frac{C_{i,\text{o}}\rho_\text{o}[k]k_\text{ro}}{\mu_\text{o}} \left(\nabla p_\text{o} - \rho_\text{o}g\,\nabla D\right)\right] + \nabla \cdot \left[\frac{C_{i,\text{g}}\rho_\text{g}[k]k_\text{rg}}{\mu_\text{g}} \left(\nabla p_g - \rho_g g\,\nabla D\right)\right]
$$

$$
+ \nabla \cdot \left[\frac{C_{i,\text{w}}\rho_\text{w}[k]k_\text{rw}}{\mu_\text{w}} \left(\nabla p_\text{w} - \rho_\text{w}g\,\nabla D\right)\right] - C_{i,\text{o}}\tilde{q}_{\text{m,o}} - C_{i,\text{g}}\tilde{q}_{\text{m,g}} - C_{i,\text{w}}\tilde{q}_{\text{m,w}}
$$

$$
= \frac{\partial}{\partial t} \left[\phi \left(C_{i,\text{o}}\rho_\text{o}S_\text{o} + C_{i,\text{g}}\rho_\text{g}S_\text{g} + C_{i,\text{w}}\rho_\text{w}S_\text{w}\right)\right], \tag{13.41}
$$

where:

$C_{i,\text{o}}, C_{i,\text{g}}, C_{i,\text{w}}$ represent the *mass fractions* of the component i in, respectively, the oil, gas and water phase – in other words, the mass of component i per unit mass of liquid hydrocarbon, gaseous hydrocarbon and water phases respectively.

and:

$\tilde{q}_{\text{m,o}}, \tilde{q}_{\text{m,g}}, \tilde{q}_{\text{m,w}}$ represent the *mass flow rates* of oil, gas and water (*under reservoir conditions*) extracted or injected per unit volume of porous medium.

a) Moderately Volatile Oil

Since there is no need to use a fully compositional treatment of the flow equations for moderately volatile oils, we can conveniently represent the reservoir behaviour in terms of *three components*:

– *oil* under standard conditions of 0.1013 MPa and 288.2 K (stock tank oil);
– *gas* under standard conditions;
– *water* under standard conditions.

The exchange of mass between the *three phases* (gas, liquid hydrocarbon and water) is modelled using the respective solubilities:

$R_\text{s} =$ m^3 of gas under standard conditions that are dissolved in a
volume of liquid hydrocarbon in the reservoir that corresponds
to 1 m^3 of stock tank oil;

$R_\text{sw} =$ m^3 of gas under standard conditions that are dissolved in a
volume of water in the reservoir that corresponds to 1 m^3 of
water under standard conditions.

These were already defined in Chap. 2. We also need to define a new term:

$R_\text{v} =$ m^3 of stock tank oil dissolved in a volume of reservoir gas
that corresponds to 1 m^3 of gas under standard conditions.

The introduction of R_v, whose variation with pressure can be evaluated experimentally, allows for the vaporisation of the oil *component* in the gas *phase*.
We will also use the customary nomenclature:

$\rho_{\text{o,sc}}, \rho_{\text{g,sc}}, \rho_{\text{w,sc}}$ the densities of the oil, gas and water components under standard conditions.

Now we can calculate the mass fractions $C_{i,j}$ which appear in Eq. (13.41).

Gas Phase

From 1 sm^3 of gas of density $\rho_{\text{g,sc}}$, and R_v m^3 of stock tank oil of density $\rho_{\text{o,sc}}$, we obtain B_g m^3 of gas of density ρ_g under reservoir conditions. Therefore:

- the mass fraction of the gas *component* in the gas *phase* is:

$$C_{\text{g,g}} = \frac{\rho_{\text{g,sc}}}{\rho_{\text{g}} B_{\text{g}}},$$
(13/3.1a)

- the mass fraction of the oil *component* in the gas *phase* is:

$$C_{\text{o,g}} = \frac{R_{\text{v}} \rho_{\text{o,sc}}}{\rho_{\text{g}} B_{\text{g}}}.$$
(13/3.1b)

Liquid Hydrocarbon Phase

From 1 m³ of stock tank oil of density $\rho_{\text{o,sc}}$, and R_{s} sm³ of gas of density $\rho_{\text{g,sc}}$, we obtain B_{o} m³ of oil of density ρ_{o} under reservoir conditions. Therefore:

- the mass fraction of the gas *component* in the hydrocarbon liquid *phase* is:

$$C_{\text{g,o}} = \frac{R_{\text{s}} \rho_{\text{g,sc}}}{\rho_{\text{o}} B_{\text{o}}}.$$
(13/3.2a)

- the mass fraction of the oil *component* in the hydrocarbon liquid *phase* is:

$$C_{\text{o,o}} = \frac{\rho_{\text{o,sc}}}{\rho_{\text{o}} B_{\text{o}}}.$$
(13/3.2b)

Water Phase

The solubility $C_{\text{w,g}}$ of water in gas is considered to be zero, or negligible, as is the solubility $C_{\text{w,o}}$ of water in oil, and $C_{\text{o,w}}$ of oil in water.

From 1 m³ of water of density $\rho_{\text{w,sc}}$ under standard conditions, and R_{sw} sm³ of gas of density $\rho_{\text{g,sc}}$, we obtain B_{w} m³ of water of density ρ_{w} under reservoir conditions. Therefore:

- the mass fraction of the gas *component* in the water *phase* is:

$$C_{\text{g,w}} = \frac{R_{\text{sw}} \rho_{\text{g,sc}}}{\rho_{\text{w}} B_{\text{w}}}.$$
(13/3.3a)

- the mass fraction of the water *component* in the water *phase* is:

$$C_{\text{w,w}} = \frac{\rho_{\text{w,sc}}}{\rho_{\text{w}} B_{\text{w}}}.$$
(13/3.3b)

By definition we have:

$$\rho_{\text{g}} = \frac{\rho_{\text{g,sc}}}{B_{\text{g}}}$$
(2.8c)

and the relationship between mass flow rate $\tilde{q}_{\text{m}}$ *under reservoir conditions* and volumetric flow rate $\tilde{q}_{\text{v}}$, *under standard conditions*, is expressed by:

$$\tilde{q}_{\text{v,g}} = \frac{\tilde{q}_{\text{m,g}}}{\rho_{\text{g}} B_{\text{g}}},$$
(13.39a′)

$$\tilde{q}_{\text{v,o}} = \frac{\tilde{q}_{\text{m,o}}}{\rho_{\text{o}} B_{\text{o}}},$$
(13.39b′)

$$\tilde{q}_{\text{v,w}} = \frac{\tilde{q}_{\text{m,w}}}{\rho_{\text{w}} B_{\text{w}}}.$$
(13.39c′)

If we introduce the mass fractions as expressed in Eqs. (13/3.1), (13/3.2) and (13/3.3) into Eq. (13.41) we get:
For the gas component (i = g)

$$\nabla \cdot \left[\frac{\rho_{\text{g,sc}} R_{\text{s}} \rho_{\text{o}} [k] k_{\text{ro}}}{\rho_{\text{o}} B_{\text{o}} \mu_{\text{o}}} (\nabla p_{\text{o}} - \rho_{\text{o}} g \nabla D) \right]$$

$$+ \nabla \cdot \left[\frac{\rho_{\text{g,sc}} \rho_{\text{g}} [k] k_{\text{rg}}}{\rho_{\text{g}} B_{\text{g}} \mu_{\text{g}}} (\nabla p_{\text{g}} - \rho_{\text{g}} g \nabla D) \right]$$

$$+ \nabla\cdot\left[\frac{\rho_{g,sc}R_{sw}\rho_w[k]k_{rw}}{\rho_w B_w \mu_w}\left(\nabla p_w - \rho_w g \nabla D\right)\right]$$

$$- \left[\frac{\rho_{g,sc}R_s}{\rho_o B_o}\tilde{q}_{m,o} + \frac{\rho_{g,sc}}{\rho_g B_g}\tilde{q}_{m,g} + \frac{\rho_{g,sc}R_{sw}}{\rho_w B_w}\tilde{q}_{m,w}\right]$$

$$= \frac{\partial}{\partial t}\left[\phi\left(\frac{\rho_{g,sc}R_s}{\rho_o B_o}\rho_o S_o + \frac{\rho_{g,sc}}{\rho_g B_g}\rho_g S_g + \frac{\rho_{g,sc}R_{sw}}{\rho_w B_w}\rho_w S_w\right)\right].\tag{13/3.4a}$$

For the oil component (i = o)

$$\nabla\cdot\left[\frac{\rho_{o,sc}\rho_o[k]k_{ro}}{\rho_o B_o \mu_o}\left(\nabla p_o - \rho_o g\ \nabla D\right)\right]$$

$$+ \nabla\cdot\left[\frac{\rho_{o,sc}R_v\rho_g[k]k_{rg}}{\rho_g B_g \mu_g}\left(\nabla p_g - \rho_g g\ \nabla D\right)\right]$$

$$- \left[\frac{\rho_{o,sc}}{B_o \rho_o}\tilde{q}_{m,o} + \frac{\rho_{o,sc}R_v}{B_g \rho_g}\tilde{q}_{m,g}\right]$$

$$= \frac{\partial}{\partial t}\left[\phi\left(\frac{\rho_{o,sc}}{\rho_o B_o}\rho_o S_o + \frac{\rho_{o,sc}R_v}{\rho_g B_g}\rho_g S_g\right)\right].\tag{13/3.4b}$$

For the water component (i = w)
Since $C_{w,o}$ and $C_{w,g} = 0$ we have:

$$\nabla\cdot\left[\frac{\rho_{w,sc}\rho_w[k]k_{rw}}{\rho_w B_w \mu_w}\left(\nabla p_w - \rho_w g\ \nabla D\right)\right]$$

$$- \frac{\rho_{w,sc}}{\rho_w B_w}\tilde{q}_{m,v} = \frac{\partial}{\partial t}\left(\phi\frac{\rho_{w,sc}}{\rho_w B_w}\rho_w S_w\right).\tag{13/3.4c}$$

Using Eqs. (13.39), the source term in Eq. (13/3.4a) can be written as:

$$- \rho_{g,sc}\left[R_s\tilde{q}_{v,o} + \tilde{q}_{v,g} + R_{sw}\tilde{q}_{v,w}\right] = -\rho_{g,sc}(\tilde{q}_{v,g})_{tot},$$

where the total gas production rate $(\tilde{q}_{v,g})_{tot}$ is the sum of the gas liberated from solution in oil, the free gas, and the gas liberated from solution in water.

Similarly, in Eq. (13/3.4b), the source term can be written as:

$$-\rho_{o,sc}\left[\tilde{q}_{v,o} + R_v\tilde{q}_{v,g}\right] = -\rho_{o,sc}(\tilde{q}_{v,o})_{tot},$$

where the total oil production rate $(\tilde{q}_{v,o})_{tot}$ is the sum of oil produced as oil and the oil separated from the produced gas.

Consequently, Eqs. (13/3.4a, b and c) become, after substitution and simplification:
Gas:

$$\nabla\cdot\left[\frac{R_s[k]k_{ro}}{B_o \mu_o}\left(\nabla p_o - \rho_o g\ \nabla D\right)\right]$$

$$+ \nabla\cdot\left[\frac{[k]k_{rg}}{B_g \mu_g}\left(\nabla p_g - \rho_g g\ \nabla D\right)\right]$$

$$+ \nabla\cdot\left[\frac{R_{sw}[k]k_{rw}}{B_w \mu_w}\left(\nabla p_w - \rho_w g\ \nabla D\right)\right]$$

$$- (\tilde{q}_{v,g})_{tot} = \frac{\partial}{\partial t}\left[\phi\left(\frac{R_s S_o}{B_o} + \frac{S_g}{B_g} + \frac{R_{sw}S_w}{B_w}\right)\right].\tag{13/3.5a}$$

Stock tank oil:

$$\nabla\cdot\left[\frac{[k]k_{ro}}{B_o \mu_o}\left(\nabla p_o - \rho_o g\ \nabla D\right)\right]$$

$$+ \nabla\cdot\left[\frac{R_v[k]k_{rg}}{B_g \mu_g}\left(\nabla p_g - \rho_g g\ \nabla D\right)\right] - (\tilde{q}_{v,o})_{tot}$$

$$= \frac{\partial}{\partial t}\left[\phi\left(\frac{S_o}{B_o} + \frac{R_v S_g}{B_g}\right)\right].\tag{13/3.5b}$$

Water:

$$\nabla \cdot \left[\frac{[k]k_{\text{rw}}}{B_{\text{w}}\mu_{\text{w}}} \left(\nabla p_{\text{w}} - \rho_{\text{w}}g \; \nabla D \right) \right] - \tilde{q}_{\text{v,w}}$$

$$= \frac{\partial}{\partial t} \left(\phi \frac{S_{\text{w}}}{B_{\text{w}}} \right). \tag{13/3.5c}$$

Equations (13/3.5a, b and c) are used to simulate the behaviour of reservoirs containing moderately volatile oil, characterised by the parameter R_{v}.

Medium-Heavy Oils (Black Oils)

With medium-heavy oils (called "black oils" because of their dark colour), it is generally assumed that none of the stock tank oil exists as a vapour component in the reservoir gas phase. *This is equivalent to saying that $R_{\text{v}} = 0$.*

In other words, we are assuming that the gaseous phase is of constant composition, so that mass is exchanged between the three components in the following directions only:

  - gas dissolves in (or is liberated from) oil;
  - gas dissolves in (or is liberated from) water.

Therefore:

$$C_{\text{g,g}} = \frac{\rho_{\text{g,sc}}}{\rho_{\text{g}} B_{\text{g}}}, \qquad C_{\text{o,g}} = 0, \qquad C_{\text{w,g}} = 0,$$

$$C_{\text{g,o}} = \frac{R_{\text{s}}\rho_{\text{g,sc}}}{\rho_{\text{o}} B_{\text{o}}}, \qquad C_{\text{o,o}} = \frac{\rho_{\text{o,sc}}}{\rho_{\text{o}} B_{\text{o}}}, \qquad C_{\text{w,o}} = 0,$$

$$C_{\text{g,w}} = \frac{R_{\text{sw}}\rho_{\text{g,sc}}}{\rho_{\text{w}} B_{\text{w}}}, \qquad C_{\text{o,w}} = 0, \qquad C_{\text{w,w}} = \frac{\rho_{\text{w,sc}}}{\rho_{\text{w}} B_{\text{w}}}.$$

Note that the mass fractions of the individual components in the three phases are identical to those in the preceding case of a volatile oil, with the exception that:

$$C_{\text{o,g}} = 0 \quad \text{instead of} \quad C_{\text{o,g}} = \frac{R_{\text{v}}\rho_{\text{o,sc}}}{\rho_{\text{g}} B_{\text{g}}}.$$

Consequently, we simply have to set $R_{\text{v}} = 0$ in Eq. (13/3.5b) to obtain the equations used to simulate the behaviour of reservoirs containing medium-to-heavy oil.

These are:

Gas:

$$\nabla \cdot \left[\frac{[k]k_{\text{rg}}}{\mu_{\text{g}} B_{\text{g}}} \left(\nabla p_{\text{g}} - \rho_{\text{g}}g \; \nabla D \right) \right] + \nabla \cdot \left[\frac{[k]k_{\text{ro}} R_{\text{s}}}{\mu_{\text{o}} B_{\text{o}}} \left(\nabla p_{\text{o}} - \rho_{\text{o}}g \; \nabla D \right) \right]$$

$$+ \nabla \cdot \left[\frac{[k]k_{\text{rw}} R_{\text{sw}}}{\mu_{\text{w}} B_{\text{w}}} \left(\nabla p_{\text{w}} - \rho_{\text{w}}g \; \nabla D \right) \right] - (\tilde{q}_{\text{v,g}})_{\text{tot}}$$

$$= \frac{\partial}{\partial t} \left[\phi \left(\frac{S_{\text{g}}}{B_{\text{g}}} + \frac{R_{\text{s}} S_{\text{o}}}{B_{\text{o}}} + \frac{R_{\text{sw}} S_{\text{w}}}{B_{\text{w}}} \right) \right]. \tag{13/3.6a}$$

Stock tank oil:

$$\nabla \cdot \left[\frac{[k]k_{\text{ro}}}{\mu_{\text{o}} B_{\text{o}}} \left(\nabla p_{\text{o}} - \rho_{\text{o}}g \; \nabla D \right) \right] - \tilde{q}_{\text{v,o}} = \frac{\partial}{\partial t} \left(\frac{\phi S_{\text{o}}}{B_{\text{o}}} \right). \tag{13/3.6b}$$

Water:

$$\nabla \cdot \left[\frac{[k]k_{\text{rw}}}{\mu_{\text{w}} B_{\text{w}}} \left(\nabla p_{\text{w}} - \rho_{\text{w}}g \; \nabla D \right) \right] - \tilde{q}_{\text{v,w}} = \frac{\partial}{\partial t} \left(\frac{\phi S_{\text{w}}}{B_{\text{w}}} \right). \tag{13/3.6c}$$

Note that Eqs. (13/3.6) are the same as Eqs. (13.38) derived by a completely different method in Sect. 13.8 using the "black oil" model defined in Sect. 13.3.3.

$$\Diamond \; \Diamond \; \Diamond$$

Exercise 13.4

Derive Eq. (13.77) (in Sect. 13.12.1 of the text) for the z-axis component of the transport term of the flow equation, for the case of a three-dimensional numerical model.

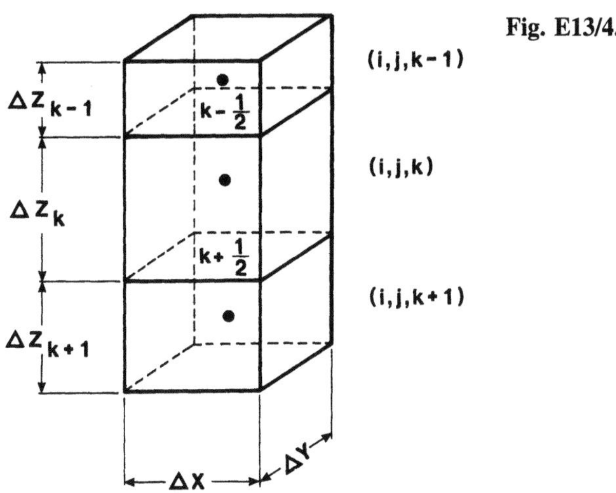

Fig. E13/4.1

Solution

Consider a series of three adjoining blocks (Fig. E13/4.1) in the z-direction, whose centres have coordinates $(i, j, k - 1)$, (i, j, k) and $(i, j, k + 1)$, and whose heights are Δz_{k-1}, Δz_k and Δz_{k+1}.

All three blocks have the same cross-sectional area $\Delta x\, \Delta y$ perpendicular to the z-axis.

From Eq. (13.67a) in the main text, the vertical interblock permeability at the interface $(k - 1/2)$ between blocks $(i, j, k - 1)$ and (i, j, k) is:

$$(k_z)_{i,j,k-1/2} = \frac{(k_z)_{i,j,k-1}(k_z)_{i,j,k}}{(k_z)_{i,j,k-1}\Delta z_k + (k_z)_{i,j,k}\Delta z_{k-1}} \left(\Delta z_{k-1} + \Delta z_k\right).$$

Similarly, at the interface $(k + 1/2)$ between blocks (i, j, k) and $(i, j, k + 1)$ it will be:

$$(k_z)_{i,j,k+1/2} = \frac{(k_z)_{i,j,k}(k_z)_{i,j,k+1}}{(k_z)_{i,j,k}\ \Delta z_{k+1} + (k_z)_{i,j,k+1}\ \Delta z_k} \left(\Delta z_k + \Delta z_{k+1}\right).$$

The potentials at the centres of the three blocks are:
Block (i,j,k-1):

$$\Phi_{i,j,k-1} = \frac{p_{i,j,k-1}}{\rho} - gz_{i,j,k-1}.$$

Block (i,j,k):

$$\Phi_{i,j,k} = \frac{p_{i,j,k}}{\rho} - gz_{i,j,k}.$$

Block (i,j,k+1):

$$\Phi_{i,j,k+1} = \frac{p_{i,j,k+1}}{\rho} - gz_{i,j,k+1}.$$

Darcy's law for linear flow states that:

$$q = -\frac{kA\rho}{\mu}\frac{\Delta\Phi}{\Delta L}.$$

The volume of fluid passing from block $(i, j, k - 1)$ to block (i, j, k) in time Δt is therefore:

$$q_{i,j,k-1/2} = \frac{(k_z)_{i,j,k-1/2}\Delta x\,\Delta y}{\mu\dfrac{\Delta z_{k-1} + \Delta z_k}{2}} \left[\left(p_{i,j,k-1} - \rho gz_{i,j,k-1}\right) - \left(p_{i,j,k} - \rho gz_{i,j,k}\right)\right]\Delta t,$$

where the distance between the two block centres is $1/2(\Delta z_{k-1} + \Delta z_k)$.

Similarly, across the interface between blocks (i, j, k) and $(i, j, k + 1)$ we have:

$$q_{i,j,k+1/2} = \frac{(k_z)_{i,j,k+1/2}\Delta x\ \Delta y}{\mu\dfrac{\Delta z_k + \Delta z_{k+1}}{2}} \left[\left(p_{i,j,k} - \rho gz_{i,j,k}\right) - \left(p_{i,j,k+1} - \rho gz_{i,j,k+1}\right)\right]\Delta t.$$

If we now define:

$$(T_z)_{i,j,k-1/2} = \frac{2(k_z)_{i,j,k-1/2}\Delta x\ \Delta y}{\mu(\Delta z_{k-1} + \Delta z_k)},$$

$$(T_z)_{i,j,k+1/2} = \frac{2(k_z)_{i,j,k+1/2}\Delta x\ \Delta y}{\mu(\Delta z_k + \Delta z_{k+1})}$$

and note that:

$$z_{i,j,k} - z_{i,j,k-1} = 1/2(\Delta z_{k-1} + \Delta z_k),$$

$$z_{i,j,k+1} - z_{i,j,k} = 1/2(\Delta z_k + \Delta z_{k+1}).$$

The difference between the inflow and outflow rates for block (i, j, k) in the z-direction will be:

$$(\Delta q_{\mathrm{v}})_{i,j,k} = q_{i,j,k-1/2} - q_{i,j,k+1/2}$$

$$= (T_z)_{i,j,k-1/2}\left[p_{i,j,k-1} - p_{i,j,k} + \rho g\frac{\Delta z_{k-1} + \Delta z_k}{2}\right]\Delta t$$

$$- (T_z)_{i,j,k+1/2}\left[p_{i,j,k} - p_{i,j,k+1} + \rho g\frac{\Delta z_k + \Delta z_{k+1}}{2}\right]\Delta t.$$

Now defining:

$$\Delta_z T_z \Delta_z p = (T_z)_{i,j,k+1/2}\left[p_{i,j,k+1} - p_{i,j,k} - \frac{\rho g}{2}(\Delta z_k + \Delta z_{k+1})\right]$$

$$- (T_z)_{i,j,k-1/2}\left[p_{i,j,k} - p_{i,j,k-1} - \frac{\rho g}{2}(\Delta z_{k-1} + \Delta z_k)\right], \qquad (13/4.1)$$

we can write this as:

$$(\Delta q_{\mathrm{v}})_{i,j,k} = \Delta_z T_z \Delta_z p \times \Delta t,$$

from which:

$$(\Delta q_{\mathrm{m}})_z = \rho \Delta_z T_z \Delta_z p \times \Delta t,$$

which is the same as Eq. (13.73).

Equation (13/4.1) is identical to Eq. (13.77) in the text, so the objective of this exercise has been met.

◊ ◊ ◊

Exercise 13.5

Derive the finite difference Eqs. (13.99a,b,c) for polyphasic flow corresponding to the differential Eqs. (13.38a,b,c).

Solution

For the discretisation of the porous medium, we can refer to the block dimensions and symbols used in Fig. 13.13 of the text, not forgetting that this in fact described a two-dimensional system.

We define the potential $\Phi_{i,j,k}$ at the centre of block (i, j, k) as:

$$\Phi_{i,j,k} = \int_0^{p_{i,j,k}} \frac{\mathrm{d}p}{\rho} - gD_{i,j,k},$$

and the depth of the block centre as $D_{i,j,k}$.

The generalised Darcy equation states that:

$$q = -\frac{[k]k_\mathrm{r}A\rho}{\mu}\ \mathrm{grad}\ \Phi.$$

Therefore, the flow rate, *measured at standard p and T*, of component f (f = g,o,w – gas, oil, water) passing from block $(i - 1, j, k)$ into block (i, j, k) in the x-axis direction will be:

$$q_{\mathrm{in},x} = \frac{\rho_f(k_x)_{i-1/2}k_{rf}\Delta y_j\Delta z_k}{\mu_f B_f}\ \frac{\Phi_{f,i-1} - \Phi_{f,i}}{\dfrac{\Delta x_{i-1} + \Delta x_i}{2}}, \qquad (13/5.1)$$

where $(k_x)_{i-1/2}$ is defined in Eq. (13.67a) of the text, μ_f and B_f are, respectively, the viscosity and formation volume factor of fluid f at reservoir p and T, k_{rf} is the relative permeability to fluid f, and Φ_f is the potential of the fluid f, and is a function of position and time.

Since we are investigating flow in the x-direction, the indices j and k (y- and z-directions) have been omitted from Eq. (13/5.1) for simplification.

In Eq. (13/5.1), if we expand the potential term as defined above, and ignore the variation of ρ_f for small values of Δp, we get:

$$q_{in,x} = \frac{2(k_x)_{i-1/2}k_{rf}\Delta y_j \Delta z_k}{\mu_f B_f(\Delta x_{i-1} + \Delta x_i)}\left[\left(p_{f,i-1} - p_{f,i}\right) - \rho_f g\left(D_{i-1} - D_i\right)\right]. \tag{13/5.1'}$$

Similarly, the flow rate of component f out of block (i, j, k) along the x-axis is:

$$q_{out,x} = \frac{2(k_x)_{i+1/2}k_{rf}\Delta y_j \Delta z_k}{\mu_f B_f(\Delta x_i + \Delta x_{i+1})}\left[\left(p_{f,i} - p_{f,i+1}\right) - \rho_f g\left(D_i - D_{i+1}\right)\right]. \tag{13/5.2}$$

The interblock transmissibilities are defined as:

$$(T_{f,x})_{i-1/2} = \frac{2(k_x)_{i-1/2}k_{rf}\Delta y_j \Delta z_k}{\mu_f B_f(\Delta x_{i-1} + \Delta x_i)},$$

$$(T_{f,x})_{i+1/2} = \frac{2(k_x)_{i+1/2}k_{rf}\Delta y_j \Delta z_k}{\mu_f B_f(\Delta x_i + \Delta x_{i+1})}.$$

Consequently, the change *per unit time* in the volume of fluid f *(expressed at standard p and T)* in block (i, j, k) as a result of flow in the x-direction – in other words, *the x-component of the transport term* – will be:

$$(\Delta q_f)_x = q_{in,x} - q_{out,x} = (T_{f,x})_{i-1/2}\left[\left(p_{f,i-1} - p_{f,i}\right) - \rho_f g\left(D_{i-1} - D_i\right)\right]$$
$$- (T_{f,x})_{i+1/2}\left[\left(p_{f,i} - p_{f,i+1}\right) - \rho_f g\left(D_i - D_{i+1}\right)\right].$$

If we define the operator:

$$\Delta_x T_{f,x}(\Delta_x p_f - \rho_f g \Delta_x D) = (T_{f,x})_{i+1/2}[(p_{f,i+1} - p_{f,i}) - \rho_f g(D_{i+1} - D_i)]$$
$$- (T_{f,x})_{i-1/2}[(p_{f,i} - p_{f,i-1}) - \rho_f g(D_i - D_{i-1})],$$

we have, finally:

$$\left(\Delta q_f\right)_x = \Delta_x T_{f,x}\left(\Delta_x p_f - \rho_f g \Delta_x D\right). \tag{13/5.3a}$$

In a similar manner, we define:

Transmissibility in the y-direction

$$\left(T_{f,y}\right)_{j-1/2} = \frac{2(k_y)_{j-1/2}k_{rf}\,\Delta x_i\,\Delta z_k}{\mu_f B_f(\Delta y_{j-1} + \Delta y_j)},$$

$$\left(T_{f,y}\right)_{j+1/2} = \frac{2(k_y)_{j+1/2}k_{rf}\,\Delta x_i\,\Delta z_k}{\mu_f B_f(\Delta y_j + \Delta y_{j+1})}.$$

Transmissibility in the z-direction

$$\left(T_{f,z}\right)_{k-1/2} = \frac{2(k_z)_{k-1/2}k_{rf}\Delta x_i\,\Delta y_j}{\mu_f B_f(\Delta z_{k-1} + \Delta z_k)},$$

$$\left(T_{f,z}\right)_{k+1/2} = \frac{2(k_z)_{k+1/2}k_{rf}\Delta x_i\,\Delta y_j}{\mu_f B_f(\Delta z_k + \Delta z_{k+1})}.$$

and the operators:

$$\Delta_y T_{f,y}\left(\Delta_y p_f - \rho_f g \Delta_y D\right) = \left(T_{f,y}\right)_{j+1/2}\left[\left(p_{f,j+1} - p_{f,j}\right) - \rho_f g\left(D_{j+1} - D_j\right)\right]$$
$$- \left(T_{f,y}\right)_{j-1/2}\left[\left(p_{f,j} - p_{f,j-1}\right) - \rho_f g\left(D_j - D_{j-1}\right)\right]$$

and:

$$\Delta_z T_{f,z}\left(\Delta_z p_f - \rho_f g \Delta_z D\right) = \left(T_{f,z}\right)_{k+1/2}\left[\left(p_{f,k+1} - p_{f,k}\right) - \rho_f g\left(D_{k+1} - D_k\right)\right]$$
$$- \left(T_{f,z}\right)_{k-1/2}\left[\left(p_{f,k} - p_{f,k-1}\right) - \rho_f g\left(D_k - D_{k-1}\right)\right].$$

The components of the transport term for fluid f (*expressed under standard conditions*) along the x-, y- and z-directions will therefore be:

$$\left(\Delta q_f\right)_x = \Delta_x T_{f,x}\left(\Delta_x p_f - \rho_f g \Delta_x D\right), \tag{13/5.3a}$$

$$\left(\Delta q_f\right)_y = \Delta_y T_{f,y}\left(\Delta_y p_f - \rho_f g \Delta_y D\right), \tag{13/5.3b}$$

$$\left(\Delta q_f\right)_z = \Delta_z T_{f,z}\left(\Delta_z p_f - \rho_f g \Delta_z D\right). \tag{13/5.3c}$$

The *transport term* in the flow equation – the change in volume per unit time (measured under standard conditions) of the component f in block (i, j, k) due to the movement of fluid in the three coordinate directions x, y and z – is therefore:

$$\left(\Delta q_f\right)_{\mathrm{tr}} = \left(\Delta q_f\right)_x + \left(\Delta q_f\right)_y + \left(\Delta q_f\right)_z$$
$$= \sum_s \Delta_s T_{f,s}\left(\Delta_s p_f - \rho_f g \Delta_s D\right) \qquad (s = x, y, z). \tag{13/5.4}$$

Now, the transport term in the differential flow equation is:

$$\nabla \cdot \left[\frac{[k]k_{rf}}{\mu_f B_f}\left(\nabla p_f - \rho_f g\ \nabla D\right)\right] \equiv \sum_s \Delta_s T_{f,s}\left(\Delta_s p_f - \rho_f g \Delta_s D\right),$$

where the right-hand term is the finite difference formulation of the left-hand term.

For the *accumulation term* for component f in the block, we have:

V_{b} : block volume $= \Delta x_i\ \Delta y_j\ \Delta z_k$,

ϕ : average block porosity,

S_f : average saturation of the phase containing component f in the block,

B_f : formation volume factor of the phase containing component f.

The change of volume (*at standard p and T*) of component f in block (i, j, k) during the time step Δt is therefore:

$$\left(\Delta q_f\right)_{\mathrm{acc}} = V_{\mathrm{b}}\Delta_t\left(\frac{\phi S_f}{B_f}\right), \tag{13/5.5}$$

where the operator Δ_t represents the change in the value of the term to which it is applied, in time Δt.

The *source term* expresses the flow rate (*under standard conditions*) at which component f is extracted from, or injected into, block (i, j, k). This is negative $(-q_f)$ for extraction, positive $(+q_f)$ for injection.

Since all terms for all components *are expressed under standard conditions* (where the density of component f is $\rho_{f,\mathrm{sc}}$), *the mass balance equation* for component f in block (i, j, k) can be written as a *volumetric balance equation* by dividing through by $\rho_{f,\mathrm{sc}}$:

$$\Delta t \times \sum_s \Delta_s T_{f,s}\left(\Delta_s p_f - \rho_f g\ \Delta_s D\right) \pm q_f\ \Delta t = V_{\mathrm{b}}\Delta_t\left(\frac{\phi S_f}{B_f}\right). \tag{13/5.6}$$

In the case of medium and heavy oils (black oils) the *components* are:

- *gas*, of constant composition, at standard p and T
- *stock tank oil* at standard p and T
- *water* at standard p and T.

while the *phases* are:

- *gas phase*, of constant composition independent of the pressure (the oil component is assumed to have zero volatility in the gas phase)

- *hydrocarbon liquid phase*, consisting of the oil component plus that part of the gas component which has dissolved in the oil [defined by the solubility $R_s(p)$, which is expressed as sm^3 of dissolved gas per sm^3 of oil].
- *water phase*, consisting of the water component, plus any dissolved gas [defined by the solubility $R_{sw}(p)$ of gas in water].

We can ignore the solubility of gas in water (usually very small), but the gas component still moves through the reservoir partly as a gas phase, partly in the hydrocarbon liquid phase as gas dissolved in oil.

The corresponding transport term can be calculated as follows. Because the flow rate of each component is expressed under standard conditions, the flow rate of the gas component transported in solution in the oil component will be:

$$\left(q_g\right)_o = R_s q_o,$$

where: $R_s = \dfrac{\text{m}^3 \text{ of gas at standard } p \text{ and } T}{\text{m}^3 \text{ of stock tank oil}}$

and $\quad q_o =$ oil transport term, expressed under stock tank conditions.

From Eq. (13/5.1′), therefore:

$$
\begin{aligned}
\left(q_{g,in,x}\right)_o &= R_s q_{o,in,x} \\
&= \frac{2(k_x)_{i-1/2}k_{ro}\,\Delta y_j\,\Delta z_k}{\mu_o B_o(\Delta x_{i-1} + \Delta x_i)} R_s \left[\left(p_{o,i-1} - p_{o,i}\right) - \rho_o g\left(D_{i-1} - D_i\right)\right],
\end{aligned}
$$

and from Eq. (13/5.2):

$$
\begin{aligned}
\left(q_{g,out,x}\right)_o &= R_s q_{o,out,x} \\
&= \frac{2(k_x)_{i+1/2}k_{ro}\Delta y_j\,\Delta z_k}{\mu_o B_o(\Delta x_i + \Delta x_{i+1})} R_s \left[\left(p_{o,i} - p_{o,i+1}\right) - \rho_o g\left(D_i - D_{i+1}\right)\right].
\end{aligned}
$$

Consequently, recalling the transmissibilities defined earlier:

$$
\begin{aligned}
\left(\Delta q_{g,x}\right)_o &= \left(q_{g,in,x}\right)_o - \left(q_{g,out,x}\right)_o \\
&= \left(T_{o,x}\right)_{i-1/2} R_s \left[\left(p_{o,i-1} - p_{o,i}\right) - \rho_o g\left(D_{i-1} - D_i\right)\right] \\
&\quad - \left(T_{o,x}\right)_{i+1/2} R_s \left[\left(p_{o,i} - p_{o,i+1}\right) - \rho_o g\left(D_i - D_{i+1}\right)\right].
\end{aligned}
$$

By defining the operator:

$$
\begin{aligned}
\Delta_x R_s T_{o,x}\,(\Delta_x p_o - \rho_o g \Delta_x D) \\
= \left(T_{o,x}\right)_{i+1/2} R_s \left[\left(p_{o,i+1} - p_{o,i}\right) - \rho_o g\left(D_{i+1} - D_i\right)\right] \\
- \left(T_{o,x}\right)_{i-1/2} R_s \left[\left(p_{o,i} - p_{o,i-1}\right) - \rho_o g\left(D_i - D_{i-1}\right)\right],
\end{aligned}
$$

we can write this as:

$$\left(\Delta q_{g,x}\right)_o = \Delta_x R_s T_{o,x}(\Delta_x p_o - \rho_o g \Delta_x D).$$

Treating the components of the transport term in the y- and z-directions in a similar manner, the full transport term for the gas component dissolved in the oil component becomes:

$$\left(\Delta q_g\right)_o = \sum_s \Delta_s R_s T_{o,s}(\Delta_s p_o - \rho_o g\,\Delta_s D) \qquad (s = x, y, z). \tag{13/5.7}$$

We now have all the necessary terms to write the volumetric balance equation (equivalent to the mass balance equation, as we saw earlier), for each component in the system. All terms are written relative to unit time.

a) Gas component at 15°C and 0.1013 MPa

   (free gas transport term)

   +(dissolved gas transport term)

   −(produced free gas flow rate)

   −(associated gas flow rate from produced oil)

   =(change per unit time in the free gas accumulation term)

   +(change per unit time in the dissolved gas accumulation term)

Therefore:

$$\sum_s \Delta_s T_{g,s}\left(\Delta_s p_g - \rho_g g \Delta_s D\right) + \sum_s \Delta_s R_s T_{0,s}(\Delta_s p_0 - \rho_0 g \Delta_s D)$$

$$- \left(q_{g,sc}\right)_{free} - R_s q_{0,sc} = \frac{V_b}{\Delta t}\left[\Delta_t\left(\frac{\phi S_g}{B_g}\right) + \Delta_t\left(\frac{R_s \phi S_0}{B_0}\right)\right]. \tag{13/5.8a}$$

b) Oil component under stock tank conditions

   For the "stock tank oil" component, the volumetric balance equation is much simpler because it is only transported in one flowing phase – the "hydrocarbon liquid".

   From Eq. (13/5.6) we therefore have:

$$\sum_s \Delta_s T_{0,s}\left(\Delta_s p_0 - \rho_0 g \Delta_s D\right) - q_{0,sc} = \frac{V_b}{\Delta t}\Delta_t\left(\frac{\phi S_0}{B_0}\right). \tag{13/5.8b}$$

c) Water component under standard conditions

   For the component "water at standard conditions", Eq. (13/5.6) becomes:

$$\sum_s \Delta_s T_{w,s}\left(\Delta_s p_w - \rho_w g \Delta_s D\right) - q_{w,sc} = \frac{V_b}{\Delta t}\Delta_t\left(\frac{\phi S_w}{B_w}\right). \tag{13/5.8c}$$

   Equations (13/5.8a,b,c) are the same as Eqs. (13.99 a,b,c) in the text, and the derivations are complete.

$$\diamond \ \diamond \ \diamond$$

Exercise 13.6

Express the general transport term $\Delta_s T_{f,s}(\Delta_s p_f - \rho_f g \Delta_s D)$ in fully implicit form (third-order implicit as described in Sect. 13.14.3.4) at the $(k+1)$th iteration, assuming the kth iteration term has been evaluated.

Solution

For simplicity, we will limit the derivation to the x-axis component of the transport term. The y- and z-components can then be obtained by simple substitution of the appropriate indices.

   In Exercise 13.5 we developed Eq. (13/5.3a), using the index f to indicate a component type (f = g,o, or w – gas, oil, or water):

$$\Delta_x T_{f,x}\left(\Delta_x p_f - \rho_f g \Delta_x D\right) = \left(T_{f,x}\right)_{i+1/2}\left[\left(p_{f,i+1} - p_{f,i}\right) - \rho_f g \left(D_{i+1} - D_i\right)\right]$$

$$- \left(T_{f,x}\right)_{i-1/2}\left[\left(p_{f,i} - p_{f,i-1}\right) - \rho_f g \left(D_i - D_{i-1}\right)\right]. \tag{13/5.3a}$$

   Using the superscripts $(k+1)$ and k to represent the $(k+1)$th and kth iteration, respectively, and with:

$$\delta T_{f,x} \quad = T_{f,x}^{k+1} - T_{f,x}^k$$

$$\text{and} \quad \delta p_f = p_f^{k+1} - p_f^k,$$

we have:

$$\Delta_x T_{f,x}^{k+1}\left(\Delta_x p_f^{k+1} - \rho_f g \Delta_x D\right)$$

$$= \Delta_x\left(T_{f,x}^k + \delta T_{f,x}\right)\left[\Delta_x\left(p_f^k + \delta p_f\right) - \rho_f g \ \Delta_x D\right], \tag{13/6.1}$$

where D is independent of the iteration level since it is not a function of time.

Equation (13/6.1) can be expanded using Eq. (13/5.3a):

$$\Delta_x \left(T_{f,x}^k + \delta T_{f,x} \right) \left[\Delta_x \left(p_f^k + \delta p_f \right) - \rho_f g \Delta_x D \right]$$

$$= \left[\left(T_{f,x}^k \right)_{i+1/2} + \left(\delta T_{f,x} \right)_{i+1/2} \right] \left[\left(p_{f,i+1}^k - p_{f,i}^k \right) + \left(\delta p_{f,i+1} - \delta p_{f,i} \right) - \rho_f g \left(D_{i+1} - D_i \right) \right]$$

$$\quad - \left[\left(T_{f,x}^k \right)_{i-1/2} + \left(\delta T_{f,x} \right)_{i-1/2} \right] \left[\left(p_{f,i}^k - p_{f,i-1}^k \right) + \left(\delta p_{f,i} - \delta p_{f,i-1} \right) - \rho_f g \left(D_i - D_{i-1} \right) \right]$$

$$= \left(T_{f,x}^k \right)_{i+1/2} \left[\left(p_{f,i+1}^k - p_{f,i}^k \right) - \rho_f g \left(D_{i+1} - D_i \right) \right]$$

$$\quad - \left(T_{f,x}^k \right)_{i-1/2} \left[\left(p_{f,i}^k - p_{f,i-1}^k \right) - \rho_f g \left(D_i - D_{i-1} \right) \right]$$

$$\quad + \left(T_{f,x}^k \right)_{i+1/2} \left(\delta p_{f,i+1} - \delta p_{f,i} \right) - \left(T_{f,x}^k \right)_{i-1/2} \left(\delta p_{f,i} - \delta p_{f,i-1} \right)$$

$$\quad + \left(\delta T_{f,x} \right)_{i+1/2} \left[\left(p_{f,i+1}^k - p_{f,i}^k \right) - \rho_f g \left(D_{i+1} - D_i \right) \right]$$

$$\quad - \left(\delta T_{f,x} \right)_{i-1/2} \left[\left(p_{f,i}^k - p_{f,i-1}^k \right) - \rho_f g \left(D_i - D_{i-1} \right) \right]$$

$$\quad + \left(\delta T_{f,x} \right)_{i+1/2} \left(\delta p_{f,i+1} - \delta p_{f,i} \right) - \left(\delta T_{f,x} \right)_{i-1/2} \left(\delta p_{f,i} - \delta p_{f,i-1} \right) . \tag{13/6.2}$$

The terms:

$$\left(\delta T_{f,x} \right)_{i+1/2} \left(\delta p_{f,i+1} - \delta p_{f,i} \right)$$

$$\text{and } \left(\delta T_{f,x} \right)_{i-1/2} \left(\delta p_{f,i} - \delta p_{f,i-1} \right)$$

being products of two infinitesimal quantities, are second-order infinitesimal and can be ignored.

The first two terms on the right side of the equality in Eq. (13/6.2) are none other than the transport terms already evaluated in the preceding iteration k.

Therefore:

$$\Delta_x T_{f,x}^{k+1} \left(\Delta_x p_f^{k+1} - \rho_f g \Delta_x D \right) = \Delta_x T_{f,x}^k \left(\Delta_x p_f^k - \rho_f g \Delta_x D \right)$$

$$\quad + \left(T_{f,x}^k \right)_{i+1/2} \left(\delta p_{f,i+1} - \delta p_{f,i} \right) - \left(T_{f,x}^k \right)_{i-1/2} \left(\delta p_{f,i} - \delta p_{f,i-1} \right)$$

$$\quad + \left(\delta T_{f,x} \right)_{i+1/2} \left[\left(p_{f,i+1}^k - p_{f,i}^k \right) - \rho_f g \left(D_{i+1} - D_i \right) \right]$$

$$\quad - \left(\delta T_{f,x} \right)_{i-1/2} \left[\left(p_{f,i}^k - p_{f,i-1}^k \right) - \rho_f g \left(D_i - D_{i-1} \right) \right] . \tag{13/6.3}$$

From Exercise 13.5 we have that:

$$\left(T_{f,x} \right)_{i+1/2} = 2 k_{rf} \frac{(k_x)_{i+1/2}}{\mu_f B_f} \frac{\Delta y_j \, \Delta z_k}{\Delta x_i + \Delta x_{i+1}}$$

with similar expression for $(T_{f,x})_{i-1/2}$

Since:

$$k_{rf} = k_{rf}(S_f)$$

and

$$\frac{(k_x)_{i+1/2}}{\mu_f B_f} = f(p),$$

we get:

$$\delta T_{f,x} = \frac{\partial T_{f,x}}{\partial S_f} \delta S_f + \frac{\partial T_{f,x}}{\partial p} \delta p. \tag{13/6.4}$$

Given that:

$$\frac{\mathrm{d}}{\mathrm{d}p} \left(\frac{k}{\mu_f B_f} \right)$$

is usually very small, the term containing δp in Eq. (13/6.4) can be ignored. For the interblock relative permeability we use the value corresponding to the upstream block (see Sect. 13.14.1), on the assumption that the direction of flow is from block $(j-1)$ to block $(j+1)$, so that:

$$
\begin{aligned}
\left(\delta T_{f,x}\right)_{i+1/2} &= \frac{\partial}{\partial S_f}\left(T_{f,x}\right)_{i+1/2}\delta S_f \\[2mm]
&= \left[\frac{(k_x)_{i+1/2}}{\mu_f B_f}\right]^k \frac{2\Delta y_j\,\Delta z_k}{\Delta x_i + \Delta x_{i+1}}\left(\frac{dk_{rf}}{dS_f}\right)^k \delta S_{f,i} \\[2mm]
&= \alpha_{i+1/2}^k \delta S_{f,i}.
\end{aligned}
\tag{13/6.5a}
$$

Similarly:

$$
\left(\delta T_{f,x}\right)_{i-1/2} = \alpha_{i-1/2}^k \delta S_{f,i-1}.
\tag{13/6.5b}
$$

Note that $\alpha_{i+1/2}^k$ and $\alpha_{i-1/2}^k$ are known quantities, since they consist of parameters calculated in the preceding, kth, iteration.

Using Eqs. (13/6.5a,b), Eq. (13/6.3) becomes:

$$
\begin{aligned}
\Delta_x T_{f,x}^{k+1}&\left(\Delta_x p_f^{k+1} - \rho_f g \Delta_x D\right) \\[2mm]
&= \Delta_x T_{f,x}^k\left(\Delta_x p_f^k - \rho_f g \Delta_x D\right) \\[2mm]
&\quad + \left(T_{f,x}^k\right)_{i+1/2}\left(\delta p_{f,i+1} - \delta p_{f,i}\right) - \left(T_{f,x}^k\right)_{i-1/2}\left(\delta p_{f,i} - \delta p_{f,i-1}\right) \\[2mm]
&\quad + \alpha_{i+1/2}^k \delta S_{f,i}\left[\left(p_{f,i+1}^k - p_{f,i}^k\right) - \rho_f g\left(D_{i+1} - D_i\right)\right] \\[2mm]
&\quad - \alpha_{i-1/2}^k \delta S_{f,i-1}\left[\left(p_{f,i}^k - p_{f,i-1}^k\right) - \rho_f g\left(D_i - D_{i-1}\right)\right].
\end{aligned}
\tag{13/6.6}
$$

In this equation, *all the terms on the right side of the equality*, with the exception of:

$$
\begin{array}{ccc}
\delta p_{f,i+1} & \delta p_{f,i} & \delta p_{f,i-1} \\
& \delta S_{f,i} & \delta S_{f,i-1}
\end{array}
$$

were already evaluated in the kth iteration.

After rearranging Eq. (13/6.6) we get:

$$
\Delta_x T_{f,x}^{k+1}\left(\Delta_x p_f^{k+1} - \rho_f g \Delta_x D\right)
$$

$$
= \Delta_x T_{f,x}^k\left(\Delta_x p_f^k - \rho_f g \Delta_x D\right)
$$

$$
+ \left(T_{f,x}\right)_{i+1/2}^k \delta p_{f,i+1} \tag{1}
$$

$$
- \left[\left(T_{f,x}\right)_{i+1/2}^k + \left(T_{f,x}\right)_{i-1/2}^k\right]\delta p_{f,i} \tag{2}
$$

$$
+ \left(T_{f,x}\right)_{i-1/2}^k \delta p_{f,i-1} \tag{3}
$$

$$
+ \alpha_{i+1/2}^k\left[\left(p_{f,i+1}^k - p_{f,i}^k\right) - \rho_f g\left(D_{i+1} - D_i\right)\right]\delta S_{f,i} \tag{4}
$$

$$
- \alpha_{i-1/2}^k\left[\left(p_{f,i}^k - p_{f,i-1}^k\right) - \rho_f g\left(D_i - D_{i-1}\right)\right]\delta S_{f,i-1}. \tag{5}
$$

$$
\tag{13/6.7}
$$

This presentation distinguishes the five unknowns.

We now add the components of the flow term in the y- and z-directions to Eq. (13/6.7) and get an equation in the seven unknowns δp_f (one for each of the seven blocks in contact) and the four unknowns $\delta S_{f,i-1,j,k}, \delta S_{f,i,j,k}, \delta S_{f,i,j-1,k}, \delta S_{f,i,j,k-1}$ (for the four "upstream" blocks in the flow equation).

Once again, the values of all the coefficients in the terms appearing in Eq. (13/6.7) are known, having been calculated in the preceding iteration. This *linear* equation is the transport term for block

(i, j, k) in the linearised and discretised system of equations describing the reservoir behaviour, *in the fully implicit form*.

This system can be solved using one of the methods described in the text to obtain values of δp_f and δS_f for each block. Then:

$$p_f^{k+1} = p_f^k + \delta p_f$$

$$S_f^{k+1} = S_f^k + \delta S_f$$

It now remains only to consider the case where, when choosing the independent variables (for instance: p_o, S_o, S_w; see Sect. 13.10.3), p_f is expressed as the sum of the pressure measured in a different fluid and the corresponding capillary pressure.

For example:

$$p_{\rm w} = p_{\rm o} - P_{\rm c,ow}.$$

In this case:

$$p_w^{k+1} = p_o^k + \delta p_o - P_{\rm c,ow}\left(S_o^k + \delta S_o\right).$$

Since:

$$P_{\rm c,ow}\left(S_o^k + \delta S_o\right) = P_{\rm c,ow}\left(S_o^k\right) + \left(\frac{{\rm d}P_{\rm c,ow}}{{\rm d}S_o}\right)^k \delta S_o$$

$$= P_{\rm c,ow}\left(S_o^k\right) + \beta\left(S_o^k\right)\delta S_o,$$

we have:

$$p_w^{k+1} = p_o^k - P_{\rm c,ow}\left(S_o^k\right) + \delta p_o - \beta\left(S_o^k\right)\delta S_o$$

$$= p_w^k + \delta p_o - \beta\left(S_o^k\right)\delta S_o.$$

In the calculation of the transport term in the manner of Eq. (13/6.2), terms like:

$$\delta T_{\rm w} \times \delta p_{\rm o}$$

$$\text{and} \quad \delta T_{\rm w} \times \delta S_{\rm o}$$

are infinitesimal to the second-order, and can be ignored.

◇ ◇ ◇

14 Forecasting Well and Reservoir Performance Through the Use of Decline Curves and Identified Models

14.1 Introduction

Chapter 13 made the point very clearly that the use of numerical models must be preceded by a rigorous investigation into the internal structure – the "geological model" – of the reservoir. Generally speaking, this sort of study is only justifiable where the reservoir is extremely large or highly complex, or where the production process is to be modified in some way – by the drilling of in-fill wells, or the implementation of water injection, for instance.

When no changes to the development program are planned, the behaviour of the reservoir, and individual wells, can be modelled by more simplistic methods which do not require a detailed knowledge of the reservoir structure. These methods are based on the forward extrapolation of past reservoir behaviour by means of *production decline curves*, or the *identification of the reservoir model* using systems theory to analyse production data (flow rates and flowing pressures) gathered over the life of the reservoir.

The decline curve method is a long-established approach to the prediction of reservoir and well behaviour. Identified models – an Italian contribution to reservoir engineering – are a very recent innovation. These two methods will be described in broad terms in the following sections.

14.2 Production Decline Curves

14.2.1 Conditions for the Use of Decline Curves

The decline curve method dates back to the very dawn of petroleum engineering[1,5]. In essence, the method is based on the forward extrapolation of the flow rate/time curve established over a sufficient period of production history. Of course, the method can only provide valid results *if the producing conditions remain unaltered throughout the entire period under consideration* (past history + period of forecast).

The kind of alterations that would render the decline curve approach unusable include: the recent drilling and completion of new wells (or wells planned for the near future), well stimulation, a change in the production mechanism (such as switching oil wells from natural production to artificial lift, or implementing water injection in a naturally producing reservoir), etc. Generally speaking, then, decline curves are most applicable in reservoirs that are in the advanced stages of production, with wells producing at their deliverability (Sect. 13.20.3.1), either naturally or on artificial lift.

The majority of the oil fields in the USA are of this type. The wells produce at very low rates (from several to a few tens of m^3/day), and are maintained at their deliverability for a fixed number of days per month on a reservoir-by-reservoir, or even well-by-well, basis, specified by local regulatory bodies such as the well-known TRRC (Texas Rail Road Commission). Under these conditions, the decline curve method can be used to obtain a reliable estimate of the volume of oil that will be produced through to abandonment of the reservoir, to estimate the remaining oil reserves, and to forecast how the production from the reservoir or individual wells will vary with time. Decline curves, then, are a semi-empirical approach to the estimation of oil reserves, which supplement the volumetric method (Chap. 4) and the material balance method (Chap. 10).

14.2.2 The Characteristics of Decline Curves

We will define the instantaneous rate at which the production is declining, per unit production rate, as:

$$D(t) = -\frac{1}{q}\frac{dq}{dt}. \tag{14.1}$$

Field data show that $D(t)$ is a function of the nth power of the instantaneous production rate itself:

$$D = Kq^n. \tag{14.2a}$$

n is generally between 0 and 1, but in some cases[10] can exceed 1.

For a given well or reservoir, K and n will not change as long as the producing conditions remain unaltered, as described in Sect. 14.2.1.

Except for the case where $n = 0$, therefore, D varies over the life of the reservoir as a power of the instantaneous production rate.

Combining Eqs. (14.1) and (14.2a), we get:

$$-\frac{dq}{q^{n+1}} = K\,dt. \tag{14.3}$$

For $n \neq 0$, Eq. (14.3) can be integrated from initial conditions ($t = 0$, $q = q_i$) to time t ($t = t$, $q = q_t$) to give:

$$\frac{1}{n}\left(q_t^{-n} - q_i^{-n}\right) = Kt \tag{14.4a}$$

The initial decline rate D_i at $t = 0$ is defined as:

$$D_i = Kq_i^n, \tag{14.2b}$$

so that Eq. (14.4a) becomes:

$$q_t = q_i(1 + nD_i t)^{-1/n} \quad \text{for } n \neq 0. \tag{14.5a}$$

For the case where $n = 0$, Eq. (14.2a) reduces to:

$$D = K = \text{const.} \tag{14.2c}$$

and integrating Eq. (14.3) we get:

$$\ln\frac{q_i}{q_t} = Dt, \tag{14.4b}$$

from which:

$$q_t = q_i\,e^{-Dt} \quad \text{for } n = 0. \tag{14.5b}$$

Equations (14.5a,b) enable us to calculate the production rate at any time for the two cases $n \neq 0$ and $n = 0$.

The cumulative oil production N_p can be obtained by the straightforward integration of $q_t(t)$ from zero to t.

For $n = 0$, we have:

$$N_p = \int_0^t q_t \, dt = q_i \int_0^t e^{-Dt} \, dt$$

$$= \frac{q_i}{D} \left(1 - e^{-Dt}\right) = \frac{q_i}{D} \left(1 - \frac{q_t}{q_i}\right), \tag{14.6}$$

so that:

$$N_p = \frac{q_i - q_t}{D} \quad \text{for } n = 0. \tag{14.7a}$$

For $n \neq 0$ and $n \neq 1$, we have:

$$N_p = \int_0^t q_t \, dt = q_i \int_0^t (1 + nD_i t)^{-1/n} \, dt$$

$$= \frac{q_i}{(n-1)D_i} \left[(1 + nD_i t)^{(n-1)/n} - 1\right] \tag{14.8}$$

which, substituting from Eq. (14.5a), becomes:

$$N_p = \frac{q_i}{(n-1)D_i} \left[\left(\frac{q_i}{q_t}\right)^{n-1} - 1\right] \quad \text{for } n \neq 0 \text{ and } \neq 1. \tag{14.7b}$$

For $n = 1$, Eq. (14.7b) takes the undefined form $\infty \times 0$. In this case, we return to the definition of N_p and introduce Eq. (14.5a) with $n = 1$:

$$N_p = \int_0^t q_t \, dt = q_i \int_0^t (1 + D_i t)^{-1} \, dt$$

$$= \frac{q_i}{D_i} \ln (1 + D_i t) . \tag{14.9}$$

Referring again to Eq. (14.5a) with $n = 1$, this simplifies to:

$$N_p = \frac{q_i}{D_i} \ln \frac{q_i}{q_t} \quad \text{for } n = 1. \tag{14.7c}$$

Equations (14.7a,b,c) can be used to estimate the volume N_p of oil that will have been produced by the time the production rate has decreased from its current value q_i to some future value q_t. q_t is usually taken as the minimum rate at which it is still considered economical to keep the well or reservoir on production.

Of course, we cannot forecast the flow rate [Eqs. (14.5)] or the cumulative production [Eqs. (14.7)] without knowing n and D_i (or D). The methods for determining them from the production history will be described next.

Historically, the following types of decline curve have been defined:

- exponential, or constant percentage ($n = 0$, so that $D = \text{const}$);
- hyperbolic ($n \neq 0$ and $\neq 1$);
- harmonic ($n = 1$).

The terms "hyperbolic" and "harmonic" refer to the behaviour of q_t as a function of time, which, from Eq. (14.5a) with $n \neq 0$ and $\neq 1$, and with $n = 1$, is expressed

Table 14.1. *Principal equations for the three types of decline curve*

Parameter	Exponential	Hyperbolic	Harmonic
		Decline curve type	
n	0	$\neq 0$ and $\neq 1$	1
$q_t(t)$	$q_t = q_i e^{-Dt}$	$q_t = q_i(1 + nD_i t)^{-1/n}$	$q_t = \dfrac{q_i}{1 + D_i t}$
$N_p(t)$	$N_p = \dfrac{q_i}{D}\left(1 - e^{-Dt}\right)$	$N_p = \dfrac{q_i}{(n-1)D_i}\left[(1 + nD_i t)^{(n-1)/n} - 1\right]$	$N_p = \dfrac{q_i}{D_i}\ln\left(1 + D_i t\right)$
$N_p(q)$	$N_p = \dfrac{q_i - q_t}{D}$	$N_p = \dfrac{q_i}{(n-1)D_i}\left[\left(\dfrac{q_i}{q_t}\right)^{n-1} - 1\right]$	$N_p = \dfrac{q_i}{D_i}\ln\dfrac{q_i}{q_t}$

respectively as:

$$\left(\frac{q_i}{q_t}\right)^n - 1 = nD_i t \quad n \neq (0; 1) \tag{14.10a}$$

$$\frac{1}{q_t} - \frac{1}{q_i} = \frac{D_i}{q_i}t \qquad n = 1. \tag{14.10b}$$

The relevant equations describing q_t and N_p for exponential, hyperbolic and harmonic decline are presented in Table 14.1.

14.2.3 Practical Applications of the Decline Curve Method

If we intend to forecast the behaviour of a reservoir (or well), we must have access to data from a sufficient length of production history. It is first essential to make sure that none of the producing conditions outlined in Sect. 14.2.1 have been altered during this period, for this would invalidate the decline curve method.

Secondly, it is assumed that these very same producing conditions will be maintained over the forecast period.

The historical production data are used to establish the type of decline curve that best describes the behaviour of the reservoir or well, and the appropriate value of D_i or D. This is achieved by plotting:
–the curve described by Eq. (14.4b) on semilog axes:

$$\ln\frac{q_i}{q_t} = f(t), \tag{14.11a}$$

as illustrated in Fig. 14.1.
–the curve described by Eq. (14.5a) with $n = 1$ on Cartesian axes:

$$\frac{q_i}{q_t} = f(t), \tag{14.11b}$$

as illustrated in Fig. 14.2.

If the data in the semilog plot [Eq. (14.11a)] lie on a straight line trend, the decline curve is of the exponential type ($n = 0$). The slope of the straight line is equal to D/2.302 in semilog coordinates.

If, on the other hand, it is the Cartesian plot [Eq. (14.11b)] which exhibits the linear trend, the decline curve is harmonic ($n = 1$) and the slope of the line is equal to D_i.

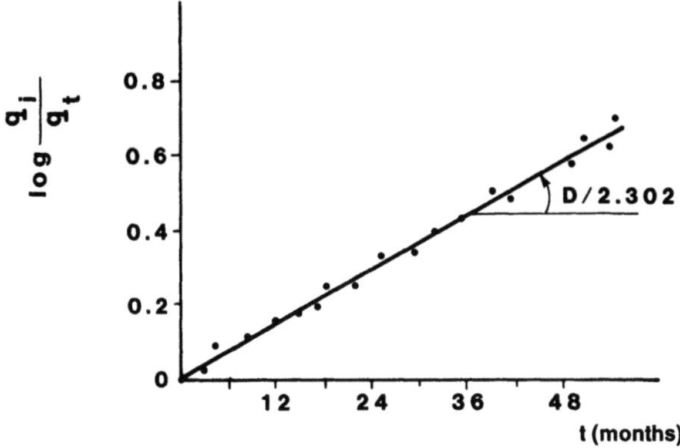

Fig. 14.1. *Exponential (constant percentage) decline curve*

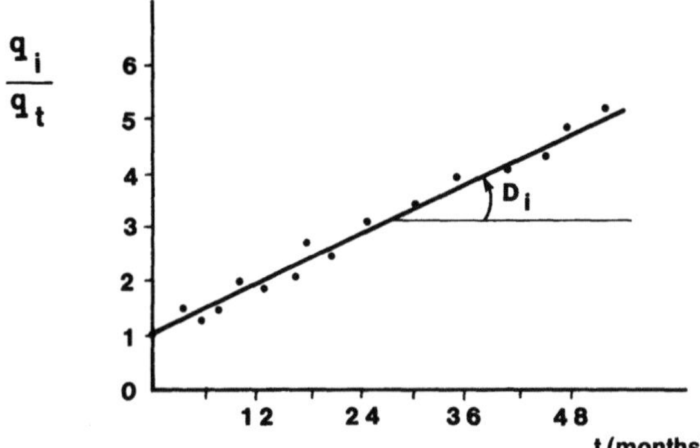

Fig. 14.2. *Harmonic decline curve*

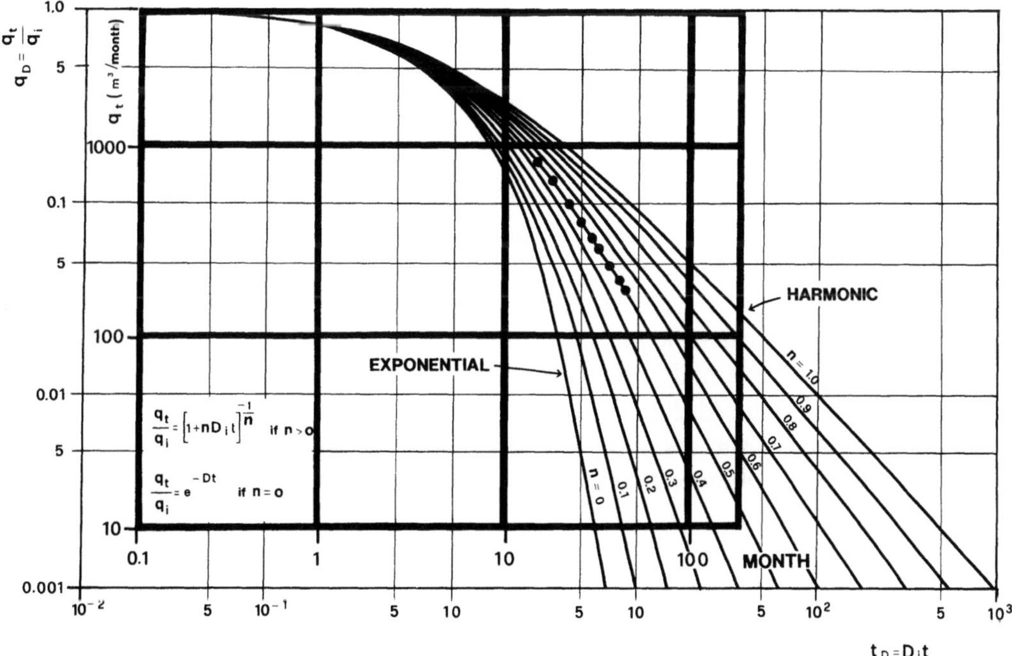

Fig. 14.3. *Example of the use of Fetkovich's type curves[6] to determine the value of n and D_i for a hyperbolic decline. (From Ref. 6, 1980, Society of Petroleum Engineers of AIME. Reprinted with permission of the SPE)*

If no straight line trend is observed on either of these two plots, the decline curve is hyperbolic ($n \neq 0$ and $\neq 1$). In this case, n and D_i can be determined from Eq. (14.5a) by a non-linear regression technique using a programmable calculator or computer.

Fetkovich[6] presented a graphic approach to the determination of these parameters, using type curves. The type curves illustrated in Fig. 14.3 are simply graphic representations of Eq. (14.5a), plotted as $\log q_t/q_i$ versus $\log D_i t$ for different values of n.

A log-log plot of the data from the production history ($\log q_t$ versus $\log t$) is overlain on the type curves and shifted in the x- and y-directions until a match is obtained against one of the curves. This match curve corresponds to the value of n, and D_i can be calculated from $D_i = (x$-value on the type curve)/(corresponding x-value on the field data plot).

Once n and D_i or D have been determined, it is a straightforward matter to use the equations listed in Table 14.1 to predict the behaviour of the production rate q_t, the cumulative volume $N_p(t)$ of oil produced, and the additional oil that can be produced before q_t declines to its economic limit q_{ab}.

14.3   Identified Models

14.3.1   The Reservoir and Aquifer as a Single Dynamic System

In Chap. 13, we saw that the use of numerical models to study and forecast reservoir behaviour depended on a knowledge of:

- the equations describing the thermodynamic and mass transport processes occurring in the reservoir;
- geometry and internal structure of the reservoir + aquifer system;
- the fluid dynamic and thermodynamic characteristics of the system consisting of the reservoir rock + aquifer + pore fluids.

In the 1980s, the Faculty of Engineering of the University of Bologna[2,3,7] undertook the development of a totally different approach, which had first been proposed in the 1960s by Rowan and Clegg[11] and Rowan and Warren[12]. The reservoir and aquifer in contact are treated as a single hydrodynamic system about which no knowledge of physical structure is required, but whose behaviour is defined only by its response to external stimuli – usually consisting of cumulative volumes of injected and/or produced fluids.

In other words, the reservoir + aquifer system is considered to be a "black box" for which the relationships between input parameters $u_i(t)$ (applied stimuli) and output parameters $y_i(t)$ (responses to the stimuli) are derived by applying a *technique of identification* to the time-ordered sequences of $u_i(t)$ and $y_i(t)$ data measured over the production history of the reservoir.

The input/output relationships obtained through this identification procedure are then used to predict the behaviour of the reservoir in response to a sequence of applied input parameters representing a future production program. In other words, after the *system* has been identified, its reaction to the *process* to which it is subjected (the *production* process in the case of a petroleum reservoir) can be numerically simulated.

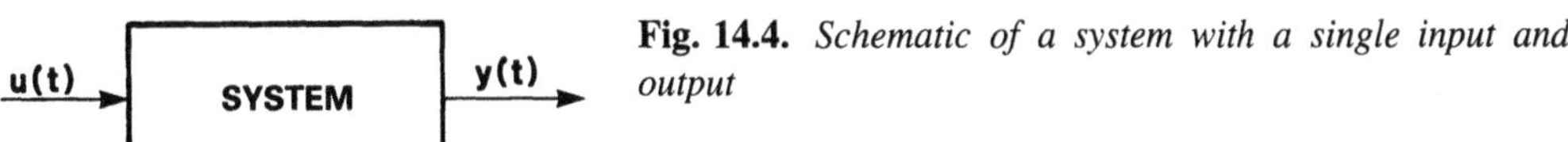

Fig. 14.4. *Schematic of a system with a single input and output*

For simplicity, the discussion will be limited to systems with *a single input and a single output* (Fig. 14.4) where, for example, the following selections have been made:

Input	Output
- cumulative volume of oil (gas) produced	- average reservoir pressure
- cumulative volume of water injected	- cumulative volume of oil produced
- rate of steam injection	- produced water/oil ratio

The technique of identification for the construction of dynamic models presents a number of advantages, and disadvantages, relative to the classical numerical modelling approach. An obvious advantage is that it does not require any knowledge of the internal structure of the reservoir, or of the physical relationships governing its behaviour. A significant disadvantage lies in the fact that *it only allows forecasting of those parameters which were used in the identification phase.* For example, it would not be possible to predict the extent of the water influx $W(t)$ (Chap. 9) in a model for which identification had been based on the *flow rate/pressure* record from the production history.

The technique of identification represents an important part of systems theory. The next section will review some of the basic concepts of the theory, highlighting those which are of practical significance to the modelling of reservoirs.

14.3.2 The Basics of Systems Theory

14.3.2.1 Definitions

Systems can be classified according to the nature of the relationship between input and output:

- *Algebraic systems:* here, the output is an instantaneous function of the input, in the sense that any variation at the input will provoke an immediate response at the output.
- *Dynamic systems:* the output is a function not only of the instantaneous input, but also of the past history of the system up to the instant the input is applied. In other words, dynamic systems "remember" their past. Oil and gas reservoirs are typical dynamic systems – their "memory" consists of the distribution of pressure and fluid saturation resulting from their production history.
- *Linear systems:* the principle of superposition applies here: the output resulting from a number of inputs applied simultaneously is equal to the sum of the individual outputs that would be obtained by applying each input signal in turn.
- *Non-linear systems:* in this type of system, the principle of superposition does *not* apply (see Duhamel's theorem in Sect. 5.7). Many natural systems are

non-linear, but their behaviour is generally difficult to model, and it is often
expedient to employ linear systems whose behaviour is found to be "equivalent"
over an acceptable range of operating conditions.

- *Time-invariant systems:* the characteristics of these systems do not change with
  time. For a specified set of initial conditions, the response to given input signal
  will be the same at any time.
- *Time-varying systems:* the characteristics of these systems change with time,
  so that the response to a given stimulus will be different at different times. Oil
  and gas reservoirs in contact with an aquifer are time-varying systems, because
  the hydrocarbon-bearing volume of reservoir rock varies with time owing to
  the encroachment of water from the aquifer.
  To reduce the complexity of the problem, time-varying systems tend to be
  represented by means of a succession of time-invariant systems, each with
  slightly different characteristics.

An alternative classification is based on the methods used to describe how the
input and output signals vary with time:

- *Continuous systems:* the input signal $u(t)$ and output $y(t)$ are continuous functions of time (Fig. 14.5A).
- *Discrete (or sampled signal) systems:* the input and output signals are defined
  only at discrete times, which are multiples of a fixed *sampling interval* Δt
  (Fig. 14.5B).

The conventional notation is:

$$u(k\Delta t) = u(k)$$
$$k = 0, 1, 2, \ldots, N \tag{14.12}$$
$$y(k\Delta t) = y(k)$$

*When studying reservoir behaviour by the identified model technique, discrete
models are used.* Input and output parameters are defined at constant time steps
(every month, or 3 months, for instance).

A system is considered *stable* when a finite output is obtained for *any* finite
input, regardless of its shape. If this is not the case, the system is *unstable.*

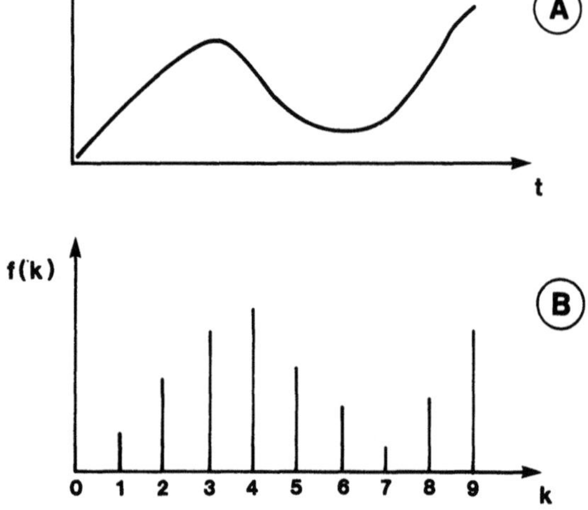

Fig. 14.5. *Signal forms for continuous (A)
and discrete (B) systems*

14.3.2.2 The "z-Transform"

Remember that for a *continuous function* $f(t)$, *finite for all t* and defined for $t \geq 0$, the *Laplace transform*[13] is the integral:

$$\mathcal{L}[f(t)] = \int_0^\infty f(t)e^{-st}\, dt, \tag{14.13}$$

where the parameter $s = \sigma + j\omega$ appearing inside the integral is a complex variable, defined in such a manner that the integral itself converges to a finite value.

The Laplace transform is widely used in systems theory because it facilitates the inversion of a problem in "real space", described by a differential equation in the independent variable *time*, into a problem in "virtual (Laplace) space", described by an algebraic equation in the complex variable s, which is usually easier to deal with.

In conventional notation:

$$\mathcal{L}[f(t)] = F(s). \tag{14.14}$$

As illustrated in Fig. 14.5B, in a discrete system, input and output signals are sampled at regular time intervals Δt. If $f^*(t)$ is the sampled signal, we have:

$$f^*(t) = \sum_{0}^{\infty}{}_k f(k\Delta t)\delta(t - k\Delta t), \tag{14.15}$$

where $\delta(t - k\Delta t)$ = unit impulse applied at the instant $t = k\Delta t$.

In a discrete system, the integral in Eq. (14.13) is replaced by a summation:

$$\mathcal{L}[f^*(t)] = \mathcal{L}\left[\sum_{0}^{\infty}{}_k f(k\Delta t)\delta(t - k\Delta t)\right]$$

$$= \sum_{0}^{\infty}{}_k f(k\Delta t)e^{-ks\Delta t} = F^*(s). \tag{14.16a}$$

Defining:

$$z = e^{s\,\Delta t}, \tag{14.16b}$$

we have, using the notation of Eq. (14.12):

$$F(z) = F^*(s)_{z=e^{s\Delta t}} = \sum_{0}^{\infty}{}_k f(k)\, z^{-k}, \tag{14.17}$$

where the complex variable z may only assume values for which the summation in Eq. (14.17) converges to finite values.

$F(z)$ *is referred to as the z-transform*[9] *of* $f^*(t)$ for values of $f^*(t)$ sampled at constant time intervals. The z-transform plays the same role in the analysis of discrete systems as the Laplace transform for continuous systems. The z-transform has a number of important properties, which are summarised below.

Linear Property

The z-transform is a linear function: as a consequence, the principle of superposition can be applied to it.

If $g^*(t)$ is a linear function of n sampled signals $f_1^*(t)$, $f_2^*(t)$, ..., $f_n^*(t)$, so that:

$$g^*(t) = a_1 f_1^*(t) + a_2 f_2^*(t) + \ldots + a_n f_n^*(t) \tag{14.18a}$$

then:

$$G(z) = a_1 F_1(z) + a_2 F_2(z) + \ldots + a_n F_n(z), \tag{14.18b}$$

where $G(z)$, $F_1(z)$, $F_2(z)$, ..., $F_n(z)$ are the z-transforms of $g^*(t)$, $f_1^*(t)$, $f_2^*(t)$, ..., $f_n^*(t)$ respectively.

Delay Property

Suppose we wish to delay a signal $f_1^*(t)$ by m time steps Δt. If $f_2^*(t)$ is the delayed signal, then for every value of k we can write (Fig. 14.6):

$$f_2^*(k) = f_1^*(k - m). \tag{14.19}$$

Applying the z-transform to Eq. (14.19) we will get:

$$F_2(z) = \sum_{0}^{\infty} {}_k f_1^*(k - m)\, z^{-k}. \tag{14.20}$$

If $\qquad k - m = n,$

so that $\qquad k = m + n,$

then $\quad F_2(z) = \sum_{0}^{\infty} {}_n f_1^*(n)\, z^{-m-n}$

$$= z^{-m} \sum_{0}^{\infty} {}_n f_1^*(n)\, z^{-n} = z^{-m} F_1(z) \tag{14.21a}$$

and, inversely:

$$F_1(z) = z^m F_2(z). \tag{14.21b}$$

Equations (14.21a,b) demonstrate two important rules:

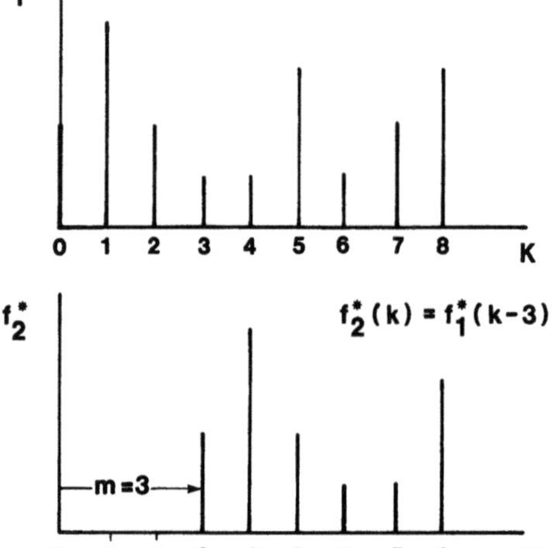

Fig. 14.6. *Base sampled signal $f_1^*(k)$, and signal delayed by three time steps $f_2^*(k) = f_1^*(k - 3)$*

– the z-transform of a signal which is to be delayed by m time steps can be obtained by simply multiplying the z-transform of the base-signal by z^{-m}.
– the z-transform of a signal which is to be advanced by m time steps can be obtained by multiplying the z-transform of the base-signal by z^m.

z^{-m} is the *delay operator for m time steps*.

14.3.3 Discrete Dynamic Systems

The dynamic system was defined in Sect. 14.3.2.1 as one which retained a "memory" of its past life. Suppose the single input/single output system in Fig. 14.4 is a discrete dynamic system. We will suppose also that *this system is linear*; this means that the output signal $y(t)$ sampled at time t will be a linear function of the instantaneous input $u(t)$, *and* of a combination of all input and output signals handled by the system during the period over which its "memory" extends. The length of the system's memory is m time steps for both input and output signals. Therefore:

$$y(t) = a_o u(t) + a_1 u(t-1) + a_2 u(t-2) + \ldots + a_m u(t-m)$$
$$+ b_1 y(t-1) + b_2 y(t-2) + \ldots + b_m y(t-m) \tag{14.22a}$$

or, in a more concise form:

$$y(t) = \sum_{0}^{m} {}_j a_j u(t-j) + \sum_{1}^{m} {}_j b_j y(t-j). \tag{14.22b}$$

We will now calculate the z-transform of Eq. (14.22). Applying the principle of superposition, and the delay operator described in Sect. 14.3.2.2, we can write:

$$Y(z) = a_0 U(z) + \frac{a_1}{z} U(z) + \frac{a_2}{z^2} U(z) + \ldots + \frac{a_m}{z^m} U(z)$$
$$+ \frac{b_1}{z} Y(z) + \frac{b_2}{z^2} Y(z) + \ldots + \frac{b_m}{z^m} Y(z) \tag{14.23a}$$

or, after rearranging:

$$(-z^m + b_1 z^{m-1} + b_2 z^{m-2} + \ldots + b_{m-1} z + b_m) Y(z)$$
$$= -(a_0 z^m + a_1 z^{m-1} + a_2 z^{m-2} + \ldots + a_{m-1} z + a_m) U(z), \tag{14.23b}$$

which leads to:

$$Y(z) = -\frac{a_0 z^m + a_1 z^{m-1} + a_2 z^{m-2} + \ldots + a_{m-1} z + a_m}{-z^m + b_1 z^{m-1} + b_2 z^{m-2} + \ldots + b_{m-1} z + b_m} U(z)$$
$$= G(z) U(z). \tag{14.23c}$$

Equation (14.23c) expresses the relationship between the z-transform of the input signal $U(z)$ and the z-transform of the output signal $Y(z)$ in a linear dynamic system with one input and one output, and memory length m.

The rational function $G(z)$ appearing in this equation is the *system transfer function*. The order m of the polynomial in the denominator of $G(z)$ is the *order* of the system. Note that the order of a discrete system always corresponds to the number of time steps over which the system remembers its past history.

The values of z satisfying the algebraic equation:

$$-z^m + b_1 z^{m-1} + b_2 z^{m-2} + \ldots + b_{m-1} z + b_m = 0 \tag{14.24}$$

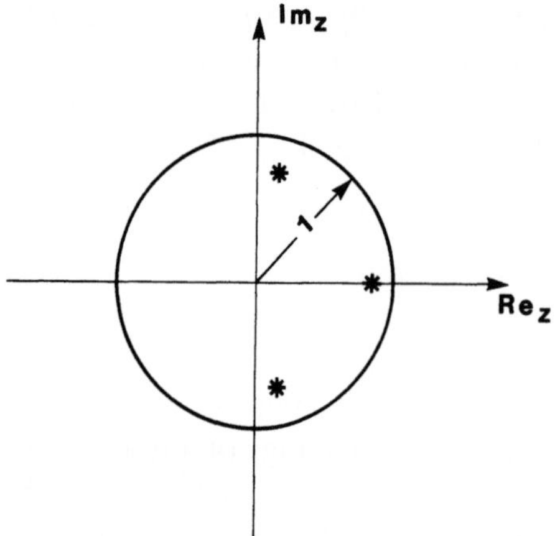

Fig. 14.7. *Possible pole positions in complex space for a stable discrete system*

[obtained by equating the denominator of $G(z)$ to zero] are the *poles* of the transfer function. $G(z) = \infty$ when z is a pole.

An algebraic equation of order m has, of course, m solutions. Some are real, some complex, and each is a pole of $G(z)$. It can be shown that *a discrete system is stable* (by which we mean that any finite input signal will produce a finite output signal) *when all of its poles have a modulus less than* 1. This is illustrated graphically in Fig. 14.7: in complex space, all the poles must lie within a circle of unit radius for the system to be stable.

14.3.4 The Model Identification Procedure

14.3.4.1 Ordering of the Model

The first step in the process of model identification is to determine its *structure*, in other words, the *order m* of the model best describing the behaviour of the discrete system. m, remember, corresponds to the number of equally spaced time steps spanned by the "memory" of the system and, therefore, of the appropriate identified model.

We will assume for the moment that m is known. The output signal $y(k+m)$ at time $(k+m)\Delta t$ will be a linear function of the input signals $u(k+m), u(k+m-1), \ldots, u(k+1), u(k)$ at times $(k+m)\Delta t, (k+m-1)\Delta t, \ldots, (k+1)\Delta t, k\Delta t$; and of the output signals at times $(k+m-1)\Delta t, (k+m-2)\Delta t, \ldots, (k+1)\Delta t, k\Delta t$. Therefore:

$$y(k+m) = a_0 u(k+m) + a_1 u(k+m-1) + \ldots + a_{m-1} u(k+1) + a_m u(k)$$
$$+ b_1 y(k+m-1) + \ldots + b_{m-1} y(k+1) + b_m y(k).$$

$$(14.25)$$

There are $(2m+1)$ coefficients in Eq. (14.25) – more precisely: $(m+1)$ coefficients of the u terms, and m of the y terms. Since we know the input signals $u(t)$ and the output signals $y(t)$, we can solve for these $(2m+1)$ coefficients by writing Eq. (14.25) $(2m+1)$ times, corresponding to each of the times $(k+m)$, $(k+m+1), \ldots, (k+3m)$. This will produce a set of $(2m+1)$ equations which can be expressed in matrix form as follows:

$$
\begin{bmatrix} y(k+m) \\ y(k+m+1) \\ \vdots \\ y(k+3m) \end{bmatrix} =
\left[\begin{array}{ccc|ccc}
u(k+m)\ldots & u(k) & & y(k+m-1)\ldots & y(k) \\
u(k+m+1)\ldots & u(k+1) & & y(k+m)\ldots & y(k+1) \\
\vdots & \vdots & & \vdots & \vdots \\
u(k+3m)\ldots & u(k+2m) & & y(k+3m-1)\ldots & y(k+2m)
\end{array}\right]
$$

$$
\times \begin{bmatrix} a_0 \\ a_1 \\ \vdots \\ a_m \\ b_1 \\ b_2 \\ \vdots \\ b_m \end{bmatrix}. \tag{14.26}
$$

The $(2m+1)$th order matrix appearing in the second term of Eq. (14.26) must be non-singular (that is, with a non-zero determinant) in order for it to be inverted and the coefficient vector $a_0, a_1, \ldots, a_m, b_1, b_2, \ldots, b_m$ calculated.

Let us now see what would happen if we wrongly assumed a system which is *physically* of order m to be of order $(m+1)$. The coefficient matrix in Eq. (14.26) would be of order $(2m+3)$, with $(m+2)$ coefficients to be calculated for u and $(m+1)$ for y:

$$
\left[\begin{array}{ccc|cccc}
u(k+m+1)\ldots & u(k) & & y(k+m) & y(k+m-1)\ldots & y(k) \\
u(k+m+2)\ldots & u(k+1) & & y(k+m+1) & y(k+m)\ldots & y(k+1) \\
u(k+m+3)\ldots & u(k+2) & & y(k+m+2) & y(k+m+1)\ldots & y(k+2) \\
\vdots & \vdots & & \vdots & \vdots & \vdots \\
u(k+3m+3)\ldots & u(k+2m+2) & & y(k+3m+2) & y(k+3m+1)\ldots & y(k+2m+2)
\end{array}\right]. \tag{14.27}
$$

Now, in fact, this is a physical system of order m, and not $(m+1)$ as assumed initially. Referring to Eq. (14.25), note that the first column on the right of the partition in Eq. (14.27) [headed by the term $y(k+m)$] is a linear combination of all the others, with the exception of the very first column in the matrix [headed by the term $u(k+m+1)$]. In this situation, since it has a zero determinant, the matrix cannot be inverted. This is rigorously true for any assumed system order n which is greater than the true order m.

As a consequence, the set of Eqs. (14.26), written for order $n > m$, become meaningless. This suggests a simple method for determining the order of a system.

If the square matrix in Eq. (14.27) is of the ith order, and its determinant is $A(i)$, we simply need to calculate $A(i)$ for successive values of $i = 2r+1$, with $r = 1, 2, 3, \ldots$ and so on. If n is the first value of i for which $A(i) = 0$, the order m of the system will be:

$$
m = \frac{n-3}{2}. \tag{14.28}
$$

This, then, will be the order of the identified model describing the system.

For this method to work, the input and output signals whose sampled values are used to derive the determinant $A(i)$ must be free of disturbances. Otherwise it is difficult to ascertain the precise value of i at which $A(i)$ goes to zero. When disturbances are present in the signals, the PPCRE (Predicted PerCent Reconstruction Error)[8] approach is used. The PPCRE is defined as 100 times the root mean square of the difference between the "predicted reconstruction" (the past history as simulated by the model) and the system output. The value of the PPCRE is calculated for increasing values of the assumed order i of the system. The lowest value of i above which the PPCRE does not decrease significantly is taken as the order m of the system.

14.3.4.2 Determination of the Parameters of the Model

Once we have decided on the order m of the system, the model parameters – coefficients $a_0, a_1, \ldots, a_m$; $b_1, b_2, \ldots, b_m$ appearing in Eq. (14.22) – can be determined by the least squares method (at least as far as single input/single output systems are concerned). If N is the number of ordered pairs of input/output values $[u(t), y(t)]$ available from the past history, we can write Eq. (14.22) $(N-m)$ times and get $(N-m)$ linear equations in the $(2m+1)$ parameters appearing in Eq. (14.22).

Since it is practically always the case that:

$$(N-m) > (2m+1), \tag{14.29}$$

these parameters can be evaluated applying a linear regression based on least squares to the $(N-m)$ equations.

We now know the order of the model and its parameters, so it is completely *identified*.

14.3.4.3 Evaluation of Model Stability

To assess the stability of an identified model, we calculate the poles. These were introduced in Sect. 14.3.3 as the roots of the algebraic equation:

$$-z^m + b_1 z^{m-1} + b_2 z^{m-2} + \ldots + b_{m-1} z + b_m = 0. \tag{14.24}$$

The values of m and the coefficients $b_1, b_2, \ldots, b_{m-1}, b_m$ have already been obtained in the identification phase (Sects. 14.3.4.1 and 14.3.4.2). There are, of course, m roots $z_1, z_2, \ldots, z_m$ to Eq. (14.24). These can be all real, mixed real and conjugate complex or, for even values of m, all conjugate complex.

For model stability (Sect. 14.3.3), all poles must lie inside a circle of unit radius in the complex plane (Fig. 14.7).

14.3.5 Practical Example: Identification of a Gas Reservoir with an Aquifer

14.3.5.1 Initial Considerations

In the identification phase it is common practice to assume that the *system* (or *process*) is *linear and time-invariant* as described in Sect. 14.3.2. In reality, reservoir systems and the development processes they undergo are often intrinsically non-linear and time-varying.

We will consider the simple example of a gas reservoir. If we assume it to be a closed reservoir – that is, not in contact with an aquifer – the relationship between the cumulative production G_p and the mean reservoir pressure $\bar{p}$ is expressed by the well-known equation:

$$\frac{\bar{p}}{\bar{z}} = \frac{p_i}{z_i} - K G_p, \tag{10.6c}$$

where K is a constant and:

    p_i is the initial reservoir pressure,

    $\bar{z}$, z_i are the gas compressibility factors at reservoir temperature, at $\bar{p}$ and p_i respectively.

Since z is a non-linear function of p (Sect. 2.3.1; Fig. 2.4), G_p is a non-linear function of $\bar{p}$. For a closed gas reservoir, a model in which the input signal is the cumulative production G_p and the output signal is the mean reservoir pressure $\bar{p}$ is therefore intrinsically non-linear.

In the presence of a supporting aquifer, the well-known relationship between G_p and $\bar{p}$ is as follows:

$$\frac{\bar{p}}{\bar{z}} = \frac{p_i}{z_i} \frac{1 - \dfrac{G_p}{G}}{1 - \dfrac{W_e}{V_{g,o}}}, \tag{10.15}$$

where:

    G is the initial volume of gas in the reservoir, measured under standard conditions,

    $V_{g,o}$ is the initial volume occupied by the gas in the reservoir,

    W_e is the volume of water that has entered the reservoir up to time t. This corresponds to the cumulative production G_p.

G, $V_{g,o}$, p_i and z_i are constants, while W_e is a function of t and, as seen in Chap. 9, of the past history of $\bar{p}(t)$.

Since W_e increases with time, the term $[1 - (W_e / V_{g,o})]$ appearing in the denominator of Eq. (10.15) will be a decreasing function of t. Therefore, for a gas reservoir with aquifer, a model having cumulative production G_p as the input signal and the mean reservoir pressure $\bar{p}$ as the output *will be both non-linear and time- varying*. The *same* variation ΔG_p in the input signal can produce *any number of different* output variations $\Delta \bar{p}$, depending on the time [and, consequently, the influx $W_e(t)$ of water into the reservoir] at which ΔG_p is applied. In particular, if after a certain period the reservoir is shut in ($\Delta G_p = 0$), the pressure will start to rise ($\Delta \bar{p} < 0$) because water entering the reservoir from the aquifer ($\bar{p}$ being $< p_i$) recompresses the gas.

Obviously, then, in this case the model, which is intrinsically linear and time-invariant, will have to be "adapted" to suit a situation which is not rigorously either of these. The following example, taken from a real reservoir, demonstrates how this adaptation is carried out.

14.3.5.2 *The Minerbio Gas Reservoir*

The subject of this case study[4] is the C pool of the Minerbio reservoir, discovered in 1956. The C pool, lying at a depth of approximately 1300 m, held initial reserves

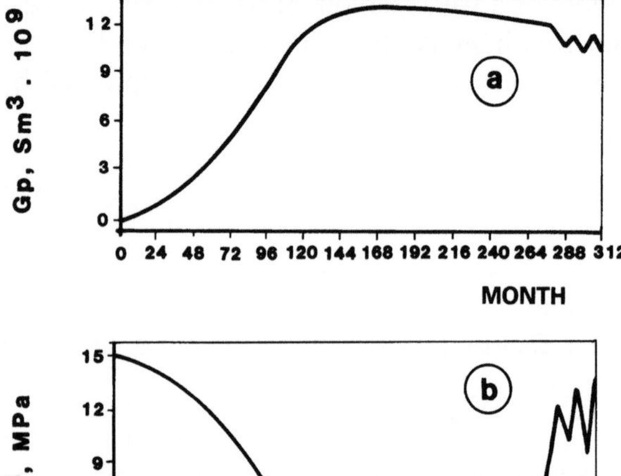

Fig. 14.8. *Records of the cumulative production $G_p(t)$ (curve **a**) and mean reservoir pressure $p(t)$ (curve **b**) for pool C of the Minerbio gas reservoir. (From Ref. 4. Reprinted with permission of the Institut Français du Pétrole)*

of 14.2 billion sm^3 of gas, and was in contact with a flanking aquifer of limited extent. Production began in February 1959, and was terminated in the summer of 1972 after producing 12.8 billion sm^3. The reservoir was then converted for underground gas storage with the drilling of additional wells and the installation of the necessary surface equipment (compressors, increased gas treatment facilities).

At the end of this phase, approximately 1.1 billion sm^3 of gas were injected to create the "cushion" necessary to ensure that the "peaks" of demand made on the gas supply in the winter months could be met.

The first annual storage cycle was started in June 1982, with massive injection of gas during the summer season and extraction during the winter. Overall, we have a total of 312 months of production history for this reservoir. Figure 14.8 shows the variation of $G_p(t)$ and $\bar{p}(t)$ over this period.

The history consists of three phases:

0–162 months: 12.8 billion sm^3 of gas produced. $\bar{p}$ declined from 15.05 MPa to approximately 3 MPa.

163–282 months: conversion to storage and injection of approximately 1.1 billion sm^3 of gas. $\bar{p}$ increased to 7.8 MPa, mainly through the influx of water from the aquifer.

283–312 months: 2.5 cycles of storage and production (three summers and two winters), with corresponding oscillations in G_p and $\bar{p}$.

For the identification process, the history was sampled at intervals of $\Delta t = 3$ months, thereby providing 105 pairs of input/output values $(G_p, \bar{p})$.

14.3.5.3 Determination of the Order of the Model

The sampled input and output signals are affected by noise caused by:

– errors in the measurement of G_p and the static pressure p_{ws} in individual wells,

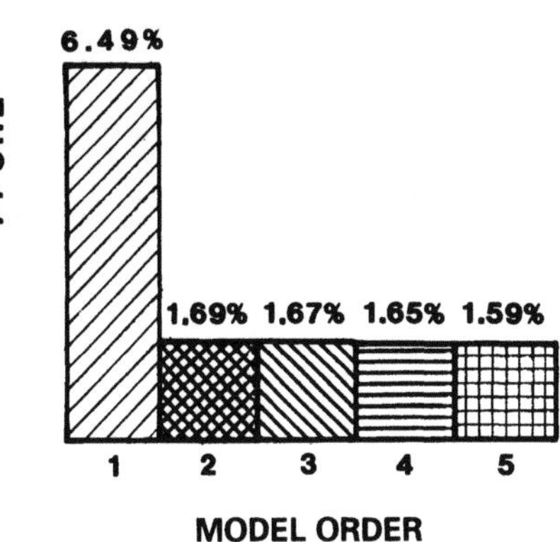

Fig. 14.9. *The PPCRE as the model order is increased. Minerbio gas reservoir, pool C*

- the calculation of the mean reservoir pressure $\bar{p}$ from measurements of p_{ws} which were not exactly simultaneous in different wells,
- sampling values of $(G_p, \bar{p})$ at regularly spaced time intervals, when the actual measurements – particularly $\bar{p}$ – were in fact made irregularly, depending on production requirements.

When determining the order m of the model, then, it is preferable to use the criterion of the minimum PPCRE (Sect. 14.3.4.1), which allows for the effects of noise in the input and output signals, rather than the alternative criterion of matrix singularity.

Figure 14.9 shows the PPCRE calculated for the set of 105 sampled signal pairs, for model order m increasing from 1 to 5. There is no doubt that the PPCRE does not decrease appreciably for $m > 2$. We will therefore represent the production process by means of an identified model of order 2.

14.3.5.4 Investigation of the Conditions for Time Invariance

If a linear time-invariant identified model is used to describe a process which might be intrinsically neither linear nor time-invariant (Sect. 14.3.5.1), it may prove impossible to simulate the reservoir behaviour accurately. In addition, the model could very rapidly become unstable (Sect. 14.3.2.1). This will become readily apparent when an attempt is made to simulate the past history of the reservoir. Consequently, it is essential to establish whether or not the process is time-invariant, by comparing its behaviour at different times.

To do this, a "window" of width r time steps is slid one step at a time along the time axis through the N sampled signal pairs $(G_p, \bar{p})$. This produces $(N - r + 1)$ subsets, each containing r values. The parameters for an identified model of order m are then calculated for each subset, as described in Sect. 14.3.4.2. m was determined from the whole set of N available signal pairs (Sect. 14.3.5.3), and in this case is equal to 2. The poles of the $(N - r + 1)$ identified models corresponding to these subsets can then be obtained (Sect. 14.3.4.3).

If the process being analysed is truly time-invariant, all the poles will be the same. Otherwise, if the poles are scattered in the complex plane, the process is time-varying.

The distribution of poles for the Minerbio C pool is shown in Fig. 14.10. The window was set at a width of 20 time steps, and was slid one step at a time between the 1st and 105th sample. This produced 86 data subsets and identified models, each

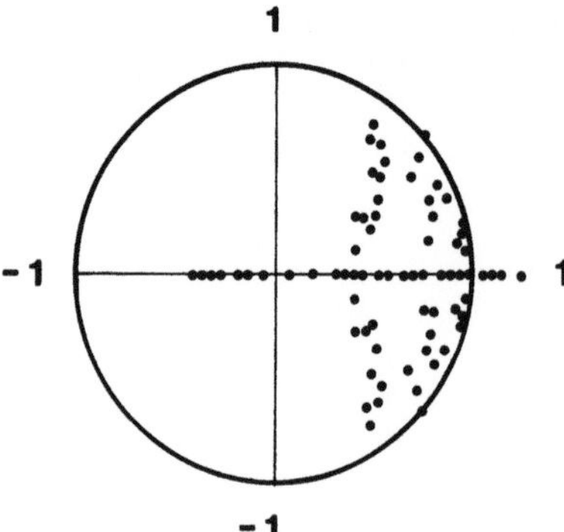

Fig. 14.10. *Distribution of poles in the complex plane, and analysis of time invariance conditions, for the identified models derived from the past history of the Minerbio C pool*

consisting of 20 time steps. Note that some of the poles fall outside the circle of unit radius in Fig. 14.10, indicating that the associated models are unstable.

By looking closely at the distribution of poles with time in the complex plane, it is possible to discern time intervals where the poles lie more or less in the same place. In each of these intervals, the process can be considered quasi-time-invariant, and can therefore be reasonably represented by an identified model.

The quasi-time-invariant intervals for the Minerbio C pool were[4]:

0–120 months: the pool was on continuous production,
121–267 months: the pool was initially on production, was then shut in, then received a small amount of gas,
231–312 months: the major part of the gas cushion was injected, and 2.5 storage/production cycles were completed (three summers, two winters).

The three intervals in which the system is quasi-time-invariant correspond to three successive and quite distinct sets of producing conditions for the pool. The analysis of the time invariance of the process therefore has a definite physical significance.

The parameters of the second-order identified models pertaining to each of the quasi-time-invariant intervals must next be determined, such that the process can be replicated. This is done by the method described in Sect. 14.3.4.2.

Note that the width r of the window must be chosen according to certain criteria. It must not be too long, so as not to mask possible time variance; nor too short, so as not to diminish the significance of poles calculated for the sequence of identified models.

A theoretical analysis of the problem establishes that the minimum window width, r_{min}, is related to the order m of the model by:

$$r_{min} = 3m + 2. \tag{14.30}$$

Values of $r > r_{min}$ are always used in practice. Models used for reservoirs are typically of order $2 \sim 3$, and windows will usually span 12 to 22 sampled input/output pairs.

There is an alternative approach to the investigation of time invariance. A window is defined with a certain initial width and position. In successive passes, its width is incremented by a fixed number of sample times, keeping the position of the lower end fixed. Each successive pass therefore has a window of increased width, and the corresponding models will describe the *average* behaviour of the system over a progressively longer interval of time. In the intervals where the model is unstable, *clustering* of the poles – regrouping them inside the circle of unit radius – will impose time invariance. This should be followed by a post-clustering optimisation of the model parameters so as to be able to reproduce the reservoir behaviour over its production history.

14.3.5.5 Validation of the Identified Model

In order to assess the effectiveness of a model at predicting the future behaviour of a process, it is common practice to "construct" (or, in the present case, to identify) the model using only *part* of the historical data, and to let the model "predict" the remaining part.

If the behaviour "predicted" by the model matches the historical data to an acceptable degree, the model can be considered *validated*. This procedure is of course applicable to identified models.

In the case of the Minerbio C pool, three identified models must be considered, corresponding to three successive periods of the production history (derived in Sect. 14.3.5.4). The identification of the appropriate model[4] was based on the ordered signal pairs $(G_p, \bar{p})$ from the early part of each of these periods. The models derived were then used to "predict" the behaviour of the reservoir[4] over the remaining parts of their respective periods.

More precisely:

Model	Identification period (months)	Predictive period (months)
a	0–57	58–120
b	120–177	178–267
c	231–285	286–312

The results are shown in Fig. 14.11. The match is very good over periods *a* and *c*, but only fair over period *b*.

Model *b* corresponds to three successive and quite distinct mechanisms during the one interval of time: production of gas, shut-in, and injection of gas. The model identification was based on the signals from the production and shut-in phases, and was used to predict the reservoir behaviour during the injection of the gas cushion. Because of the extreme difference between the conditions prevailing during identification and prediction, it was difficult to define a model that could both match the "past" and predict the "future" closely.

This sort of problem is quite common even with numerical models (Chap. 13). For example, a numerical model that has been validated by history matching to data from the primary production phase of a reservoir is very often unable to replicate the behaviour observed when the reservoir subsequently produced by water injection.

Models *a* and *c* in Fig. 14.11 can be considered completely validated. Model *c* was in fact used successfully to predict the reservoir behaviour during the subsequent gas storage and production cycles.

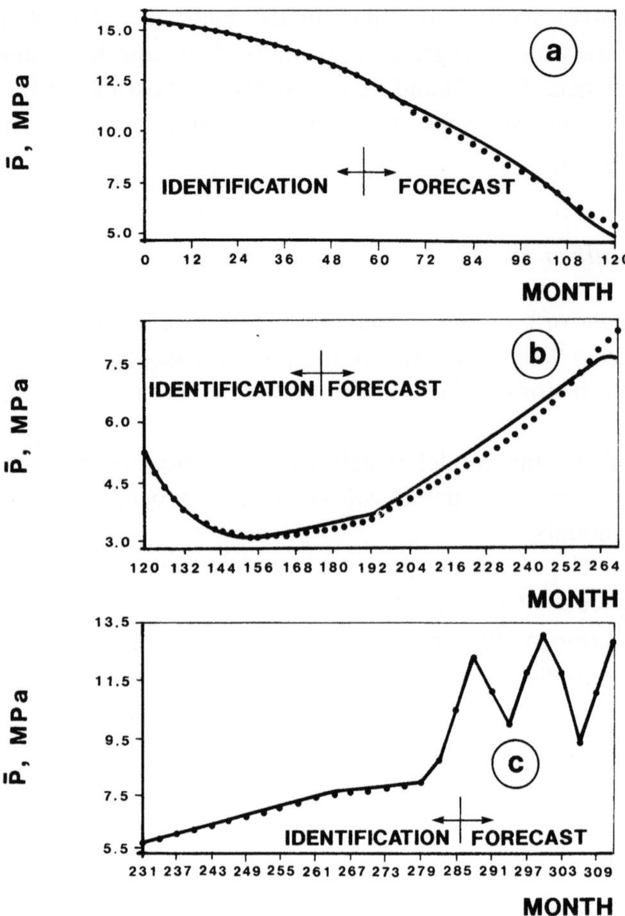

Fig. 14.11a–c. *Model identification and validation in three different intervals of the production history of the Minerbio gas reservoir, pool C. Solid lines: field data; dotted lines: data simulated by the model. (From Ref. 4. Reprinted with permission of the Institut Français du Pétrole)*

References

1. Arps JJ (1945) Analysis of decline curves. Trans, AIME 160: 228–247
2. Beghelli S, Chierici GL, Gottardi GA, Guidorzi RP, Terragni F (1987) An identification approach for oil recovery processes. Proc, In ASME Conf on Modelling and simulation, Cairo, March 2–4, vol 3c, pp 11–22
3. Chierici GL, Gottardi G, Guidorzi RP (1981) Identified models for gas storage dynamics. Soc Pet Eng J April: 151–159; Trans, AIME 271
4. Chierici GL, Gottardi GA, Guidorzi RP (1986) Forecasting gas reservoir behavior with identified models. Rev Inst Fr Pét 41 (5): 637–647
5. Cutler WW Jr (1924) Estimation of underground oil reserves by well production curves. Bull US Bur Mines, 91: 228–240
6. Fetkovich MJ (1980) Decline curve analysis using type curves. J Pet Technol June: 1065–1077; Trans, AIME 269
7. Gottardi G, Guidorzi RP, (1978) Experimental results in the identification of some water-driven Po River Valley gas reservoirs. Proc, MECO '78, Athens, pp 1108–1113
8. Guidorzi RP, Losito MP, Muratori T (1982) The range error test in the structural identification of linear multivariable systems. IEEE Trans on Automatic control AC-27 (5): 1044–1054
9. Jong MT (1982) Methods of discrete signal and system analysis. McGraw-Hill, New York
10. Long DR, Davis MJ (1988) A new approach to the hyperbolic curve. J Pet Technol July: 909–912
11. Rowan G, Clegg MW (1963) The cybernetic approach to reservoir engineering. Paper SPE 727 presented at the 38th SPE Annu Fall Meet, New Orleans, Oct 6–9

12. Rowan G, Warren JE (1967) A systems approach to reservoir engineering – optimum development planning. J Can Pet Technol July – Sept: 84–96
13. Widder DV (1946) The Laplace transform. Princeton University Press, Princeton, NJ

EXERCISES

Exercise 14.1

The AX-1 reservoir has been producing oil for over 30 years from the Bartlesville (Oklahoma) sands at a depth of 300-400 m. Between 1985 and 1988, 42 wells were on production.

The field can be operated economically as long as production is in excess of 1 m^3/day per well. Eight in-fill wells were drilled in early 1989, and another 25 wells were stimulated with acid. This resulted in an increase in oil production. Table E14/1.1 is a record of the daily production from the reservoir between January 1985 and December 1989.

Determine what the incremental cumulative production would be at reservoir abandonment following the drilling of the in-fill wells and the stimulation of the others.

Solution

By plotting ln q_0 versus time (Fig. E14/1.1) we can see clearly that the production data in Table E14/1.1 lie on two distinct straight line trends corresponding to before and after the interventions of January 1989 (8 new wells on production, 25 wells stimulated).

These lines have the general form:

$$q_0(t) = q_i e^{-Dt}. \tag{14.5b}$$

The two curves describing the decline in productivity are therefore exponential (or constant percentage decline).

Table E14/1.1. *Production data for the AX-1 reservoir*

	Months	Wells on production	Oil rate (m^3/day)
1985–Jan.	0	42	210
April	3	42	198
July	6	42	187
Oct.	9	42	177
1986–Jan.	12	42	167
April	15	42	157
July	18	42	149
Oct.	21	42	141
1987–Jan.	24	42	133
April	27	42	125
July	30	42	118
Oct.	33	42	112
1988–Jan.	36	42	106
April	39	42	100
July	42	42	94
Oct.	45	42	89
1989–Jan.	48	50	150
April	51	50	136
July	54	50	124
Oct.	57	50	113
Dec.	60	50	103

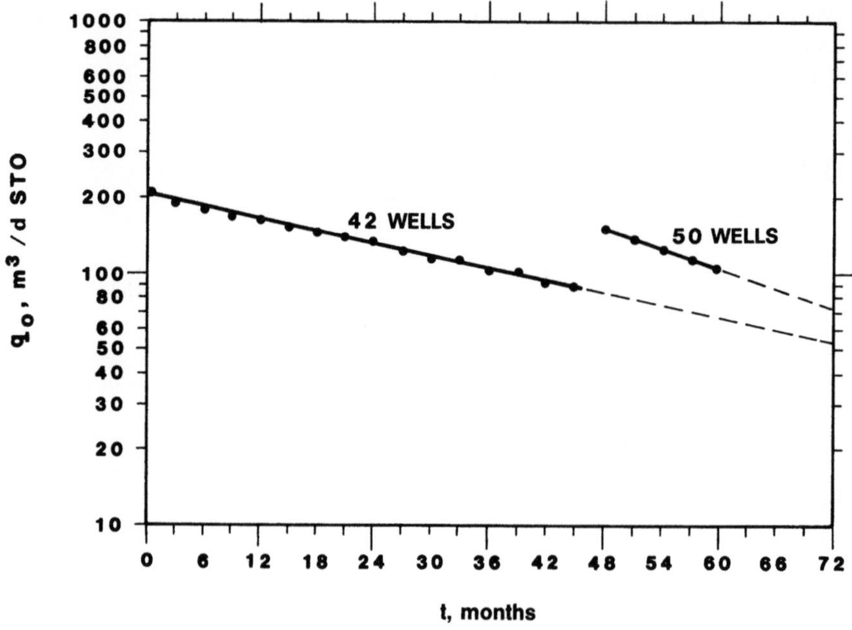

Fig. E14/1.1

The coefficient D is calculated by a least squares linear regression on the production data.
To be exact:

For months 0 to 45:

$$q_0(t) = 210\,e^{-0.01909\,t} \tag{14/1.1}$$

For months 48 to 60:

$$q_0(t) = 150\,e^{-0.03132(t-48)}, \tag{14/1.2}$$

where:

$q_0(t)$ = the daily oil production rate from the reservoir (m³/day STO),

t = time in months.

The cumulative oil production between month 48 (when the 8 new wells were put on production and 25 wells were stimulated) and abandonment of the reservoir at its economic limit is calculated from the equation:

$$N_p = \frac{q_i - q_t}{D}, \tag{14.7a}$$

where:

q_i = reservoir production rate at month 48,

q_t = reservoir production rate at abandonment = 1 m³/day per well,

   = 42 m³/day when there are 42 wells;

   = 50 m³/day when there are 50 wells.

The production rate at month 48 in the case of 42 wells is calculated using Eq. (14/1.1) with $t = 48$. The result is q_i (42 wells, 48 months) = 84 m³/day STO.
Therefore:

from month 48 to reservoir abandonment

$$N_p(50\text{wells}) = \frac{150 - 50}{0.03132} = 3193 \text{ m}^3\text{STO}$$

$$N_p(42\text{wells}) = \frac{84 - 42}{0.01909} = 2200 \text{ m}^3\text{STO}.$$

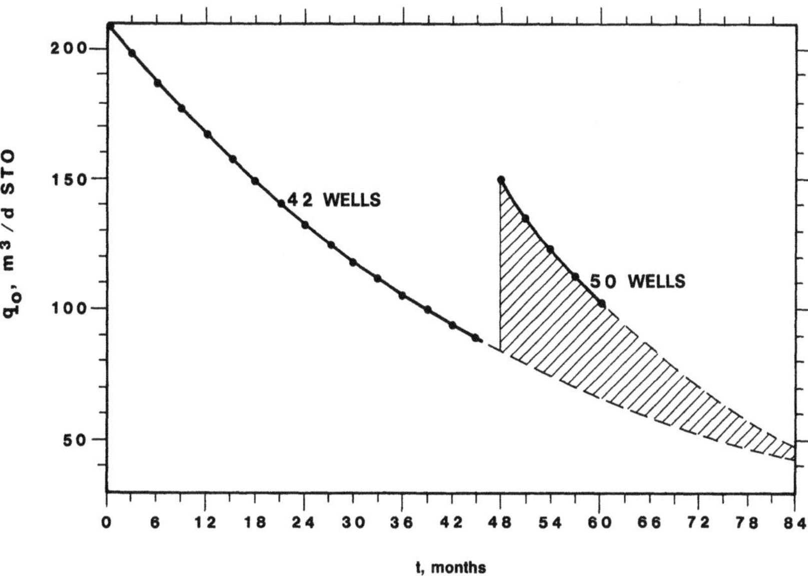

Fig. E14/1.2

Solving Eqs. (14/1.1) and (14/1.2) for t, with $q_o(t)$ equal to the economic limit at abandonment we get:

reservoir is abandoned at:

  $t = 84.3$ months with 42 wells;

  $t = 83.1$ months with 50 wells.

From this we can conclude that the "revamping" operation of drilling 8 new wells and stimulating 25 others will achieve a significant increase in oil recovery – nearly 1000 m³ – for almost the same abandonment time (83 months instead of 84).

In all likelihood, this gain is due largely to the drilling of the in-fill wells, which provided improved connectivity with parts of the reservoir not adequately drained by the original 42 wells.

The shaded area in Fig. E14/1.2 [between the two curves $q_o(t)$ corresponding to the production from 42 wells and 50 respectively] depicts the extent of the improvement in recovery. The curves were extrapolated forward in time using Eq. 14/1.1 ($t > 45$ months) and Eq. 14/1.2 ($t > 60$ months).

The revenue realised from the 1000 m³ of extra production should, of course, be compared with the cost of the revamping operation to verify that the project is economically viable.

◇ ◇ ◇

Fig. E10/2

15 Techniques for Improving the Oil Recovery

15.1 Introduction

We have seen in preceding chapters that not all of the hydrocarbon contained in a reservoir can be extracted. The basic *resource*, therefore, cannot be converted entirely to *reserves*: part of the resource is left behind as a *remaining oil saturation* (ROS).

The oil that remains consists of oil trapped in the pore space of the rock that has been invaded by displacing fluid (S_{or}), plus oil distributed in isolated sections of rock that have been bypassed by the displacing fluid.

A statistical study of oil reservoirs that have been totally depleted by natural drive processes (*primary recovery*) and by gas or water injection (*secondary recovery*) found that the average percentage of oil recovered was only 33.6% (see Sect. 12.3 for the recovery factor, E_R).

The main causes of this poor recovery are: the existence of surface tension forces between non-miscible oil and water or gas, resulting in an immobile saturation of residual oil S_{or} in the pores; and heterogeneities in the reservoir rock, which give rise to a volumetric invasion efficiency E_V which is considerably less than unity.

Not surprisingly, the quantities of oil left behind when production is terminated are enormous, as we shall see in Sect. 15.2. It is a staggering thought that *if we could double the recovery factors for all existing reservoirs, it would be equivalent to discovering again as many new reservoirs as have been discovered to date.*

The recovery factor, then, is of tremendous technical and commercial significance for the exploration and production sectors (E&P) of the petroleum industry. As a consequence, since the early 1950's a huge effort has been devoted to the task of achieving a higher recovery factor by economically viable methods.

The following sections of this chapter will deal with both the technical and economic aspects of this very important subject.

15.2 The Current Status of the World's Oil Reserves

With the discoveries that have been announced over the last two decades or so in the Arabian Gulf area (Saudi Arabia, The Emirates, Iran, Iraq, Kuwait, etc.)[3], the world's oil reserves as of 1 January 1992 (the quantity of oil that could be extracted from all the known reservoirs of the world at that date) were estimated[99] to be 156×10^9 m^3 STO. Of this, some 66% belonged to the OPEC countries[149]. If we add to this the 106×10^9 m^3 STO of oil already extracted up to the end of 1991[97], we obtain a figure of 262×10^9 m^3 STO for the total quantity of recoverable oil originally in place. Assuming an average recovery factor of 33.6%, the actual volume of oil originally present in the world's reservoirs prior to discovery and exploitation amounts to some 780×10^9 m^3 STO.

The difference – 520×10^9 m^3 – between *resource* and *reserves* is the quantity of oil which will be left in the ground when all primary and secondary production has been terminated. *This represents some 150 years of continued production at current levels.*

To the reservoirs already discovered we should, of course, add those yet to be discovered, both within areas that have already been explored, and in totally unexplored areas such as the deep oceans, the Arctic and Antarctic, the Amazon central Africa, large parts of the CIS (ex-USSR) and of the People's Republic of China.

At a broad estimate, exploration geologists reckon[10,29,60,97,130] there might be resources of some $180 \sim 240 \times 10^9$ m^3 STO in reservoirs still to be discovered, which means $60 \sim 80 \times 10^9$ m^3 STO of reserves amenable to primary and secondary recovery.

The situation presented in Table 15.1 can be summarised as follows:

- the world oil reserve as of 1 January 1992 was of the order of 226 billion m^3 STO, of which approximately one-third is in reservoirs not yet discovered,
- some 660 billion m^3 STO will remain in the ground after recovery by primary and secondary processes has been accomplished. Roughly one fifth of this is in reservoirs not yet discovered.

If even a quarter of the remaining oil could be produced economically, it would satisfy world demands for a further 50 years at current rates of consumption.

Geographically, a major part of the remaining oil is, of course, concentrated in the more mature petroleum producing areas – areas from which a high percentage of the original reserves have already been extracted, and where there are, consequently, a large number of depleted or nearly depleted reservoirs.

The USA and Canada are of particular interest in this context. Both countries have already been very thoroughly explored, and are producing from a large number of predominantly low-rate wells[2] (Fig. 15.1). Obviously, the probability of finding new reserves is significantly smaller than in less "mature" countries where the petroleum industry is not so well developed.

Table 15.1. *Worldwide petroleum resources and reserves*[97,120,130] *(as of 1 January, 1992)*

| | Resources | Reserves | | | Oil remaining |
| | | Original | Produced by 1/1/92 | Unproduced by 1/1/92 | in the reservoir[a] |
	(10^9 m^3)	(10^9 m^3)	(10^9 m^3)	(10^9 m^3)	(10^9 m^3)
In known reservoirs	780	262	106	156	518
In undiscovered reservoirs	180–240	60–80	–	60–80	120–160
Total	960–1020	322–342	106	216–236	638–678

[a] At abandonment, after reservoir exploitation by primary and secondary recovery processes

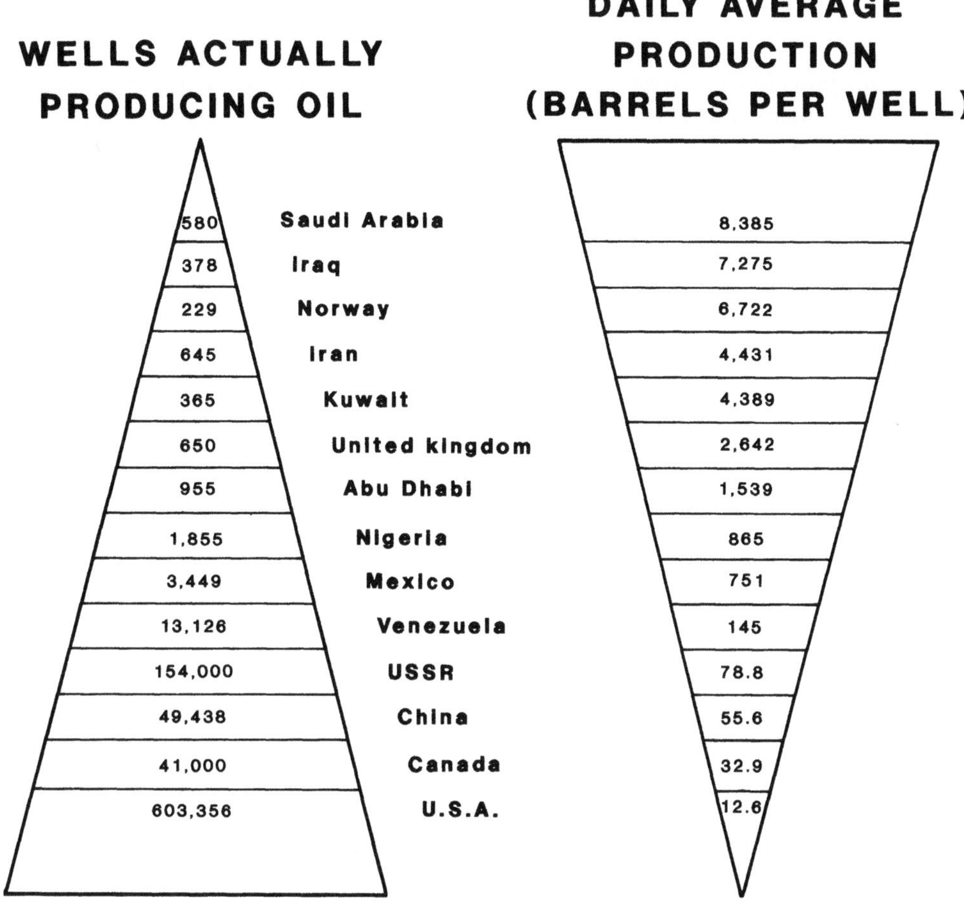

Fig. 15.1. *Number of wells and their average daily production in 14 of the principal oil-producing countries (from data for 1989 reported in Ref. 96). Data since 1989 have not been included, in view of the recent situation in two of the major producers, Kuwait and Iraq*

The costs of exploration for "replacement barrels" – discovering reserves each year sufficient to match the country's annual production – are extremely high. Thus we find that a major investment in the research, development and implementation of techniques for improving oil recovery is mainly focused in these two countries[54].

15.3 Quantity and Distribution of Oil Remaining in the Reservoir

15.3.1 Introduction

The practical and economic feasibility of any improved recovery process obviously depends on a knowledge of the quantity of oil still present in the reservoir at the time the process is planned to start – be it at reservoir abandonment, or earlier in the life cycle.

But this is not the whole story: we also need reliable information on the areal distribution of the oil, and on how it is divided between residual oil (S_{or}) contained in pore space that has been swept during primary and secondary recovery, and bypassed oil contained in isolated pockets of rock that have not been completely swept because of reservoir heterogeneities ($E_V =$ volumetric efficiency).

As we shall see later, in fact, processes designed to improve the displacement efficiency at the pore level (reducing S_{or}) are completely different from – and sometimes even opposite to – those dedicated to improving the volumetric efficiency of displacement (increasing E_V).

It goes without saying that there is no point in considering a process to reduce an S_{or} which is already small when most of the remaining oil is in poorly swept parts of the reservoir. Nor is it sensible to attempt to improve E_V when most of the oil remains in the pore space already swept by water or gas.

A surprising proportion of the unsuccessful attempts at enhanced recovery failed because of the application of a process ill-adapted to the "type" of remaining oil (high S_{or} or low E_V). However, it is only in the last decade, following a number of post mortem analyses on reservoirs where enhanced recovery had failed, that this simple fact has been fully appreciated.

It is now a prerequisite for the reservoir engineer, before designing any improved ·recovery program whatsoever, to focus his attention first and foremost on both the quantity and the distribution of oil remaining in the reservoir (ROS).

15.3.2 Methods for Evaluating the Overall Quantity of Remaining Oil

15.3.2.1 The Material Balance Method

This is the simplest, but also the most inaccurate, method of evaluating the total volume N_r of oil remaining in the reservoir, under standard conditions.

If N is the original volume of oil in place (Chap. 10) as estimated by the volumetric method (Chap. 4), and N_p is the cumulative volume of oil produced, both expressed under standard conditions, then quite simply:

$$N_r = N - N_p \tag{15.1}$$

The value of N_r calculated in this way suffers principally from the uncertainty[138] inherent in N: if the estimate of N was optimistic, N_r will be too large, if pessimistic, N_r will be too small.

Furthermore, the material balance method provides no information whatsoever about spatial distribution of the remaining oil throughout the reservoir.

15.3.2.2 Single Well Tracer Test

The single well tracer test (SWTT) provides an estimate of the *average* ROS over a region of about 1 m radius around the well. An idea of the *areal* distribution (*not* vertical) of the ROS can therefore be built up by running an SWTT in each well.

The test consists of injecting a pad of *primary tracer* (an ester such as methyl, ethyl or iso-propyl acetate) down the well. This is followed by a certain volume of water which displaces the tracer outwards into the formation (Fig. 15.2).

The primary tracer is soluble both in the water and, at least partially, in the remaining oil. If the tracer concentration in water is C_w, and C_o in oil, the *equilibrium partition coefficient K* is defined as the ratio:

$$K = \frac{C_o}{C_w}. \tag{15.2}$$

After injection, the well is shut in for an appropriate period of time. The fraction of the primary tracer that remains dissolved in water undergoes a partial hydrolysis,

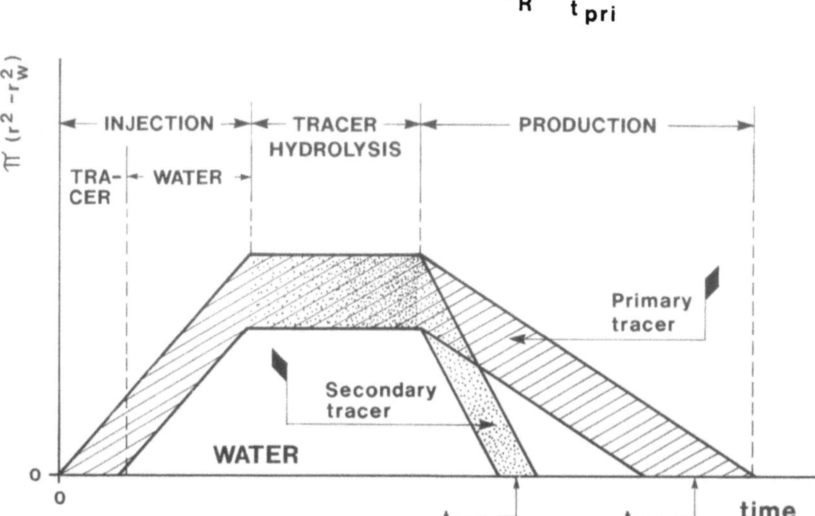

Fig. 15.2. *The Single Well Tracer Test (SWTT). Schematic of the radial distribution with time of the primary and secondary tracer. (From Ref. 83, 1973, Society of Petroleum Engineers of AIME. Reprinted with permission of the SPE)*

in which it is decomposed slowly into the organic acid and alcohol that made up the ester. One of the reactions is, for example:

$$CH_3COO-C_2H_5 + H_2O \Rightarrow CH_3COOH + C_2H_5OH$$
$$\textit{ethyl acetate} \qquad\qquad\qquad \textit{acetic acid} \quad\ \textit{ethyl alcohol} \qquad (15.3)$$

The alcohol by-product of the reaction – the *secondary tracer* – remains in solution in the water, and is insoluble in oil. The acid will react with any carbonates in the rock and disappear in the form of calcium and magnesium salts.

After a time sufficient to allow well-advanced (but not total) hydrolysis of the primary tracer, the well is put back on production and the concentration of primary and secondary tracers appearing in the produced water are measured (Fig. 15.3). No oil is produced, of course, since it is immobile.

The bell shape of the curves (tracer concentration)/(cumulative volume of produced water) is caused by dispersion (described in Sect. 12.7.2).

Note that the primary tracer peak always lags behind that of the secondary tracer. This is a chromatographic phenomenon caused by the fact that the primary tracer is partially soluble in the oil remaining in the rock through which it passes, while the secondary tracer is insoluble.

If u_w is the velocity of the water, the velocity u_p of the molecules of primary tracer will be a fraction F_R of u_w:

$$u_p = F_R u_w, \qquad\qquad\qquad\qquad (15.4)$$

where F_R is the "delay factor"[83]. It represents the fraction of time the molecules spend in the water phase; $(1 - F_R)$ is therefore the fraction of time spent in the oil phase.

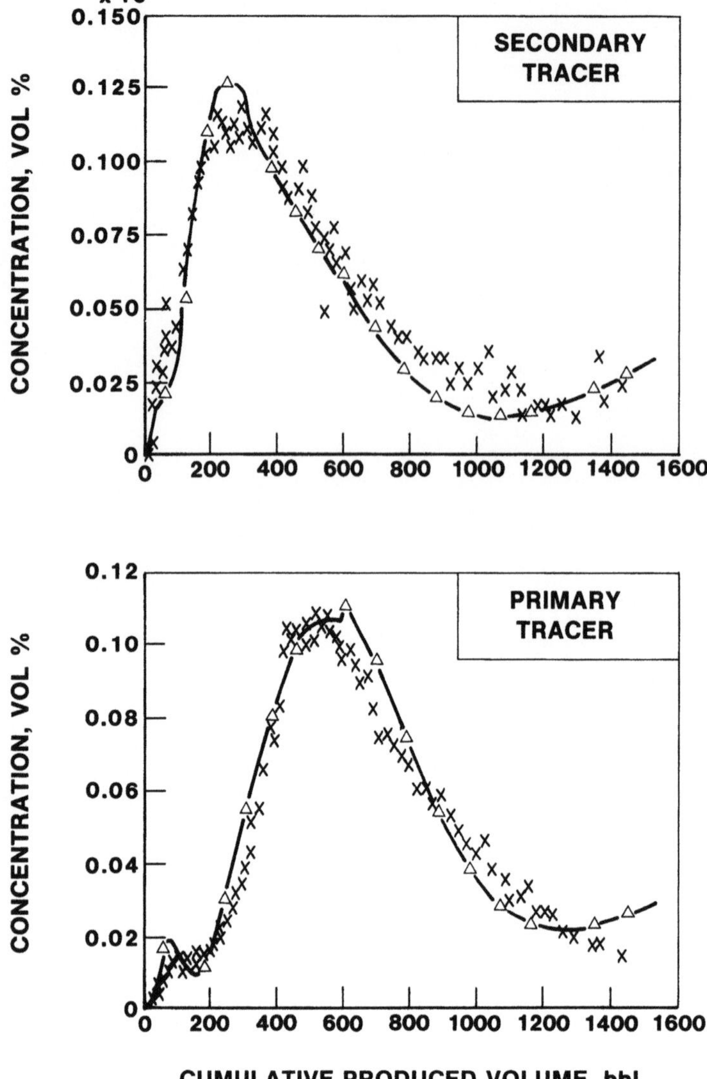

Fig. 15.3. *Variation in the concentrations of primary and secondary tracers in the produced water during a single well tracer test (SWTT). (From Ref. 75, 1982, Society of Petroleum Engineers of AIME. Reprinted by permission of the SPE)*

If u_w is sufficiently small, the distribution of *primary* tracer between the water and remaining oil phases will correspond to an equilibrium condition[75,83] in which:

$$\frac{\text{mass dissolved in remaining oil}}{\text{mass dissolved in water}}$$

$$= \frac{C_o\text{ROS}}{C_w(1-\text{ROS})} = K\frac{\text{ROS}}{1-\text{ROS}} = \frac{1-F_R}{F_R} \tag{15.5}$$

Therefore:

$$F_R = \frac{1-\text{ROS}}{1+(K-1)\text{ROS}} \tag{15.6a}$$

Since the *secondary* tracer is insoluble in oil, $C_o = 0$. From Eq. (15.2), then, $K = 0$, and from Eq. (15.6), $F_R = 1$.

The molecules of secondary tracer therefore move with the same velocity u_w as the water phase. If u_s is the secondary tracer velocity:

$$u_s = u_w. \tag{15.7}$$

Since the paths taken by the molecules of primary and secondary tracer are the same, we have:

$$\frac{u_p}{u_s} = \frac{u_p}{u_w} = F_R = \frac{\text{time to arrival of secondary tracer peak}}{\text{time to arrival of primary tracer peak}}$$

$$= \frac{\text{cumulative volume of water produced at arrival of secondary tracer peak}}{\text{cumulative volume of water produced at arrival of primary tracer peak}}$$

$$\tag{15.6b}$$

The value of F_R calculated from Eq. (15.6b) is entered in Eq. (15.6a) to get a first approximation of the ROS.

More sophisticated methods of interpreting the SWTT have been published[32], using numerical models to simulate the observed response.

Laboratory tests are run under reservoir conditions to evaluate the equilibrium partition constant K and the shut-in time necessary to allow the primary tracer to be adequately hydrolysed. A knowledge of these parameters also helps in choosing the appropriate primary tracer for that particular test.

15.3.3 Determination of the Distribution of Remaining Oil Throughout the Reservoir

There are four principal methods[9,15,82] by which reservoir engineers can assess the spatial distribution of the remaining oil:

1. Analysis of downhole cores
2. Resistivity logs
3. Pulsed neutron logs (PNL)
4. Nuclear magnetism logs[5] (NML)

Some techniques are still experimental; others, such as those based on the measurement of the dielectric properties of the rock, and on gamma-active tracers, are now in field test phase.

Cores can only be recovered from wells during the drilling phase. In a producing field, they must be obtained from in-fill wells whenever they are drilled.

Resistivity logging is limited to open hole – the presence of a steel casing shortcircuits the measuring currents. However, if necessary, special observation wells can be drilled and cased through the hydrocarbon-bearing intervals with a non-conductive plastic or fibre glass liner. This will permit induction-type electrical logs to be run.

Pulsed neutron logs (PNL) can be used to measure the ROS in open hole or holes cased with steel, plastic or fibre glass.

The nuclear magnetism log (NML) can only be run in open hole, or through non-metallic casings. Provided the magnetic properties of the drilling mud and flushed zone water are correctly controlled, it is possible to measure the ROS directly.

The following sections outline the principles of each of the methods listed above. For more detailed reading, please refer to specialised literature on the topic[50].

15.3.3.1   Retrieval and Analysis of Downhole Cores

In Sect. 3.2 we saw that the value of S_o measured in a laboratory on a core cut with a conventional core barrel, or using a rubber-sleeved device, bears little or no relation to the actual S_o existing in the reservoir at the point where the core was taken.

There are several reasons for this:

- the core may be invaded by mud filtrate while being cut. S_o in the downhole core, therefore, is likely to be closer to a swept zone S_{or} in the case of a water-based mud (especially in a water-wet rock), or to S_{iw} in the case of an oil-based mud (especially in an oil-wet rock). For this reason, a water-based mud is to be preferred when coring in a depleted interval where S_{or} is to be determined.
- as the core is brought to the surface, gas is liberated from the oil, taking a fraction of the remaining oil saturation with it.

Invasion by well-bore fluids is difficult to avoid. (A low-invasion core bit has recently been introduced, but is not yet in widespread use.) However, the loss of oil through the liberation and expansion of dissolved gas can be prevented by using specialised core-cutters:

- *pressure core barrel:* the core is maintained at bottom-hole pressure at all times. The subsequent laboratory measurement of S_{or} requires special equipment and expertise.
- *sponge core barrel:* the oil exuded by the core on its way to the surface is collected in an absorbent sleeve made of a sponge of resinous material surrounding the core.

In both cases, as far as the estimation of the remaining oil saturation is concerned, the value of S_o measured from the core is still the oil saturation S_{or} after invasion by mud filtrate, rather than that to be found in the reservoir.

15.3.3.2   Resistivity Logs

The electrical resistivity $R_{t,R}$ of the virgin formation beyond the zone invaded by mud filtrate can be measured by means of a number of logging techniques[71,72]: the dual induction log (DIL) or a combination of induction logs of different depths of investigation (PHASOR); focused resistivity logs such as the dual laterolog (DLL) and the earlier LL7 and LL3; or a simultaneous combination of DIL and DLL.

The water saturation in the uninvaded formation some distance from the well-bore has a well-defined relationship to the resistivity $R_{t,R}$. In a "clean" reservoir (containing no dispersed or stratified shales), this is expressed by the *Archie equation*:

$$S_w = \phi^{-m} \left(\frac{R_{w,R}}{R_{t,R}} \right)^{1/n} \tag{15.8a}$$

where:

ϕ = connected porosity (decimal fraction);

$R_{w,R}$ = electrical resistivity of the formation water;

m = cementation factor;

n = saturation exponent.

The local porosity can be obtained from the analysis of porosity logs (Sect. 3.2), previously calibrated against core measurements.

m and n depend on the microscopic structure of the rock, and should be evaluated from core tests for each sedimentary unit in the reservoir. Cluster analysis (Sect. 3.6.3) will provide a properly correlated and statistically valid interpretation.

$R_{w,R}$ can be estimated from the SP (spontaneous, or self, potential) log[72]. In injection wells, it can of course be measured directly from an injection water sample.

With Eq. (15.8a), S_w beyond the zone flushed by mud filtrate can be calculated *point by point*. Since $S_o = 1 - S_w$, an oil saturation profile $S_o(z)$ can therefore be constructed over the thickness of the formation.

In order to obtain greater precision in the estimation of the saturation, the need to know ϕ and m can be eliminated by the use of a technique known as "log-inject-log" (LIL). For resistivity logging, this is of course restricted to open-hole. A cased-hole LIL method based on pulsed neutron logging is described in the next section.

A log is first made of the formation resistivity profile $R_{t,R}$. A pad of solvent which is miscible with oil and water is then injected into the oil-bearing interval. This is followed by a volume of water of known resistivity $R_{w,I}$.

Sufficient solvent and water are injected to displace all of the oil from the region of investigation of the logging device, so that $S_w = 1$.

The resistivity profile is then logged again. If $R_{o,I}$ is the resistivity measured at any point, then from Eq. (15.8a):

$$1 = \phi^{-m} \left(\frac{R_{w,I}}{R_{o,I}} \right)^{1/n} \tag{15.8b}$$

By combining Eqs. (15.8a, b), we can estimate the water saturation seen by the first logging pass:

$$S_w = \left(\frac{R_{w,R}}{R_{w,I}} \frac{R_{o,I}}{R_{t,R}} \right)^{1/n} \tag{15.8c}$$

This methodology can be extended to the estimation of the ROS corresponding to when the reservoir has been fully swept by water (the minimum ROS achievable by primary or secondary recovery) by the addition of an extra injection cycle, in a technique referred to as ILIL (inject-log-inject-log).

The formation is first flushed with a suitable volume of high salinity water of resistivity $R_{w,I}$, with the objective of displacing all the movable oil away from the wellbore region so that only the remaining oil saturation will be seen by the logging tool. The LIL sequence described above is then performed, using water of the same resistivity $R_{w,I}$ for the second injection cycle as for the first.

Since in this case $R_{w,R} = R_{w,I}$, Eq. (15.8a) reduces to:

$$1 - \text{ROS} = \left(\frac{R_{o,I}}{R_{t,I}} \right)^{1/n} \tag{15.8d}$$

Note that, apart from the two log readings, Eq. (15.8d) only requires the saturation exponent n to calculate the ROS.

15.3.3.3 Pulsed Neutron Logs (PNL)

The logging techniques based on high energy pulsed neutron emission were described in Sect. 8.5.4.

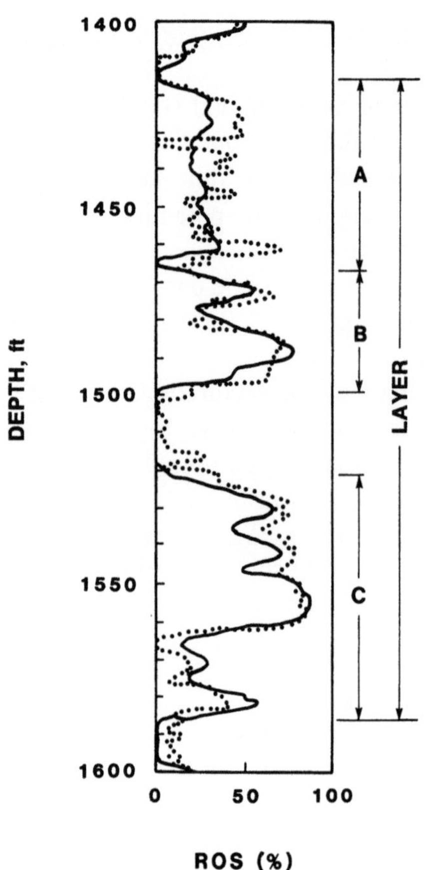

Fig. 15.4. *Comparison between ROS estimates derived from a carbon/oxygen (C/O) log and a deep-reading induction log in the same well. (From Ref. 15, 1988, Society of Petroleum Engineers of AIME. Reprinted with permission of the SPE)*

They fall into two categories:

- *gamma-ray spectrometry log* (the GST gamma spectrometry tool, the MSI-C/O carbon/oxygen tool): high energy (14 MeV) neutrons emitted from a pulsed source interact with atomic nuclei in the surroundings. From the energy spectrum of gamma radiation resulting from inelastic interactions between neutrons and matter, a number of elements can be identified and their relative abundance estimated. Two of these elements are carbon and oxygen. The carbon/oxygen ratio (COR) log (see Sect. 8.7.3), which can be recorded in open hole or through casing, responds to the concentration of carbon in oil, and oxygen in water, in the pore space.

It can therefore be used to measure the ROS. In Fig. 15.4, ROS profiles derived from a resistivity log and a carbon/oxygen ratio log run in the same well are compared.

- *thermal decay time log* (TDT, PDK-100, TMD): high energy neutrons emitted from the same type of pulsed source as is used for gamma spectrometry logging are rapidly slowed down by successive collisions with atomic nuclei until they reach thermal energy. They are then annihilated by capture interactions with certain elements, particularly chlorine. The tool measures the rate of decay, through capture, of the population of thermalised neutrons. This varies with the abundance of chlorine in the formation and is therefore directly related to the water saturation S_w.

Pulsed neutron capture logging also lends itself to the log-inject-log technique[142] – or LIL(PNL). After the first TDT log has been run, water whose salinity is higher than that of the formation water is injected in sufficient quantity to displace all of the original water from the near-well-bore region investigated by the tool. A second TDT log is then run.

If $\Sigma_{w,1}$ is the neutron capture cross-section of the formation water, $\Sigma_{w,2}$ is that of the injected water, and $\Sigma_{TDT,1}$ and $\Sigma_{TDT,2}$ are the first and second log readings, we can derive from Eq. (8.15) that, at any point:

$$S_w = \frac{\Sigma_{TDT,2} - \Sigma_{TDT,1}}{\Sigma_{w,2} - \Sigma_{w,1}} = 1 - \text{ROS} \qquad (8.15a)$$

This equation is only valid if the ROS was not altered by the injection of the higher salinity water, and if the injection effectively displaced all of the formation water from the region investigated by the logging tool.

15.3.3.4 Nuclear Magnetism Log (NML)

Both oil and water contain a high percentage of hydrogen. The nuclear magnetism log (NML) measures the density of hydrogen atoms (or, more precisely, of protons) present in the pore fluids, but only in the *mobile* fraction.

Some of the pore fluid is bound to the rock by electrical forces (water in shales), or by van der Waals forces (the film of water or oil adhering to the grain surfaces in water-wet or oil-wet rocks respectively). This fraction of fluid does not contribute to the NML signal, and the log is essentially a measurement of the "free fluid index" (FFI). For example, in a preferentially water-wet rock containing only water and oil ($S_g = 0$), the FFI is a measure of the quantity $\phi(1 - S_{iw})$.

Special procedures (described later) allow the contribution from the water phase to be eliminated from the signal, so that the NML measures ϕS_o. In parts of the reservoir swept by water, this is the same as $\phi \times$ ROS.

The NML can therefore be used to determine effective porosity and remaining oil saturation in the reservoir rock around the well-bore. Below is a very basic description of the principles upon which the NML measurement is based.

The nucleus of the hydrogen atom is a single proton. The proton, with a mass approximately 1836 times heavier than the electron and a positive electrical charge, spins about its own axis with an intrinsic angular momentum **s**. In quantum mechanics it can be shown that the modulus of **s** is $\sqrt{s(s+1)}h/2\pi$, where s is the spin quantum number and h is Planck's constant (6.625×10^{-34} joule $\cdot$ seconds).

The spin quantum number of a proton can be either $+1/2$ or $-1/2$. This means it can only rotate about itself in one direction or the opposite. In a magnetic field $\boldsymbol{H}$, the spin vector will therefore align either parallel or antiparallel to $\boldsymbol{H}$. Protons whose spins have aligned parallel to the field have a lower potential energy than those aligned antiparallel. At equilibrium, the distribution of protons between the two energy levels (Boltzmann distribution) is such that the lower energy protons (parallel to H) are in a small majority. This gives rise to a nuclear magnetisation M in the same direction as H, and proportional to the local proton density. M is therefore proportional to the local density of hydrogen atoms – but only of those atoms present in the molecules of the *mobile* oil and water.

Normally, in the absence of any other magnetic fields, M aligns itself in the direction and sense of the Earth's field H_e. Now suppose a strong polarising magnetic field H_p of constant magnitude and direction is superposed on H_e, perpendicular to it. At equilibrium, there will be a slight prevalence of proton spin axes oriented parallel to H_p, with the result that the nuclear magnetisation vector M will gradually realign itself with H_p.

The approach of M to its equilibrium value $M_o = \chi_o H_p$ (where χ_o is the static spin susceptibility) is described by the *spin lattice relaxation time* T_1. After a time $t_p > 6T_1$, the magnetisation M has attained at least 99.8% of its equilibrium value M_o.

In the NML tool, the polarising field H_p is generated by a coil carrying a DC current of several amps (over 1 kW of power consumption). The field is applied for a time t_p sufficient for the nuclear magnetisation M to reach M_o, and is then very abruptly shut off. The spins are unable to respond instantaneously to this sudden change, and for a time they are left still aligned perpendicular to H_e. The nuclear magnetisation M is therefore subjected to a torque $M \wedge H_e$ which is perpendicular to the plane containing H_e and M. The couple invokes an angular displacement of the vector M, so that its rotation about H_e describes a cone with H_e as its axis. This is illustrated in Fig. 15.5.

This phenomenon, *proton free precession*, is described quantitatively by Larmor's theory. The rotational frequency f_L of M about H_e (its Larmor frequency) is:

$$f_L = \gamma |H_e|, \tag{15.9}$$

where γ is the gyromagnetic ratio of the proton:

$$\gamma = 4.2576 \times 10^7 \ \text{Hz Tesla}. \tag{15.10}$$

Since the value of H_e depends on the longitude, latitude and depth of the point of measurement, so does the Larmor frequency. It lies somewhere between 1.3 and 2.6 kHz in typical oil wells. Because of its precessional motion, the component

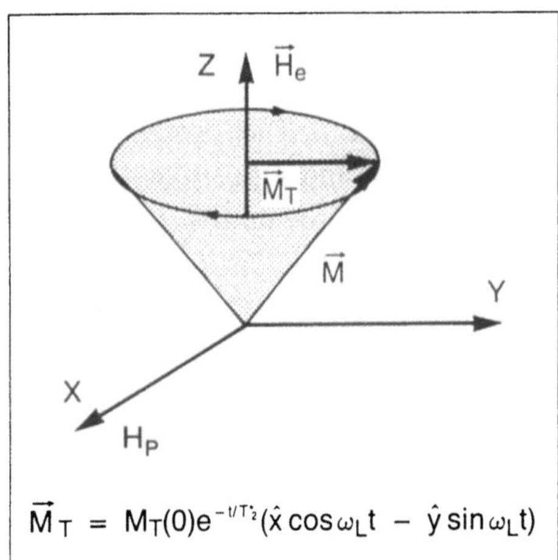

Fig. 15.5. *Schematic of the Larmor free precession of the proton magnetisation vector about the Earth's magnetic field. (From Ref. 113, 1989, Society of Petroleum Engineers of AIME. Reprinted with permission of the SPE)*

M_T shown in Fig. 15.5 perpendicular to H_e will generate a sinusoidal emf (the *free precession signal*) of frequency f_L. Its amplitude – typically a fraction of a μV – will be proportional to the density of the protons constituting the hydrogen atoms not bound to the grain surfaces by the electrical or van der Waals forces described earlier. The free precession signal is measured by the same coil used to generate the polarising magnetic field H_p, and transmitted to surface.

As the polarising magnetic field cannot, in practice, be turned off instantaneously, much of the precessional signal can be lost. To compensate for this, the NML polarising current is allowed to oscillate at the Larmor frequency for a few cycles before being turned off. This "ringing" field keeps the proton spins aligned perpendicular to H_e until H_p has completely died away.

The amplitude of the free precession signal decays exponentially, with a time constant T_2^*. This phenomenon is called free induction decay (FID), or nuclear magnetic relaxation (Fig. 15.6). It is caused by microscopic spin-spin interactions among protons, characterised by the transverse spin-spin relaxation time constant T_2; and by inhomogeneities in the Earth's magnetic field. The following inequalities hold true: $T_2^* < T_2 \le T_1$.

The signal at the beginning of the free precession period is contaminated by transients induced by turning off the polarising current, and this initial part of the cycle must be ignored. The signal amplitude at time zero, needed to get the free fluid index (FFI) or free fluid porosity (ϕ_f), can, however, be obtained by backward extrapolation of the signal envelope recorded later in the relaxation period (Fig. 15.6). The decay time constant T_2^* is calculated simultaneously. Some authors claim that the rock permeability can be derived from T_2^*, but this viewpoint is by no means universally accepted.

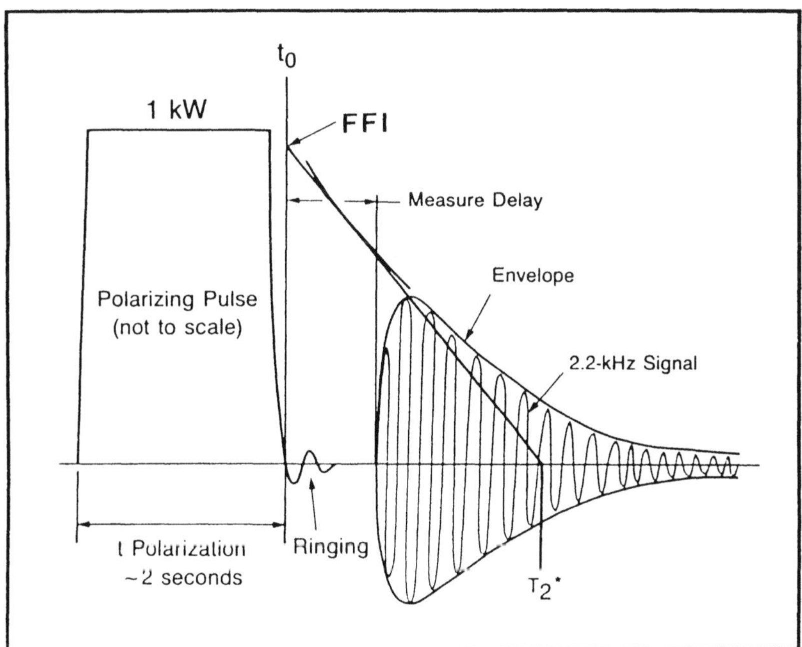

Fig. 15.6. *NML signal: Larmor free precession sine wave, with amplitude decaying exponentially with time constant T_2^*. Estimation of the FFI (signal amplitude at t_0). (From Ref. 72. Reprinted courtesy of Schlumberger)*

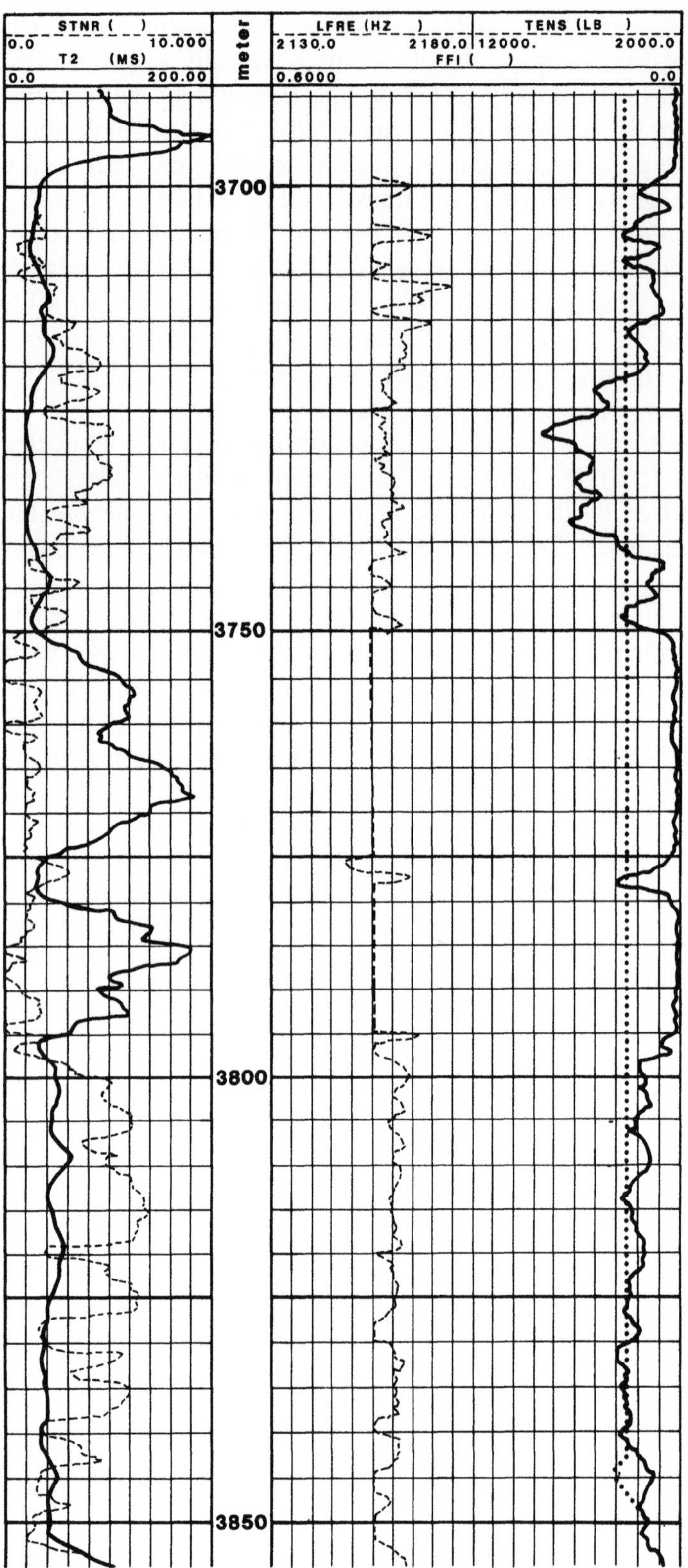

Fig. 15.7. *An example of an NML log run in open hole. FFI free fluid index; LFRE Larmor frequency;* T_2 *nuclear magnetic relaxation time constant* (T_2^*)

The basic NML log, then, consists of the recording versus depth of the FFI, T_2^*, and the Larmor frequency f_L (Fig. 15.7). It can only be run in open hole, or through nonmetallic casing – a metal casing will completely shield the pick up coil from the formation precessional signal.

The signal originating from oil or water in the mud in the well bore can be eliminated by the addition of magnetite, which is paramagnetic. This causes the protons to realign themselves with H_e almost instantaneously when the field H_p is switched off, so that they do not contribute to the FFI measurement. The FFI can therefore be calibrated directly in terms of $\phi(1 - S_{iw})$.

The NML can be used *to measure the remaining oil saturation directly*. To achieve this, the mobility of the protons in the formation water is increased by the injection of an aqueous solution of a paramagnetic salt such as EDTA-Mn (the manganese salt of ethyl-diamino-tetracetic acid). The precessional signal from the protons in the water now decays too rapidly to contribute to the FFI measurement, which now consists almost entirely of the signal from the oil present in the pore space. *The FFI is therefore now a direct measure of the* ROS.

Since a uniform distribution of the EDTA-Mn solution over the hydrocarbon-bearing interval is desirable, this operation is best performed in open hole.

With this procedure, the NML becomes *the only logging tool that can measure the* ROS *directly*, rather than inferring it from an indirect measurement.

15.4 ARM – Advanced Reservoir Management: Improvement of the Oil Recovery Factor by Optimisation of Water or Gas Injection

15.4.1 ARM for Water Injection

15.4.1.1 The Effect of Well-Spacing on Oil Recovery by Water Injection

In the days before reservoir engineering had become a recognised science, a simple rule known as Cutler's law[31] stated that, from the evidence of field observations, the recovery factor was inversely proportional to the spacing between producing wells.

Figure 15.8, from a publication by van Everdingen and Kriss[85], shows the validity of this relationship in the Slaughter Field, Texas, where the various leases have been developed with a variety of well spacings.

With the advances made in reservoir engineering in the 1930s and 1940s, Cutler's law was dismissed as "too empirical" and "unscientific", and was not corroborated by theoretical developments of at the time.

In fact, in these early years, the lack of adequate computing capabilities (numerical models, hardware, etc.) forced the theoreticians to treat the reservoir as if it were homogeneous and isotropic, particularly as far as the permeability was concerned. Without such simplifications, analytical solution of the flow and continuity equations would not have been possible.

Intuitively, in a truly homogeneous and isotropic reservoir, the recovery factor should not depend on the distribution of the producing wells. This explains why, at least until the end of the 1970s, the notion prevailed that the recovery factor should be more or less independent of well spacing.

On 27 August 1979, van Everdingen wrote his famous letter to the then US President Jimmy Carter, to the US Department of Energy, and to the heads of all

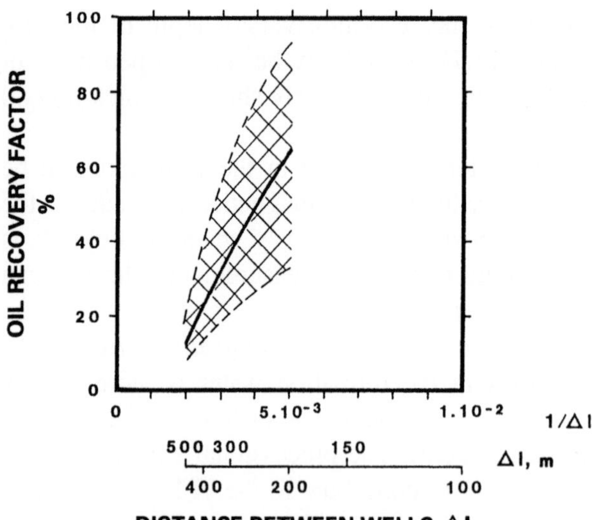

Fig. 15.8. *Slaughter Field, Texas. Relationship observed between the oil recovery factor and the reciprocal of the well spacing in a number of leases. (From Ref. 21. Reprinted with permission of the Institut Français du Pétrole)*

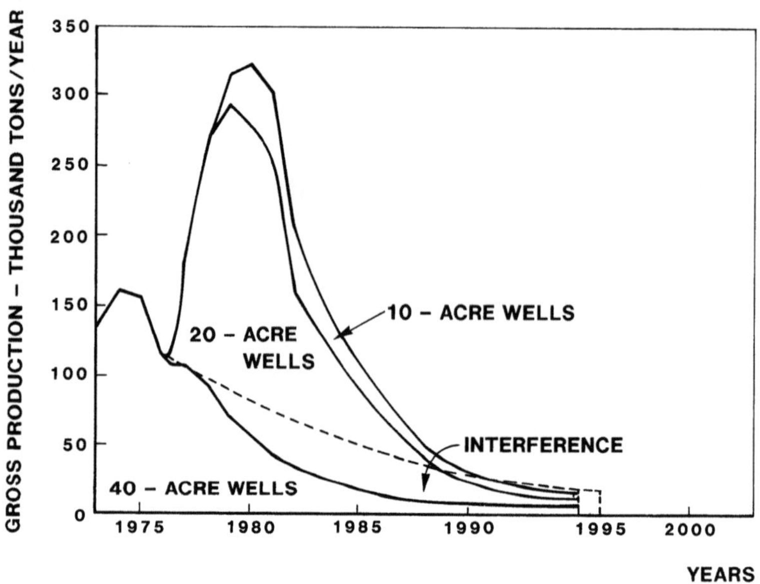

Fig. 15.9. *Increase in oil production rate and recovery obtained by infill drilling. Robertson Clearfork Unit, Gaines County, Texas. (From Ref. 7, 1983, Society of Petroleum Engineers of AIME. Reprinted with permission of the SPE)*

the major oil companies, pointing out that, on the evidence of a wealth of field data, *the recovery factor did indeed depend on the well spacing*. This controversial announcement was backed up by a detailed technical publication[85] which followed in 1980.

An increase in the oil recovery factor has been achieved in numerous reservoirs[140] by reducing the average distance between wells through infill drilling[95]. The results presented in Fig. 15.9 illustrate this very clearly.

Van Everdingen's opinion has been the subject of much debate, and is not universally accepted, particularly by aficionados of enhanced oil recovery (EOR). The EOR approach, based on the injection of "exotic" fluids such as steam, carbon dioxide, surfactants or polymers into the reservoir, will be described later in this chapter.

Nonetheless, infill drilling is now used throughout the world, and generally produces positive results. In Table 15.2 you will see that more than 70% of the new reserves discovered in Texas[43] between 1973 and 1982 was attributed to infill drilling. In fact, the new wells simply allowed the recovery of oil from parts of developed reservoirs that existing wells were not able to drain effectively.

According to the report by Fisher[43] of a study undertaken by US consultants Lewin and Associates, *of the current US oil resource* not recoverable by conventional methods, *approximately 16% – some 13 billion m³ STO – could be extracted by infill drilling* (Fig. 15.10), *on condition that the location of these wells is based on a precise study of the internal structure (or "architecture") of the reservoirs concerned.*

Table 15.2. *Oil reserves discovered in Texas between 1973 and 1982*[43]

	Reserves (10^6 m³)	% of total
New reservoirs	103.4	11
New pools in existing fields	95.4	10
Infill and development wells	696.4	73
Tertiary recovery	38.8	4
Prolongation of productive life of existing fields	22.3	2
	956.3	100

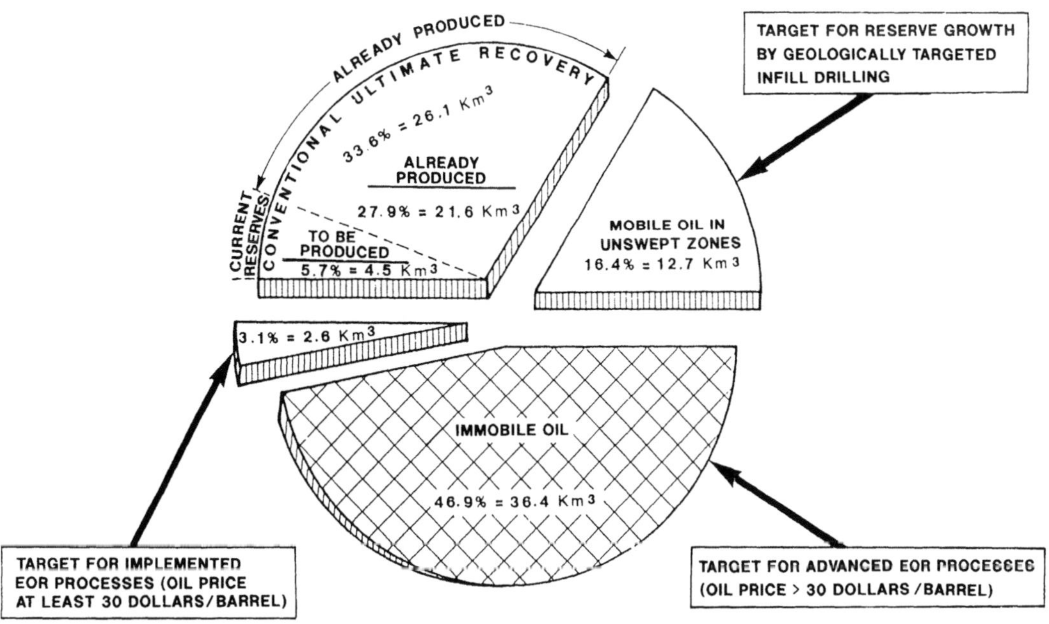

Fig. 15.10. *Additional oil reserves recoverable from fields in the USA by infill drilling, assuming the wells are located according to precise mapping of the internal reservoir structure. This is compared with the reserves remaining at the end of 1987, and with the quantity of oil that can be recovered by EOR. (Ref. 43)*

15.4.1.2   Why Infill Drilling Can Lead to an Increase in Oil Recovery

The reason for the improvement in oil recovery resulting from the drilling of infill wells becomes quite obvious when the reservoir is considered in terms of its real internal structure, and not as some idealised homogeneous isotropic medium. Owing to the nature of sediment deposition, reservoirs are often made up of a number of overlying "sedimentary units" or "zones", few of which are ever continuous over the entire area of the reservoir. These, in turn, consist of lenses of various sizes and differing petrophysical characteristics. The extent of any particular lens depends on its sedimentary history – most importantly, the local energy and type of flow of the water or air transporting the sediment. The higher the energy of deposition, the smaller the areal extent of the lens. This is clearly illustrated in the diagram from Weber[88] presented in Fig. 13.27.

Lenses of permeable and porous rock are in contact vertically and laterally with lenses of low permeability or porosity: this heterogeneity on a megascopic scale means that hydraulic continuity through the reservoir rock can only be assumed with confidence over an area whose extent is a function of the mean lens size. Consequently, the probability that a reservoir is continuous between two wells (the reservoir *connectivity*) increases as the distance between wells is decreased.

To produce any oil at all from an individual lens, there must of course be at least one well penetrating it, in the case of primary production, or a producer-injector pair in the case of secondary production by waterflooding[101]. In a heterogeneous reservoir, the percentage volume of rock contributing to production (the volumetric efficiency E_V; Sect. 12.5) therefore increases with the well density, as more lenses are intercepted. This explains the frequently observed increase in the oil recovery factor when infill wells are drilled.

The example from two reservoir units in Texas[6,7] shown in Fig. 15.11 demonstrates how the continuity, or connectivity, of the pay varies with well spacing. This diagram was derived from a geological study of the reservoirs performed after the well spacing had been reduced to 10 acres/well (average distance between wells: 200 m) by infill drilling.

It shows that in the Means San Andres reservoir, by doubling the number of wells to reduce the well spacing from 400 m (40 acres/well) to 285 m (20 acres/well), the connectivity (and E_V) were increased from 28% to 42%. In other words, with wells

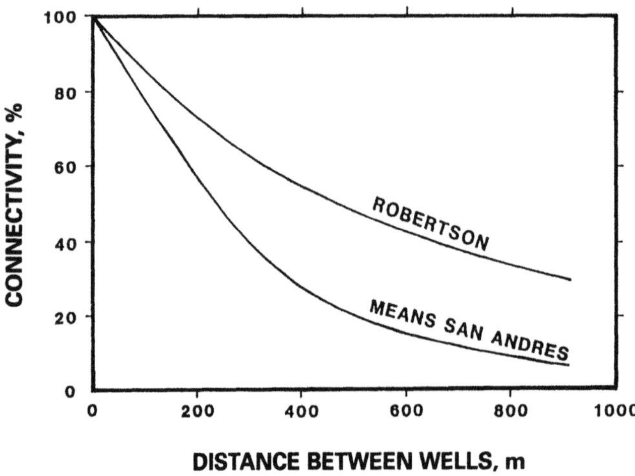

Fig. 15.11. *Connectivity between wells as a function of their separation. Robertson Unit and Means San Andres Unit, Texas. (From Ref. 21. Reprinted with permission of the Institut Français du Pétrole*

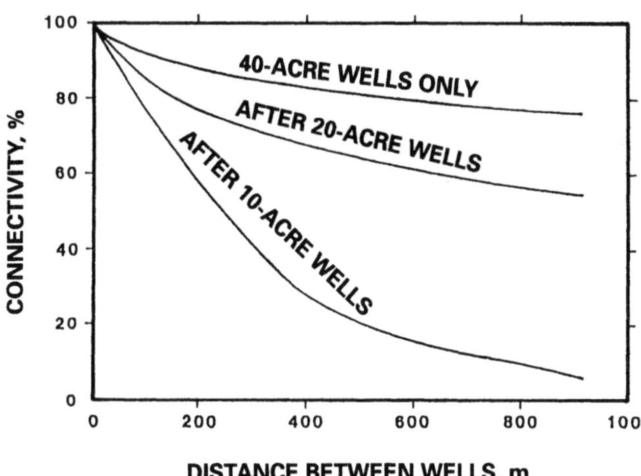

Fig. 15.12. *Means San Andres unit, Texas. Initial distribution of interwell connectivity estimated from data obtained from wells drilled with a 40 acres/well spacing, and the subsequent improvements indicated by new data following infill drilling to 20 and 10 acres/well. (From Ref. 7, 1983, Society of Petroleum Engineers of AIME. Reprinted with permission of the SPE)*

drilled 400 m apart, only 28% of the reservoir volume was being drained; with twice as many wells, and a mean interwell distance of 285 m, an extra 14% of the volume was drained. The oil recovery factor is of course increased proportionately: in this case by $42/28 = 1.5$ — a 50% improvement. The extra recovery from the other reservoir in Fig. 15.11 — the Robertson — was less impressive at 20%, but significant nevertheless.

In the light of the above discussion, we should not be surprised to learn that, over the vast complex of reservoirs found in East Texas, with an average well spacing of less than 5 acres/well (140 m between wells), and producing by water injection for over 40 years, *the recovery factor to date is estimated*[84] *to be 80%!*

Effective mapping of the distribution of reservoir connectivity with distance depends of course on the amount and quality of the available data. For instance, in the case of the Means San Andres featured in Fig. 15.11, geological studies performed, firstly, after the initial development of the field with a 40 acres/well spacing, after infill drilling to 20 acres/well, and then to 10 acres/well, led to revision of the connectivity vs distance relationship as shown in Fig. 15.12.

Note how, as more information becomes available, the heterogeneity of the reservoir becomes more apparent. This fact should always be borne in mind when evaluating reservoir connectivity with large well spacings.

15.4.1.3 Optimising Infill Well Location by ARM Techniques

While it is often true that the drilling of infill wells results in increased oil recovery, the drilling and completion of these wells represents a major financial investment. Basic economic sense dictates that infill drilling can only be considered viable when the net discounted profit (extra oil revenue minus cost of infill wells) is positive. Therefore, the *location* of the infill wells is of prime importance. Their placement must be carefully optimised so that, for a given number of wells, a maximum increase in reservoir connectivity (and, therefore, of volumetric efficiency E_V) is achieved. This is a very significant factor in the planning of advanced reservoir management (ARM)[21]. From a perusal of some of the recent technical publications[104,118,119,141,145,147,151,153,155], it will be readily apparent that the major oil companies involved in exploration and production have recognised the importance of optimising infill drilling, and are committing considerable resources to researching the subject.

The current low price of oil ($16–$18/barrel at the end of 1994) is forcing operators to reduce their development costs per barrel of oil produced, and their investment in potential reservoirs. This requires careful attention to every aspect of reservoir management, from the initial discovery right through to reservoir abandonment.

A fundamental prerequisite for the use of ARM techniques is an accurate and detailed knowledge of the internal structure of the reservoir rock. Until a few years ago this was believed to be a problem lying purely in the domain of reservoir geology. As such, it was considered sufficient to analyse all the lithological information obtained during drilling (cores, logs), from seismic surveys (especially 3D), and the study of outcrops, viewed within the framework of the basin typology (paleogeography, sedimentary history, tectonics, diagenesis). This is, of course, all "static" information – distribution of lithology, possible presence and location of faults, variation in porosity throughout the reservoir, and so on – and sheds little light on the possible "dynamic" behaviour of the reservoir once fluid movement is initiated by the extraction of oil, gas and water.

The "static" model constructed from these data is used to determine the location of development and delimiting wells, at least up to the moment that the reservoir is put on production. Experience from the study of a large number of producing reservoirs has shown that this initial static model must be continually modified in line with the response of the reservoir to the production of fluids. In addition to the conventional measurements of production rates, bottom-hole shut-in and flowing pressures, pressure drawdown and buildup tests, data must be acquired specifically with a view to studying the *dynamic* behaviour of the reservoir: continuity of the pay, its connectivity (the fraction of the reservoir volume in communication with at least one well) as a function of well density, the distribution of horizontal and vertical permeability between wells, the movement of the water/oil or gas/oil contacts with time. Interwell testing is particularly useful in this respect.

The resulting model, which must of course still respect the initial geological information, is the "dynamic" model. This will be used to predict, by numerical simulation, the results that could be obtained from any course of action that might be undertaken to improve production, and return on investment, from the reservoir. This might be anything from working over one or a group of wells, or the *drilling of vertical, extended reach, "designer", or horizontal, infill wells*, or a decision to implement a conventional improved recovery process (water or gas injection, both non- miscible phases), or a major reassignment of production and injection wells within the existing well pattern, to deciding to initiate crestal gas injection or enhanced oil recovery (covered later in this chapter).

The information required to set up the geological, or static, model is obtained from:

- the basinwide study of paleogeography, sedimentology, tectonics and diagenesis;
- study of reservoir rock outcrops[62,116,119,124], with the evaluation of the continuity of the various lithologies (see Fig. 13.27) as well as their porosity and permeability;
- sedimentological, petrographic and petrophysical data from all cores recovered from the reservoir rock, quantitative interpretation of the logs recorded in all

the wells of the field, correlation of log and core data and reservoir zonation using cluster analysis[37] (Sect. 3.6.3);

- interpretation of seismic surveys (especially 3D[33,156]) in lithological and petro-physical terms
- assessment of the interwell distribution of lithology by means of crosswell tomography[51,126], using connectivity mapping techniques[167].

Construction of the static model also brings probabilistic techniques into play to process the available data (Sect. 13.15) and, of course, the personal expertise of the technicians involved. The validity of the resulting model is directly dependent on the density of data points. This topic is illustrated very clearly in a recent publication[109].

In order to go from the static model to the dynamic, the following specialised measurements must be taken *systematically* (i.e. at regular intervals) *throughout the entire productive life of the reservoir*. These are, in addition to the conventional routine measurements of oil, water and gas flow rates and bottom-hole pressures, made in every well.

- PLT logs (Sect. 8.3) in all producing wells, to identify fluid entry points and the types of fluid, and to establish the production profile across the producing interval. In injection wells, it is sufficient to run the flowmeter alone, or in combination with a temperature survey.
- pulsed neutron capture (PNL) logs[123], gamma-ray spectrometry (GRS) logs[133] (described in Sect. 8.7), combination neutron-density logs (CNL + FDC) (Sect. 8.6), or the nuclear magnetism log (NML) (Sect. 15.3.3.4): these all have the capability, under the appropriate conditions, to distinguish the water/oil or gas/oil contact even in intervals of pay not being produced in a particular well[42,110,154].
- field wide measurement of the static reservoir pressure in all wells, to identify regions that are being poorly drained[106].
- pulse tests[61,122] (Sect. 6.12), interference tests[53,105] (Sect. 6.11) and tracer tests[12,47,137] (Sect. 12.7.2) between groups of wells to determine the rock characteristics in the region between wells (permeability, storability ϕc_t), as well as the communication between wells.
- vertical pulse tests (Sect. 6.12) to determine the vertical permeability of the reservoir in the region surrounding each well, and to assess the potential for crossflow between layers.

The transition from the initial static model to the final dynamic model is achieved by an iterative procedure[108] during which all data on the response of the reservoir to the production of fluids are incorporated as soon as they become available. This process of iteration, or "feedback", is flow-charted in Fig. 15.13.

The numerical model constructed from the initial static model is run in "history match" mode (Sect. 13.19) to replicate the real behaviour of the reservoir observed at selected times during its producing life to date – in particular: pressure, WOR and GOR at each well, and the position of the displacing fluid front. This requires successive refinement of the model parameters until a match is obtained. *Modifications must in all cases remain consistent with the geological description incorporated in the initial static model*. There is a well-defined hierarchy[144] for parameter modification: for example, the movement of a permeability barrier

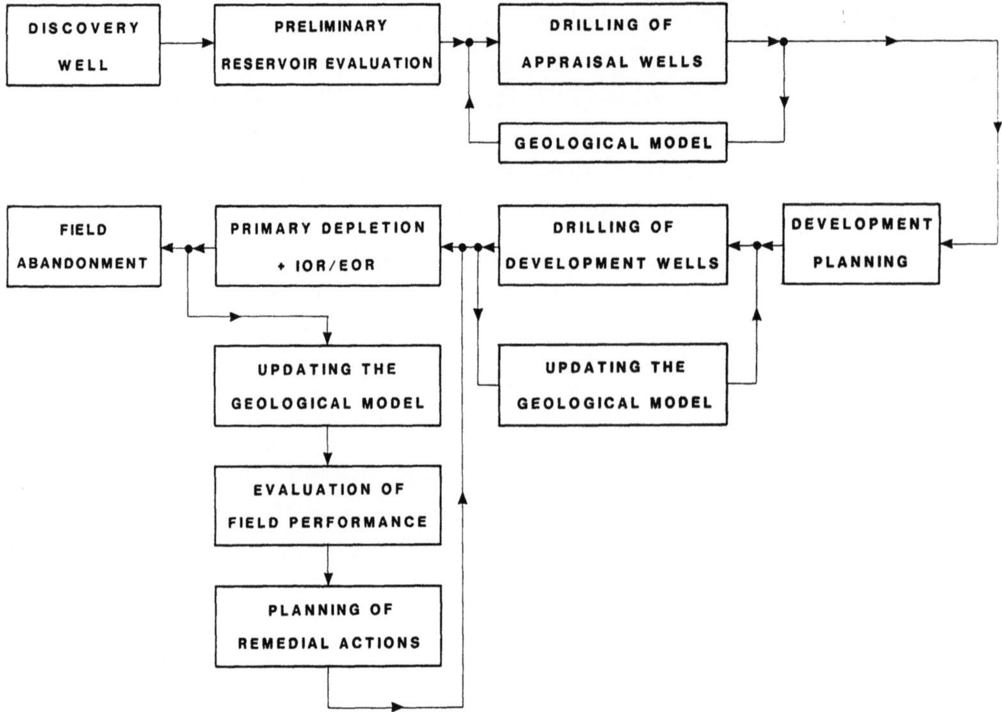

Fig. 15.13. *Reservoir management: the iterative (or "feedback") approach. (From Ref. 108. Reprinted with permission of Elsevier Science Publishers)*

requires considerably more field evidence to justify it than does a simple modification of the permeability of a layer.

An extended flow test[112] run prior to starting production proper can speed up the process of determining the definitive dynamic model. This can be considered to have been achieved when the most recent production history can be matched with only minimal adjustment to the model.

The final numerical dynamic model is then used to predict the consequences of any modifications that might be made, at reservoir level, to the production process. In particular, it will be used to determine the optimum types of infill wells (vertical, extended reach, "designer" horizontal) and their targets, to drain oil from parts of the reservoir bypassed by the displacing fluid front.

From the above discussion, it will be clear that the effective use of ARM techniques requires continual innovation on the part of the service companies (new logging tools and log interpretation techniques[100]) and operators (exploration and production), aimed at facilitating a synergistic[100,108] approach involving geologists, geophysicists, geostatisticians, and reservoir and production engineering experts. These diverse talents should be combined in multispecialist teams[98] who will be responsible for the management of each reservoir, from its discovery through to the end of its producing life.

15.4.2 Crestal Injection of Non-Miscible Gas

15.4.2.1 Unidirectional Displacement of Oil by Gas in the Presence of S_{iw}

In 1974, Dumoré and Schols[40] published the results of some important experiments performed in the Koninklijke/Shell Exploratie en Produktie Laboratorium (KSEPL) in Rijswijk, Holland. They showed that if a core of preferentially water-wet rock was saturated with oil *at irreducible water saturation*, low velocity vertical displacement of the oil by a *non-miscible* gas injected at the top of the core would create, above the capillary transition zone, a residual oil saturation $S_{or,g}$ of almost zero, or at most a few percent.

This discovery – completely unexpected because gas and oil are *not* miscible ($\sigma_{og} \neq 0$) – has since been confirmed by the research departments of numerous oil companies, including Agip S.p.A. The phenomenon is extremely important in reservoirs producing by gas drive from a gas cap, or by crestal gas injection. It is explained by the mechanism of "film flow".

In water-wet rock, residual oil would normally remain entrapped in the middle of the pore spaces. However, if the displacement velocity is low, such that gravitational forces predominate over viscous forces, the oil instead runs over the water in the form of a film just a few molecules thick, and escapes. In this way, the oil can be completely expelled from the pores, even when they have already been invaded by gas.

15.4.2.2 Vertical Displacement of Oil by Gas in a Gravity-Stabilised Regime

As seen in Sects. 11.4.1 and 11.4.2, displacement which occurs in vertical equilibrium (VE), or, rather, in a Dietz[38] segregated flow regime, is said to be *gravity stabilised* when the displacing fluid front remains parallel to itself as it advances.

Under these conditions, water will displace oil without "tongues" of water running along the base of the layer, beneath the oil (Fig. 11.25), and gas will displace oil without the gas tonguing along the top of the layer, above the oil (Fig. 11.29).

In the case of *vertical* displacement of oil by gas (injected, or from the gas cap) in a homogeneous layer, if u_z is the Darcy velocity of the front, then Eq. (11.63) tells us that the condition for the front to be stabilised by gravity is expressed by:

$$N_{gv} \geq 1, \tag{15.11a}$$

where

$$N_{gv} = \frac{k_v k_{ro,iw}}{u_z \mu_o} g (\rho_o - \rho_g) \tag{15.11b}$$

N_{gv} is a dimensionless group representing the ratio of gravitational to viscous forces.

Gravity-stabilised displacement in a multilayered reservoir was studied by Ekrann[111].

Referring to Eq. (3.53a), note that:

$$N_{gv} = \frac{1}{N_{vg}}. \tag{15.12}$$

In *practical metric units*, where:

k_v : md

u_z : m/year

g : m/s^2

$$\rho_{\rm o}, \rho_{\rm g} : {\rm kg/m}^3$$
$$\mu_{\rm o} \quad : {\rm cP}$$

Equation (15.11b) becomes:

$$N_{\rm gv} = 3.1145 \times 10^{-5} \frac{k_{\rm v} k_{\rm ro,iw}}{u_z \mu_{\rm o}} g(\rho_{\rm o} - \rho_{\rm g}) \tag{15.11c}$$

while in *US oil-field units*, with:

$$k_{\rm v} \quad : {\rm md}$$
$$u_z \quad : {\rm ft/year}$$
$$g \quad : {\rm ft/s}^2$$
$$\rho_{\rm o}, \rho_{\rm g} : {\rm lb/ft}^3$$
$$\mu_{\rm o} \quad : {\rm cP}$$

it becomes:

$$N_{\rm gv} = 4.9888 \times 10^{-4} \frac{k_{\rm v} k_{\rm ro,iw}}{u_z \mu_{\rm o}} g(\rho_{\rm o} - \rho_{\rm g}) \tag{15.11d}$$

Remember that, according to Dietz's theory, all the mobile oil is displaced by the front, so that behind it we have $S_{\rm o} = S_{\rm or}$. Because of the low viscosity of gas, the ratio $\mu_{\rm o}/\mu_{\rm g}$ is always very large. It is therefore impossible for the fractional flow curves $f_{\rm g} = f_{\rm g}(S_{\rm g})$ to be concave upwards (Sect. 11.3.2). This means that some mobile oil will be left behind the gas front ($S_{\rm o} > S_{\rm or}$), and this should be taken into account when estimating the quantity of oil that will be produced at breakthrough. The theory developed by Richardson et al.[143] can be used to calculate the variation of the oil saturation profile with time behind the front.

15.4.2.3 Production of Oil Through Vertical Displacement by Gas Under Gravity-Dominated Conditions

Hagoort[46] and, later, Ypma[93], both scientists at KSEPL, have perfected a procedure for calculating the recovery factor of oil under vertical displacement by gas in gravity-dominated conditions ($N_{\rm gv} \gg 1$). Their method is based on the solution of the Buckley-Leverett equation [Eq. (11.19a)] by the method of characteristics (Sect. 11.3.2) for the case of vertical cocurrent flow of gas and oil. In order to obtain an analytical solution, the capillary term, which is always very small in this case, is ignored. Hagoort's analytical treatment is described in Exercise 15.1.

If we express the oil relative permeability curve using Corey's equation (Sect. 3.5.5):

$$k_{\rm ro} = k_{\rm ro,iw}(S_{\rm o}^*)^n, \tag{15.13a}$$

where

$$S_{\rm o}^* = \frac{S_{\rm o} - S_{\rm or,g}}{1 - S_{\rm iw} - S_{\rm or,g}} \tag{15.13b}$$

is the normalised oil saturation; and the volume of oil $N_{\rm p}^*$ produced at gas breakthrough as a fraction of the mobile oil contained in the reservoir above the point of breakthrough:

$$N_{\rm p}^* = \frac{N_{\rm p}}{V_{\rm R} \phi (1 - S_{\rm iw} - S_{\rm or,g})}, \tag{15.14}$$

where N_p = volume of oil produced at gas breakthrough;

V_R = volume of oil-bearing rock above the point of gas breakthrough.

Hagoort and Ypma's results are the curves[21] presented in Fig. 15.14.

These curves demonstrate that, for example, when $N_{gv} = 10$ and $n = 3$, *86% of the mobile oil* (which, practically speaking, amounts to 86% of the oil in the reservoir, since $S_{or,g} \simeq 0$) *has been produced by the time gas breakthrough occurs.* This falls to 60% if n = 6. Because the value of n depends on the pore structure of the porous medium, and cannot therefore be evaluated beforehand, an accurate determination of the oil relative permeability curves is obviously an essential component in the planning of a crestal injection program. We will next look at some of the practical aspects of Fig. 15.14.

A high value of N_{gv} [Eq. (15.11b)] can result from high vertical permeability k_v, a large fluid density difference $(\rho_o - \rho_g)$, low oil viscosity μ_o and/or, most importantly, a low vertical component to the frontal velocity $v_z = u_z/\phi$.

For example, if:

k_v = 100 md

$k_{ro,iw}$ = 0.9

ϕ = 0.25

ρ_o = 750 kg/m^3

ρ_g = 150 kg/m^3

μ_o = 0.5 cP

then from Eq. (15.11c) we can determine that:

$N_{gv} = 10$ for $u_z = 3.3$ m/year, corresponding to $v_z = 13.2$ m/year.

If the surface area of the gas/oil contact were to be 2 km^2, this frontal advance velocity would be equivalent to an oil production rate of approximately 18,000 m^3/day (6.6 million m^3/year) measured under reservoir conditions. The production rate would be lower, of course, if the vertical permeability or contact area were smaller, or the oil more viscous.

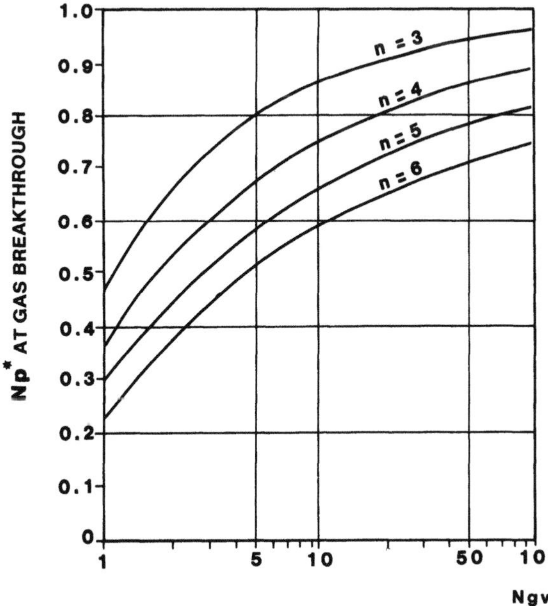

Fig. 15.14. *Fraction of mobile oil produced at gas breakthrough, as a function of the gravity/viscous force ratio and of the Corey n-exponent. (From Ref. 21. Reprinted with permission of the Institut Français du Pétrole)*

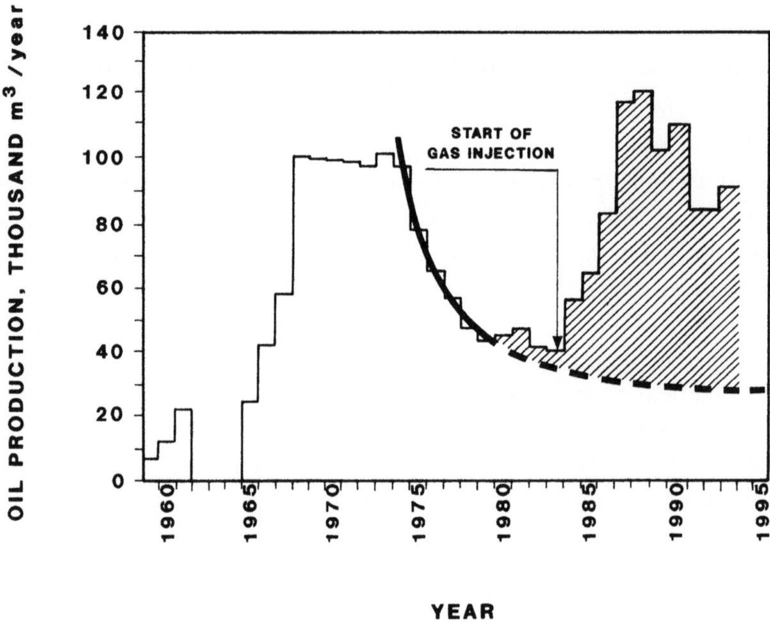

Fig. 15.15. *Production record for the Ponte Dirillo reservoir in Sicily, showing the effect of crestal gas injection under gravity-dominated conditions, initiated in 1983. Updated from Ref. 21*

The above discussion indicates that high recovery factors can be obtained, at commercial production rates, simply through crestal injection of a *non-miscible* gas. Numerous examples of the application of this technique have been reported[13,44,52,55]. There is at least one case where a field already on water injection was converted to crestal gas injection to improve recovery[66].

The sources of injection gas are several: gaseous hydrocarbon (in areas where there is no local market for natural gas); nitrogen[25] (from the fractional distillation of liquified air); combustion gases (nitrogen/carbon dioxide mixtures) from large power stations and petrochemical plants; or carbon dioxide produced, where available, from a local natural deposit. Reservoirs best suited to crestal injection will have a very thick pay section of large areal extent, containing light oil, and a high vertical permeability.

The Ponte Dirillo reservoir[23] in Sicily is an example of the application of crestal gas injection in Italy. Ponte Dirillo is a fractured carbonate of high permeability, but contains a very viscous oil. The injection of natural gas with a high CO_2 content, separated off from the oil from the nearby Gela field, resulted in an almost tripling of the oil production rate (Fig. 15.15), at the same time reducing the water influx from the underlying aquifer, which had begun to present coning problems (Sect. 12.6). Following these excellent results, it was planned to apply the same strategy to the nearby Gela field, a much larger field containing an oil very similar to that in Ponte Dirillo.

15.5 Enhanced Oil Recovery (EOR): Definition and Classification of Methods

15.5.1 Introduction

So far, in Chap. 12 and in this chapter, we have looked at methods for improving productivity and recovery based on the use of low cost "classical" fluids, such as water or non-miscible gases. These used to be referred to as "secondary recovery" processes, because at one time they were implemented only after the natural, or "primary", drive energy of the reservoir had been exhausted. At present, water injection is often started as soon as production begins, to prevent the reservoir pressure from declining (pressure maintenance), thereby sustaining well productivity. Gas and water injection are therefore now referred to as *improved oil recovery* (IOR) processes.

Since the early 1950s, there has been an ongoing program of laboratory and field research, starting with small-scale pilots, then progressing to full-size reservoirs, to develop methods for obtaining even better oil recovery factors than those achieved through improved recovery. These advanced processes, referred to as *enhanced oil recovery* or *EOR* will be described in the following pages.

EOR is based on the injection of more "exotic" – and correspondingly more expensive – fluids than water or non-miscible gases. Their purpose is to improve either or both of the parameters upon which the final oil recovery factor $E_{R,o}$ depends (Chap. 12):

$E_{D,o}$:the microscopic displacement efficiency (at pore level)

$E_{V,o}$:the volumetric efficiency of the displacement front

where:

$$E_{R,o} = E_{D,o}E_{V,o}.$$

Generally speaking, these methods are brought into play when the benefits obtained from improved, or "secondary", recovery have been exhausted. For this reason, EOR is sometimes referred to as "tertiary" recovery, even if, as in some cases, it is initiated in the primary phase at the onset of production (for example, the Athabasca tar sands have been produced by steam injection or in situ combustion right from the start).

In the USA, the extra oil produced by EOR enjoyed certain tax concessions not available for conventional production or improved oil recovery methods. To take advantage of this, the EOR process used on a given reservoir must be certified by a regulatory body – usually the US Dept. of Energy (US DOE).

In Energy Regulation 10 CFR 212.78c issued by the US DOE in June 1979, the following are recognised as EOR processes:

1. Miscible gas flooding
2. Steam drive ("steam flooding")
3. Micellar solution flooding
4. In situ combustion
5. Polymer flooding
6. Cyclic steam stimulation ("steam soak" or "huff'n'puff")
7. Alkaline (or "caustic") flooding
8. Carbonated waterflooding.

Unfortunately, in many articles which have been published on oil recovery, "improved" and "enhanced" processes have been erroneously grouped under the one title of "enhanced", causing a good deal of confusion. The classification "enhanced" is in fact strictly limited to the processes cited in the above list.

15.5.2 Types of Enhanced Oil Recovery Methods

Enhanced recovery processes can be classified according to a number of different criteria. One criterion is how increased recovery is achieved: by improving the microscopic displacement efficiency $E_{D,o}$, the volumetric efficiency $E_{V,o}$, or both. The processes are ranked by this criterion in Table 15.3. Note that:

– *miscible processes* have a significant effect on $E_{D,o}$, but none at all on $E_{V,o}$ (except, perhaps, to make it worse!),
– *thermal processes* affect $E_{V,o}$ very favourably, and $E_{D,o}$ more moderately,
– among the *chemical processes*, micellar solutions improve $E_{D,o}$ and, to a lesser extent, $E_{V,o}$, while polymer solutions only effect $E_{V,o}$. Relatively little improvement in either efficiency is obtained with alkaline solutions.

Another criterion for the classification of EOR processes is the nature of the chemical and physical interactions between the injected fluid, the oil, and the reservoir rock. This is the most commonly used method of classification, and it places the processes into three groups (Fig. 15.16):

– *Thermal processes*: the reservoir temperature is increased to reduce oil viscosity, thus improving the water/oil mobility ratio (M_{wo}). This category includes:
  – *steam injection* by:
    • *steam drive* – continuous injection of steam into injector wells, driving the oil towards producing wells;
    • *huff'n'puff* or *steam soak* – alternating steam injection and oil production in the same well.
  – *partial combustion* of the oil in situ, sometimes followed by the injection of water to recover the heat from the rock behind the combustion front.
– *Miscible processes*: based on the injection of a gas which, through a progressive exchange of mass with the oil, becomes miscible with it. This has the effect of reducing the residual oil saturation to zero.

Table 15.3. *The effect of the various EOR processes on the values of $E_{D,o}$ and $E_{V,o}$ (relative to the results achieved with water injection)*

	Process	Effect on $E_{D,o}$	Effect on $E_{V,o}$
Thermal	– steam drive	•	• • •
	– steam soak	–	• •
	– in situ combustion	• •	• • •
Miscible	– carbon dioxide	• • •	–
	– natural gas	• • •	–
	– nitrogen	• • •	–
	– combustion gases	• • •	–
Chemical	– micellar solutions	• •	•
	– polymer solutions	–	• • •
	– alkaline solutions	•	–
Legend	– no effect		
	• slight effect		
	• • fairly strong effect		
	• • • strong effect		

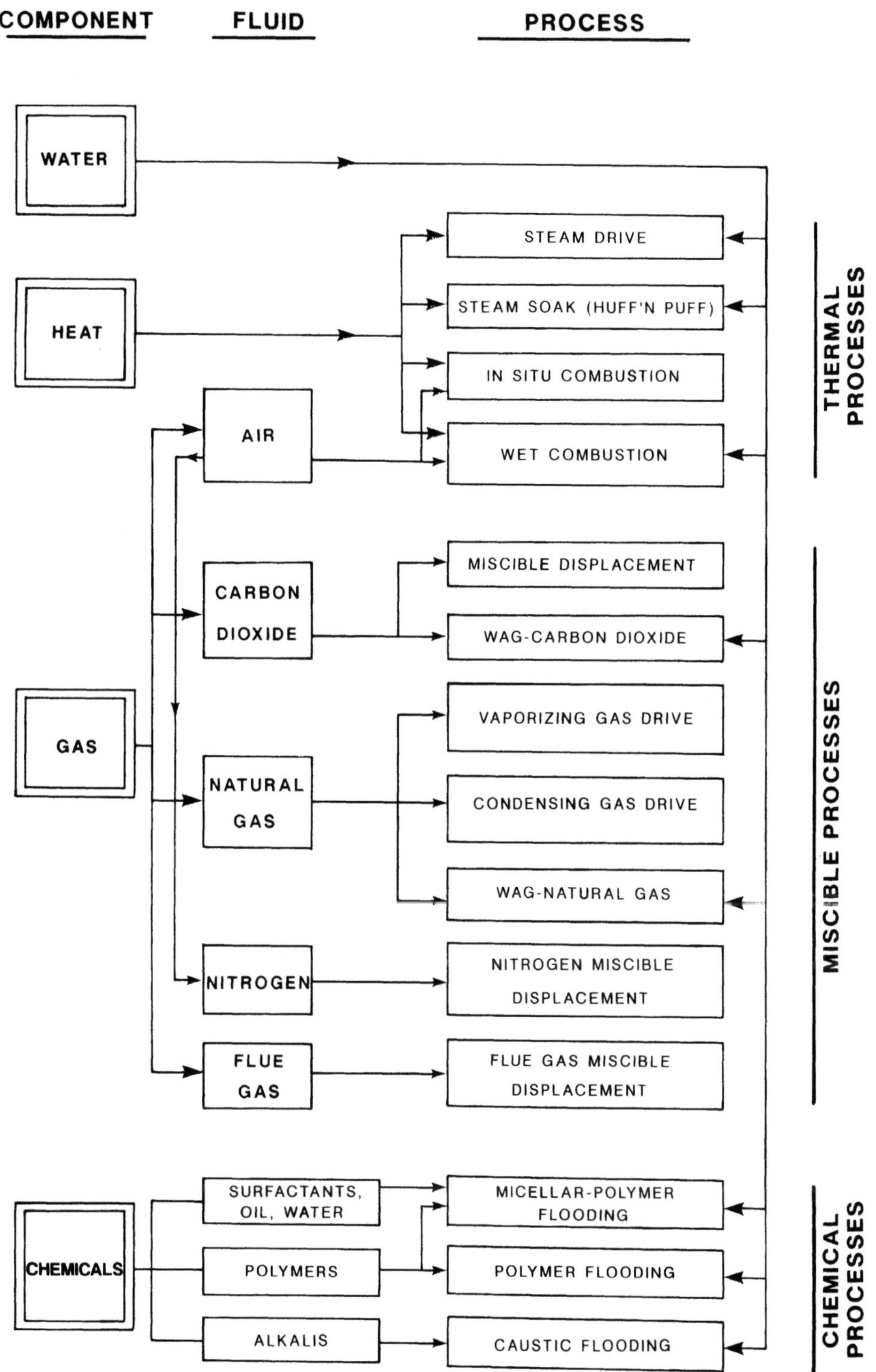

Fig. 15.16. *Classification of enhanced oil recovery processes. (From Ref. 108. Reprinted with permission of Elsevier Science Publishers)*

There are several different methods, depending on the nature of the gas injected and the manner in which the miscible front is created:

- *miscible displacement with carbon dioxide*, optionally alternated with pads of water (WAG – water alternated with gas) to reduce the mobility of the displacing fluid;

- *vaporising gas drive* – miscible displacement with high pressure gas
- *condensing gas drive* – miscible displacement with C_3-C_6-enriched gas
- *WAG with vaporising or condensing gas drive;*
- *miscible injection of nitrogen* – optionally alternated with pads of water to control mobility;
- *injection of combustion gases* – consisting mainly of N_2 and CO_2, optionally alternated with pads of water.
- *Chemical processes*: based on the injection of water containing chemical additives. This category includes:
  - *polymer flooding* (polymer solutions, or "thickened water") to reduce the water/oil mobility ratio;
  - *micellar solutions* – mixtures of surfactants/cosurfactants/water/oil, which have interfacial tensions with oil and water of almost zero, and are therefore miscible with both;
  - *alkaline solutions* – these act as surfactants with certain oils, and improve the value of $E_{D,o}$.

Each of these processes will be described briefly in the following pages. For further reading, please refer to the published material[20,50,69,74,77] on the subject.

15.6   Steam Drive

15.6.1   The Influence of Temperature on Reservoir Rock and Fluid Properties

Studies on the feasibility of thermal EOR processes showed that the heat transport characteristics of reservoir rock were almost independent of the nature of the rock, for a given set of fluid saturations (Table 15.4). Thermal recovery owes much of its success to this fact. In fact, it is sufficient to heat up one layer (by steam

Table 15.4. *Thermal properties of some rocks[a]*

Sample	Density	Specific heat	Thermal conductivity	Thermal diffusivity
	kg/m^3	$kJ/kg\ K$	$W/m\ K$	mm^2/s
Dry rock				
Fine sand	1634	0.766	0.627	0.500
Coarse sand	1746	0.766	0.557	0.415
Sandstone	2082	0.766	0.877	0.550
Silty sandstone	1906	0.846	0.692	0.431
Shale	2323	0.804	1.04	0.557
Limestone	2195	0.846	1.70	0.916
Rock saturated with water				
Fine sand	2018	1.419	2.75	0.960
Coarse sand	2082	1.319	3.07	1.12
Sandstone	2275	1.055	2.76	1.15
Silty sandstone	2114	1.206	2.60	1.02
Shale	2387	0.892	1.69	0.792
Limestone	2387	1.114	3.55	1.33

[a] Data extracted from the paper by W.H. Somerton, "Some Thermal Characteristics of Porous Rocks", published in *Trans.*, AIME *213* (1958): 374–378, and converted to SI units.

injection, for instance) for the heat to propagate into the overlying and underlying layers (unfortunately raising the temperature even beyond the top and bottom of the reservoir). The beneficial effects of injecting steam into the layer are therefore extended to other layers.

In particular, *the rock permeability has almost no effect on the thermal properties*: tight layers, which impede the flow of fluid, allow heat to pass through. Because the grains expand on heating, the porosity and permeability are reduced to some extent. The effect of temperature on wettability and, consequently, on the relative permeability curves, is less well understood. But the most important effect is the reduction in the reservoir oil viscosity.

Figure 15.17 shows that increasing the layer temperature by 100 °C (from 50 to 150 °C) reduces the oil viscosity by 99.2% (from 2500 mPa·s to about 20 mPa·s) in a 10 ° API oil (density 1000 kg/m^3). Lighter oils (higher API gravities) are less significantly affected by temperature changes: for example, the viscosity of a 20 ° API oil (934 kg/m^3) decreases by only 92.5% (from 40 mPa·s to 3 mPa·s) for the same 100 °C increase in temperature.

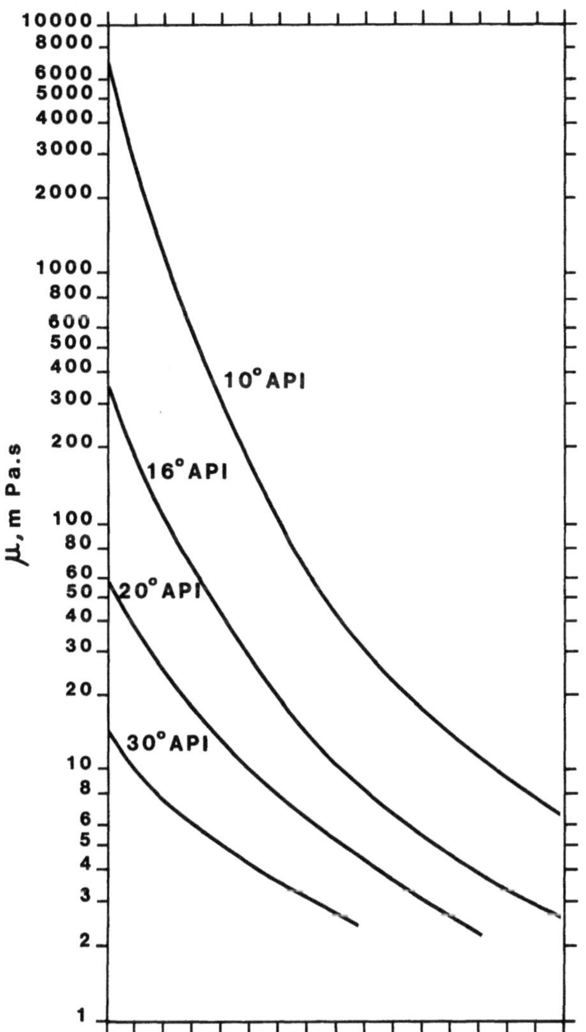

Fig. 15.17. *The effect of temperature on oil viscosity at atmospheric pressure*

Walther[50] derived an empirical equation relating the oil viscosity and temperature:

$$\log[\log(\mu_o + \alpha)] = A - B \log T, \tag{15.15}$$

where T is in Kelvin, and A, B and α are constants depending on the type of
oil.

Because the well flow rate is inversely proportional to the viscosity of the
oil (all other factors being equal; see Chaps. 5 and 6), increasing the reservoir
temperature serves not only as an EOR process, but also as means of *increasing
the productivity of the wells and reservoir*. Of course, the viscosity of water μ_w also
decreases with increasing temperature, as illustrated in Fig. 2.18 [see Eq. (2.33)].
However, the effect of temperature on water viscosity is less than on oil so that,
overall, the viscosity ratio μ_o/μ_w [and the *water/oil mobility ratio* M_{wo}, defined
in Eq. (3.52a)] *decreases with increasing formation temperature*. This is illustrated
in Fig. 15.18. Raising the formation temperature therefore has a positive effect on
the areal sweep efficiency E_A (Sect. 12.5.1) and the vertical invasion efficiency E_I
(Sect. 12.5.2), both of which increase as M_{wo} decreases.

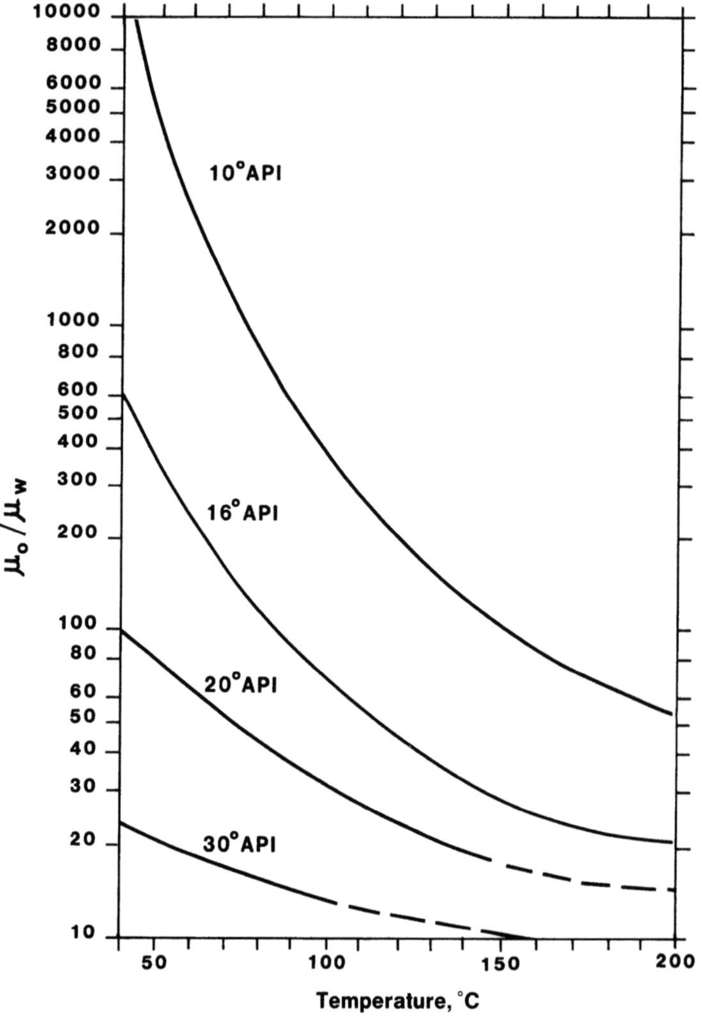

Fig. 15.18. *The effect of temperature on the oil/water viscosity ratio at atmospheric pressure*

There are other beneficial effects from heating. The increasing temperature raises the bubble point p_b of the oil, especially if it is a heavy oil with high critical temperature T_c (as is clearly shown in the phase diagram in Fig. 2.2). It also increases the volume factors of the oil and the connate water by thermal expansion. Combined with the reduction in porosity brought about by the expansion of the grains, these factors lead to an improvement in the microscopic displacement efficiency E_D.

Increasing the formation temperature, then, improves both the volumetric efficiency E_V and the microscopic displacement efficiency E_D, and so *increases the oil recovery factor* $E_R = E_V \, E_D$. The improvement will be more significant for oils which have a high initial viscosity.

Thermal EOR processes based on steam injection are therefore applied principally to heavy oils, or semi-solid bituminous deposits. For practical reasons, which will be explained in the next section, they are generally limited to depths of less than 1000 m. Steam injection is currently being tried out in some light oil reservoirs[114] which have been produced to depletion by water flooding. It is hoped that increased recovery will result through vaporisation of the residual oil, which will be then swept along in the current of steam.

15.6.2 Steam Drive Technology

Figure 15.19 is an artist's impression[5] of the continuous steam drive process. The heart of the system is the steam generator. Here, fresh (or desalinated) water, treated with suitable inhibitors, is vaporised at high pressure and temperature, the objective being to deliver steam downhole at a pressure higher than that of the reservoir, for injection into the formation.

As Fig. 15.20 shows, the higher the reservoir pressure (which usually means the deeper the reservoir), the higher the operating temperature of the boiler. The actual quantities of steam involved in this process are enormous – typically several tens of thousands of tonnes per day. There is no recycling, so all of this is left in the reservoir. This contrasts with the technology of thermo-electric power stations where, although the steam they operate with is supercritical (very high temperature), it is recycled. In this case, because of the recycling, a limited quantity of water is involved, and costly water treatment techniques can be employed to combat corrosion and erosional problems in the equipment.

Because it is not recycled, economic considerations prevent the water for EOR steam generators from being treated to the same extent. This limits the maximum operating temperature of the system and, therefore, the injection pressure. In order to overcome the problems of high temperature corrosion and erosion in the surface piping and downhole tubing, as well as atmospheric pollution from the combustion gases, the feasibility of downhole steam generators has been studied and tested.

In this type of steam generator, fuel, air (or oxygen), and water are transported downhole either simultaneously through three separate tubings, or as alternating pads down a single tubing. Combustion and steam generation occur at the bottom of the well. So far, results have been disappointing, both technically and economically.

The maximum pressure is also limited by the thermodynamics of the process.

Referring to Fig. 15.21, you will see that the latent heat of vaporisation of water (the quantity of heat given out by 1 kg of saturated vapour when it condenses at

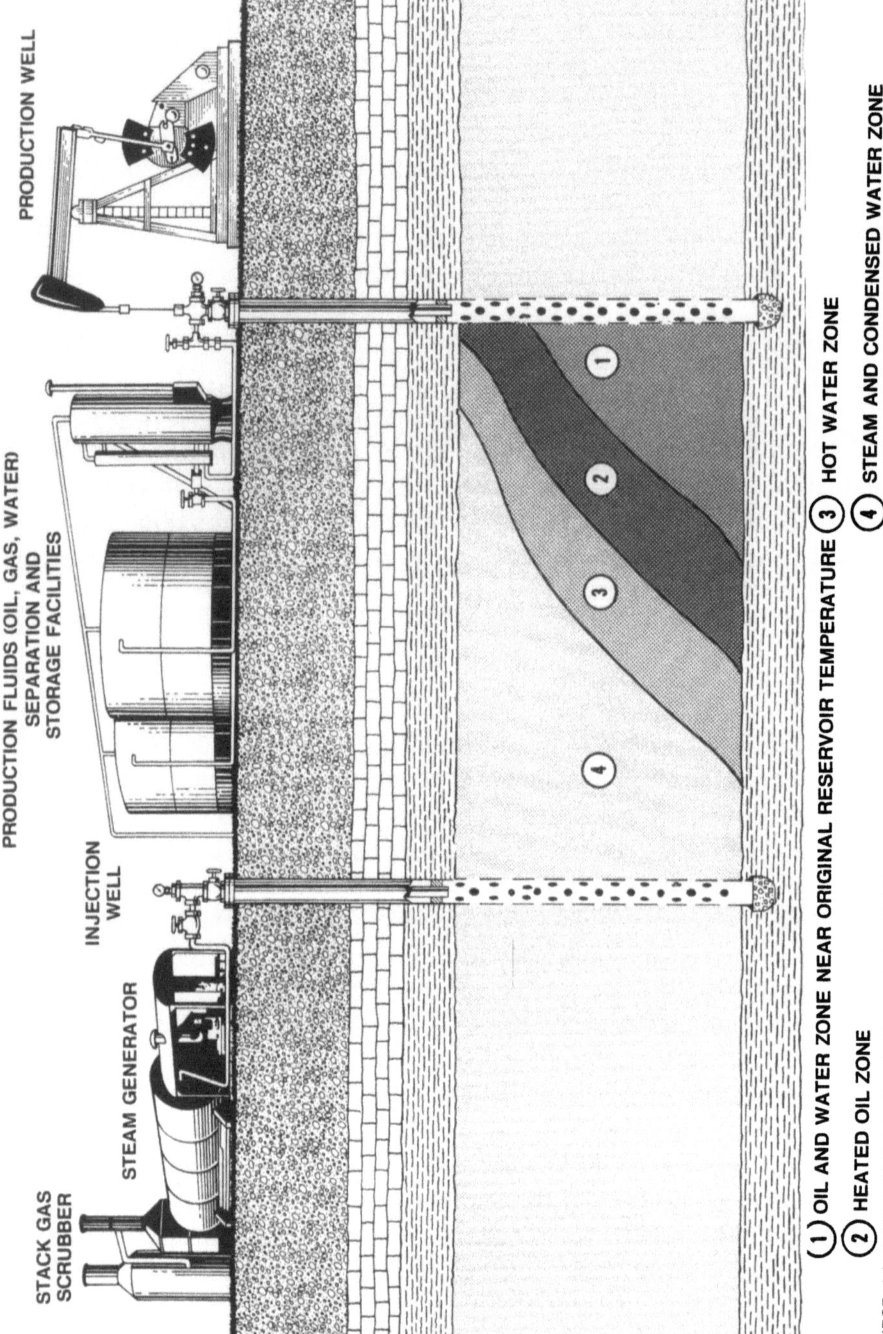

Fig. 15.19. *Artist's impression of the continuous steam injection process. (From Ref. 5. Reprinted with permission of the National Petroleum Council and the US Department of Energy)*

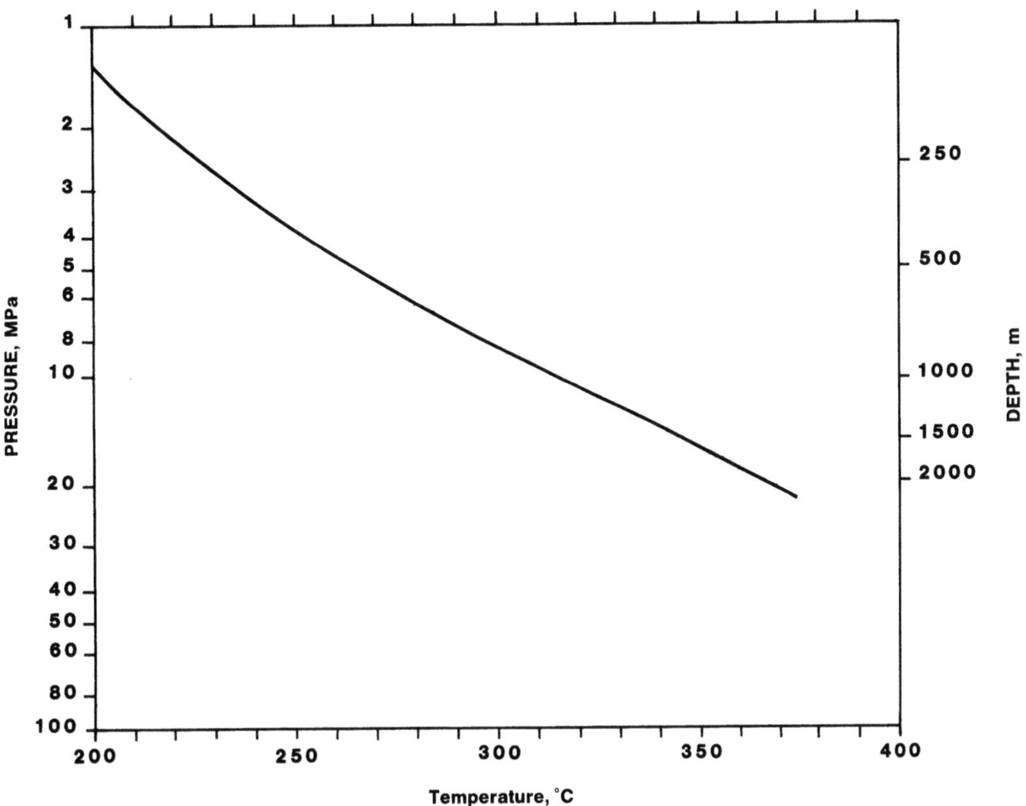

Fig. 15.20. *Variation in the boiling point of water with pressure (which is roughly proportional to reservoir depth)*

saturation pressure) decreases considerably at higher pressures (and, consequently, higher temperatures). This means that the higher the reservoir pressure, the greater the mass of vapour needed to heat up a given mass of rock.

It would be reasonable to ask at this point if perhaps it would not be adequate to use hot water instead of steam. The increased hydrostatic pressure obtained would balance out the major part, if not all, of the reservoir pressure. This technique has had little success in the past.

In fact, liquid water yields slightly less than 1 kcal/kg of heat energy when its temperature is decreased by 1 °C. Under the same conditions, saturated water vapour yields the entirety of its latent heat of vaporisation – approximately 340 kcal/kg at 8 MPa. The thermal effects of hot water injection are therefore almost non-existent compared to those resulting from steam injection.

For the reasons outlined above, the practical limit to the steam pressure that can be used for thermal EOR lies in the range 8–10 MPa, corresponding to a temperature of 295–315 °C and a maximum reservoir depth of 800–1000 m. It is only recently that steam injection at a depth of 1300 m has been reported, from the People's Republic of China[18,57].

The most commonly used boiler fuel is the oil produced from the reservoir itself. A typical boiler will use 1 t of oil to generate 13–14 t of vapour. Unused heat is dispersed with the flue gases. If 13–14 t of vapour produce less than 1 t of oil, therefore, the system will run with a negative *energy balance*.

Taking into account amortisation of equipment, running costs and taxation, the *economic balance* goes negative when the *average* ratio of steam injected/oil

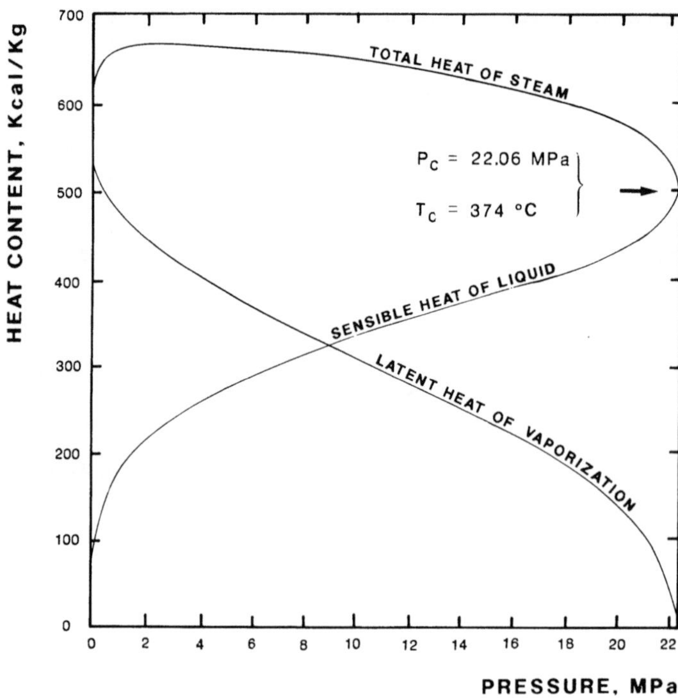

Fig. 15.21. *Latent heat of vaporisation and sensible heat of liquid water, and total heat of saturated steam as a function of pressure. (From Ref. 50. Reprinted with permission of the Interstate Oil Compact Commission)*

produced (the steam/oil ratio, SOR) over the life of the project exceeds 5–7 t of steam per tonne of oil.

The technique of *cogeneration* has seen increasing use over the last few years as a means of improving the economics of steam injection for EOR. Cogeneration plants are two-stage systems producing electrical energy in the first stage, and high pressure steam, suitable for injection, in the second. The thermal efficiency of a cogeneration plant is substantial. The commercial value of the electricity produced can be offset against the cost of steam production. The steam is essentially a by-product from the generator and is therefore significantly cheaper than from a conventional plant. Locally produced oil, natural gas or coal can be used as fuel, according to availability and cost. Natural gas burns more cleanly than oil or coal and is therefore less of a pollutant.

There are two basic types of cogeneration. In one type, the fuel is burnt in a high pressure boiler, and the steam is fed to a turbine which drives an electrical generator. The steam is then piped from the turbine outlet to the injection wells and into the reservoir. This process can be fuelled by coal as well as oil or natural gas.

The second type uses a gas turbine, fuelled by oil or natural gas, to generate electricity. The exhaust gases, which reach temperatures of at least 200 °C, are fed to a boiler which generates steam at a sufficiently high pressure for injection.

With a large installation, special techniques for improving the efficiency of the process can be considered economically viable: heat recuperators, combustion of carbon under pressure in fluidised beds, etc. The treatment and disposal of combustion gases is a persistent problem with all steam generation plants. In recent years, this has become the subject of increasingly stringent environmental pollution standards.

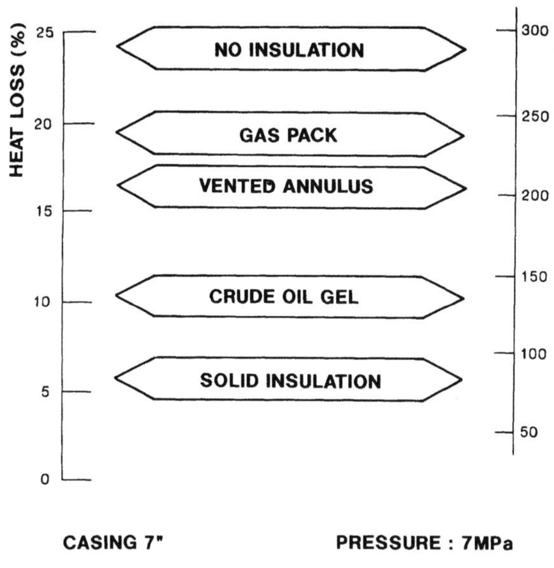

Fig. 15.22. *Example of how different insulating materials in the tubing-casing annulus influence the overall heat loss and casing temperature. (From Ref. 50. Reprinted with permission of the Interstate Oil Compact Commission)*

Heat loss from the steam in transit to the wellhead and then down the tubing poses a further problem. To combat this, the surface piping is insulated with a jacket of magnesia or similar material.

In the well, the steam usually passes down the tubing. The tubing-casing annulus is either left empty, filled with a low conductivity oil-based gel, or, most effective of all, the tubing is encased in a solid insulating jacket.

The lower part of the tubing string is anchored by one or several special packers which can accommodate the differential thermal expansion between tubing and casing.

Figure 15.22 compares the effectiveness of the different methods of reducing heat dissipation from the tubing. In all cases, a considerable amount of heat escapes from the well to the overburden before the steam reaches the reservoir itself.

In order to estimate the temperature, pressure, quality and enthalpy of the steam actually entering the reservoir, the overall heat losses from the well must be quantified. Computer models originally developed for the study of geothermal wells[24] *producing* steam can be easily adapted for this purpose.

15.6.3   Problems Encountered in the Reservoir

Because of the heat lost on the way down the well, the fluid entering the reservoir consists of a mixture of vapour and water that has condensed out – the steam is said to have a quality of less than 100%. Once in the formation, the latent heat of vaporisation of the steam as it undergoes (complete or partial) condensation is transmitted directly to the rock. The equilibrium temperature of the formation can be calculated from the enthalpy balance between the injected fluid, rock and reservoir fluid.

The oil is in fact displaced by the condensed water, so this is an *immiscible* displacement, with M_{wo} better than that for cold water injection (Sect. 15.6.1), which itself always has $M_{wo} > 1$. The displacement process can therefore be studied by the same methods as for water injection, described in Chaps. 11 and 12. In the absence of gravitational forces, the fluid distribution in the reservoir should be as illustrated in Fig. 15.23.

Now the steam has a much lower density than the oil or water in the reservoir (Table 15.5), and so tends to segregate to the upper part of horizontal or gently dipping layers, where it moves in parallel with the oil ("gravity overriding"). Owing to its relatively low viscosity (Table 15.5), the steam moves faster than the oil and may quite rapidly break through at the producing wells. This short circuit increases the fraction of condensed water produced with the oil, resulting in a lowering of the oil/steam ratio (OSR).

For the same reasons, steam will also advance preferentially along zones or through regions of the reservoir with high permeability. This reduces the volumetric efficiency $E_{V,o}$ (Sect. 12.5). The high steam/water mobility ratio can also give rise to frontal instability through fingering (Sect. 11.3.6). The advance of the steam front through the oil reservoir is monitored by measuring the steam/oil

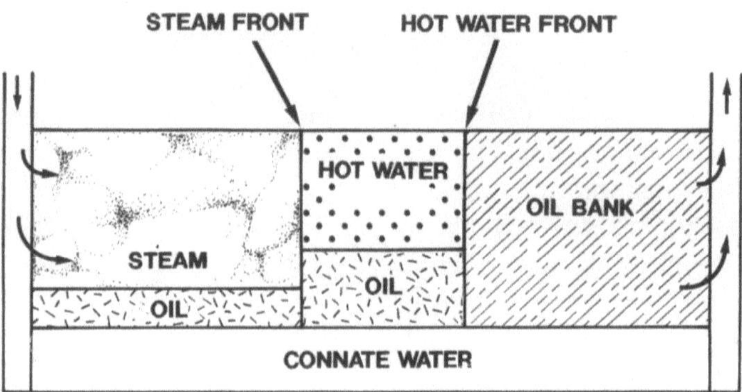

Fig. 15.23. *Continuous steam injection. Idealised fluid distribution assuming no gravitational effects*

Table 15.5. *Density and viscosity of saturated steam*

Pressure (kg/cm²abs)	(MPa)	Temperature (°C)	Density (kg/m³)	Viscosity (mPa·s)
1	0.098	99.1	0.578	0.0120
5	0.490	151.1	2.63	0.0141
10	0.981	179.0	5.00	0.0153
20	1.96	211.4	9.80	0.0167
30	2.94	232.8	14.7	0.0175
40	3.92	249.2	19.6	0.0182
50	4.90	262.7	25.0	0.0187
60	5.88	274.3	30.3	0.0192
70	6.86	284.5	35.7	0.0196
80	7.85	293.6	41.7	0.0200
90	8.83	301.9	47.6	0.0203
100	9.81	309.5	55.6	0.0206
150	14.7	340.5	91.0	0.0219

ratio (SOR) produced at each well. Pay zones that have not been perforated can be monitored by running pulsed neutron (PNL) surveys over the full reservoir interval[129]. In steeply dipping formations, steam overriding can be eliminated by operating under gravity-stabilised conditions (Sect. 11.4.2), or gravity-dominated conditions (Sect. 15.4.2.3).

A technique relatively new to the petroleum industry to combat the effects of the high steam mobility, involves increasing the steam viscosity through the formation of a foam[11,107,125]. This is achieved by injecting surfactants and inert gases (nitrogen or natural gas[41]) continuously with the steam. This creates foams which behave as non-Newtonian fluids and have viscosities several tens of times higher than pure steam. This results in a reduction in gravity overriding because it takes longer for the steam to segregate to the top of the layer, and eliminates the undesirable consequences of a high steam mobility.

The reservoir behaviour during steam drive is usually predicted using numerical modelling[26]. This is a very complex task which suffers from the difficulty of describing the simultaneous fluid dynamic and thermal aspects of the process correctly, and in recent years there has been a tendency to use scale models (Sect. 13.1), which are better suited than numerical models to the simulation of the various phenomena associated with steam injection[78]. For an initial feasibility study, however, it is quite adequate to use analytical methods[86] or empirical correlations based on data from reservoirs which have been produced by steam drive[63].

15.6.4 Summary

The use of steam injection for EOR started in 1959-1960 with pilot projects which Shell and other participants conducted in the Schoonebek field in Holland, the Mene Grande field in Venezuela, and the Yorba Linda field in California[58]. Since then, continuous steam injection, and the alternative "huff'n'puff" process described in the next section, have become the most important EOR processes in the whole world[102].

In 1992, steam injection was used[136] in 117 fields in the USA, 3 in Canada, 38 in Venezuela, 19 in the CIS (ex-USSR), 7 in West Germany, and several other fields mainly in Colombia, Indonesia and Trinidad. In the USA, 455 000 bbl/day (26.3 million m^3/year) of oil were produced by steam injection (continuous and alternating) in 1992. This amounts to about 60% of the oil produced in the USA by EOR, but is still only 6.2% of its total production. In terms of world oil production, steam injection was responsible for some 820 000 bbl/day (47.6 million m^3/year) – about 1.4% of the total – in 1991.

Cumulative experience of the behaviour of reservoirs produced under continuous steam injection[18] has allowed a number of ground rules to be established for the process to be effective. The following conditions are indicated:

- reservoir rock   : sands and sandstones, or carbonates with only
                          primary porosity
- reservoir thickness   : at least 10 m
- net pay   : more than 50% of reservoir thickness
- angle of dip   : above 10°
- porosity   : over 20%

- permeability : better than 100 md, *with no secondary permeability (fractures)*
- oil saturation at the : not less than 40%
  start of the process
- oil API gravity : less than 25° API (> 904 kg/m^3)
- oil viscosity : in excess of 20 mPa·s (cP)
  (the oil must, however, be mobile)

In addition, the reservoir rock should not be too heterogeneous, and there should be connectivity between the injection and production wells (Sect. 15.4.1.2).

In Sect. 15.6.2 it was stated that the reservoir depth should be less than 1000 m. Reservoirs shallower than 300 m pose an additional problem: because of their low overburden pressure, there is a risk of hydraulic fracturing when injecting steam at the required rate. This results in loss of cement integrity around the casing, and a reduction in the volumetric efficiency of the steam drive.

15.7 Huff'n'puff, or Steam Soak

15.7.1 Origins of the Method

In the early 1960s, while a continuous steam drive project was under way in the Tia Juana reservoir, alongside Lake Maracaibo, steam broke through to the surface through a fracture that had formed in the overlying strata (see Sect. 15.6.4). Injection was shut down to stop this unexpected eruption, and the steam already injected was bled off through an injection well. To the surprise of the engineers supervising the project[34], the steam and vapour returns contained considerable quantities of oil, even though the well in question had not been producing oil before injection was started, despite penetrating the oil-bearing interval.

This chance observation led to the development of the technique of alternating steam injection and oil production in the same well, known as "steam soak" or "huff'n'puff". This is widely used in highly viscous oil reservoirs which are shallow enough to allow steam injection to take place (Sect. 15.6.2).

Strictly speaking, *steam soak is a stimulation treatment rather than an enhanced oil recovery technique.* It leads to an increase in the oil production rate (sometimes quite appreciable) and therefore accelerates recovery, *but does not increase the percentage of oil recovered.*

15.7.2 Steam Soak Technology

The conventional steam soak process is a succession of cycles, each consisting of:

- a steam injection phase;
- a predetermined shut-in period (one or several weeks) to allow thermal equilibrium to be established;
- a production phase, when a mixture of oil and condensed water is produced, with a declining production rate. When it falls below a certain threshold value, a new cycle is started.

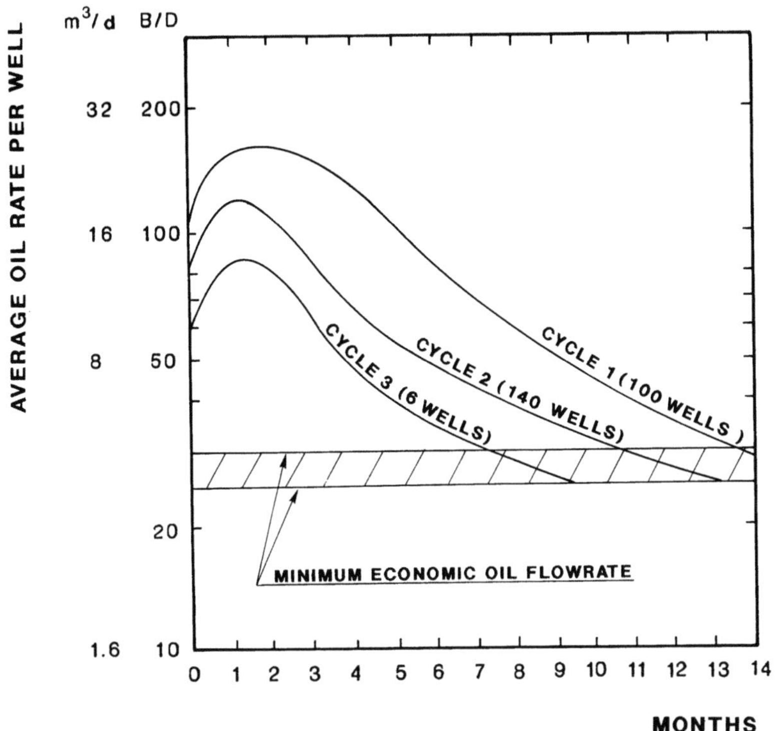

Fig. 15.24. *Results achieved after successive cycles of steam injection/oil production in the Huntington Beach field, California*

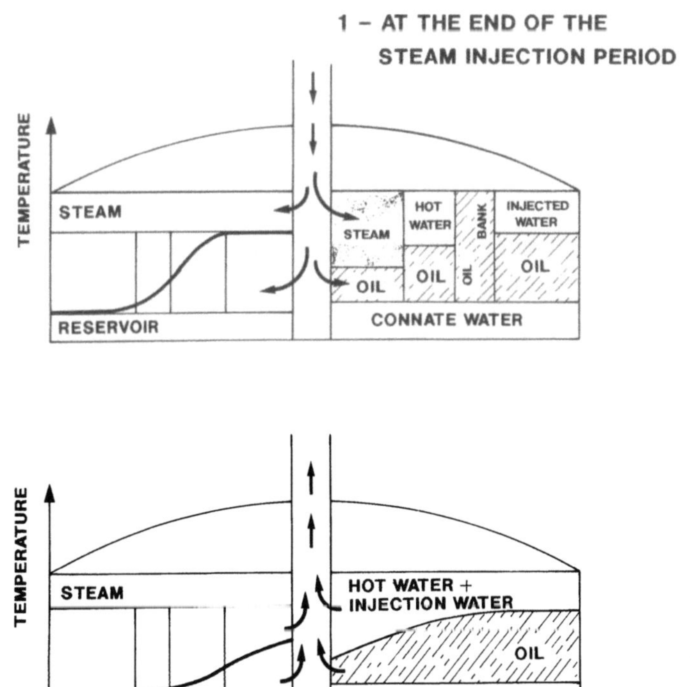

Fig. 15.25. *Fluid and temperature distribution around the well after the steam injection and oil production phases of the huff'n'puff process*

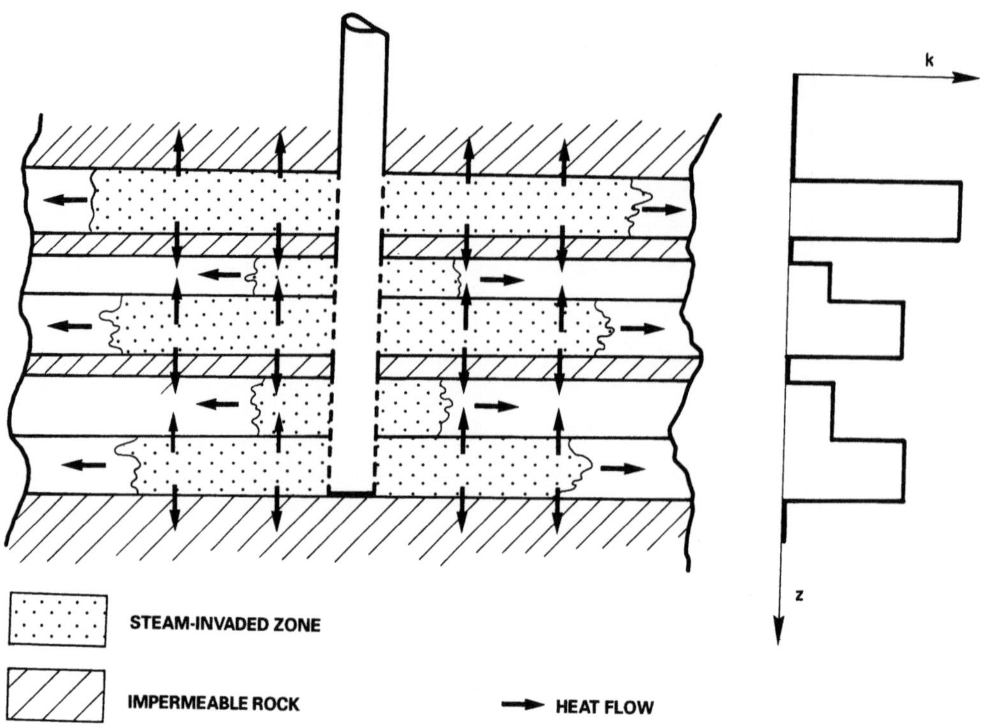

Fig. 15.26. *The diffusion of heat across impermeable strata in a strongly heterogeneous rock undergoing steam injection*

The yield from each cycle decreases until, after a certain number of cycles, no further oil is extracted (Fig. 15.24).

The process is quite simply explained. During the injection phase, the rock surrounding the well is heated up. During the production phase, this heat is transmitted to the oil as it passes through (Fig. 15.25). As in the case of steam drive, the ability of the heat to diffuse to the over- and underlying layers across low permeability strata means that oil will be heated in parts of the reservoir that are not directly affected by the steam injection process (Fig. 15.26). This tends to reduce the negative effects of reservoir heterogeneity. The heating of the oil reduces its viscosity (Sect. 15.6.1), resulting in an improvement in well productivity which is roughly inversely proportional to viscosity.

15.7.3 Reservoir Engineering Considerations

The rock gets heated out to a radius which can be estimated using the equations proposed by van Lookeren[86].

To a first approximation[34] we have:

$$r_\mathrm{h} = \left[\frac{\Delta H \, m_\mathrm{s}}{\pi h_\mathrm{s} \rho_\mathrm{r} C_\mathrm{r} (T_\mathrm{s} - T_\mathrm{i})} \right]^{1/2} \tag{15.16}$$

with: C_r: specific heat of the rock (kJ/kg °C)

        h_s: thickness of rock affected by the steam injection (m)

     ΔH: decrease in enthalpy of steam between T_s and T_i (kJ/kg)

      m_s: mass of steam injected during the cycle (kg)

r_h: radius of the heated region (m)
T_i: rock temperature at the beginning of the cycle (°C)
T_s: temperature of injected steam (°C)
ρ_r: density of the reservoir rock (kg/m$^{3)}$.

Still to a first approximation, we have, assuming steady state fluid flow:

$$\frac{q_{o,h}}{q_{o,c}} = \frac{\ln\dfrac{r_e}{r_w} - \dfrac{1}{2}}{\dfrac{\mu_{o,h}}{\mu_{o,c}}\left(\ln\dfrac{r_h}{r_w} - \dfrac{r_h^2}{2r_e^2}\right) + \ln\dfrac{r_e}{r_h} - \dfrac{1}{2} + \dfrac{r_h^2}{2r_e^2}} \qquad (15.17)$$

with: $q_{o,c}$: oil flow rate at the original reservoir temperature (m^3/day)

$\quad q_{o,h}$: oil flow rate after heating the rock (m^3/day)

$\quad r_e$: drainage radius (m)

$\quad r_h$: radius of the heated region (m)

$\quad r_w$: well radius (m)

$\quad \mu_{o,c}$: oil viscosity at the original reservoir temperature (mPa·s)

$\quad \mu_{o,h}$: viscosity of the heated oil (mPa·s)

As the oil production phase continues, r_h gradually decreases, with a simultaneous lowering of $q_{o,h}$. Steam soak is often used in the early stages of a steam drive to boost the output from the producing wells before the arrival of hot oil displaced by the steam front. For this reason, the production figures for oil produced by steam soak are almost always lumped with those for the oil produced by steam drive. Another application of steam soak is the removal of oil surrounding the injection wells so as to improve steam injectivity.

The conditions which favour steam soak are almost identical to those for steam drive (Sect. 15.6.4), with a preference for highly viscous oils for which well productivity without thermal stimulation would be totally uneconomical. Two accumulations fitting this category are the Faja Petrolifera of the Orinoco in Venezuela (extremely heavy oil), and the Athabasca (bituminous tar sands) of Canada.

15.8 In Situ Combustion

15.8.1 Principles of the Method

The in situ combustion process involves the partial combustion of the oil for generating heat *in the reservoir itself*. This avoids the heat loss and depth-related problems associated with steam injection. The only fluid injected is air, air enriched with oxygen or, more recently, nearly pure oxygen. There are therefore no reservoir pressure constraints to take into consideration, and the feasibility of the method is limited only by the surface compressor capabilities and the economics of the process in any particular situation. The injected air partially oxidises the oil in the reservoir, under the catalysing influence of acid sites that are present on the grain surfaces. This provokes a rise in temperature, which may become sufficient to cause the oil to combust. Downhole gas-powered burners or electrical heaters can be used in the initial phase of the process to get the combustion started. Fig. 15.27 presents a schematic[5] of the in situ process and surface equipment.

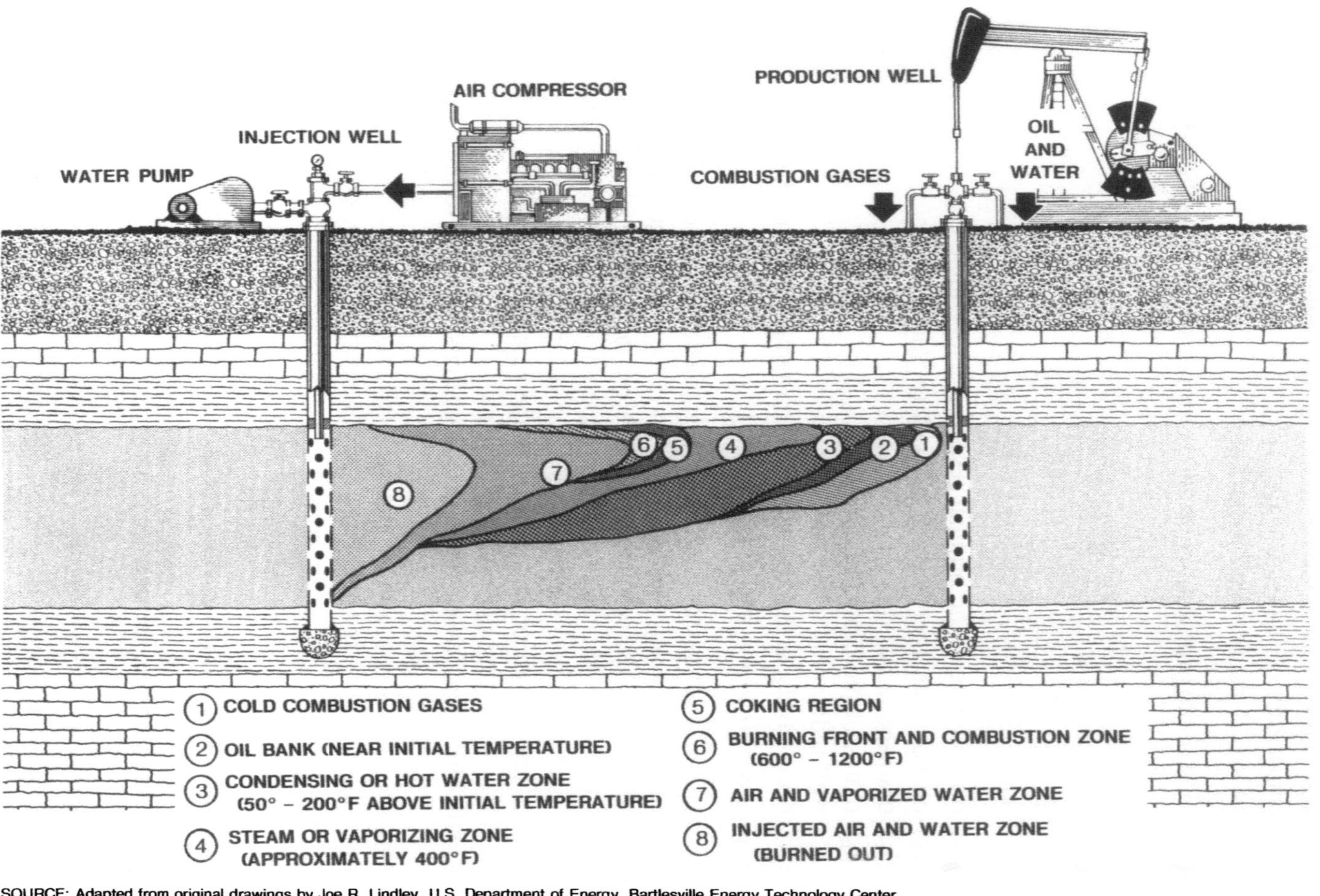

SOURCE: Adapted from original drawings by Joe R. Lindley, U.S. Department of Energy, Bartlesville Energy Technology Center.

Fig. 15.27. *Schematic of an in situ combustion process with simultaneous injection of water (wet combustion). (From Ref. 5. Reprinted with permission of the National Petroleum Council and the US Department of Energy)*

15.8.2 What Happens in the Reservoir

The following discussion applies to direct combustion, where the combustion front advances in the same direction as the flow of injected air. Inverse combustion, where the front moves in the opposite direction to the injected air, has also been tested, but without much success. The latter process is a concept that would be very interesting for highly viscous oils which are barely mobile. Unfortunately, it is very difficult to get the combustion front to propagate through the reservoir.

The fluid and temperature distribution during direct combustion[89] are shown in Fig. 15.28. In order for combustion to occur, the air must be able to flow in the reservoir – it is therefore essential that the relative permeability of the rock to gas should be non-zero. Owing to gravitational force, the air tends to travel through the upper part of the layer, and its low viscosity allows it to flow preferentially along any high permeability strata. The combustion gases (mostly CO_2, CO, SO_2 and N_2) also migrate to the top of the layer, where they heat the oil ahead of the combustion front.

In the high temperature region close to the front (Fig. 15.28), the oil is cracked into light hydrocarbons which migrate with the combustion gases towards the cooler region ahead of the combustion front, where they condense. A bank of light oil forms which displaces the reservoir oil as a miscible phase. Ahead of the front, the combustion gases also heat up the rock, reducing the oil viscosity (Sect. 15.6.1). Formation water is vaporised and contributes a steam drive component to the process.

The residual coke from the cracking of the oil acts as the fuel necessary to sustain combustion. The quantity of coke produced is the most critical factor influencing the propagation of the combustion front. Recent studies have shown that the amount of coke depends not only on the nature of the reservoir oil (particularly its density; Fig. 15.29), but on the type of rock, which acts as a catalyst to the cracking.

The presence of shales, very fine grains, or any sort of rock with a high concentration of acid sites promotes the oxidation of oil at low temperatures, with the formation of large amounts of coke. This will lead to a very hot combustion front, which may even vitrify silicious rock. The consequent loss of permeability will kill the combustion process because the flow of air to the front, and the escape of combustion gases towards the producing wells, are stifled.

At the other extreme, insufficient coke formation will cause the process to die out through lack of combustible material. To increase the chances of success, it is common practice to use a high well density to reduce the distance the combustion front has to travel between the injectors and producers. The improved performance, however, must be balanced against the added cost of drilling and completing the extra wells.

In situ combustion is characterised by a poor volumetric efficiency E_V (Sect. 12.5), because of the tendency of the hot air and combustion gases to migrate to the top of the layer (resulting in a low vertical invasion efficiency E_I), and to move preferentially along high permeability strata (low areal sweep efficiency E_A).

The relatively uniform thermal diffusivity of rock (Table 15.4) favours an even distribution of temperature over the thickness of the layer, but heat is lost to the overlying and underlying formations, which can be a serious problem if the layer is thin. The thermal efficiency of in situ combustion is very low: only about 10%

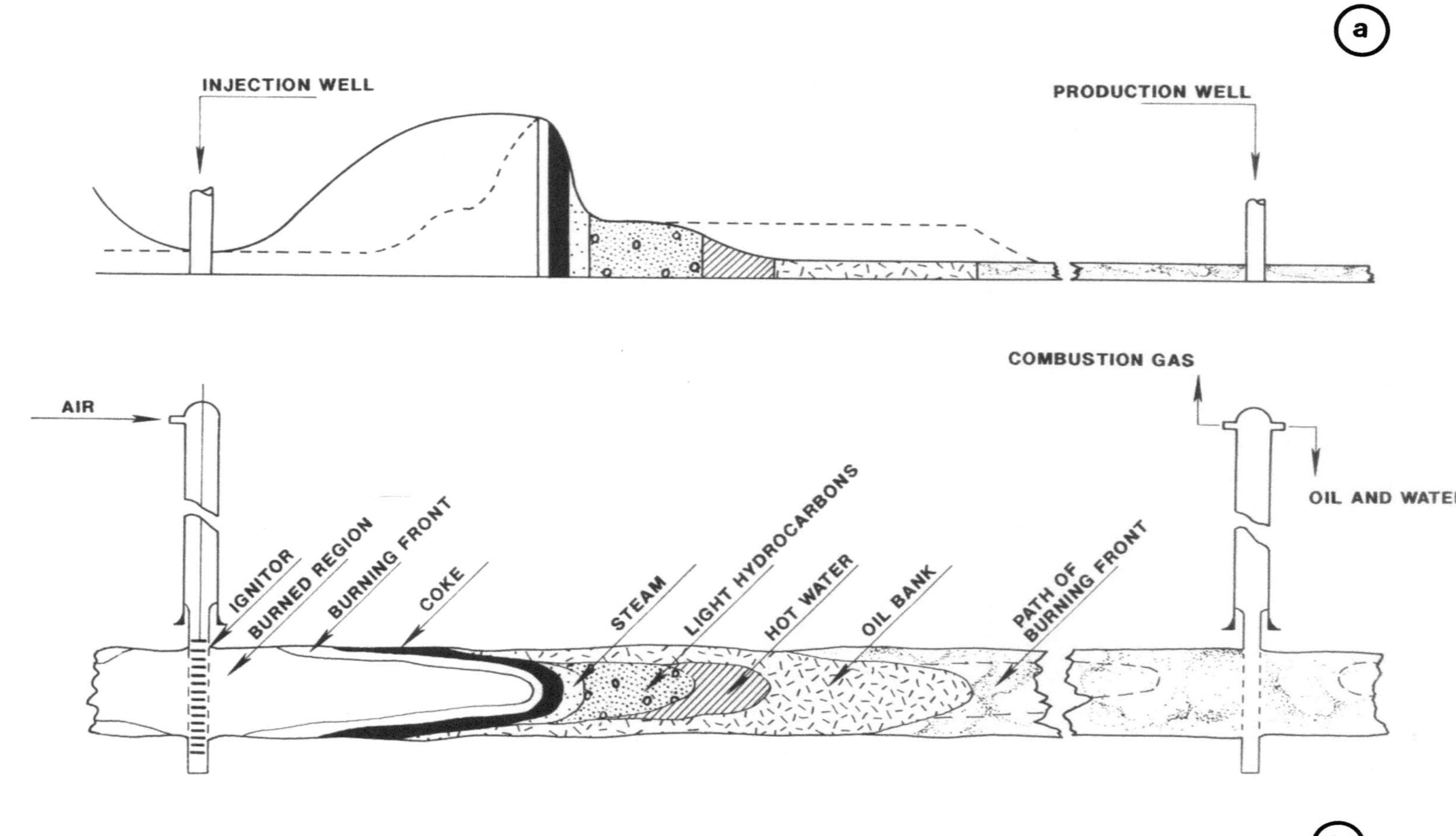

Fig. 15.28. **a** *Temperature distribution;* **b** *Fluid distribution in the reservoir during in situ combustion. (From Ref. 89, 1985, Society of Petroleum Engineers of AIME. Reprinted with permission of the SPE)*

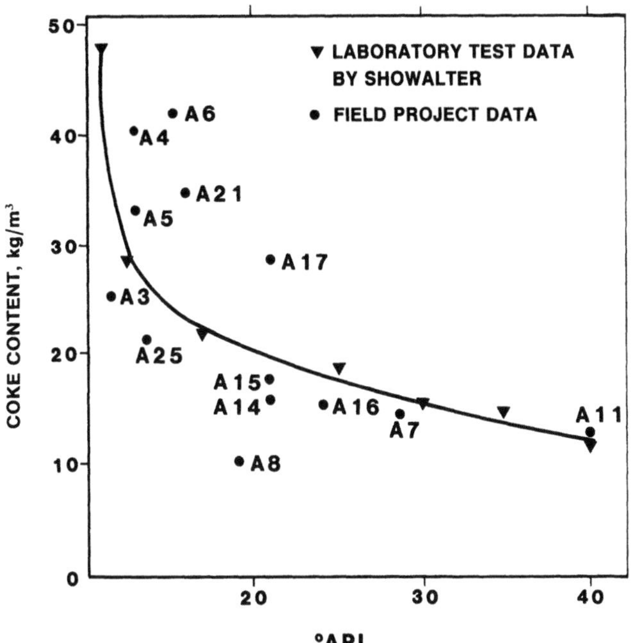

Fig. 15.29. *Quantity of coke produced in the reservoir as a function of the oil API gravity. (From Ref. 50. Reprinted with permission of the Interstate Oil Compact Commission)*

of the heat generated actually goes to heating the oil[89]; 20% is lost in the waste gases (which, in addition, create serious atmospheric pollution problems because of the presence of SO_2 and carbon monoxide); while 70% is retained in the rock behind the combustion front.

15.8.3   Wet Combustion

The heat lost to the rock behind the combustion front can be recuperated by injecting water[17], which it transforms to steam in situ. In this way, a steam drive is created (Sect. 15.6) which boosts the effectiveness of the displacement process. This widely used method is called "wet combustion". The water is injected with the air from the very start of the process, or shortly after. The temperature in the reservoir depends on the water/air ratio (WAR) (Fig. 15.30). The WAR should not be so high as to drop the temperature below the minimum needed for combustion to proceed.

Combination thermal drive (CTD) uses a low WAR and therefore sustains a high temperature front. The water is turned to superheated steam as it passes through the front, and then propagates through the reservoir as a region of limited extent carrying with it the major part of the heat from behind the front.

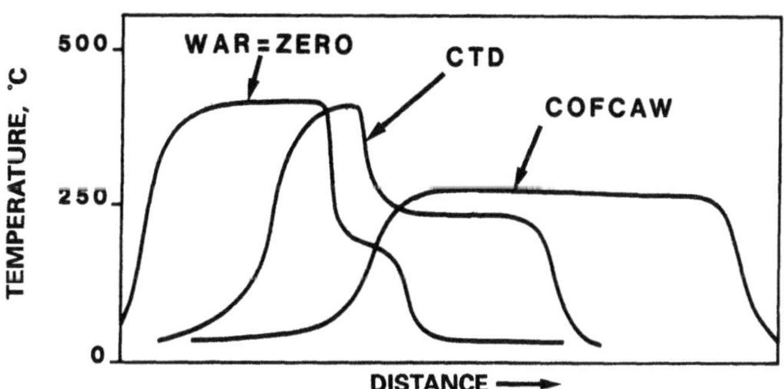

Fig. 15.30. *Wet combustion. Temperature profile through the reservoir for different water/air ratios (WAR). (From Ref. 50. Reprinted with permission of the Interstate Oil Compact Commission)*

The process of "Combination of forward combustion and waterflooding" (COFCAW) uses the maximum WAR and therefore operates at the lowest temperature. In this case, the combustion front is cooled by the saturated steam generated as the water passes through hot rock behind the front. Rather than true combustion, we now have what amounts to oxidation of residual oil at moderate temperature. There is a risk with COFCAW of a gravitational segregation of the water and air in the formation. If this occurs, there will be dry combustion in the upper part of the layer, and a simple flushing with hot water along the bottom. The process can only be properly engineered through a detailed understanding of the reservoir and the use of numerical modelling[27].

15.8.4 Injection of Oxygen or Oxygen-Enriched Air

Since only the oxygen in the air takes part in the in situ combustion, and only one-fifth of the volume of air injected is oxygen, it follows that, by injecting pure oxygen, the same amount of coke could be burnt with a smaller compressor and reduced running costs. Some additional costs will, however, be incurred: firstly, pure oxygen must be obtained by the fractional distillation of liquefied air and, secondly, the compression of pure oxygen requires special equipment. An alternative approach is to use air enriched with oxygen.

The advantage of pure oxygen injection is that the combustion gases consist almost entirely of carbon dioxide, which dissolves completely in the oil ahead of the combustion front. This results in a sharp reduction in the oil viscosity, and an increase in its mobility. The pure oxygen does, however, need to be diluted somewhat with steam to avoid the excessively high combustion temperatures that would otherwise result. This is then a form of wet combustion. This process is currently in the laboratory testing phase, and some pilot tests have been conducted in the field with promising results.

15.8.5 Summary

The technique of in situ combustion originated in the former Soviet Union in the 1930s, and development was taken up in the USA and Canada in the 1950s. To date, it has enjoyed a fairly limited success, despite the significant theoretical advantages it offers (including its applicability to deep reservoirs and highly viscous oils), owing to a number of problematic aspects. The main drawbacks are:

- low volumetric invasion efficiency;
- difficulty in propagating the front over long distances;
- erosion and corrosion in the producing wells, caused by acidic components of the waste gases;
- damage to the tubing, packer and casing, as well as to the cement sheath, as a result of thermal expansion at the high temperatures involved;
- production of water/oil emulsions which are difficult to separate, provoked by the presence of surfactants generated by the partial oxidation of oil;
- extraction of toxic components and suspended solid particles from the waste gases before they are allowed to enter the atmosphere.

In 1992, in situ combustion was employed[136] in eight fields in the USA, six in Canada, 10 in the CIS (ex-USSR), and a few fields in Rumania and Hungary. In the USA, 4700 bbl/day (about 270 000 m^3/year) of oil were recovered by in situ combustion in 1992, representing only 0.6% of its total production by EOR.

A study of the performance of the reservoirs produced to date by in situ combustion[17] has established the following optimum conditions for successful EOR:

- reservoir rock   : preferably sandstone, with only primary porosity;
- reservoir thickness : greater than 3 m;
- porosity   : over 20%;
- permeability   : better than 100 md, with no *secondary permeability (fractures)*;
- initial oil saturation : at least 35%;
- oil API gravity   : less than 30 °API (>876 kg/m^3);
- oil viscosity   : less than 1000 mPa·s (cP).

There is no theoretical limit to the depth of the reservoir. The process has in fact already been used at depths in excess of 3000 m.

15.9 Miscible Gas Flooding

15.9.1 Introduction

The displacement of oil by a miscible gas is the second most commonly used EOR process in the world, after steam drive. Suitable fluids for this process which are miscible with oil are: natural gas, LPG (liquefied petroleum gases) consisting of the propane-butane fractions of natural gas, carbon dioxide, and, in special cases, nitrogen or combustion gases.

To understand how it is possible to achieve miscibility between these gases and oil in the reservoir, we must first of all look in some detail at the phase behaviour of oil and gas. This will be covered in the following sections. We will then examine the displacement process itself, and the problems encountered.

15.9.2 Pseudo-Ternary Diagrams
for the Reservoir Oil/Injection Gas Mixture

It was explained in Sect. 2.1 that a reservoir oil is a complex mixture of hydro-carbons and non-hydrocarbons (N_2, CO_2, nitrogen and sulphur compounds). An n-component oil would require an $(n - 1)$-dimensional diagram to represent its phase behaviour in parametric form, which poses obvious problems for graphic presentation.

The conventional approach is to group the reservoir oil + injected gas system into three pseudo-components as follows:

- *oil/hydrocarbon gas mixture or oil/nitrogen*
    - non-condensing gases ($C_1 + N_2$)
    - intermediate hydrocarbons [from ethane to the hexanes (symbol C_{2-6})]
    - heavy hydrocarbons [from the heptanes onwards (symbol C_{7+})]

- *oil/carbon dioxide mixture*
    - carbon dioxide (CO_2)

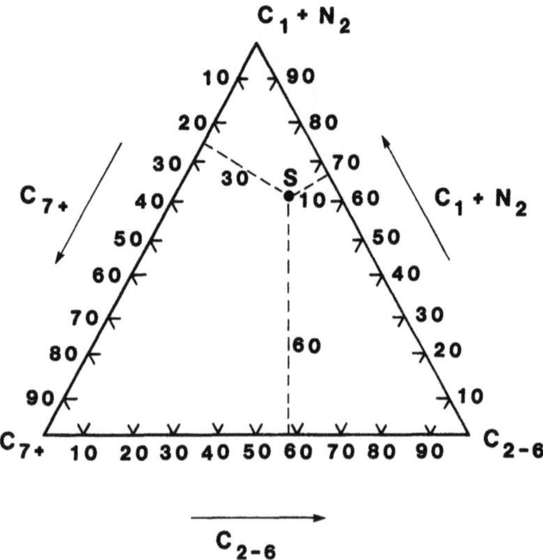

Fig. 15.31. *Triangular diagram used to represent the composition of a hydrocarbon system. The composition at point S is : $C_1 + N_2 = 60\%$; $C_{2-6} = 30\%$; $C_{7+} = 10\%$*

- hydrocarbons than can be extracted by vaporisation in the CO_2 [from methane to the hexanes (symbol C_{1-6})]
- heavy hydrocarbons (C_{7+}).

The composition of any ternary mixture can be represented by a point on a triangular diagram like Fig. 15.31. The perpendicular distance from the point to any side of the equilateral triangle is proportional to the fraction of the pseudo-component which is named in the opposite vertex. For example, in Fig. 15.31 the point S represents a mixture of 60% (C_1+N_2), 30% C_{2-6} and 10% C_{7+}.

15.9.3 Phase Behaviour of a Mixture of Reservoir Oil/Gaseous Hydrocarbon and/or Nitrogen

15.9.3.1   Use of Pseudo-Ternary Diagrams to Describe Phase Behaviour

The phase diagram for the system (C_1+N_2), C_{2-6}, C_{7+} is similar to the one shown in Fig. 15.32, drawn for a given reservoir pressure and temperature. Any mixture represented by a point lying between the curve ACB and the side (C_1+N_2) - N_{7+} is biphasic. One such mixture, M, in Fig. 15.32 will split into a vapour phase of composition V in equilibrium with a liquid phase of composition L. The line joining L and V is the *equilibrium line* or *tie line*; all mixtures that lie on this line will split into vapour and liquid phases of the same compositions V and L, but in different proportions, depending on the position of the mixture point on the tie line.

The part AC of the curve is the *dew point line*, and CB is the *bubble point line*, using the same terminology as for the phase diagram of a hydrocarbon system of fixed composition (Fig. 2.2). In the present case, however, the composition varies from point to point. Point C, where the dew point and bubble point lines meet, is the *critical point* of the system at the specified pressure and temperature (that is, the composition for which that pressure and temperature are the critical values).

As we shall see later in this Chapter, *the mixture represented at point C is miscible in any proportion with any mixture on the triangular diagram.* At constant temperature, the biphasic region decreases in size with increasing pressure (Fig. 15.33), while at constant pressure, the area increases with temperature.

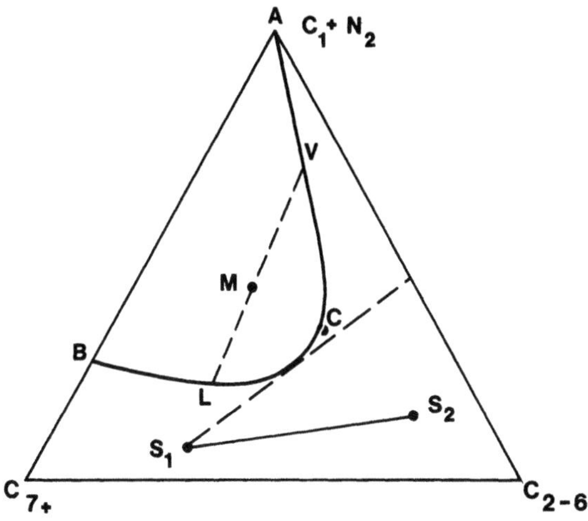

Fig. 15.32. *Example of the phase diagram for a hydrocarbon system at a given p and T, using the triangular presentation*

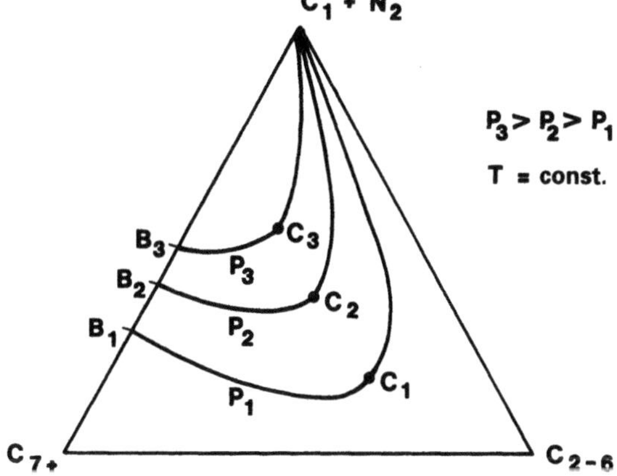

Fig. 15.33. *Variation in the biphasic region with pressure, at constant temperature*

15.9.3.2 First Contact Miscibility

Two mixtures represented as points whose connecting line does not cut the biphasic region (S_1 and S_2 in Fig. 15.32) are miscible with each other in any proportions. So if S_1 corresponds to the composition of the reservoir oil, any other mixture whose point falls below the tangent to curve BC through S_1 will be miscible with the oil. This is *first contact miscibility*.

From Fig. 15.32 it is apparent that any miscible composition must consist of a high proportion of intermediate hydrocarbons C_{2-6}. In practice, a miscible bank of LPG (a propane/butane mixture) is injected first. This is followed by the injection of dry gas, which in turn is miscible in all proportions with the LPG provided the line joining S_2 to the vertex (C_1+N_2) of the triangle does not pass through the biphasic region. The technique is not widely used, partly because of the high cost of LPG, partly because of the risk of a loss of miscibility in the reservoir. This can occur if the continuity of the LPG bank is interrupted by reservoir heterogeneity or a breakthrough of the more mobile dry gas into the oil. There is no miscibility at the direct contact between oil and dry gas.

15.9.3.3 Multiple Contact Miscibility: Condensing Gas Drive

Consider a reservoir oil with a relatively high C_{7+} fraction such as point S_1 in Fig. 15.34, into which a gas rich in intermediate hydrocarbons C_{2-6} (point S_2) is

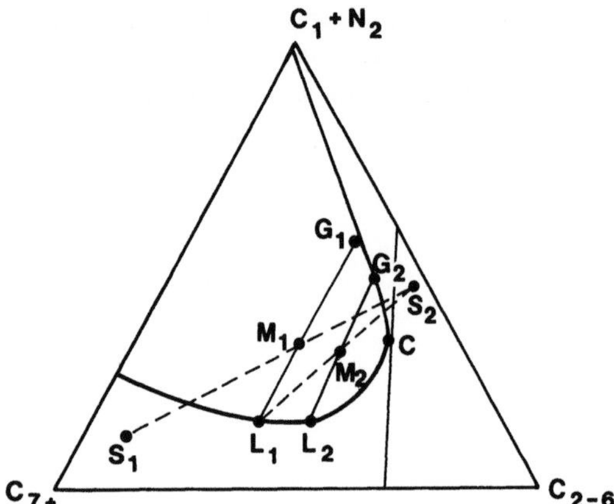

Fig. 15.34. *Triangular diagram showing the principle of condensing gas drive – the formation of a miscible front by condensation into the oil of the intermediate components of the injected gas*

injected. A mixture M_1 forms where the gas is in contact with the oil. This separates into a vapour phase G_1 and a liquid phase L_1 which is miscible with the reservoir oil. The contact between L_1 and injected gas S_2 in turn produces a mixture M_2 which separates into a vapour phase G_2 and a liquid phase L_2, which is miscible with both L_1 and the reservoir oil.

As a result of successive contacts between the injected gas and the liquid phase formed at each preceding contact, the mixture corresponding to the critical point C is eventually reached. This mixture is miscible with L_1, L_2,... and the reservoir oil because it lies on the bubble point line, and has first contact miscibility with injection gas S_2.

Looking at Fig. 15.34, you will see that, at each contact, the C_{2-6} fraction "condenses" out from the injected gas and enriches the equilibrium liquid phase, making it progressively lighter: hence the name *condensing gas drive*. This process is in fact continuous, and builds up a "bank" of liquid phase whose composition varies between that of the reservoir oil and the critical point. A particular advantage of the process is that formation of the bank will resume automatically if at any time it is interrupted because of reservoir heterogeneity or injection gas breakthrough.

From Fig. 15.34 you will appreciate that condensing gas drive can only occur if the C_{2-6} content of the injected gas is greater than (to the right of) the tangent to the biphasic boundary at the critical point. This is very significant observation. The higher the reservoir pressure, the closer the critical point is to the $(C_1 + N_2)$ - C_{7+} line on the triangular diagram. Miscibility can therefore be achieved with a lower C_{2-6} content in the injected gas at higher pressures.

As we shall see later in this chapter, there are correlations for the approximate calculation of the minimum miscibility pressure (MMP) for a specified composition of the reservoir oil and injected gas, and the reservoir temperature. The MMP can be obtained more accurately through "slim tube" laboratory testing under simulated reservoir conditions.

15.9.3.4 Multiple Contact Miscibility: Vaporising Gas Drive

In Fig. 15.35, at point S_1, we have a light ("volatile") reservoir oil which is rich in intermediate hydrocarbons C_{2-6}. Dry gas or pure nitrogen, corresponding to (C_1+N_2) or just N_2 in the diagram are injected into the reservoir.

As seen in Sect. 15.9.3.3, at the contact between the gas and oil a mixture M_1 forms which consists of a vapour phase G_1 *enriched with C_{2-6} extracted from the oil*, and a liquid phase L_1 which has first contact miscibility with the reservoir oil.

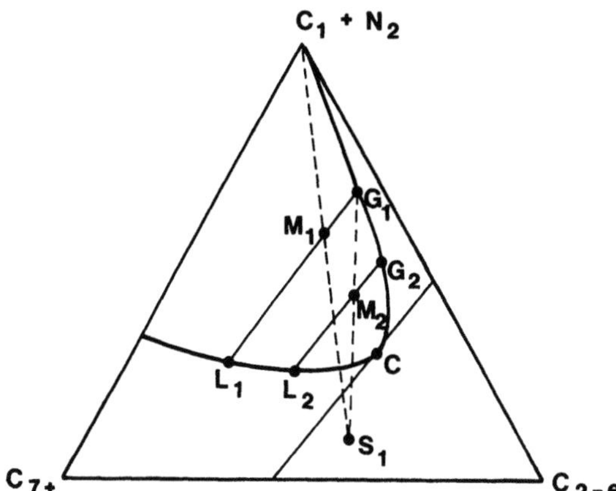

Fig. 15.35. *Triangular diagram showing the principle of vaporising gas drive process, in which the miscible bank is formed by vaporisation of the intermediate components of the oil*

Further contact between G_1 and the oil S_1 produces a mixture M_2 consisting of a vapour phase G_2, richer still in C_{2-6} extracted from the oil, and a liquid phase L_2, miscible with L_1 and the reservoir oil.

Through successive contacts between oil and a vapour which is becoming increasingly rich in C_{2-6}, the critical composition C is eventually attained. This is miscible with L_1, L_2, ... and with the reservoir oil because it falls on the bubble point line, and with G_1,G_2, ... and the injected gas because it falls on the dew point line. Because of the path by which miscibility is achieved, the process is known as *vaporising gas drive*.

In reality, it occurs in a continuous manner rather than as discrete steps as implied above, and leads to the formation of a bank of vapour phase whose composition varies between that of G_1 and the critical mixture C. The injected gas is miscible in all proportions with the bank of rich gas, and the critical mixture is in turn miscible in all proportions with L_1, L_2, ... and the reservoir oil. Miscibility is achieved, then, by means of a bank of fluid of variable composition created through a process of multiple contacts. As with condensing gas drive, if the bank is interrupted by reservoir heterogeneity, or injection gas breakthrough from behind, it can reform automatically through the process just described.

Vaporising gas drive can only occur if the critical mixture forms. For this to happen, the oil must contain a higher fraction of C_{2-6} than that represented by the line which is tangent to the biphasic envelope at the critical point. In Fig. 15.35, S_1 must lie to the right of the tangent through C. The higher the pressure, the more the biphasic region contracts towards the side (C_1+N_2) - C_{7+} on the triangular diagram (Fig. 15.33): miscibility can therefore be achieved for reservoir oils lower in C_{2-6} content. In other words, the higher the reservoir pressure, the less volatile the oil needs to be for vaporising gas drive to be feasible.

With pure nitrogen injection, vaporising gas drive is the only miscible process possible, because the N_2 obviously has no intermediate components to yield to the oil. Note that, at a given temperature and pressure, the pseudo-ternary diagram $N_2/C_{1-7}/C_{7+}$ has a much larger biphasic region than the equivalent $C_1/C_{2-6}/C_{7+}$ diagram (Fig. 15.36). For a given reservoir temperature and pressure, therefore, miscible drive using pure nitrogen requires a more volatile reservoir oil. By the same token, for a given reservoir oil, miscibilty with nitrogen occurs at a much higher pressure than with natural gas. Consequently, miscible gas flooding with nitrogen is limited to a relatively small number of deep, high pressure volatile oil reservoirs.

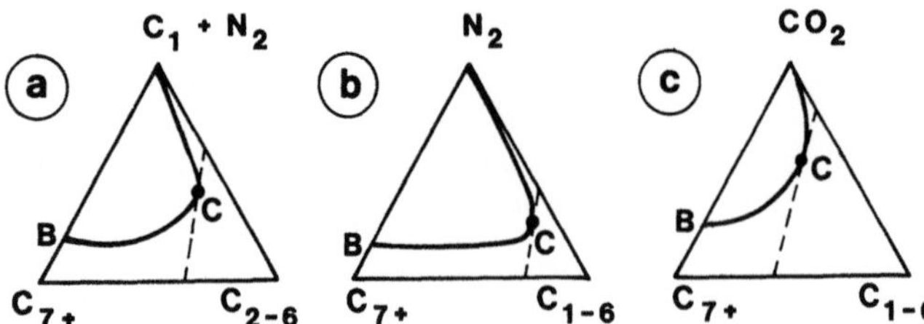

Fig. 15.36a-c. *Extent of the biphasic region, with tangent through the critical point, for the same oil at the same temperature and pressure, for three systems: a reservoir oil + natural gas; b reservoir oil + nitrogen; c reservoir oil + carbon dioxide*

On the other hand, nitrogen injection is widely used in *non-miscible flooding* for the vertical displacement of oil under gravitational or gravity-dominated conditions (Sect. 15.4.2.3). Even in this case, however, the nitrogen extracts part of the light hydrocarbon fraction of the oil, and this is produced to surface in the nitrogen returns at the producing wells[94]. The comments in Sect. 15.9.3.3 apply equally to the estimation of the minimum miscibility pressure (MMP) for vaporising gas drive.

15.9.4   Phase Behaviour of a Mixture of Reservoir Oil and Carbon Dioxide

15.9.4.1   Thermodynamic Properties of Carbon Dioxide

The critical properties of pure CO_2 are as follows:

- critical temperature   : 304.2 K = 31 °C
- critical pressure   : 7.144 MPa = 72.85 kg/cm^2
- critical density   : 467 kg/m^3
- viscosity at critical point   : 0.033335 mPa s (cP)
- deviation factor at critical point : 0.275

CO_2 cannot therefore exist as a liquid above 31 °C, at any pressure.

At pressures higher than critical, the CO_2 is said to be "supercritical". In this state, we do not observe any transition from liquid to vapour phase or vice versa (the appearance of bubbles or droplets) as the temperature crosses 31 °C – the fluid properties change continuously. For this reason, the flow of CO_2 in pipelines is usually kept under supercritical conditions as far as possible, so as to avoid the formation of gas/liquid mixtures, whose undesirable flow characteristics lead to increased energy losses, and the possible development of stratified flow regimes or slugs.

A mixture of CO_2 and hydrocarbon can exhibit a critical temperature higher than 31 °C, up to a maximum of over 50 °C (122 °F). With such a mixture, two liquid phases can exist in equilibrium with a vapour phase. The denser of the two liquid phases will be richer in hydrocarbons, while the lighter will be richer in CO_2.

We will examine the behaviour of CO_2/reservoir oil mixtures for two cases:

Type I: reservoir temperature higher than 50 °C

Type II: reservoir temperature less than 50 °C.

In both cases, vaporising gas drive is the only miscible process that can occur, because CO_2 does not contain any intermediate hydrocarbon components (C_{2-6}) that could condense into the oil. Figs. 15.37 and 15.38 show the dependence of the density and viscosity of CO_2 on temperature and pressure.

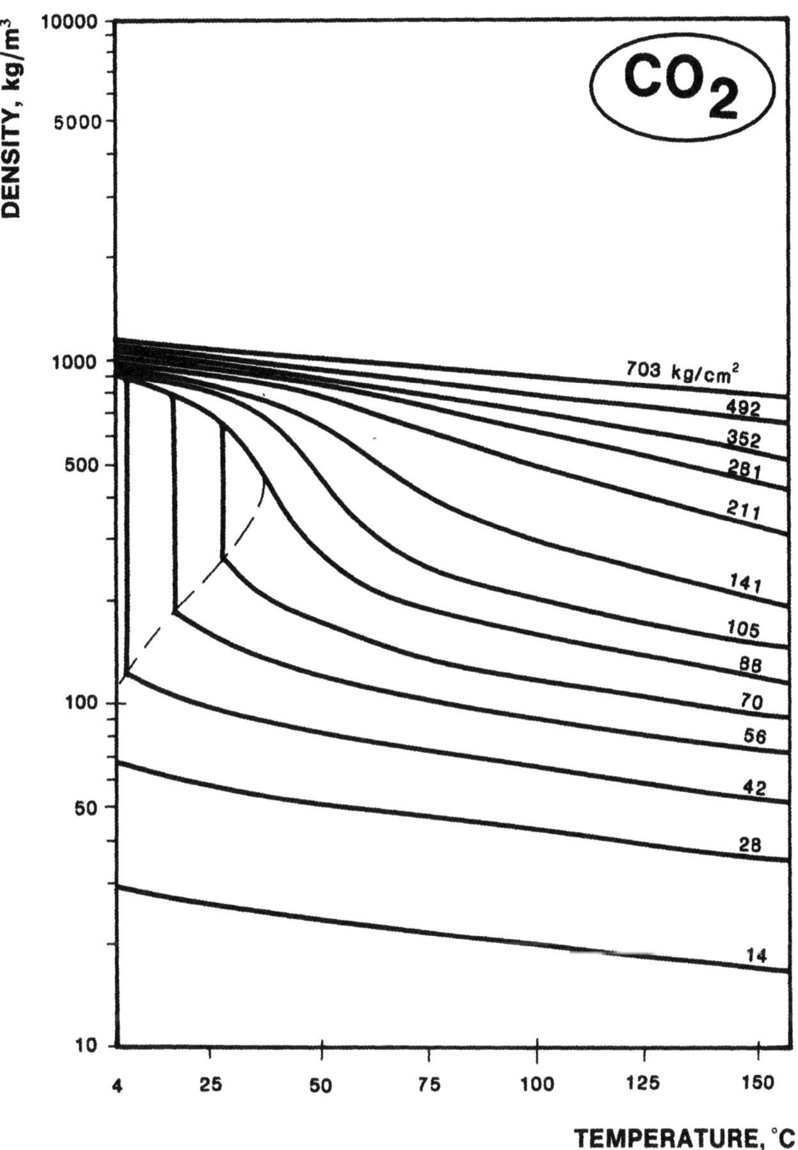

Fig. 15.37. *Density of carbon dioxide as a function of temperature and pressure*

15.9.4.2 Type I Behaviour of a Mixture of Reservoir Oil and Carbon Dioxide

Type I behaviour of a mixture of reservoir oil and CO_2 is observed when the temperature is higher than 50° C. The phase behaviour is described by a diagram which is similar to Fig. 15.35 for a vaporising gas drive using natural gas, but with the vertices of the triangle now labelled as follows:

- CO_2 : in the place of (C_1+N_2)
- extractable hydrocarbons (usually C_{1-6}) : in the place of C_{2-6}
- C_{7+} : in the place of C_{7+}

The formation of the miscible bank, and the approach to the critical point, occur in the manner described in Sect. 15.9.3.4. The process does, however, have two important additional characteristics:

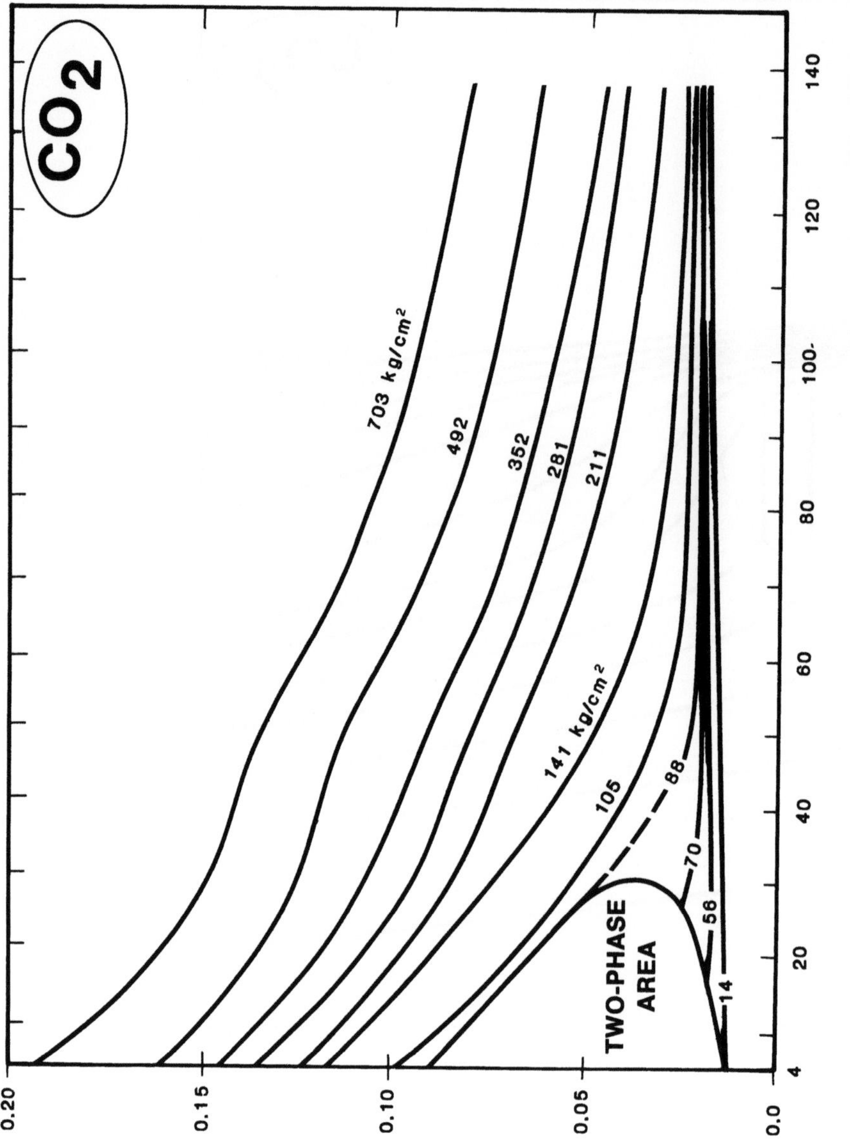

Fig. 15.38. *Viscosity of carbon dioxide as a function of temperature and pressure*

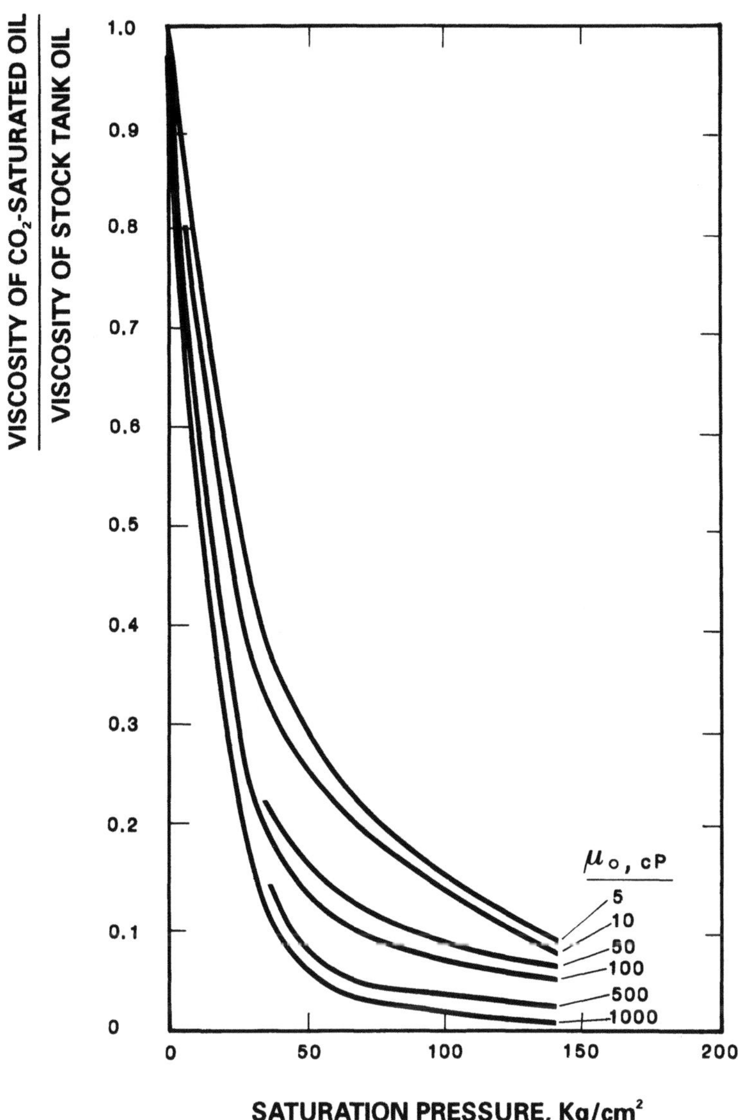

Fig. 15.39. *The influence of dissolved carbon dioxide on the oil viscosity. (From Ref. 76, 1965, Society of Petroleum Engineers of AIME. Reprinted with permission of the SPE)*

– for a given temperature, pressure and oil composition, the biphasic region ACB (Fig. 15.36) is smaller than is the case for the equivalent $(C_1+N_2)/(C_{2-6})/C_{7+}$ system. *The MMP for CO_2 injection is lower than for natural gas.*

– as the CO_2 dissolves in the oil (equilibrium points L_1, L_2, ... along the bubble point curve) there is a significant decrease in the oil's viscosity (Fig. 15.39). This greatly enhances the displacement process, and increases the limiting critical velocity u_{crit} for gravity-stabilised conditions (Sect. 11.4.2).

15.9.4.3 Type II Behaviour of a Mixture of Reservoir Oil and Carbon Dioxide

The type II behaviour of a mixture of reservoir oil and CO_2, observed when the temperature is less than 50 °C, is illustrated in Fig. 15.40. In the biphasic region ACB there is an area AA_1B_1B, corresponding to a low fraction of extractable

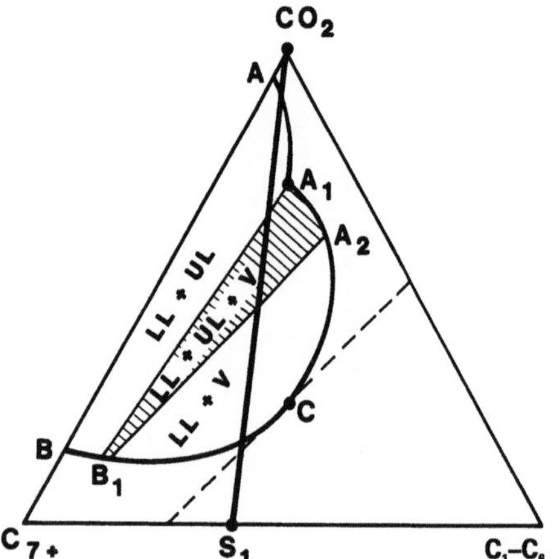

Fig. 15.40. *Phase diagram for a type II system of reservoir oil and carbon dioxide*

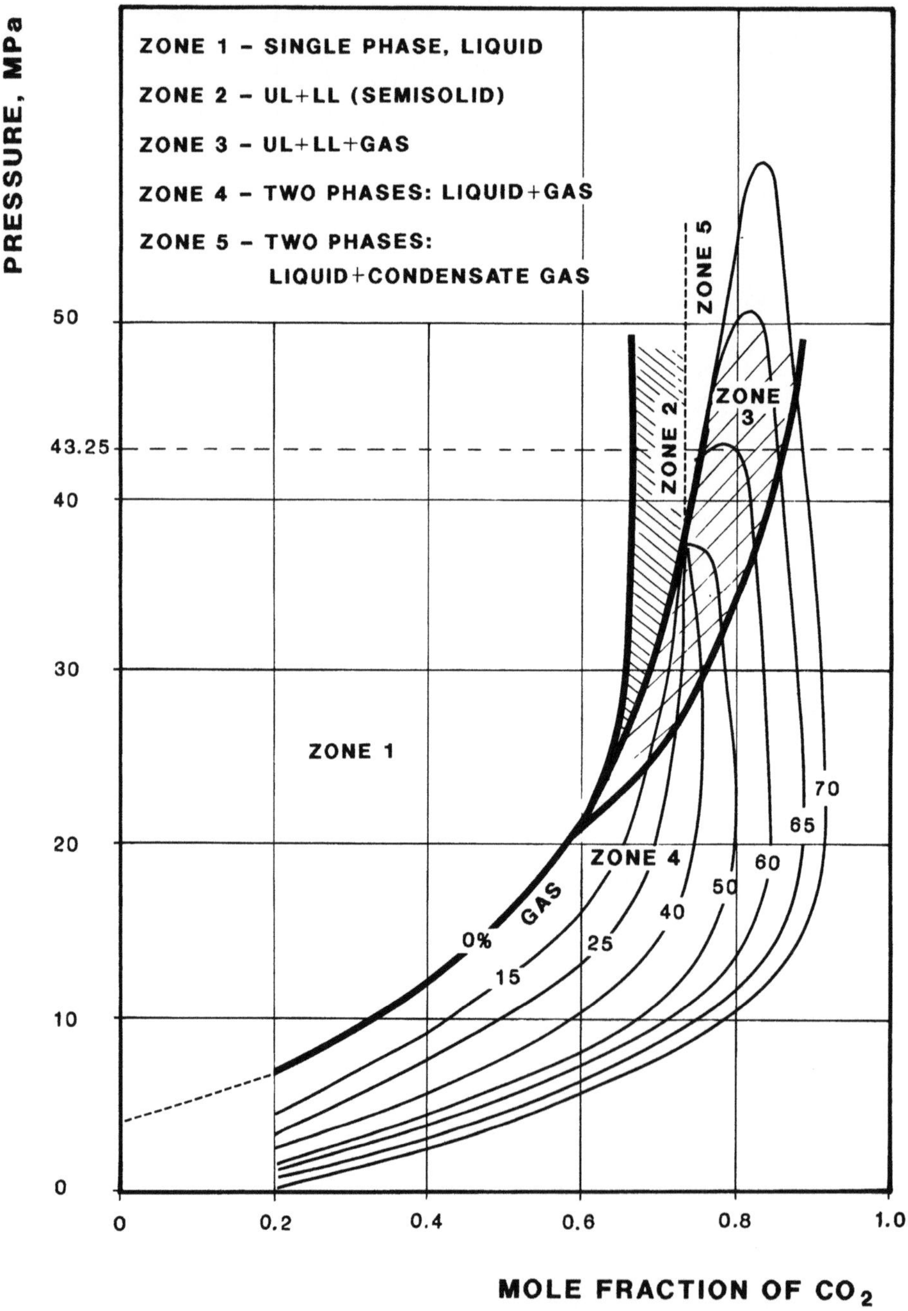

Fig. 15.41. *Phase diagram for carbon dioxide/reservoir oil from the Piropo reservoir[8] (Adriatic Sea). This is a type IIb system*

hydrocarbons, where the mixture consists of *two liquid phases*. The heavier liquid (LL) contains a high percentage of hydrocarbons, while the lighter liquid (UL) is carbon dioxide with a small hydrocarbon fraction. In the case of very heavy asphaltic oils, LL is actually semi-solid[8], and is made up almost entirely of asphaltenes.

Adjacent to this biphasic liquid region, there exists a region $A_1A_2B_1$ *in which the system is triphasic*: LL + UL + vapour phase. The phase UL disappears to the right of A_2B_1, and the system becomes a biphasic mixture consisting of LL and the vapour phase.

Miscibility occurs through extraction of the intermediate hydrocarbons, either by gaseous CO_2 or, in the region AA_2B_1B, by liquid CO_2. Once again, miscibility can only be achieved if the reservoir oil point S_1 lies below the tangent to the curve $AA_1A_2CB_1B$ at the critical point C (the dashed line in Fig. 15.40).

In some systems, the triphasic area $A_1A_2B_1$ is displaced towards the critical point by increasing pressure, and eventually disappears completely. These are referred to as type IIa systems. If the converse occurs, and the triphasic region is displaced towards the side CO_2/C_{7+} of the triangle (that is, *away* from the critical point) by increasing pressure, and eventually disappears, the system is said to be type IIb. Fig. 15.41 illustrates the behaviour of a type IIb system in terms of the CO_2 molar fraction versus pressure.

15.9.5 Correlations for the Estimation of Minimum Miscibility Pressure (MMP)

It has already been mentioned that the minimum miscibility pressure (MMP) for a reservoir oil/injection gas system can only be determined accurately through laboratory tests with a slim tube[77] at reservoir temperature and pressure. However, reasonably good estimates of MMP under reservoir conditions for a given oil and injection gas composition can be obtained using any of a number of correlations which have been derived from experimental data.

15.9.5.1 *Vaporising Gas Drive*

Stalkup[77] proposed a correlation for the calculation of the MMP specifically for vaporising gas drive:

$$\text{MMP (MPa)} = p_b \frac{X + 4.09068}{2.252354}, \tag{15.18}$$

where:

$$X = \frac{W(C_1)W_iT_r^{1/3}}{W(C_{7+})} \tag{15.19a}$$

and:

$$W(C_1) = \frac{y(C_1)M(C_1)}{M_o} \tag{15.19b}$$

$$W_i = \frac{\sum_{2}^{6}{}_i y(C_i)M(C_i) + y(CO_2)M(CO_2) + y(H_2S)M(H_2S)}{M_o} \tag{15.19c}$$

$$W(C_{7+}) = \frac{y(C_{7+})\,M(C_{7+})}{M_o}, \qquad (15.19\text{d})$$

with: p_b = bubble point pressure of the oil (MPa)
T_r = reduced temperature of the reservoir oil = T/T_c
$y(\ )$ = mole fraction of the component within the brackets
$M(\ )$ = molecular weight of the component within the brackets
M_o = molecular weight of the oil.

15.9.5.2 Condensing and Vaporising Gas Drives

Glasø's correlation[45] is valid for both condensing and vaporising gas drives:

For $M(C_{2-6}) = 34$

$$\text{MMP (MPa)} = 43.637 - 0.175M - (32.2 - 0.127M)z$$
$$+ \left[7.77 \times 10^{-15} M^{5.258} e^{31\,980zM^{-1.073}}\right](1.8T_R + 32). \qquad (15.20\text{a})$$

For $M(C_{2-6}) = 44$

$$\text{MMP (MPa)} = 37.942 - 0.133M - (55.8 - 0.188M)z$$
$$+ \left[1.172 \times 10^{-11} M^{3.730} e^{1356.7zM^{-1.058}}\right](1.8T_R + 32). \qquad (15.20\text{b})$$

For $M(C_{2-6}) = 54$

$$\text{MMP (MPa)} = 51.276 - 0.177M - (50.7 - 0.147M)z$$
$$+ \left[3.392 \times 10^{-16} M^{5.520} e^{2170.6zM^{-1.109}}\right](1.8T_R + 32), \qquad (15.20\text{c})$$

where:

$M(C_{2-6})$ = molecular weight of the C_{2-6} fraction of the reservoir oil
M = corrected average molecular weight of the C_{7+} fraction
of the oil:

$$M = 1.094 \times 10^{-14} \rho_{C_{7+}}^{5.573} \qquad (15.20\text{d})$$

z = mole fraction of C_1 in the injection gas
T_R = reservoir temperature, $^\circ$C
$\rho_{C_{7+}}$ = density of the C_{7+} fraction of the oil, kg/m^3.

15.9.5.3 Miscibility of Oil and Nitrogen

Glasø[45] proposed the following correlation for the MMP of oil/nitrogen mixtures:

For $y(C_2\text{-}C_6) < 0.28$

$$\text{MMP (MPa)} = 64.563 - 8.335 \times 10^{-2} M(C_{7+})$$
$$+ \left[7.770 \times 10^{-15} M(C_{7+})^{5.258} e^{23\,025M(C_{7+})^{-1.703}} - 0.143\right]$$
$$\times (1.8T_R + 32) - 68.5y(C_2 - C_6). \qquad (15.21)$$

For $y(C_2\text{-}C_6) > 0.28$

   a: if $M(C_{7+}) < 160$:

$$MMP\ (MPa) = 53.056 - 8.336 \times 10^{-2} M(C_{7+})$$
$$+ \left[7.770 \times 10^{-15} M(C_{7+})^{5.258} e^{23\ 025 M(C_{7+})^{-1.703}} - 0.274\right]$$
$$\times (1.8 T_R + 32) \tag{15.22a}$$

   b: if $M(C_{7+}) > 160$:

$$MMP\ (MPa) = 43.878 - 8.336 \times 10^{-2} M(C_{7+})$$
$$+ \left[7.770 \times 10^{-15} M(C_{7+})^{5.258} e^{23\ 025 M(C_{7+})^{-1.703}} - 0.143\right]$$
$$\times (1.8 T_R + 32), \tag{15.22b}$$

where, as usual:

$$y(C_2 - C_6) = \text{mole fraction of } C_{2-6} \text{ hydrocarbons in the reservoir oil}$$
$$M(C_{7+}) \quad = \text{molecular weight of the } C_{7+} \text{ fraction of the oil}$$
$$T_R \qquad\quad = \text{reservoir temperature (}^{\circ}\text{C)}.$$

15.9.5.4 Miscible Displacement with Carbon Dioxide

A very simplistic correlation for the MMP of an oil/carbon dioxide mixture was proposed by Yellig and Metcalfe[92]. In their equation, the MMP depends solely on the reservoir temperature:

$$MMP\ (MPa) = 2.23346 + 0.15824 T_R \tag{15.23}$$

where: $T_R = $ reservoir temperature ($^{\circ}$C).

The MMP predicted by Eq. (15.23) is only used if it is higher than the bubble point of the oil, otherwise the bubble point itself is used as the MMP.

Cronquist[77] derived a more elaborate correlation which took account of the composition of the oil, in addition to the temperature:

$$\log(MMP) = [0.744206 + 1.1038 \times 10^{-3} M(C_{5+}) + 0.15279 y(C_1)]$$
$$\times \log(1.8 T_R + 32) - 0.95769, \tag{15.24}$$

where the MMP is in MPa, and:

$$y(C_1) \qquad = \text{mole fraction of methane in the reservoir oil}$$
$$M(C_{5+}) = \text{molecular weight of the } C_{5+} \text{ fraction of the oil}$$
$$T_R \qquad\quad = \text{reservoir temperature (}^{\circ}\text{C)}.$$

Glasø's correlation[45] is also based on the oil composition and temperature, but is of a different form, as follows:

For $y(C_2\text{-}C_6) < 0.18$:

$$MMP\ (MPa) = 20.325 - 2.347 \times 10^{-2} M(C_{7+})$$
$$+ \left[1.172 \times 10^{-11} M(C_{7+})^{3.730} e^{786.8 M(C_{7+})^{-1.058}}\right] (1.8 T_R + 32)$$
$$- 83.6 y(C_2 - C_6). \tag{15.25a}$$

For $y(C_2\text{-}C_6) > 0.18$:

$$\text{MMP (MPa)} = 5.585 - 2.347 \times 10^{-2} M(C_{7+})$$
$$+ \left[1.172 \times 10^{-11} M(C_{7+})^{3.730} e^{786.8 M(C_{7+})^{-1.058}} \right]$$
$$\times (1.8 T_R + 32). \tag{15.25b}$$

These equations use the same nomenclature as Eqs. (15.21) and (15.22).

If there are appreciable quantities of other gases with the CO_2, the correlation of Sebastian et al.[73] can be used:

$$\frac{\text{MMP (mixture)}}{\text{MMP (pure } CO_2)} = 1.000 - 2.13 \times 10^{-2}(T_{pc} - 304.2)$$
$$+ 2.51 \times 10^{-4}(T_{pc} - 304.2)^2$$
$$- 2.35 \times 10^{-7}(T_{pc} - 304.2)^3, \tag{15.26}$$

where T_{pc} (K) is the *pseudo-critical temperature* [Eq. (2.11b)] of the mixture of CO_2 and other gases. Note that for pure carbon dioxide, $T_{pc} = 304.2$ K, in which case the right-hand side of Eq. (15.26) reduces to 1.0.

15.9.6　Flow Regimes Encountered in Miscible Gas Drive

The bank of light liquid hydrocarbons that forms through the exchange of mass between injected gas and oil (Sects. 15.9.3 and 15.9.4) is miscible with the reservoir oil at its leading edge, and with injected gas at its trailing edge. Miscibility between reservoir oil and injected gas is in fact achieved thanks to the presence of this region of light hydrocarbons, whose density and viscosity increase uniformly from the upstream to the downstream side.

Because the injected gas has a lower viscosity than "solvent" at the upstream edge of the bank, and the solvent in turn has a lower viscosity than the reservoir oil at the downstream edge, fingering is likely to occur at both edges (see Sect. 11.3.6 and Figs. 11.21 and 11.22). Fingering is a very complex phenomenon which depends on the global dispersion coefficient K_e of the displacing fluid in the displaced fluid, as well as the displacement velocity and viscosity ratio of the two fluids in contact.

K_e itself is a function of four interacting processes:
- molecular diffusion and convective dispersion within the pores;
- dispersion on a microscopic scale in the direction of flow, and perpendicular to it;
- dispersion on a macroscopic scale;
- dispersion on a megascopic scale.

The first two factors depend on the thermodynamic properties of the fluids, and on the pore structure of the rock. The last two depend on the stochastic heterogeneity of the rock and on its layering. Detailed descriptions of fingering in miscible fluids, including the equations governing the phenomenon, are provided by Chierici[19] and Stalkup[77].

Stalkup[77] includes a global treatment of fingering in horizontal porous media, in which the flow regime is characterised by the term N_{vg}^*, a modified form of the dimensionless group N_{vg} introduced in Sect. 3.5.3 to account for the ratio of viscous to gravitational forces:

$$N_{vg}^* = \frac{u\mu_o}{kg(\rho_o - \rho_s)}\frac{L}{h},$$ (15.27)

where: g = acceleration due to gravity
          h = reservoir thickness
          k = permeability of the porous medium
          L = distance between injector and producing well
          u = Darcy velocity of the fluid
          μ_o = oil viscosity
          ρ_o, ρ_s = densities of the oil and solvent.

In the region between wells, any of four possible flow regimes can be observed across the reservoir thickness. Each exhibits a different volumetric efficiency E_V (Sec. 12.5) at the moment of gas breakthrough at the producing well, as follows:

Regime I: occurs when N_{vg}^* is extremely low, with a single tongue of gas riding over the top of the oil. The shape of the tongue and the value of E_V depend on the local value of N_{vg}^*.

Regime II: occurs at higher N_{vg}^* values than regime I. The shape of the tongue and the value of E_V are independent of N_{vg}^*.

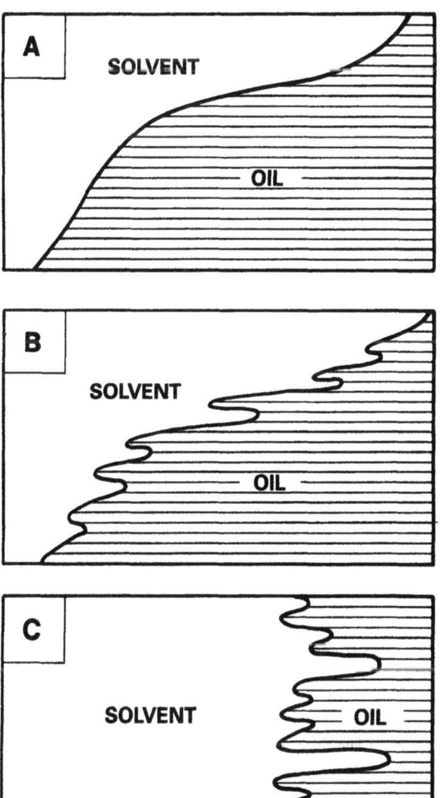

Fig. 15.42A-C. *The types of fingering that can occur in the miscible displacement of oil by gas:* **A** *Regimes I and II;* **B** *Regime III* **C** *Regime IV (From Ref. 77, 1983, Society of Petroleum Engineers of AIME. Reprinted with permission of the SPE)*

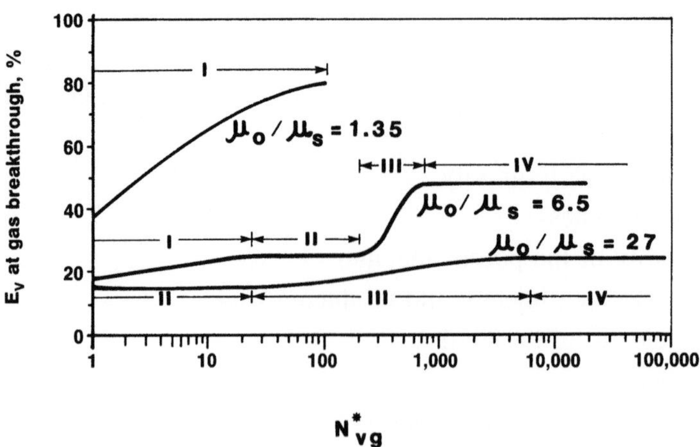

Fig. 15.43. *Volumetric efficiency at gas breakthrough for the different flow regimes, as a function of the ratio of viscous to gravitational forces, N^*_{vg}, for different values of μ_o/μ_s. (From Ref. 77, 1983, Society of Petroleum Engineers of AIME. Reprinted with permission of the SPE)*

Regime III: at even higher values of N^*_{vg}, secondary fingering starts beneath the upper tongue of gas (Fig. 15.42B). The shape of the front depends on N^*_{vg}, and E_V increases with N^*_{vg}.

Regime IV: multiple fingering occurs at very high values of N^*_{vg} (Fig. 15.42C). This is a true "viscous fingering". E_V increases with N^*_{vg}.

The behaviour of these flow regimes is clearly illustrated in Fig. 15.43.

15.9.7   Displacement Under Gravity-Stabilised Conditions

In dipping strata it becomes possible for the displacement front to be stabilised by gravity (Sect. 11.4.2). This is particularly important for miscible gas drive, as it can prevent the formation of the gas tongue that travels over the top of the oil, and eliminate, or at least reduce significantly, the tendency to fingering.

In the case of the immiscible displacement of oil by water, the critical velocity $(u_t)_{crit}$, below which gravity stabilisation is possible, is:

$$(u_t)_{crit} = \frac{k}{\dfrac{\mu_o}{k_{ro,iw}} - \dfrac{\mu_w}{k_{rw,or}}} (\rho_w - \rho_o)\, g\, \sin\theta \qquad\qquad 11.61a$$

where θ is the inclination of the formation relative to the horizontal.

Now, in the case of miscible displacement, both relative permeabilities are unity. If we represent the density and viscosity of oil by ρ_o and μ_o, and of the solvent displacing it by ρ_s and μ_s, Eq. (11.61a) reduces to:

$$(u_t)_{crit} = \frac{\rho_o - \rho_s}{\mu_o - \mu_s}\, k\, g\, \sin\theta \qquad\qquad (15.28)$$

Remember that in miscible gas drive, both ρ and μ vary continuously across the contact zone between the two fluids, as a direct consequence of the process by which miscibility occurs. From work specifically on miscible displacement, Dumoré[39] identified the velocity u_{st} below which gravity stabilisation of the front would be possible:

$$u_{st} = \left(\frac{d\rho}{d\mu}\right)_{min} kg \sin\theta \qquad (15.29)$$

Assuming that ρ varies linearly with the concentrations of the two mixing fluids as follows:

$$\rho = \rho_s C_s + (1 - C_s)\rho_o \qquad (\rho_s < \rho_o) \qquad (15.30a)$$

with C_s = concentration of solvent in the mixture; and that $\ln\mu$ also varies linearly with C_s:

$$\ln\mu = C_s \ln\mu_s + (1 - C_s)\ln\mu_o \qquad (\mu_s < \mu_o) \qquad (15.30b)$$

Eq. (15.29) becomes:

$$u_{st} = \frac{\rho_o - \rho_s}{\mu_o(\ln\mu_o - \ln\mu_s)}kg \sin\theta \qquad (15.31)$$

Therefore:

$$u_{st} = (u_t)_{crit}\frac{1 - \dfrac{\mu_s}{\mu_o}}{\ln\dfrac{\mu_o}{\mu_s}} \qquad (15.32)$$

Equation (15.31) defines the maximum Darcy velocity (which determines the maximum flow rate) that can be sustained in a reservoir under miscible gas drive before gas overriding and fingering start to develop. An example of miscible gas drive with CO_2 under gravity-stabilised conditions is described in a work by Palmer et al.[67].

15.9.8 Miscible Gas Drive Technology

If the cost of the gas to be injected for miscible gas drive is high (as is the case with LPG, CO_2, or gas enriched with C_{2-6} to achieve miscibility), only a small "pad" will actually be injected, followed by less costly dry gas or nitrogen, and then, quite often, water.

Estimation of the minimum pad volume that will still maintain miscibility over the entire interwell distance is particularly difficult, and specialised numerical models[28,132] are necessary to simulate the displacement process. To give an idea of the orders of magnitude involved, the minimum volume of an LPG pad is approximately 5–10% of the pore volume (PV) of the region to be swept; for CO_2 or enriched gas, it is nearer to 25 –50% of the PV.

Gas has a tendency to advance preferentially along zones of high permeability because of its high mobility. As a consequence, the vertical invasion efficiency (E_I, Sect. 12.5.2) can be extremely low, with a correspondingly poor oil recovery factor. Two techniques are used to counteract this undesirable characteristic of gas drive.

The first method, WAG (water alternated with gas) involves injecting alternate pads of water and miscible gas[128]. The presence of water at $S_w > S_{iw}$ in the gas zone reduces the relative permeability to gas (Sect. 3.5.2.4), thereby decreasing its mobility. One drawback[49] with the WAG technique is that the water tends to segregate under gravity once it is in the formation, and the whole process is liable

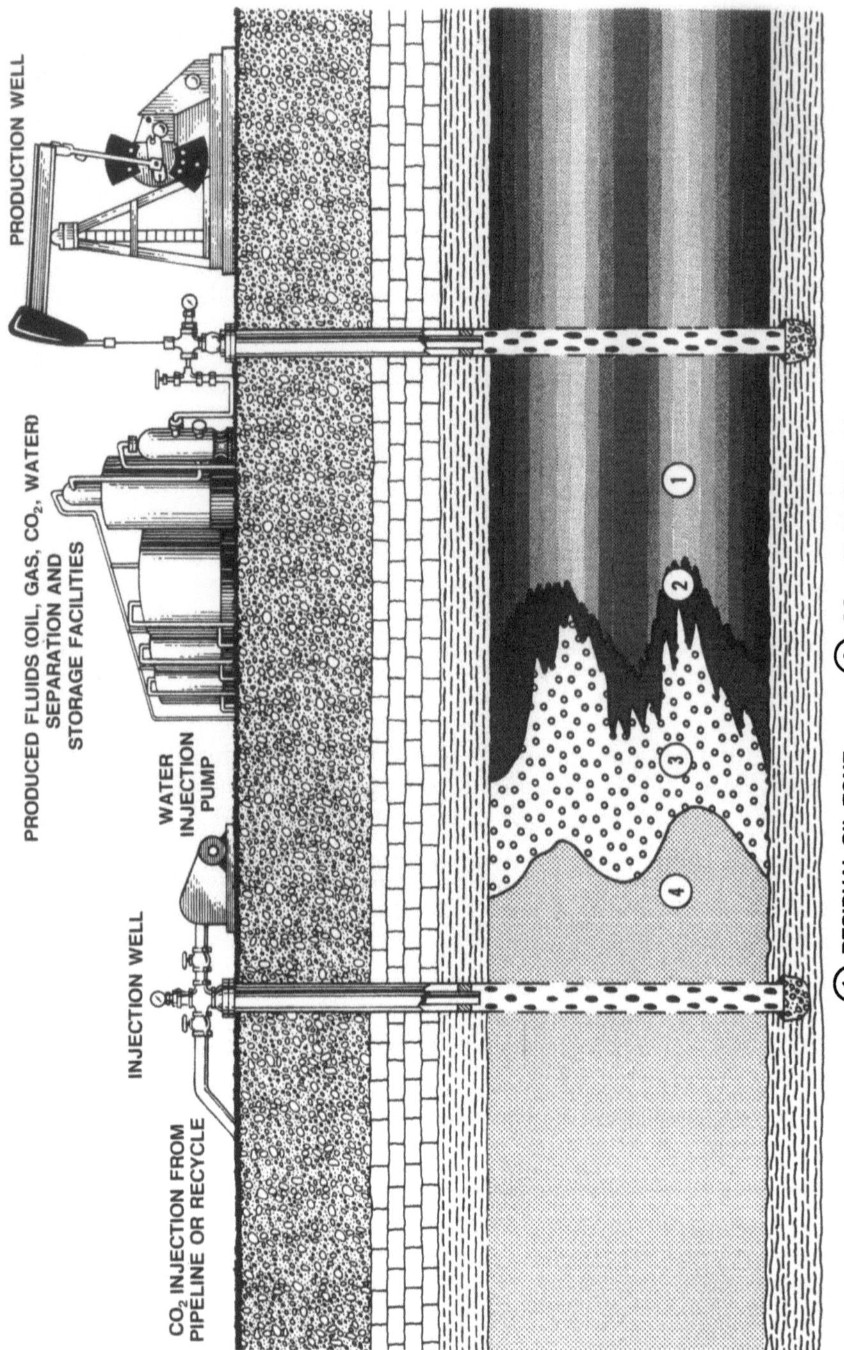

SOURCE: Adapted from original drawings by Joe R. Lindley, U.S. Department of Energy, Bartlesville Energy Technology Center.

Fig. 15.44. *Artist's impression of a miscible gas drive with carbon dioxide alternating with water (WAG). (From Ref. 5. Reprinted with permission of the National Petroleum Council and the US Department of Energy)*

to end up as a simple waterflood in the lower part of the reservoir, and a miscible gas drive in the upper part.

The second approach is the generation of foam[36,48] in the reservoir. These foams are non-Newtonian fluids, and possess a much higher viscosity (lower mobility) than the gas. The foam can be produced by two methods:

- small quantities of water and surfactant are injected simultaneously with the gas, to create a foam which remains stable over a sufficiently long period of time.
- special polymers (poly-αolephines, fluor derivatives of organic compounds of tin, etc.[48]), injected with the gas, can produce a foam even in the absence of water. This process has been under study for the particular case of miscible displacement with CO_2, and is currently being field-tested.

Figure 15.44 is an artist's impression[5] of the WAG miscible gas drive process.

Miscible gas drive with CO_2 entered a boom period in the USA in the early 1980s[134,146]. With a plentiful and cheap supply of the gas from natural sources, particularly in Colorado and Wyoming, and transport by interstate pipeline to the oil-producing areas of the southern states and the Pacific coast, there are at the present time 52 fields on CO_2 miscible drive.

The use of miscible drive with LPG or enriched gas has, on the other hand, declined to only 25 fields, because of the high costs involved. The produced CO_2 and intermediate hydrocarbons C_{2-6} are separated from the oil, processed as necessary, and then reinjected. This recycling reduces operating costs considerably.

There are a number of operational problems associated with the use of CO_2: in particular, it induces corrosion and, in the case of oils with a high asphaltene content, the precipitation of semi-solid asphalt deposits in the production tubing and surface lines. Corrosion is minimised by the use of inhibitors and special steels, while any buildup of asphalt is cleared periodically by flushing the affected pipework with hot solvents.

15.9.9 Immiscible Flooding and Huff'n'puff Using Carbon Dioxide

There are two drive processes which are based on the use of CO_2 at pressures *below* the MMP, which means that the gas is in *immiscible* phase. Although they do not qualify as EOR processes, it is worth mentioning them at this stage for completeness. They are:

- CO_2 immiscible flooding;
- CO_2 huff'n'puff.

Both processes were developed quite recently, as a result of the ready availability of CO_2 in Texas, Oklahoma, Louisiana and California (Sect. 15.9.8).

CO_2 immiscible flooding is conventional gas injection (usually crestal), but with the advantage that the dissolution of CO_2 in the oil reduces its viscosity (Fig. 15.39) and increases[76] its volume factor B_o (Fig. 15.45). Both these changes contribute to improved oil recovery, even though the process does not take place under miscible conditions.

CO_2 huff'n'puff, on the other hand, *is a means of stimulating well productivity*. It increases the rate of oil recovery, but does not improve the recovery factor itself. The method is the same as for steam soak (Sect. 15.7) with, of course, CO_2 in

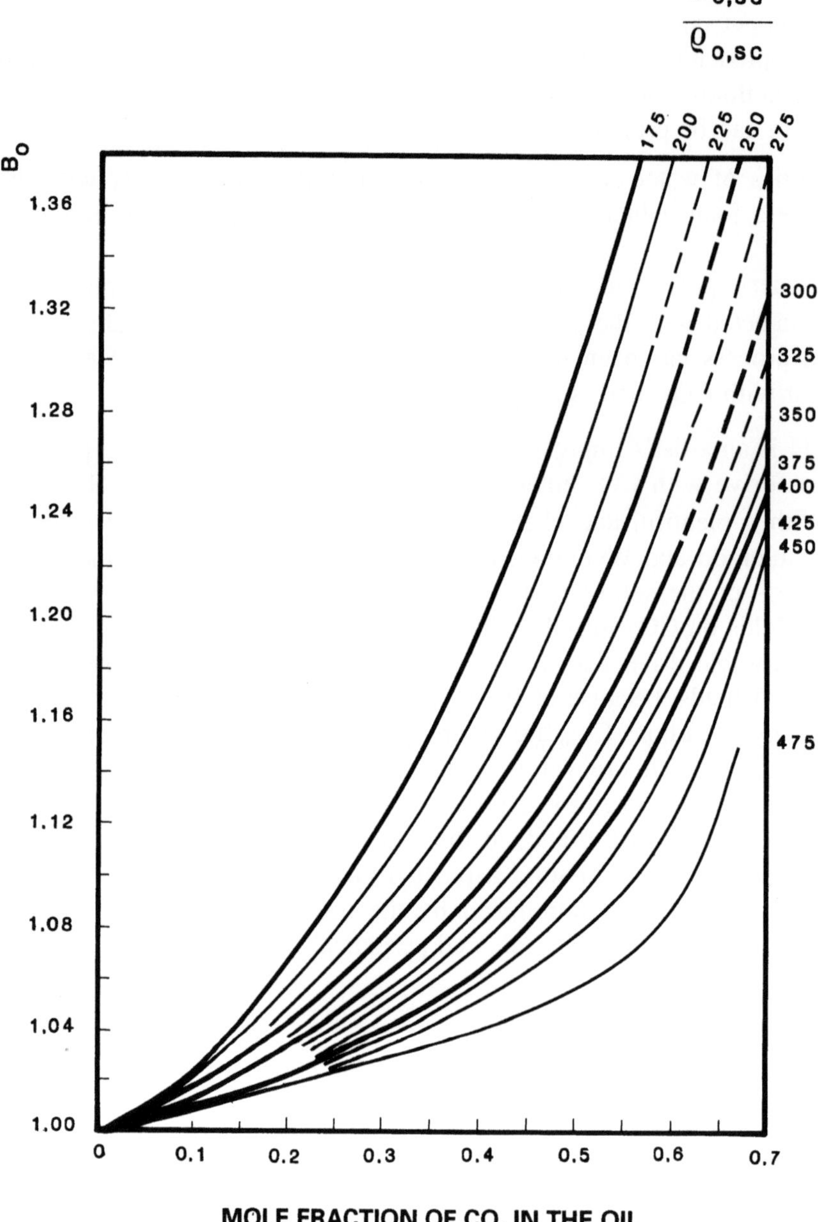

Fig. 15.45. *Increase in the volume factor of oil with the mole fraction of dissolved carbon dioxide. Each curve corresponds to a different ratio of oil molecular weight to oil density ($M_{o,sc}/\rho_{o,sc}$) under standard conditions. (From Ref. 76, 1965, Society of Petroleum Engineers of AIME. Reprinted with permission of the SPE)*

the place of steam. It is essential that the oil be undersaturated (Sect. 2.2) under reservoir conditions, and that the vertical permeability of the reservoir should not be too high: in particular, there must be no fracturing in the vicinity of the well.

After a period of CO_2 injection, the well is shut in to allow equilibrium to be reached between the oil and the CO_2. The injected CO_2 dissolves in the oil close to the well bore and reduces its viscosity μ_o, thereby increasing the productivity index J of the well (Sect. 5.8). The well is put back on production until most of the oil that has been in contact with the CO_2 has been produced, by which time J

will have deteriorated to close to its original value. The injection/shut-in/production cycle is then resumed.

High vertical permeability, or the presence of natural fractures, allows the CO_2 to migrate rapidly to the top of the reservoir without coming into any significant contact with the oil. In this case, little or no reduction in viscosity will occur. In addition, no improvement in productivity will be achieved if the reservoir oil is saturated, or very nearly saturated, because no CO_2 will be able to dissolve.

CO_2 huff'n'puff has the advantage that it can be used in deeper reservoirs than steam soak, but it is always less efficient. In fact, only the oil coming into direct contact with the CO_2 undergoes any improvement in its characteristics. Steam, on the other hand, heats up all of the rock around the well bore, regardless of permeability, and actual contact with the oil is not required.

15.9.10 Summary

Table 15.6 is a 1992 review[136] of the fields in the USA in which miscible gas drives have been used, showing the annual oil production for each type. The annual production figure represents 36.9% of the oil extracted by EOR methods in the USA, but it is still only 3.8% of the total US oil production. Outside the USA, two very large-scale miscible gas drives using natural gas rich in C_{2-6} have been in operation for 15–20 years in Hassi-Messaoud (Algeria) and the Intisar D field (Libya). Smaller-scale drives are under way in Canada, the CIS, Hungary and Abu Dhabi. The total oil produced by EOR using miscible gas drive (36.2 million m^3/year[120]) represents barely 1% of world production.

Drawing on experience gained from fields produced by miscible gas drive, it has been established that the following conditions are desirable for the process to be effective:

- reservoir rock   : any lithology, but not fractured;
- reservoir thickness   : thin, if the formation is not dipping;
                                   not important in dipping formations;
- angle of dip   : preferably steeper than 10°;
- porosity   : better than 20%;
- permeability   : not a critical parameter, provided it allows reasonable
                                   injectivity;
- initial oil saturation : at least 30%;
- reservoir depth   : > 600 m for LPG injection

Table 15.6. *Miscible gas drives in the USA in 1992*

Gas injected	Number of fields	Annual production (millions of m^3)
Carbon dioxide	52	6.56
Hydrocarbon	25	8.41
Nitrogen	7	1.31
TOTALS	84	16.28

> 1500 m for condensing gas drive
> 2500 m for vaporising gas drive and nitrogen injection
> 1000 m for CO_2 injection;

– oil API gravity   : greater than 30° API (density less than 825 kg/m^3);
– oil viscosity   : less than 10 mPa. s (cP) under reservoir conditions .

15.10 Polymer Flooding

15.10.1 Introduction

As seen in Sects. 11.3 and 12.5 reducing the ratio between the viscosities of the reservoir oil and the injected water (μ_o/μ_w) led to an increase in the oil recovery factor $E_{R,o}$ through two mechanisms:

– modification of the fractional flow curve $f_w(S_w)$ of the water [defined in Eq. (11.11)], with a gradual transition of the curve shape from concave downwards, through S-shaped, to concave upwards, as μ_o/μ_w decreases. This results in a lower WOR for a given percentage of oil recovered;

– an increase in the areal efficiency E_A (Sect. 12.5.1) and the vertical invasion efficiency E_I (Sect. 12.5.2) as μ_o/μ_w decreases, for a given well distribution and reservoir heterogeneity.

One way to reduce μ_o/μ_w, described in Sect. 15.6.1, is to increase the temperature of the reservoir. However, the same result can be achieved without altering the temperature, by increasing the water viscosity μ_w through the addition of suitable polymers. These polymer solutions, usually at concentrations of less than 1000–1500 ppm (1.0–1.5 kg/m^3 of water), are referred to as *"thickened water"*, and the associated drive process is called *polymer flooding*.

Moreover, when appropriate chemicals are added, these solutions have the property of *gelling* when injected into the reservoir. This produces a semi-permanent modification of the permeability profile – permeability is reduced in the most permeable zones, thereby reducing reservoir heterogeneity.

15.10.2 Polymers Used

There are two broad categories of polymer used in the preparation of thickened water:

– *synthetic polymers*: high molecular weight partially hydrolysed polyacrylamides (HPAMs); copolymers of acrylamide and 2-acrylamide-2 methyl propane sulphonate (AM/AMPS): only HPAMs are used in practice;
– *biopolymers*: xantham gums (polysaccharides), hydroxyethyl cellulose (CMHEC), glucan, guar gum: all of high molecular weight, obtained from the fermentation of natural substances rich in glucides.

The general structural formula of *polyacrylamides* is illustrated in Fig. 15.46A. They are treated with alkaline solution to hydrolise a portion of the amide groups:

$$- CONH_2 + NaOH \rightarrow -COO^- + Na^+ + NH_3. \tag{15.33}$$

This produces the partially hydrolysed polyacrylamide (Fig. 15.46B).

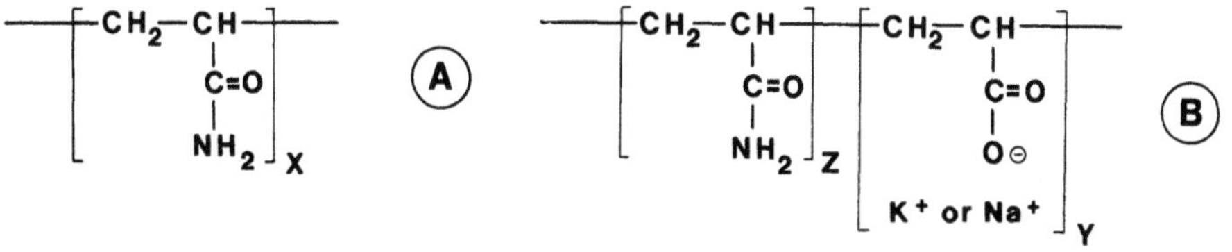

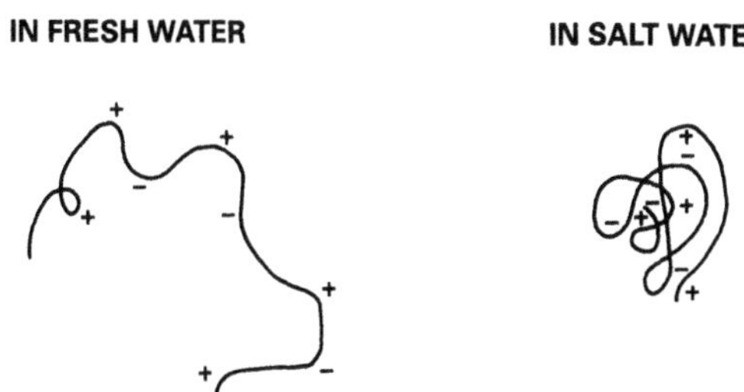

Fig. 15.46. *Structural formula of polyacrylamide or PAM (**A**) and partially hydrolysed polyacrylamide or HPAM (**B**)*

IN FRESH WATER   **IN SALT WATER**

Fig. 15.47. *Partially hydrolysed polyacrylamide (HPAM): "distended" form in fresh water, and "balled up" form in saline water*

The negative charge on the carboxyl group causes the molecules of polymer to adhere to the positive charges present on the rock grain surfaces. The resulting partial plugging of the pore channels reduces the permeability. In addition, the polymer solution is more viscous than water because the long polymer molecules cannot pass easily through the pore channels. In fact, the rheological behaviour of these polymer solutions is non-Newtonian, because their viscosity is a function of the velocity of flow (or, more precisely, the shear rate). This will be described in more detail in Sect. 15.10.3.

The action of polyacrylamides is therefore twofold:

- a permanent reduction of the rock permeability, which is due to the adsorption of the polymer molecules. The *"residual resistance factor"* is defined as the ratio of permeabilities to water after and before HPAM treatment.
- an increase in the viscosity of the solution relative to the pure solvent (water).

Both factors contribute to a reduction in the mobility ratio, M_{wo}, (Sect. 3.5.2.4) of the polymer solution displacing the oil. As more and more of the polymer is adsorbed onto the pore walls with the advance of the displacement front, the concentration of the solution and, with it, its viscosity, begin to decrease. Eventually, it is just as if pure water were being injected. This is one of the main limitations of the process.

A second drawback lies in the fact that in saline water, which is electrically conductive, the polyacrylamide molecules have a tendency to ball up on themselves (Fig. 15.47). As bundles, they pass more easily from pore to pore than the long molecular chains, and therefore offer less resistance to flow.

Biopolymers have a much more complex and rigid molecular structure (Fig. 15.48) than PAMs. They also carry a smaller electrical surface charge, and are therefore adsorbed to a lesser extent by the rock. Any permanent alteration to the permeability is therefore much less significant than with HPAMs, and they are able to travel further through the formation. Also, because of their more rigid molecular

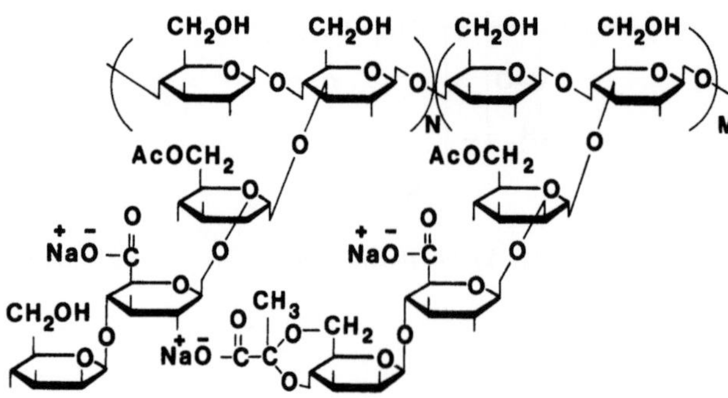

Fig. 15.48. *Structural formula of a xanthan molecule*

structure, they have less of a tendency to ball up in electrically conductive solutions. They therefore retain their viscosifying action even when dissolved in saline water.

Unfortunately, biopolymers provide ideal nutrition for many types of bacteria which, if present in the reservoir, would break up the molecules, and quite possibly destroy them completely. The viscosifying action of the biopolymers would therefore cease, or be seriously degraded, and the polymer drive would end up as a simple waterflood. Even HPAM polymers are prone to bacterial degradation under certain conditions, but to a much lesser extent. Therefore, the polymer solution is injected as a sequence of pads, each treated with a different biocide to prevent the bacteria from becoming resistant to any particular type.

Polymer degradation can also occur through the breaking up of the molecules by mechanical action as the solution passes at high velocity through restrictions in the surface equipment (filters, valves, etc.), through perforations and screens, if present, and the reservoir rock in the near-well bore region. This places an upper limit on the injection rates that can be used, and prolongs pumping time.

To improve their performance in saline water, and to reduce bacterial attack and mechanical degradation, numerous families of polymers – from totally synthetic to biopolymers copolymerised with synthetic molecules[59] – have been studied. However, they are generally still too costly to consider for field use.

15.10.3   Rheology of Polymer Solutions

15.10.3.1   Overview of Rheology

Two surfaces S_1 and S_2 shown in Fig. 15.49 are parallel, a distance d apart, and both of area S. A fluid whose rheological properties are to be determined is placed between them.

For simplicity, we assume that S_1 is fixed, while S_2 is made to move at a uniform velocity v parallel to S_1, by a force F acting in the direction of motion. The two surfaces remain parallel at all times. The fluid between the plates is subjected to a

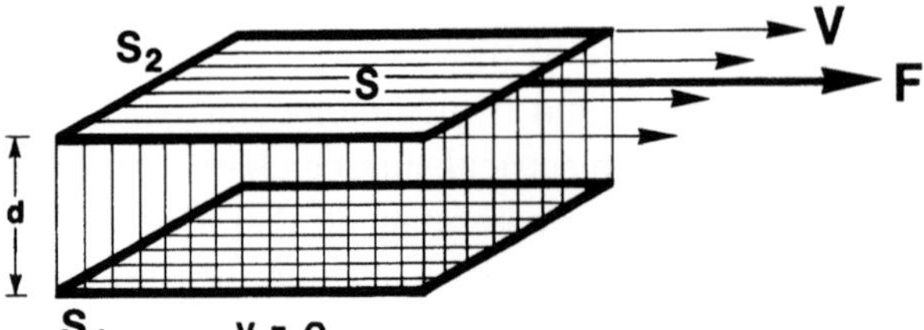

Fig. 15.49. *Geometry used to define the shear rate and shear stress*

shear stress. The shear stress per unit surface area is:

$$\tau = \frac{F}{S}.$$ (15.34a)

τ is in N/m^2 = Pascals (Pa).

The *shear rate* $\dot{\gamma}$ is the velocity gradient (assumed constant) across the fluid:

$$\dot{\gamma} = \frac{v}{d}.$$ (15.34b)

$\dot{\gamma}$ is in (m/s) $\times$ (1/m) = s^{-1}.

If $\dot{\gamma}$ is small enough for the molecular motion to correspond to laminar flow, the most general form of the relationship between τ and $\dot{\gamma}$ is:

$$\tau = a + m(\dot{\gamma})^n.$$ (15.35)

There are several special cases:

Newtonian fluids:

$$a = 0, \quad n = 1, \quad \text{so that:} \quad \tau = \mu\dot{\gamma}.$$ (15.36)

μ, the fluid viscosity, is independent of $\dot{\gamma}$ (Fig. 15.50). $\mu(= \tau/\dot{\gamma})$ is expressed in Pa·s.

Non-Newtonian fluids (all fluids which are not Newtonian):
  – *Bingham fluids*:

$$a \neq 0 \quad n = 1, \quad \text{so that:} \quad \tau = a + m\dot{\gamma}.$$ (15.37)

A minimum force per unit area is required to start the fluid moving. This is *thixotropy* and the Bingham type of fluid is said to be *thixotropic*.

  – *Power-law fluids*:

$$a = 0, \quad n \neq 1, \quad \text{so that:} \quad \tau = m(\dot{\gamma})^n.$$ (15.38)

The power-law fluid model was proposed by Ostwald, de Waele and Nuttig. There are two cases (Fig. 15.50):

if $n < 1$, the fluid is *pseudo − plastic*

if $n > 1$, the fluid is *dilatant*.

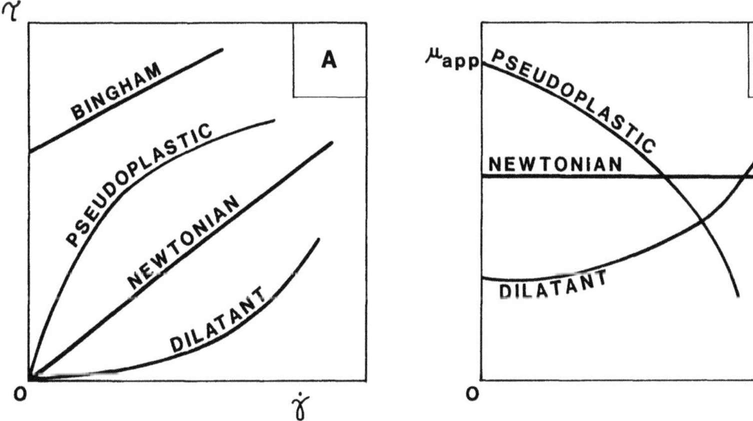

Fig. 15.50A,B. *Shear stress τ vs shear rate $\dot{\gamma}$ and apparent viscosity μ_{app} vs, shear rate $\dot{\gamma}$ for Newtonian and various types of non-Newtonian fluids (pseudoplastic, dilatant and Bingham)*

In both cases, the fluids have an *apparent viscosity* defined as:

$$\mu_{app} = \frac{\tau}{\dot{\gamma}} = m(\dot{\gamma})^{n-1} \tag{15.39}$$

Figure 15.50 illustrates the dependence of μ_{app} on the shear rate:

pseudoplastic fluids: μ_{app} decreases as $\dot{\gamma}$ increases (shear thinning)

dilatant fluids: μ_{app} increases as $\dot{\gamma}$ increases.

15.10.3.2 Polymer Solutions

Partially hydrolysed polyacrylamide (HPAM) and biopolymer solutions used in EOR are all *non-Newtonian fluids*, and behave as pseudo-plastics. Figure 15.51 is the rheogram[20] $\mu_{sol}/\mu_{H_2O} = f(\dot{\gamma})$ obtained with a rotational viscosimeter (no porous medium present) for a typical polyacrylamide at a concentration of 0.25 kg/m^3 in water at salinities between 0 and 20 kg/m^3. Note how μ_{sol}/μ_{H_2O} decreases as the shear rate $\dot{\gamma}$ rises, and how water salinity has a strong adverse effect on the

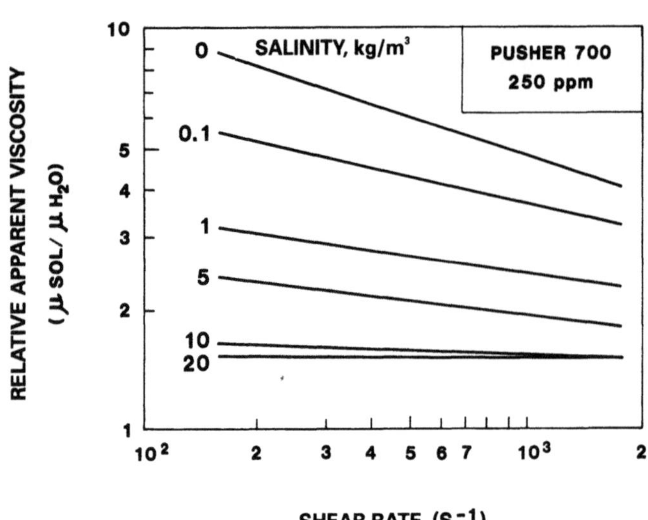

Fig. 15.51. *Rheograms for polyacrylamide solutions at a concentration of 250 ppm in different brines. (Ref. 20)*

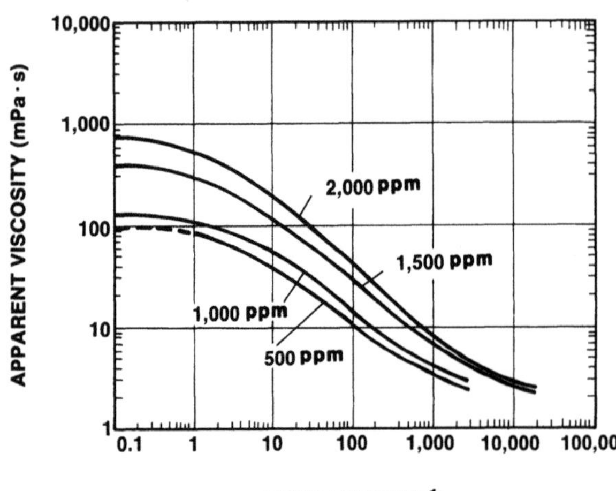

Fig. 15.52. *Rheograms for xanthan biopolymer solutions at concentrations between 500 and 2000 ppm in water with a salinity of 100 kg/m^3. (Ref. 14)*

viscosifying action of the polymer because of the tendency of its molecules to ball up in electrically conductive water.

Figure 15.52 is a rheogram[14] $\mu_{app} = f(\dot{\gamma})$, again from a rotational viscosimeter, for xanthan biopolymer solutions with concentrations between 0.5 and 2 kg/m^3 in water of about 100 kg/m^3 salinity. The salinity was made up of approximately 90% sodium chloride and 10% calcium and magnesium chlorides. In this case, the viscosity remains high, regardless of water salinity.

Rheograms of $\mu_{app} = f(\dot{\gamma})$ can also be derived from measurements of flow through porous media, in the form of cores. In this case the value of τ is deduced from the pressure gradient, and the shear velocity is calculated from the following equation[74]:

$$\dot{\gamma} = C \frac{u}{\sqrt{k/\phi}} \tag{15.40}$$

where u is the Darcy velocity of the polymer solution, $\sqrt{k/\phi}$ is the tortuosity term (Sect. 3.5.4), and C is a coefficient which can be derived by calibrating against rotational viscosimeter measurements.

The non-Newtonian behaviour of polymer solutions used in EOR complicates the calculation of pressure drops in the reservoir, especially in the near-well bore region where, because of the steep fluid velocity gradient, μ_{app} varies significantly with distance from the well axis. The same problem exists in the analysis of injection well tests (Sect. 6.10), particularly the pressure fall-off test when injection has been terminated.

15.10.4 Improvement of the Permeability Profile with Polymer Gels

Section 12.5 explained how reservoir rock heterogeneity, especially high permeability intercalations which act as thief zones by preferentially absorbing the displacing fluid, can seriously reduce the volumetric efficiency E_V, resulting in a lower percentage oil recovery. One of the techniques in use to establish a more uniform advance of the displacement front is to reduce the permeability of thief zones by making the polymer solution gel *in situ* after it has preferentially penetrated the high permeability sections. The semi-solid gel reduces the diameter of the pore channels, restricting flow, or preventing it completely.

Almost all polymer solutions used for EOR can be gelled in the reservoir. The following description applies to the polyacrylamides. In the presence of trivalent ions (Cr^{3+} and Al^{3+} in particular), a partially hydrolysed polyacrylamide will form stable compounds through ionic bonding[131] (Fig. 15.53). Their very rigid structures make them semi-solids, called *gels*. The gelling process should of course occur only in the formation, and should extend some distance from the well, so that as much as possible of the high permeability layer will subsequently be bypassed by the injected fluid. This is commonly achieved by creating the necessary reticulating ions in the formation itself. For this to happen, two substances must be added to the polymer solution: potassium dichromate ($Kr_2Cr_2O_7$ – a salt of *hexavalent* chromium), and a reducing agent such as thiourea or sodium bisulphite. Note that *the hexavalent chromium has no reticulating action on the polyacrylamide*.

The reduction of the Cr^{6+} to Cr^{3+} occurs through to the following reactions:

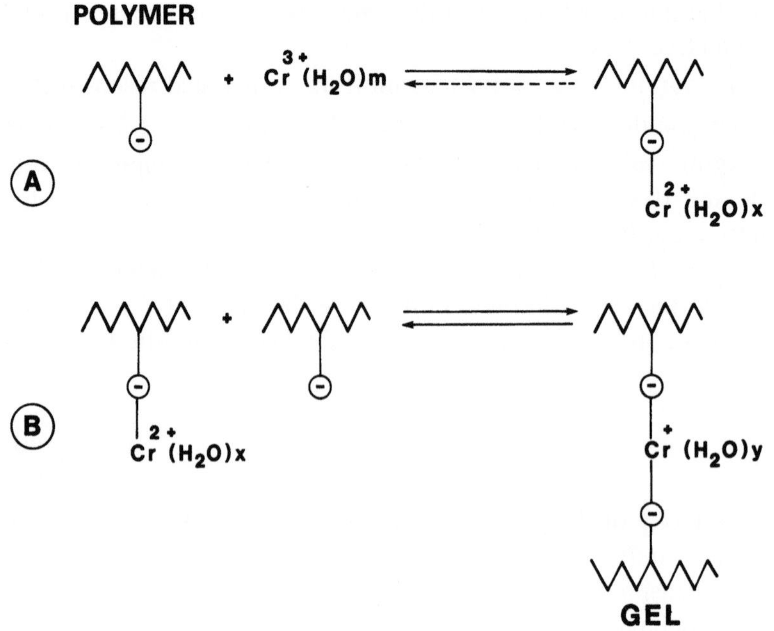

POLYMER

Fig. 15.53A,B. *Stages in the gelling of a polyacrylamide by the reticulating action of trivalent chromium ions*

With thiourea:

$$-Cr_2O_7^{2-} + 6 \left[H_2N - \overset{\overset{\displaystyle S}{\|}}{C} - NH_2 \right] + 8\ H^+ \longrightarrow 2\ Cr^{3+}$$

$$+ 3 \left[\begin{array}{c} HN \\ H_2N \end{array} \!\!\!> C - S = S - C <\!\!\! \begin{array}{c} NH \\ NH_2 \end{array} \right] + 7\ H_2O \qquad (15.41a)$$

With bisulphite:

$$-Cr_2O_7^{2-} + 3\ HSO_3^- + 5H^+ \longrightarrow 2\ Cr^{3+} + 3\ SO_4^{2-} + 4\ H_2O \quad (15.41b)$$

The kinetics of the reduction of the Cr^{6+} to Cr^{3+} and the reticulation of the polymer can be controlled by the pH, and concentrations of the polymer, dichromate and reducing agent. The gel can therefore be retarded to allow adequate penetration of the high permeability strata before it starts to set up.

Retardation can also be achieved by using as reticulating agent a compound of Cr^{3+} which releases the trivalent ion at a controlled rate[150]. HPAM gels will not form, or will quickly become unstable, at high temperatures or in high salinity environments. Special biopolymers do exist[148] that will form stable gels under these extreme conditions.

Plain water/gas surfactant foams are an alternative approach to combating the undesirable effects of high permeability strata. These foams, however, have a much shorter life than the reticulated polymer gels.

15.10.5 Polymer Technology

Polymers used for EOR are available in the form of powder or concentrated aqueous solution. The polymers used are quite fragile, and can easily be degraded

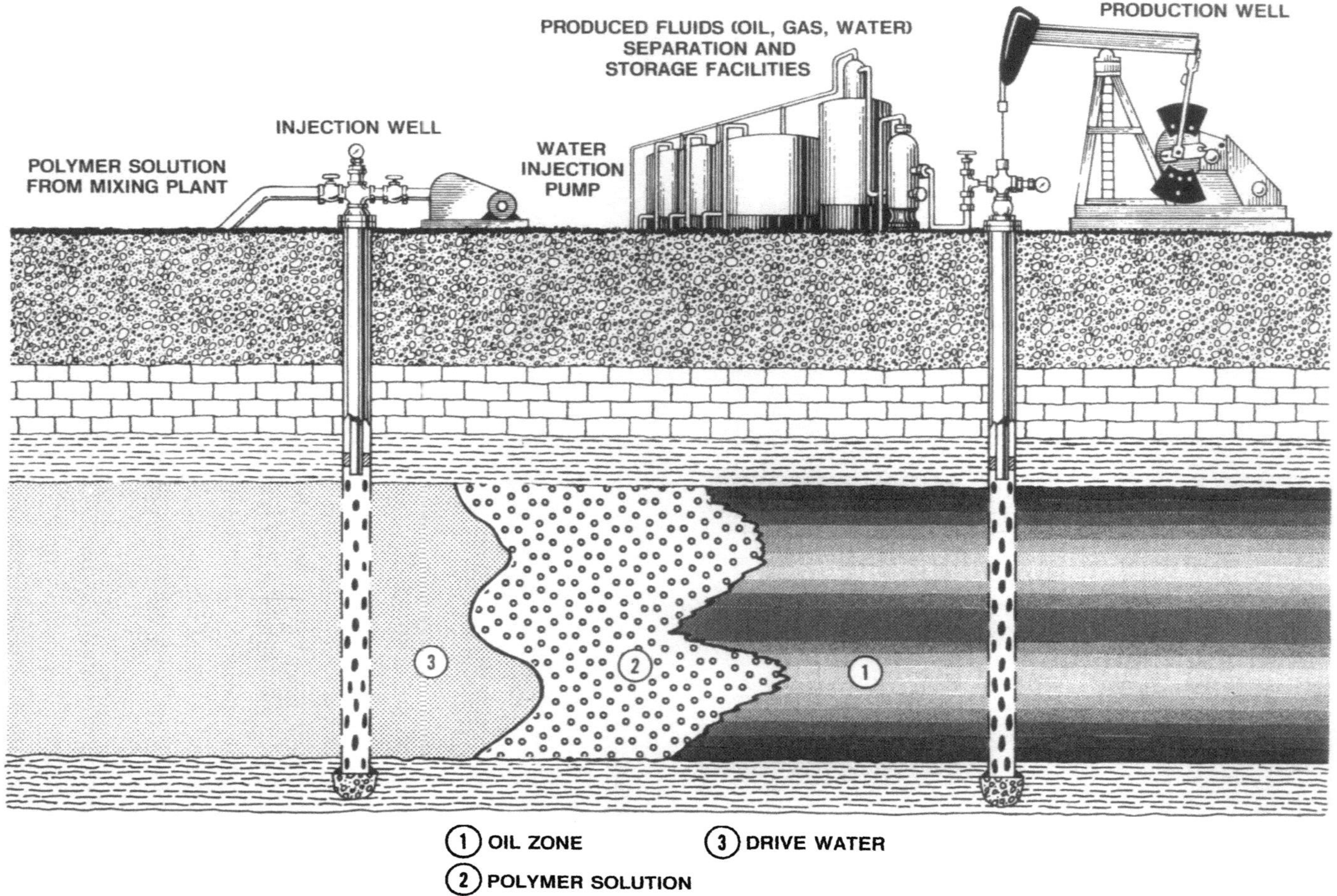

SOURCE: Adapted from original drawings by Joe R. Lindley, U.S. Department of Energy, Bartlesville Energy Technology Center.

Fig. 15.54. *Artist's impression of a polymer flood. (From Ref. 5. Reprinted with permission of the National Petroleum Council and the US Department of Energy)*

by mechanical action, which breaks up the molecules. For this reason, concentrated aqueous solutions[64] – output directly from the manufacturing process – are preferred, even though the transport costs are much higher than for the pure powder. The time and mixing velocity required for the pre-job preparation of the final polymer solution are greatly reduced, so there is less risk of mechanical degradation.

The solution then has to be filtered very carefully or, in the case of biopolymers, treated with special enzymes, so as to eliminate all solid particles in suspension. These would otherwise cause clogging when injected into the formation. Finally, successive doses of different biocides are added to combat microbial attack in the formation.

The polymer solution is usually injected in slugs of quite substantial size (0.5 to 1 times the pore volume of the region of interest), followed by neat "chase" water. Because the water is of higher mobility than the polymer solution, there is the risk of preferential advance and fingering (Sect. 11.3.6) of the water into the polymer bank. The last polymer slug is therefore usually followed by a "tail" whose concentration decreases from that of the polymer at the front end, to zero at the back. In this way, the viscosity is graded smoothly towards that of the chase water, avoiding any sharp mobility contrast. Figure 15.54 is an artist's impression[5] of the polymer flood process.

15.10.6 Summary

Despite its poor efficiency (on average, the quantity of extra oil recovered is only about 4% of the oil in place at the start of the polymer flood), displacement with "thickened water" has been widely used in the USA (especially in the period 1984–1988), particularly in reservoirs that are already on conventional waterflood.

The attraction of the method has been more fiscal than technical, however. The cost of converting from water injection to polymer injection (equipment costs for preparation of the solution, plus the polymer itself) is modest in itself, and, in addition, was more than offset by substantial tax concessions. This is because the injection of a polymer solution is classed as an EOR process by the US DOE (Sect. 15.5.1), and was eligible for a healthy reduction in taxation on the revenue from the produced oil. Other EOR processes tend to require a much greater investment in equipment and have higher running costs, and so do not offer the same obvious financial advantage.

In 1992, polymer flooding was used[136] in 49 fields in the CIS, 23 in the USA, 3 in Canada, 1 in France and 3 in Germany. The USA produced about 2000 bbl/day (115 000 m^3/year) of oil by polymer flooding in 1992, which amounts to some 0.25% of the oil produced by all EOR methods. The average production rate from fields under polymer flood was about 85 bbl/day, which is only 5000 m^3/year!

It is widely believed – although statistical evidence is quite scarce – that polymer flooding is best implemented during the early stages of a waterflood, switching from water to polymer solution shortly after the start. In this way, rather than having to correct the irregular frontal advance of the water that has already been injected (and having to produce the large quantities of water already in the reservoir), a uniform front is established from the outset.

Other favourable conditions are:

– reservoir rock   : preferably sand or sandstone, *not fractured*

$$\begin{aligned}
&- \text{permeability} && : \text{better than 50 md} \\
&- \text{initial oil saturation} && : \text{at least 10\%} \\
&- \text{reservoir temperature} && : \text{less than 80\,°C for polymer stability} \\
&- \text{oil viscosity} && : \text{less than 100 mPa·s (cP).}
\end{aligned}$$

15.11 Micellar-Polymer Flooding

15.11.1 Introduction

In Sect. 15.9 we saw how miscibility between injection gas and reservoir oil could be established through the formation, by a process of multiple contacts, of an intermediate buffer of liquid hydrocarbon which is miscible with oil at the leading edge and with gas at the trailing edge. One serious drawback to this method, however, is that the gas has a much higher mobility than the miscible buffer ahead of it (Sect. 15.9.8). Preferential advance and fingering of the gas can break up the whole displacement front, causing the process to degenerate into a simple immiscible gas drive.

The situation is further complicated by the tendency of the gas (being lighter than the oil) to migrate upwards and to move along the top of the formation, leading to an even more irregular frontal advance. In principle, this could be avoided by using a displacing fluid which was miscible with both water and oil (or had a very low interfacial tension with them; Fig. 3.26), was more viscous than both of them, and had a density lying somewhere between the two. Such a fluid would be ideal for EOR: on the one hand, it would displace all of the residual oil from the pore spaces ($E_D = 1$); on the other, it would achieve a very high volumetric efficiency E_V.

One process which comes close to this ideal, although it suffers from a number of technical difficulties, is *micellar-polymer flooding*. Firstly, a relatively small slug of *micellar solution* is injected. This is followed by a much larger buffer of *aqueous polymer solution* (described in Sect. 15.10), and then by neat chase water. Micellar solutions are extremely complex and the methodology has not yet been perfected.

The following sections are only intended to provide a basic introduction to the nature and behaviour of micellar solutions and their application to EOR. More detailed discussions are available in specialised literature[74,90].

15.11.2 Water/Oil/Surfactant Micellar Solutions

A "micellar solution" consists of very fine droplets of water dispersed in oil, or of oil in water, produced using suitable surfactants, with the addition, if necessary, of cosurfactants and solutions of various salts. The following are the most commonly used families of surfactants:

- polyethoxylate-carboxylates of aliphatic alcohols
- polyethoxylate-sulphonates of aliphatic alcohols
- alkyl phenol polyethoxylate-carboxylates
- alkyl phenol polyethoxylate-sulphonates
- alpha-olephine sulphonates
- sulphonates of hydrocarbon fractions (not used much these days because of their incompatibility with brine)

The molecules of these surfactants are characterised by a hydrophilic "head" (COO^- and SO_3^- groups with a strong affinity for water) and lipophilic "tail" (hydrocarbon chains, with or without benzene rings, with a strong affinity for oil). The surfactant molecules orient themselves at the water/oil interface with the head in the water phase and the tail in the oil phase, forming an interfacial film (together with any cosurfactant molecules) which encourages the dispersion of:

1. aggregates of water molecules in a *continuous oil phase* (Fig. 15.55A), or:
2. aggregates of hydrocarbon molecules in a *continuous water phase* (Fig. 15.55B).

These aggregates, or *"micelles"*, are between 20 and 100 Å (0.002 –0.01 μm) in diameter. To the eye, a micellar solution appears to be a transparent, homogeneous liquid.

Now, if we start out with a micellar solution where the continuous (or "external") phase is water, and we add some oil, the oil is progressively ingested into the oleic nuclei of the micelles, causing them to swell, reaching as much as 1000 Å (0.1 μm). At this stage, we have a (water-external) *microemulsion*. If we continue to add oil, the microemulsion becomes unstable, and will eventually

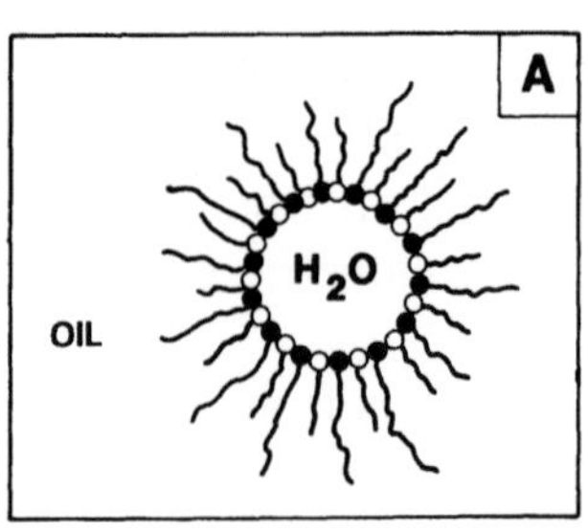

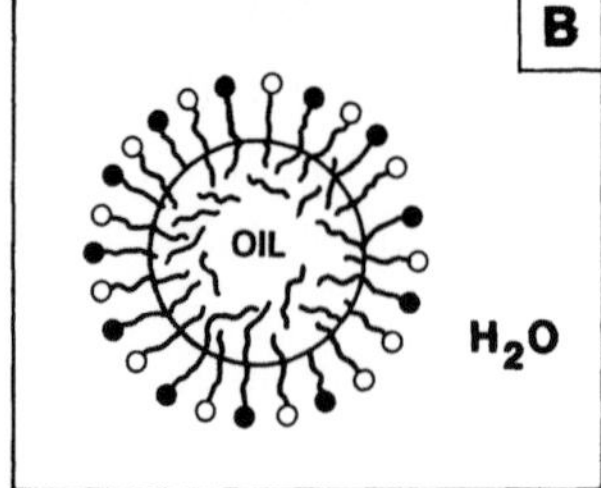

Fig. 15.55A,B. *Water-in-oil micelles (oil external phase) and oil-in-water micelles (water external phase) produced by the action of surfactants and cosurfactants at the interfaces. (From Ref. 74. Reprinted with permission of Academic Press Inc)*

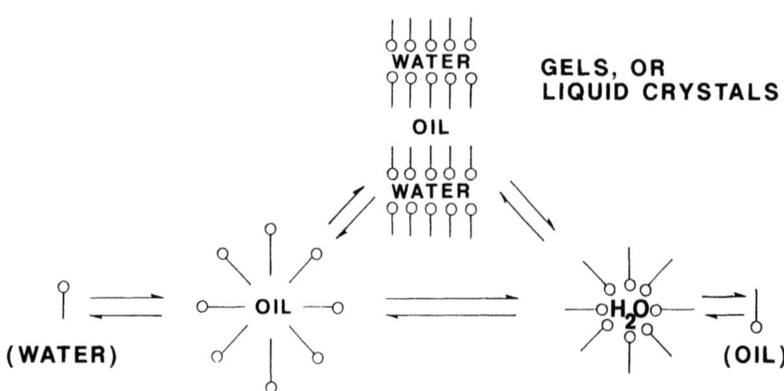

Fig. 15.56. *Transformation[90] of a micellar solution with an external water phase to one with an external oil phase, passing through intermediate gel and liquid crystal stages*

transform into a suspension of micelles containing water molecule nuclei, in a continuous external oil phase (Fig. 15.56). Sometimes an intermediate gel (or liquid crystal) stage is passed through between these two extremes.

The transformation is called *"phase inversion"*. The formation of intermediate gels or liquid crystals usually imparts an opalescence to the solution. Initially electrically conductive while the external phase was water, the solution has now become non-conductive because the external phase is oil. In fact, the moment of inversion can be identified by monitoring the electrical conductivity.

It is important to note that *phase inversion is reversible*. In other words, if water is added to an oil-external micellar solution, it will revert to a water-external micellar solution.

15.11.3 Phase Behaviour of Water/Oil/Surfactant + Cosurfactant Systems

The phase behaviour of a water/oil/surfactant + cosurfactant system is described by means of pseudo-ternary diagrams, using a technique similar to the one already described in Sect. 15.9.2 for injection gas/reservoir oil systems.

The three vertices of the equilateral triangle represent the three "pure" components: water (W), oil (O), and the surfactant plus cosurfactant mixture (T). Winsor[90] proposed a terminology in which a W/O/T system forming a micellar solution could be classed as one of three types: type $II^{(-)}$, type $II^{(+)}$ and type III.

15.11.3.1 Type $II^{(-)}$ Systems

The phase diagram for a type $II^{(-)}$ system is illustrated in Fig. 15.57a. All compositions corresponding to points in the region between the sides TW and TO and the *curve* WO are monophasic micellar solutions, and they are all miscible with one another. In some parts of this region gels or liquid crystals can exist.

Any mixture whose composition lies between the side WO and the curve WO will split into two phases whose compositions correspond to the two ends of the appropriate tie line, shown in Fig. 15.57a. One phase is almost pure oil (O), the other is a water-rich micellar solution. The reservoir fluid, consisting of oil and water, is represented by a point somewhere on the side WO of the triangle.

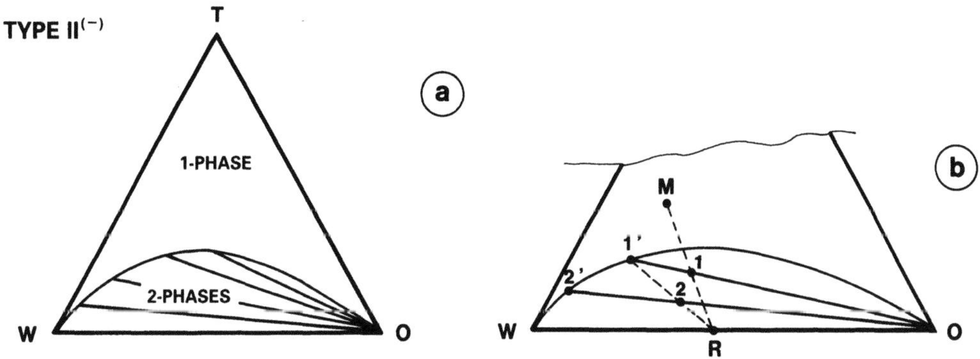

Fig. 15.57a,b. *Triangular diagram for a type $II^{(-)}$ mixture of water/oil/surfactant.* **a** *phase behaviour:* **b** *successive contacts between a microemulsion (M) and reservoir fluid (R)*

Suppose M is an injected micellar solution, and R is the reservoir fluid in a particular case (Fig. 15.57b). When the two come into contact, the mixture (point 1) will split into a microemulsion (1') and oil (O). At the contact between the microemulsion (1') and the reservoir fluid (R), the mixture (2) will split into new microemulsion (2') and oil (O). This process of successive contacts continues until it reaches the vertex W of the triangle. The system is called type II$^{(-)}$ because the tie lines have *negative* slopes in the biphasic region beneath curve WO.

15.11.3.2   Type II$^{(+)}$ Systems

The phase diagram for a type II$^{(+)}$ system is illustrated in Fig. 15.58a. This system is similar to the type II$^{(-)}$ in that all compositions lying within the area bounded by the sides TW and TO of the triangle and the *curve* WO exist as monophasic micellar solutions, miscible one with the other. Gels or liquid crystals also sometimes occur in parts of this region. Likewise, any mixture whose composition lies between the side WO and the curve WO will split into two phases whose compositions correspond to the two ends of the appropriate tie line, shown in Fig. 15.58a. However, in this case, one phase is almost pure water (W), the other is an oil-rich micellar solution.

In Fig. 15.58b, if M represents the composition of an injected micellar solution, and R is the reservoir fluid, then when the two come into contact the mixture (point 1) will split into a microemulsion (1'), and water (W) containing almost no surfactant. At the new contact between the microemulsion (1') and the reservoir fluid (R), the mixture (2) will split into a different microemulsion (2') and, again, almost pure water (W). This process of successive contacts continues until the vertex O of the triangle is reached. The system is called type II$^{(+)}$ because the tie lines have *positive* slopes in the biphasic region beneath curve WO.

15.11.3.3   Type III Systems

Figure 15.59a is the phase diagram for a type III system. All compositions corresponding to points above the curves WVO are monophasic micellar solutions. Gels or liquid crystals can exist in parts of this region. Any composition that lies in the left lobe, between the chord WV and the curve WV, behaves as a *biphasic* type

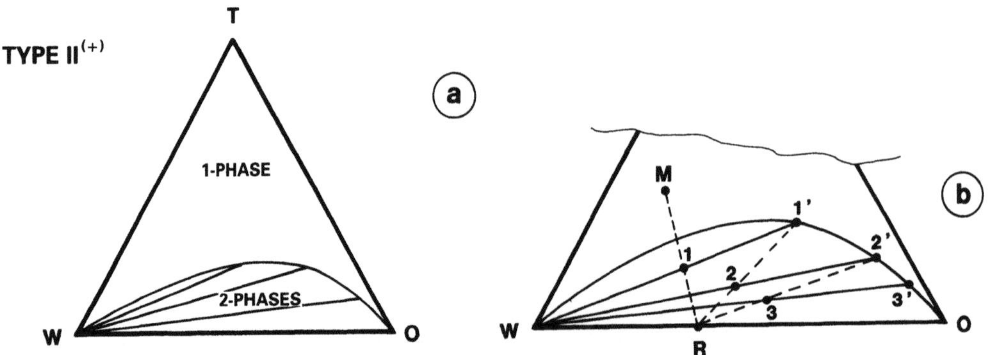

Fig. 15.58a,b. *Triangular diagram for a type II$^{(+)}$ mixture of water/oil/surfactant.* **a** *phase behaviour;* **b** *successive contacts between a microemulsion (M) and reservoir fluid (R)*

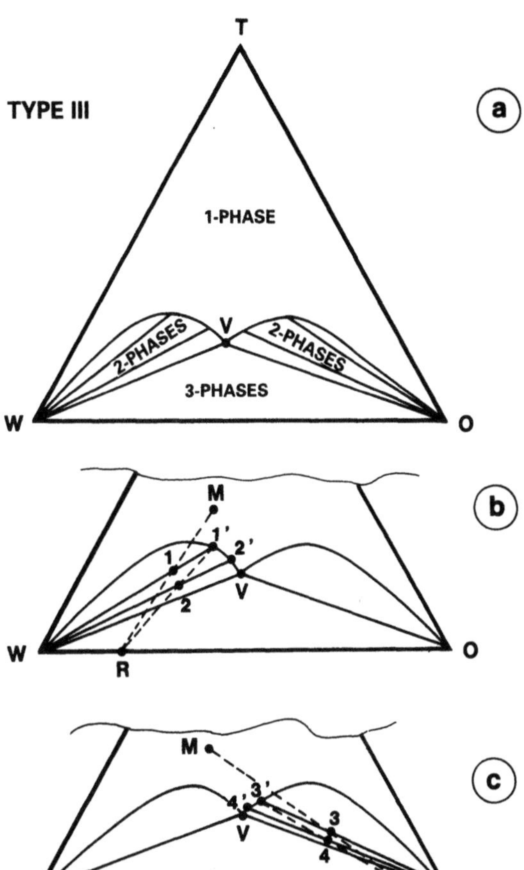

Fig. 15.59a-c. *Triangular diagram for a type III mixture of water/oil/surfactant.* **a** *phase behaviour;* **b** *successive contacts between water-rich reservoir fluid (R) and microemulsion (M);* **c** *successive contacts between oil-rich reservoir fluid (R) and microemulsion (M)*

$II^{(+)}$ system. It will split into two phases – *water* and microemulsion – as indicated by the tie line in Fig. 15.59. Any composition that lies in the right lobe, between the chord VO and the curve VO, behaves as a *biphasic* type $II^{(-)}$ system, splitting into *oil* and microemulsion as indicated by the tie line.

Finally, points lying in the triangle WVO exhibit *triphasic* behaviour, the three phases being reservoir water, reservoir oil (both almost free of surfactant), and a microemulsion of composition V, which we will call M_V.

As we shall see shortly, the interfacial tension σ_{VW} between water and M_V is the *lowest* of all the interfacial tensions between water and any of the type $II^{(+)}$ microemulsions that lie along the *curve* WV. Similarly, the interfacial tension σ_{VO} between oil and M_V is the *lowest* of all the interfacial tensions between oil and any of the type $II^{(-)}$ microemulsions that lie along the *curve* VO. From this it follows that M_V *is the microemulsion best suited to the miscible displacement of reservoir oil and water.*

An important aspect of the type III system, illustrated in Figs. 15.59b, c, is that whatever the relative proportions of water and oil in the reservoir fluid (R), and whatever the composition of the micellar solution (M) that has been injected, *the process of successive contacts always results in a microemulsion whose composition at equilibrium ends up at point V.*

Type III systems (so called because there is a three-phase region in the triangle) therefore achieve *the best microscopic displacement efficiency* E_D, the displacement of the oil by the microemulsion occurring when the interfacial tension is at a

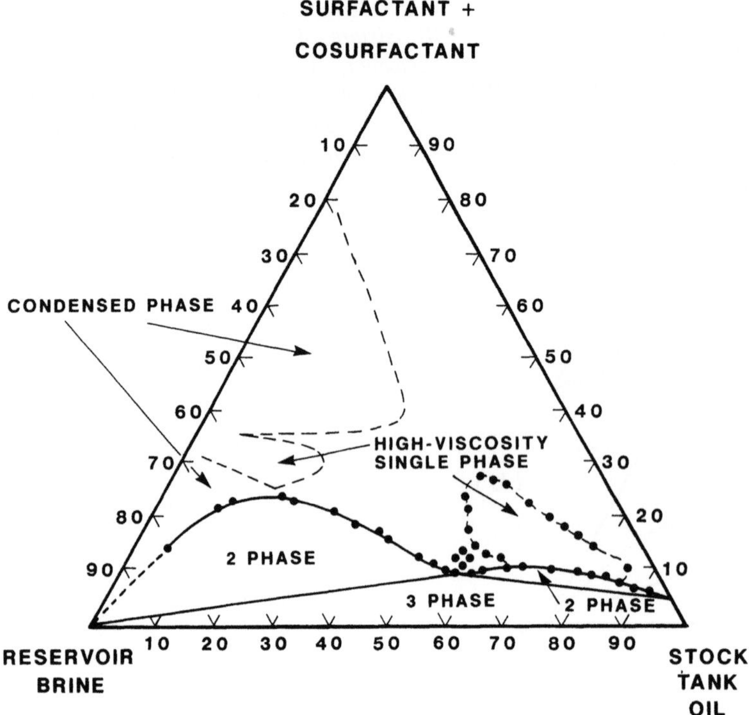

Fig. 15.60. *Triangular phase diagram from the Cortemaggiore field: reservoir oil / formation water / surfactant (an alkyl phenol polyethoxylate-carboxylate) + cosurfactant. (Ref. 14)*

minimum. The actual value is a function of the properties of the surfactant (+ cosurfactant) used, of the nature of the reservoir oil, and of the water salinity.

Figure 15.60 is the triangular diagram[14] of a type III system consisting of the reservoir oil from the Cortemaggiore field + water of 100 kg/m^3 salinity + an alkyl phenol polyethoxylate-carboxylate.

15.11.4 The Effect of Water Salinity on the Properties of the System

The phase behaviour of a water/oil/surfactant + cosurfactant system is very dependent on the salinity of the water (Fig. 15.61). The water may already be part of the system from the outset, or may be introduced into the system during the displacement process.

For a given oil and a given surfactant-cosurfactant mixture, an increase in water salinity will change the behaviour of a type II$^{(-)}$ system to type III and, eventually, to type II$^{(+)}$. This will happen in reverse if the salinity is decreased. This means that for any reservoir (characterised by type of oil and temperature) and any surfactant-cosurfactant mixture, there is an *optimum water phase salinity*, at which the system behaves as type III. Ideally, of course, this should coincide with the salinity of the formation water when the micellar buffer first arrives.

The optimum salinity can be estimated from laboratory tests conducted at reservoir temperature[127]. Samples of reservoir oil, with surfactants and cosurfactants already added, are mixed with equal volumes of brine, at increasing salinities. Once equilibrium has been established – which may take several days – the volumes of each phase (microemulsion, water and oil) present in each mixture are measured.

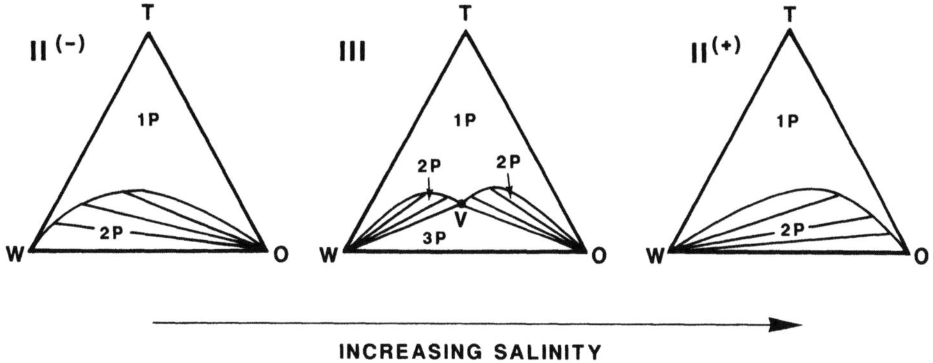

Fig. 15.61. *The effect of water salinity on the phase behaviour of a water/oil/surfactant system*

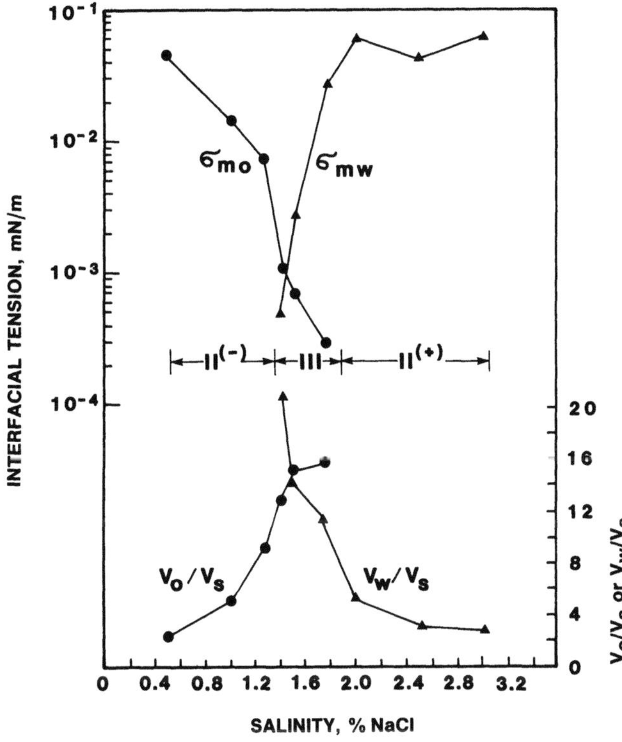

Fig. 15.62. *Influence of brine salinity on the proportions of oil/surfactant (V_o/V_s) and water/surfactant (V_w/V_s) in the microemulsion, and on the interfacial tensions between the microemulsion and oil (σ_{mo}) and water (σ_{wo}). (From Ref. 74. Reprinted with permission of Academic Press Inc., and R.D. Reed and R.N. Healy)*

Assuming that all the surfactant is in the microemulsion, the following quantities are calculated:

- oil solubility factor: V_o/V_s (15.42a)

- water solubility factor: V_w/V_s (15.42b)

where:

V_o = volume of oil in the microemulsion

V_w = volume of water in the microemulsion

V_s = volume of surfactant in the microemulsion.

Both V_o/V_s and V_w/V_s reach maximum values when the system becomes type III. Thus, it is possible to identify the *optimum brine salinity* for that particular oil/surfactant system and reservoir temperature (Fig. 15.62). Figure 15.62 also illustrates that the interfacial tensions σ_{mo} between the microemulsion and the oil, and σ_{mw} between the microemulsion and water, also reach minimum values at the optimum salinity, corresponding to type III behaviour. This relationship with salinity is one of several critical factors in the effective design of micellar solutions for EOR.

If the reservoir has already been produced by water injection, the salinity of the formation water will certainly vary from place to place. Now, even a small change in the salinity of water coming into contact with the microemulsion can radically change its behaviour[127], shifting it from type III to type II$^{(+)}$ or, worse, type II$^{(-)}$. This results in an increase in σ_{mo}, and a reduction in the microscopic displacement efficiency E_D which can be quite serious.

In principle, this can be alleviated by injecting a long preflush of brine of the desired salinity ahead of the micellar buffer, in the hope that this will sweep the entire volume of rock through which the buffer is subsequently expected to pass. This is often difficult to achieve in practice, because the water/oil mobility ratio is always larger than that of microemulsion/oil.

15.11.5   Adsorption by the Reservoir Rock (Surfactant Retention)

Since surfactant molecules are polarised, they will be adsorbed onto the reservoir rock through their interaction with the free charges which are always present on the grain surfaces and, more importantly, on the surfaces of clay particles and laminae. Furthermore, some surfactants (notably sulphonates of hydrocarbons, and those of synthetic origin containing sulphonic groups) combine with divalent Ca^{2+} and Mg^{2+} cations to form insoluble salts, and drop out from the displacement fluid. This precludes the use of these types of surfactant in carbonate reservoirs or where the formation water contains divalent ions.

The extent of the adsorption depends on the nature of the surfactant and of the rock. The bigger the specific surface area of the rock (which, to a first approximation means the smaller the grains), the greater the mass of surfactant adsorbed per unit volume of rock (the adsorption coefficient). Micellar flooding is therefore not recommended when the formation contains laminar or dispersed clays. The adsorption coefficient also depends on the manner in which the surfactant is present in the injected solution. It will be particularly high when the surfactant is simply dissolved or dispersed, but is usually lower if it is an integral part of the micelles, leaving the external phase almost surfactant-free.

A number of processes have been investigated to minimise the amount of adsorption: preflushing with low-cost chemical solutions which are strongly adsorbed; the use of low adsorption surfactants; special preparations which are able to detach surfactant that has already been adsorbed onto the rock, etc.

Micellar solutions which contain a mixture of surfactants which are intended to act synergetically are prone to the so-called chromatographic effect – the preferential adsorption of one or more of the surfactants from the mixture. This results in an altered phase diagram. In all these cases, the injected surfactants and cosurfactants, whether free or in micellar solution, will end up "trapped" by the rock. If the percentage of surfactant left in solution falls below a certain level, the whole process degenerates into a simple waterflood with a low surfactant concentration (equivalent to a high σ_{wo} immiscible drive), having little effect on the value of S_{or}. In order to ensure the continued effectiveness of the miscible displacement from the injection wells right through to the producers, it is necessary to increase the volumes of micellar solution used. The extra surfactant and cosurfactant requirement obviously incurs higher costs.

15.11.6 Micellar-Polymer Flooding Technology

Figure 15.63 is an artist's impression[5] of the micellar-polymer flooding process. The micellar solution itself is prepared in a separate unit (not shown). Locally produced oil is mixed with the required proportions of surfactant(s), cosurfactant(s) and water with optimal solids content. Where necessary, appropriate polymers are added (Sect. 15.10) to raise the viscosity of the micellar solution above that of the reservoir oil, so that the displacement can occur with a favourable mobility ratio. In some cases the micellar buffer is preceded by a large volume of water whose salinity has been optimised for the micellar solution being used (Sect. 15.11.4), and by a low-cost preflush which serves to reduce the adsorption of surfactant by the rock. Unfortunately, any improvement in conditions established by this treatment ahead of the micellar buffer will not be uniform because of the heterogeneity of the rock. The volume of micellar solution injected is usually equivalent to between 5 and 20% of the targeted pore volume. Numerical simulation can be employed to estimate the optimum buffer volume[68]. The micellar solution is followed by a long chaser of thickened water (Sect. 15.10), equivalent to some 50–80% of the pore volume, towards the end of which the polymer concentration is reduced steadily until only pure water is being injected (Sect. 15.10.5). From an economic point of view, micellar polymer flooding suffers from the fact that it requires a major front end investment (purchase of surfactant, drilling of in-fill wells), while the return, in the form of additional oil production, is spread out over a long period of time. In addition, most of the injection water remaining from the preceding waterflood (Chap. 12) has to be produced before any oil is recovered. It should be no surprise to learn that the discounted cash flow from micellar polymer flooding is often negative, even when the process itself is considered to have been successful technically.

15.11.7 Summary

Conceptually at least, micellar polymer flooding would appear to be a very promising process. By careful control of the viscosity of the injected fluid, it should be possible to attain favourable mobility ratios and achieve a high volumetric efficiency E_V. Because the front of micellar solution is miscible with the reservoir oil and water at its leading edge, and with the polymer solution at its trailing edge, it should in fact be feasible to displace all of the residual oil in the swept pore volume.

In practice, rock heterogeneity, adsorption, chemical interaction with formation water, and chromatographic effects acting on the injected fluids, have all contributed to the failure of many of the field tests conducted to date. And now even the successful floods have been rendered economic failures by the current (end 1994) low price of crude oil.

In the USA in 1992, micellar-polymer flooding was used[136] in just three fields, and was responsible for the production of only some 260 bbl/day (15 000 m^3/year).

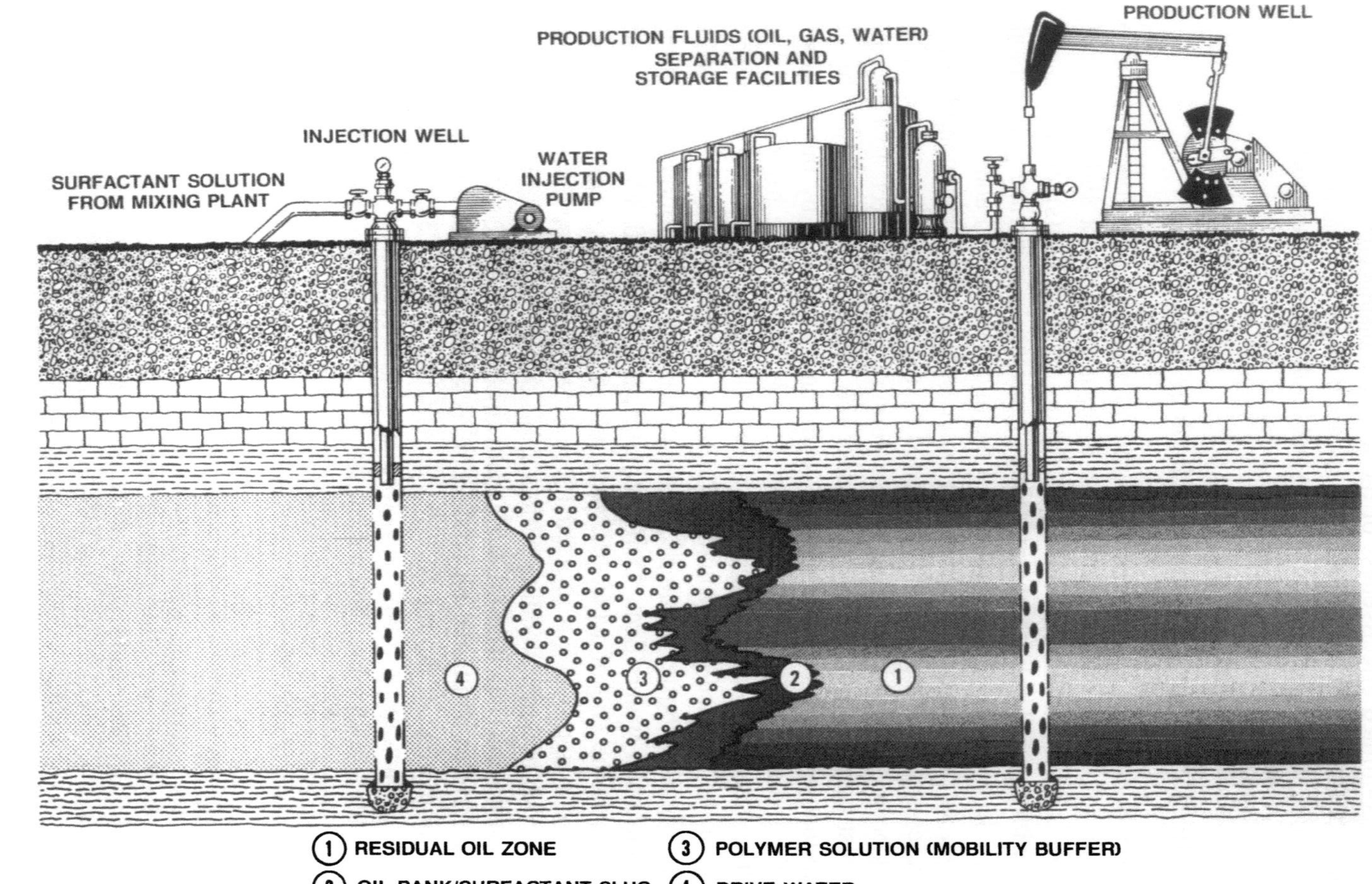

SOURCE: Adapted from original drawings by Joe R. Lindley, U.S. Department of Energy, Bartlesville Energy Technology Center.

Fig. 15.63. *Artist's impression of the displacement process in a micellar-polymer flood. (From Ref. 5 Reprinted with permission of the National Petroleum Council and the US Department of Energy)*

Still in the USA, the *highest* production rate ever achieved by this process was 2 800 bbl/day (164 000 m^3/year), from 21 fields[135] in 1984.

Outside the USA, in 1992 there was firm evidence of only one pilot EOR project using micellar solution flooding, in the Handil field in Indonesia. There are similar projects under way in the CIS[120], but here they tend to use very low concentrations of surfactant in water, which do *not* qualify the solutions as micellar.

Based on experience to date, the best conditions for micellar-polymer flooding would appear to be as follows:

    – reservoir rock   : *clay-free* sands or sandstones
    – layer thickness   : greater than 3 m
    – permeability   : better than 50 md
    – oil saturation at the
       start of the flood   : at least 30%
    – reservoir temperature : less than 90° C (necessary
                                 for surfactant stability)
    – oil viscosity   : less than 20 mPa·s (cP).

15.12   Caustic Flooding

15.12.1   Introduction

The first experimental injection of water containing alkali was based purely on empirical studies and took place in the Bradford area of Pennsylvania in 1925. The first patent on the process was taken out in 1927. Since then, the injection of alkaline solutions (*caustic* or *high-pH flooding*), has been tried with varying degrees of success in a number of reservoirs in the US – however, in 1992 there was not one caustic flood in operation.

15.12.2   The Interactions Behind the Process

Adding alkali to injection water can have a beneficial effect on the oil recovery factor only when the oil has a sufficiently high acidity. This acidity is due to the presence of molecules (usually resins and asphaltenes) containing acid radicals (mainly – COOH) formed by partial oxidisation of the hydrocarbons. The degree of acidity of a hydrocarbon is expressed as its *acid number* – the number of milligrams of caustic potash, KOH, needed to neutralise (to a pH = 7.0) one gram of oil. The acid number of the crude must be greater than 0.5 for it to be a candidate for caustic flooding.

Three mechanisms are involved in the action of the alkaline solution on the oil:

1. The alkali reacts with the acid radicals in the oil, forming molecules which are surfactants. This reduces the water/oil interfacial tension σ_{wo}.
   These surfactant molecules form at the interface between the oil and the caustic solution, but σ_{wo} will be reduced only if they diffuse into the aqueous phase. The interfacial tension varies with time – it decreases initially, then starts to increase when most of the surfactant has migrated into the aqueous phase.
2. The rock wettability is altered by the interaction between alkali and the polarised molecules adsorbed on the grain surfaces. Consequently, during the caustic

flood a variation in the surface characteristics of the rock is usually observed, tending towards increasing water-wetness.

3. A water/oil emulsion (*not* micro-emulsion) forms. In the case of heavy and viscous oils, these emulsions have a lower viscosity than the oil, which improves flow through the reservoir.

All three of these mechanisms contribute to an improvement in the displacement of oil by water, and increase the recovery factor. Unfortunately, there are a number of other mechanisms taking place at the same time which have a negative effect on the displacement process. The most significant of these involves a chemical and physical interaction between the alkali and certain minerals in the rock and formation water. The alkali reacts with divalent ions (Ca^{2+} and Mg^{2+}) to form barely soluble hydroxides:

$$2\,OH^- + Ca^{++} \rightarrow Ca(OH)_2 \tag{15.43}$$

and is therefore no longer available to react with the reservoir crude.

The increased pH also causes dispersed and laminar shales to swell to the extent that they may actually be carried in suspension in the water. Subsequent redeposition in the pore throats reduces the rock permeability. Interaction with the oil and, more importantly, with the rock, leads to a steady loss of alkali from solution, to a point where the concentration is too low to have any beneficial effect on the recovery factor.

15.12.3 Caustic Flooding Technology

Alkali solutions are usually prepared from sodium hydroxide (NaOH), a relatively cheap source, but ammonia, ammonium or sodium carbonates, and sodium orthosilicate are also suitable. (This last material precipitates calcium and magnesium salts present in the formation water). The concentration of alkali in the injection water is varied between 1 and 20 kg/m^3, depending on the nature of the oil and, in particular, the rock type and the injection water quality. Large volumes of caustic solution are injected – sometimes reaching several pore volumes.

Crudes with a suitably high acid number are usually heavy and viscous. Because of the resulting high water/oil mobility ratio M_{wo}, the volumetric efficiency E_V (Sect. 12.5) tends to be very low.

Theoretically, M_{wo} could be reduced by dissolving an appropriate polymer (Sect. 15.10) in the caustic solution to raise its viscosity. However, cost-effective polymers that can produce sufficient thickening while exhibiting long-term stability at high pH are still in the experimental phase. Currently, the caustic buffer is simply followed by a bank of polymer solution. The two are sometimes separated with a spacer of fresh water to impede degradation of the polymer. Figure 15.64 is an artist's impression[5] of a caustic flood.

15.12.4 Summary

Caustic flooding is an EOR process of minor importance. It is severely handicapped by the need for highly acidic crudes and the low sweep efficiency it achieves, with a high risk that it will degenerate into a conventional waterflood as the alkali is captured by the rock. Nevertheless, the investment required to set up the process

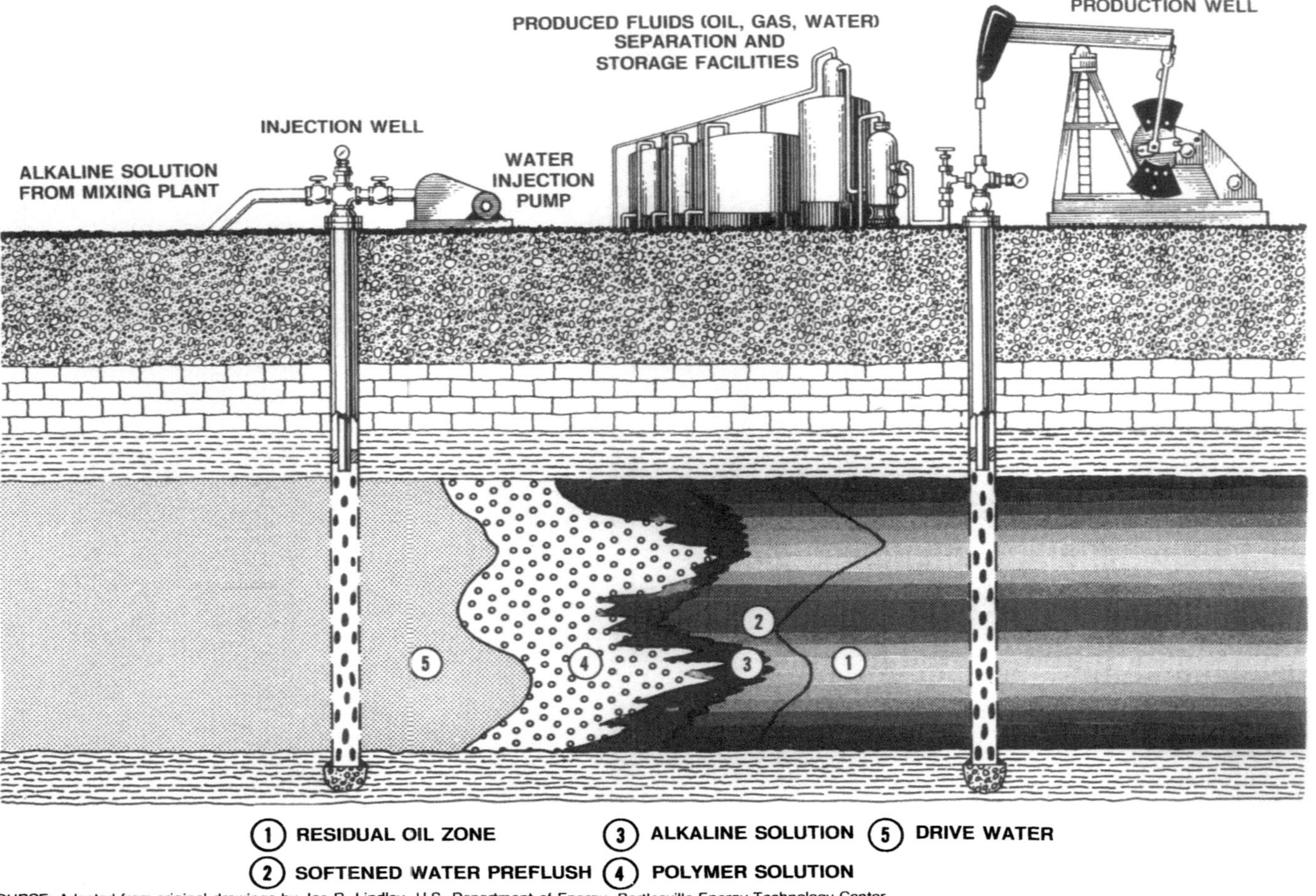

SOURCE: Adapted from original drawings by Joe R. Lindley, U.S. Department of Energy, Bartlesville Energy Technology Center.

Fig. 15.64. *Artist's impression of a caustic flood followed by polymer solution. (From Ref. 5. Reprinted with permission of the National Petroleum Council and the US Department of Energy)*

is fairly modest if water injection has already been implemented, and in the past it has enjoyed a degree of popularity in the US because it is eligible for the same fiscal incentives as polymer flooding (Sect. 15.10.6).

The most favourable conditions for caustic flooding are:

- reservoir rock   : clay-free sands or sandstones with no carbonate inclusions
- reservoir temperature   : less than 90° C
- oil density   : between 13 and 35° API (850–980 kg/m^3)
- oil viscosity   : less than 200 mPa·s (cP)
- acid number of crude   : more than 0.5.

15.13 Monitoring EOR Performance in the Reservoir

15.13.1 Classical Methods

The same classical techniques used extensively for water and immiscible gas floods to monitor the preferential direction of injected fluid flow and the movement of the displacement front (Chaps. 6, 8 and 12) can of course be employed for EOR monitoring. To recap, these techniques fall into the following categories:

1. *Nuclear logs* recorded in producing wells to determine if the displacement front has arrived, in both perforated and unperforated intervals:
   a) *for water-based displacement fluids*:
      - thermal decay time log: TDT, PDK100, TMD (Sect. 8.7.2)
      - induced gamma-ray spectrometry log: GST, RST, MSI-CO (Sect. 8.7.3; salinity indicator SIR, carbon/oxygen ratio COR)
      - chlorine log (Sect. 8.7.1)
   a) *for displacement by gas*:
      - combination density log (FDC) and compensated neutron log (CNL) (Sect. 8.6.2), or the recent Dual Burst TDT[154].
2. *Interwell testing* by pressure drawdown or buildup, especially:
      - interference tests (Sect. 6.11)
      - pulse tests (Sect. 6.12)
3. *Chemical and radioactive tracer tests*, especially "multicoloured" tests with a different tracer for each injection well (Sect. 12.7.2)

With thermal recovery (Sect. 15.6/8), temperature profiling of producing wells can provide an effective indication of the position of the thermal front, whether intervals are open to production in the well or not. Temperature logging in steam injection wells will show which parts of the pay section are taking steam. These are the well-established classical monitoring methods. There is on-going research into more sophisticated methods, and a number of experimental techniques are being field tested. EOR processes are generally costly to operate, and it is essential that the progress of the flood be carefully and accurately followed so that anomalous behaviour can be rectified promptly.

The new methods under development are evolving rapidly. The following sections will provide a broad outline of the most promising new techniques. Please refer to the list of references at the end of the chapter for more detailed reading.

15.13.2 Seismic Methods

The velocity V_p of compressional wave (P-wave) propagation in a porous medium is a function of lithology, porosity, fluid saturation and density. According to Wyllie's time-average approximation, this function can be expressed as follows[162]:

$$\frac{1}{V_p} = \frac{\phi S_w}{V_w} + \frac{\phi(1 - S_w)}{V_{nw}} + \frac{1 - \phi}{V_m}, \qquad (15.44a)$$

where V_w, V_{nw} and V_m are P-wave propagation velocities in formation water, non-water fluids saturating the pores, and rock matrix respectively. Possible non-water pore fluids are reservoir oil and/or gas, injected gas (hydrocarbon gas, carbon dioxide, nitrogen), steam, and in-situ combustion gas produced by fire flooding.

The propagation velocity of P-waves in gas is much smaller than in a liquid (oil and water in the case of a petroleum reservoir), so the presence of a gaseous phase in the pores results in an appreciable decrease in P-wave velocity. By contrast, shear wave (S-wave) velocity is hardly affected by the presence of gas in the pores.

The bulk density, ρ_b, of a porous rock depends on the nature and quantities of fluids saturating the pores in the following way:

$$\rho_b = \phi S_w \rho_w + \phi(1 - S_w)\rho_{nw} + (1 - \phi)\rho_m, \qquad (15.44b)$$

where ρ_{nw} is the density of the non-water pore fluid, and ρ_m is the density of the rock matrix. Gas density is always less than that of liquid, so the presence of a gaseous phase in the pores always results in a decrease in the bulk density of the porous medium relative to that of a fully liquid-saturated medium.

The fundamental characteristic that generates the reflection of seismic P-waves in the earth is a contrast in the acoustic impedance (a product of velocity and density) of the medium. Reflections, therefore, are generated at contacts between media of contrasting velocity, density, or both. The change in velocity and/or bulk density across the reflecting interface also modifies the amplitude of the reflected wave to a greater or lesser extent. For instance, a reflection of sizeable amplitude (a so-called bright-spot) appears below the top of an oil reservoir when a secondary gas cap is created by gas evolution from the oil, or by crestal gas injection. This allows the advance of gas/oil contact to be monitored.

Similarly, the injection of a hydrocarbon gas, carbon dioxide, or steam, into the oil column, or the accumulation in the reservoir of combustion gases from an in-situ combustion process, may result in the appearance of bright spots whose position and extension with time evidence the evolution of the displacement process.

Information about the progress of the IOR/EOR process in hand can therefore be obtained very straightforwardly by monitoring the position and lateral extent of bright spots through repeated seismic surveying. Considerably more detailed information can be extracted by processing seismic data using tomographic techniques[160]. The description of this technology is beyond the scope of this book, and only a basic discussion of the final product resulting from its application will be provided here.

Local seismic wave velocity and attenuation and interval travel times (time elapsed between the reflections at the top and bottom of a rock layer) can be computed at each point in the rock volume investigated by the seismic survey, by numerical processing of the seismic data (travel time and amplitude of each reflection) with a supercomputer. Coloured tomographic section maps of velocity and attenuation are automatically produced throughout the rock volume. Any regions showing simultaneously lower velocity and higher attenuation indicate probable invasion by gaseous fluid – hydrocarbon gas, carbon dioxide, steam, combustion gas – injected into, or produced within, the reservoir.

Before the improved/enhanced oil recovery project is started, a "base" seismic survey is run to establish the base velocity and attenuation maps for the reservoir. With each subsequent seismic survey, difference velocities[165] and attenuations[159] (the differences between base and current values) are mapped. These will allow tracking of the path of the displacing fluid, and identification of thief zones, unswept pockets of rock, impermeable layers, faults, etc.

Two different types of seismic survey can be run to monitor the evolution of an oil recovery process:

- 3D seismic surveys (fieldwide scale);
- cross-well seismic surveys (inter-well scale).

Their characteristics are described in the following pages, and some field cases presented.

15.13.2.1 Spatial Distribution of Fluid Saturations by 3D Seismic Surveys

Three-dimensional seismic reflection surveying is the most prominent technique in reservoir geophysics. For reservoir delineation and description, seismic data are used to map the structure, thickness, lateral extent and porosity distribution of reservoir rock layers. For monitoring reservoir exploitation processes, differential tomography techniques are used. Wave velocity and reflection amplitude characteristics are mapped from each 3D seismic survey run during exploitation, and the data processed by computer to obtain the differences relative to the base maps. Vertical and horizontal tomographic slice maps are then produced. Areas of reduced wave velocity and increased reflection amplitude appearing in the difference maps are indicative of regions of non-zero gas saturation – that is, regions which have been progressively swept by a gaseous phase.

Interpretation of these maps calls for the synergetic use of all additional information available – well logs, core analyses, well tests, and any general geological data. The interpretation is usually carried out by dedicated teams of geologists, petrophysicists, geophysicists and reservoir engineers. The technique of evaluating successive 3D seismic surveys run at different times throughout the life of the reservoir so as to monitor its performance (in particular, the movement of the displacement front) is referred to as "4D seismics" – the fourth dimension being time.

Because the earth is a dissipative medium, seismic resolution decreases with propagation distance, and therefore with reservoir depth. An approximate limit to spatial resolution is one-fourth of the signal's wavelength. For surface seismic data, the dominant wavelength is in the range of tens of metres, while for borehole seismic data it can be in the order of metres. The spatial detection limit of surface seismic data is a fraction of the resolution limit, and is estimated to be 1/30th the wavelength[115]. Typically, surface seismic data tend to have better vertical resolution (from 10 to 50 m) than horizontal (from 50 to 150 m).

Recent reports on the application of 3D seismic surveys in monitoring improved and enhanced oil recovery projects have been published by Pullen et al[70] (combustion project in the Athabasca tar sands), Matthews[165] (steam drive in Gregoire Lake, Alberta), Greaves and Fulp[159] (in-situ combustion project in the Holt field in Texas), and Johnstad et al[161] (pressure maintenance by miscible gas injection in the Oseberg field, Norwegian North Sea). In addition, a sequence of 3D seismic surveys has been used to monitor the irregular shape of the bottom water table as it rose during production of the huge Frigg gas field, again in the Norwegian North Sea. In the Oseberg and Frigg cases, the tomographies of gas front position versus time were cross-checked with the results of numerical field performance simulation. Reservoir parameter distribution was subsequently modified in order to obtain a satisfactory match.

15.13.2.2 Interwell Distribution of Fluid Saturations by Cross-Well Seismics

Cross-well seismic surveys are run by suspending a source of acoustic energy in the well bore, and simultaneously recording travel time and signal amplitude in a nearby well bore. By positioning the source and receivers at discrete stations in each well bore, the reservoir between the wells can be imaged. Tomographic maps of seismic velocity and reflection attenuation for the interwell region are then constructed from the data.

The source is usually lowered into one well, while a multilevel, three-component receiver string is kept stationary in the other well — sometimes by cementing it into place. If, for instance, a survey is to consist of 50 source stations, and the receiver string contains 40 geophones, then 2000 independent travel time and signal amplitude measurements will be acquired and processed to construct the tomographic map.

The interwell distances involved are usually less than the depth of the reservoir, so the path travelled by the seismic waves in cross-well seismics is much shorter than in surface seismics (where the path travelled is approximately twice the reservoir depth). Cross-well seismic signal attenuation is therefore significantly lower, and a higher dominant frequency can be recorded. As a consequence, detailed information about the vertical invasion efficiency E_I of the oil displacement process (Sect. 12.5.2) can be obtained.

As in 3D seismic tomography, a base cross-well survey is run before the oil recovery project is started, to establish the reference "blank" record. Subsequent surveys are run during the evolution of the project, and wave velocity and amplitude difference maps are produced. Their interpretation, made in conjunction with additional log and core data and numerical simulation of the recovery process, may show up possible irregularities in the path of the displacing fluid, pockets of trapped but potentially mobile oil, and other anomalies that impair the volumetric recovery efficiency.

Reports on the application of cross-well seismics to the monitoring of enhanced oil recovery projects have been published recently by Greaves et al[115], Gray and Lines[158] (a synthetic case), Justice et al[121,163] (a steam drive project), Lo et al[126] (a steam drive in the Midway Sunset field in California) and by Justice et al[164] [two projects: a WACO (water alternated with carbon dioxide) and single well cyclic steam injection].

15.13.3 Electrical and Electromagnetic Methods

A number of electrical and electromagnetic methods have been studied and pilot tested for reservoir monitoring purposes. They may provide information on resistivity variations, which could be indicative of changes in conductive fluid content[162]. Their use is still in the feasibility phase. If they prove to be applicable to the tomographic description of resistivity changes in the reservoir rock, their main use will be in mapping the spatial distribution of injected water in waterfloods[157].

The responses of electrical and electromagnetic methods to rock and fluid properties are very different from those of seismic methods, and they are inherently of lower resolution. In some reservoir monitoring situations, they could be used to complement seismic techniques. The low resolution of surface transmitter/surface receiver electrical surveys is significantly enhanced with cross-well or surface-to-borehole arrays. As with seismics, if repeated surveys are used to detect and map the area of fluid saturation change, borehole techniques may provide sufficient resolution to allow monitoring the reservoir fluid distribution.

DC (zero frequency) current is used in the electrical methods, while electromagnetic techniques use AC in two main frequency ranges: 0.1–4000 Hz for controlled source audio magnetic tellurics surveys (CSAMT), and radar frequencies of 2–200 MHz for high frequency electromagnetic methods (HFEM). The bulk of published reports on DC electrical methods can be found in technical literature from the erstwhile Soviet Union, long a leading proponent of these methods in reservoir description and monitoring[157].

15.13.3.1   DC Electrical Techniques

In DC electrical surveys, both the transmitters and receivers consist of electrodes planted directly in the earth or in a conductive fluid. A DC signal is fed into the earth through the transmitter electrodes, and the potential difference is measured between pairs of receiver electrodes.

Cross-well and borehole-to-surface surveys are both feasible. The penetration depth depends only on the geometry of the transmitter/receiver array. Increasing the transmitter/receiver separation increases the depth of investigation, but resolution deteriorates. To eliminate the shielding effect of the steel casing, it must be replaced by fibreglass or PVC liners with conductive fluid contact to the reservoir rock in the region of the borehole electrodes. While DC resistivity techniques are the simplest to describe mathematically and to model numerically, their implementation in the field has yielded mixed results. Their spatial resolution is poor and depends on the resistivity contrasts between the rock layers. Resolution is further impaired by polarisation and the severe attenuation suffered by the DC signal.

15.13.3.2   Electromagnetic (EM) Methods

15.13.3.2.1   Controlled Source Audio Magnetic Tellurics (CSAMT)

The CSAMT method is an electromagnetic geophysical technique in which a primary (or incident) electromagnetic field (EMF) is generated by an emitter consisting of a long wired dipole, earthed at both ends. The "controlled source" refers to the controlled frequencies in which the transmitter induces current in the earth, usually ranging from 2 to 2048 Hz.

The receivers consist of an ultrasensitive ferrite coil (maximum sensitivity 10^{-13} tesla) which measures the induced magnetic field. Because this is an inductive technique, it can be used through steel casing if the frequency employed is low enough[115] (less than 100 Hz). The oscillator used to control the transmitter is phase-locked to the data acquisition unit. Thus, the incident field itself can be excluded from the receiver signal by only making measurements which are not in phase with the transmitter.

In cross-well CSAMT surveys, a string of receivers is usually stationed at a number of depths in the "receiving" borehole, while the emitter is moved upwards at a constant velocity in the "emitting" borehole. This eliminates the noise that would otherwise result from the motion of these very sensitive receivers in the Earth's magnetic field.

So far, only a few examples have been published of the results obtained using the borehole-to-surface technique for reservoir characterisation[166], or for EOR monitoring[87,155]. To the author's knowledge, no example of a cross-well CSAMT exists in the petroleum literature.

15.13.3.2.2 *High-Frequency Electromagnetic (HFEM) Method*

The HFEM method uses radar frequencies in the range 2–200 MHz. Unfortunately, electromagnetic wave propagation at these high frequencies is severely attenuated in low-resistivity rocks: this limits HFEM applications to cross-well surveys where shorter distances are involved.

The propagation of high-frequency waves is governed by the wave equation, and the interpretation techniques are analogous to those used for seismic methods. The response depends on both the electrical conductivity (the inverse of resistivity) and the dielectric constant of the formation. The electrical conductivity is highly dependent on porosity, fluid conductivity and saturation, and reservoir temperature. The dielectric constant also varies considerably with formation properties.

The field equipment used in HFEM surveys consists of a transmitting antenna in one borehole that produces a signal which is detected in an adjacent borehole with a number of identical antennae. The wells must be either uncased or cased with a material that is transparent to radar frequencies (such as fibreglass or rein-forced PVC). With the transmitting antenna at some fixed depth, the attenuation of the signal is measured at a number of receiver stations in the adjacent well. The transmitter is then moved to a new position, and another set of receiver stations is recorded, and so on.

The transformation of the data into a meaningful picture of the displacement process between the two boreholes requires extensive processing. The final product is a tomographic presentation of the distribution of conductive fluid saturations in the interwell volume. Two examples of cross-well HFEM tomographs from steam floods in heavy oil sands have been presented by Witterhold et al[91] and by Greaves et al[115].

15.13.4 Summary

Of all the monitoring techniques described in this section, the most promising developments are likely to take place in seismic methods. While 4D seismics are already in field use, albeit on an experimental basis, we can expect to see significant improvements in cross-well seismic technology, both in the well-site equipment and in data processing. Electrical and electromagnetic methods would appear to have a minor chance of success, and their extensive field use is unlikely in the near future.

Table 15.7. *Applicability of the various surveying techniques to the monitoring of EOR processes and waterflooding. (From Ref. 56, 1987, Society of Petroleum Engineers of AIME. Reprinted with permission of the SPE.)*

Survey technique		EOR process							Water Injection
		Thermal		Miscible			Chemical		
		Steam	Combustion	CO_2	Hydro-carbon	Flue gas	Micellar/polymer	Polymer	
Transient pressure testing	Drawdown/buildup		Yes	Yes					Yes
	Interference	Yes	Yes						Yes
Tracer	Chemical	Yes	Yes				Yes	Yes	Yes
	Radio-active	Yes	Yes				Yes	Yes	Yes
Seismic	3D seismics	Yes	Yes		No		No	No	No
	Microseismics		Yes	No	No	No	No	No	
Electromagnetic	CSAMT	Yes	Yes		No			Yes	Yes
	HFEM	Yes	Yes		No			Yes	Yes

Note: No rating has been entered where the applicability of a technique to an EOR process is marginal, or is as yet unproven in the field

A new technique that has potential is that of "passive seismics". There is no transmitter: instead, three-component geophones suspended in the wellbore record *microseismic* signals produced in the reservoir when fluid in injected.

Thermal recovery methods commonly produce microseismic disturbances through the dilatation of the rock in the region invaded by steam or, more importantly, at the combustion front. The location of the area in which the epicentres originate can, in principle, provide an indication of the movement of the displacement front with time. This has so far only been tried in shallow reservoirs (less than 300 m), but results have been good[65]. It is hoped that this same method will also be able to track the movement of a water injection front, on the assumption that if the injected water is a lot cooler than the reservoir it will provoke microseismics by causing local contractions of the rock.

To conclude, Table 15.7 (from work by Leighton and Wayland[56] published in 1987) summarises the applicability of the various monitoring techniques described in this section to each EOR process, as well as to water injection. For each process, techniques that are only marginally successful, or need further field testing, are not rated. In 1987, cross-well seismics were still very much in the embryonic phase and have not been included in the table.

15.14   Selection of the Appropriate EOR Process for a Given Field

The preceding sections provided summaries of the optimal conditions for each EOR process. Figure 15.65 is a graphic presentation of the areas of applicability of the different processes in terms of API stock tank oil gravity and reservoir pressure[35].

EOR is not a universal solution that will benefit all reservoirs. Each situation must be evaluated thoroughly, giving due consideration to rock and fluid characteristics, the primary drive mechanism (and secondary if relevant), the local availability of injection fluids, etc. and, of course, the economic viability of any process that might appear suitable. There are fields where there is no chance of economic success

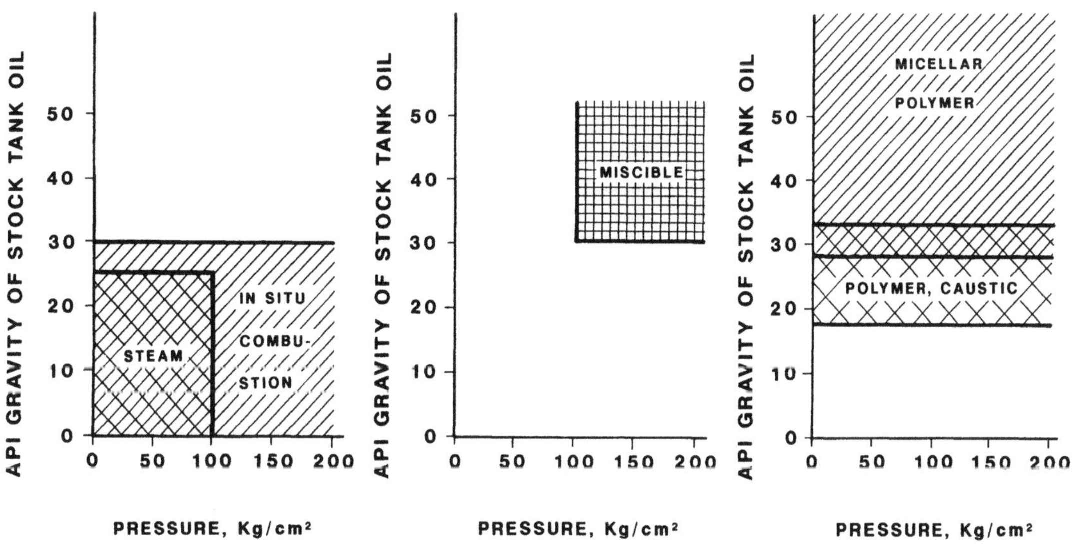

Fig. 15.65. *Applicability charts for EOR processes. (Ref. 35)*

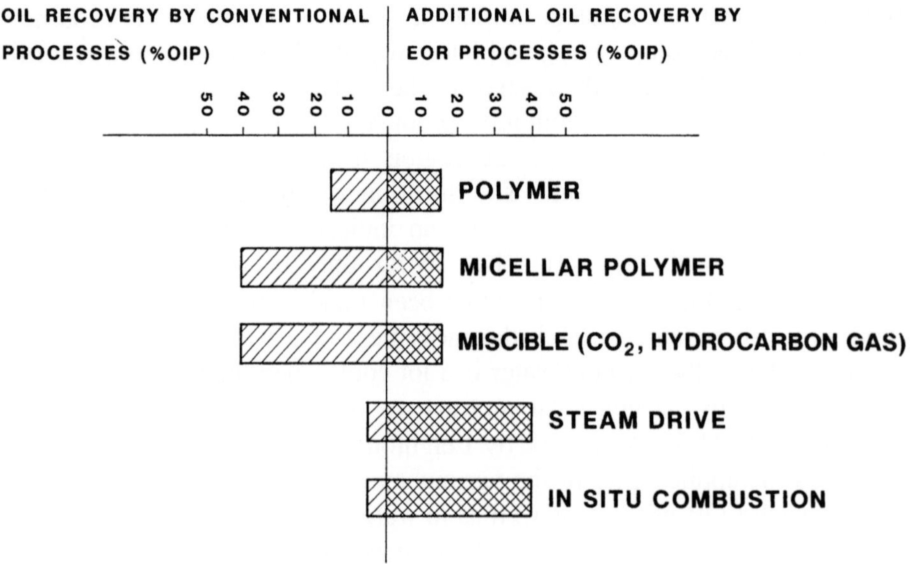

Fig. 15.66. *Gross estimates of the additional percentage oil recovery attainable by EOR, compared with recovery from primary production plus water injection, in reservoirs where each process is applicable. (Ref. 35)*

for any form of EOR, even though one or more processes might be operationally feasible.

Figure 15.66 compares[35] gross estimates of the additional percentage recovery of oil that could be obtained from each of the EOR processes, and the percentage recovery from primary production (plus water injection where applicable).

The economic success of any EOR project depends on the investment required (preparatory studies, drilling additional wells, surface facilities, etc.) and the subsequent operating expenses. Some approximate figures for the costs involved are provided in Table 15.8.

A comparison (Table 15.8) of the capital investment and operating expenses incurred for EOR processes with those of waterflooding and conventional methods[60] shows immediately just why EOR is so little used throughout the world. Note also that some 66% of the world's oil reserves are located in the Middle East[99], and this is precisely where *conventional* recovery methods are most cost-efficient.

In any case, no EOR process should ever be implemented in a field without first going through a rigorous project evaluation, which should include experimental data and numerical simulation. Below is a short list of the main components of such a preliminary evaluation:

1. Establish by employment of the various criteria for applicability, and by comparison with similar reservoirs on EOR, those recovery processes that would be suited to the reservoir in question.
2. Use numerical simulation to forecast the behaviour of the reservoir under each of the EOR processes selected in (1). This should incorporate the results of *ad hoc* fluid dynamic and thermodynamic measurements carried out on cores and samples of reservoir fluid.
3. Assess the capital investment and operating expenses associated with each method studied in (2), and the discounted profit.

Table 15.8. *Capital investment and operating expenses (1988 dollars). (Ref. 60)*

Recovery process	Capital investment	Operating expenses
	$ per bbl/day of oil produced	$/bbl
EOR		
– thermal recovery	8000–25000	15–25
– miscible flood	10000–25000	10–30+
– chemical flood		
– polymer	10000–30000	10–20
– micellar/polymer	15000–30000	20–50+
Water injection	1000–5000	2–10
Conventional production methods		
– Middle East	500–3000	1
– non-OPEC countries	3000–12000	8
WORLD AVERAGE	4000–8000	5

4. From the conclusions drawn in (2) and (3), select the process best suited to the current conditions.

5. Set up a small-scale *pilot* version of the selected EOR method in the field, to be run long enough for the response of the reservoir to be observed. Information on reservoir behaviour and the performance of the surface equipment can then be reviewed.

6. As the pilot project progresses, the numerical modelling (2) and the economic evaluation (3) can be validated and revised as necessary. The validated numerical model can then be used to forecast the behaviour of the entire field under full operating conditions, and the economic model to forecast the commercial prospects.

15.15 The Future of EOR

It is difficult – if not impossible – to make predictions about future developments in the use of EOR, at a time (end 1994) when there is an over supply of oil and the oil price has fallen to between \$16 and \$18/bbl. The rapid increase in the use of EOR that was predicted 10–12 years ago has failed spectacularly to materialise.

It would be interesting, nonetheless, to look at how EOR has evolved in the USA between 1971 and 1992. The need for enhanced recovery methods is perhaps felt more strongly in the US than in any other oil-producing country, given the very large number of fields that are by now approaching depletion, and the declining oil production associated with this. The basic data for this review are to be found in Tables 15.9, 15.10 and 15.11. The term "unconventional steam" in Table 15.11 refers to steam floods in reservoirs deeper than 750 m, or containing oil heavier than 16° API.

The most important conclusions that can be drawn from these tables are summarised as follows:

Table 15.9. *Active US EOR projects, years 1971–1992. (Refs. 97 and 135)*

	Number of projects										
	1971	1974	1976	1978	1980	1982	1984	1986	1988	1990	1992
Thermal											
Steam	53	64	85	99	133	118	133	181	133	137	117
In-situ combustion	38	19	21	16	17	21	18	17	9	8	8
TOTAL THERMAL	91	83	106	115	150	139	151	198	142	145	125
Chemical											
Micellar/polymer	5	7	13	22	14	20	21	20	9	5	3
Polymer	14	9	14	21	22	55	106	178	111	42	23
Caustic/alkaline		2	1	3	6	10	11	8	4	2	
TOTAL CHEMICAL	19	18	28	46	42	85	138	206	124	49	26
Gas											
Hydrocarbon miscible/immiscible	21	12	15	15	9	12	16	26	22	23	25
CO_2 miscible	1	6	9	14	17	28	40	38	49	52	52
CO_2 immiscible						1	18	28	8	4	2
Nitrogen miscible/immiscible					1	4	7	9	9	9	7
Flue gas miscible/immiscible					3	3	3	3	2	3	2
TOTAL GAS	22	18	24	29	30	48	84	104	90	91	88
GRAND TOTAL	132	119	158	190	222	272	373	508	356	285	239

Table 15.10. *US EOR production, years 1980–1992. (Refs. 97 and 135)*

	Barrels of oil per day						
	1980	1982	1984	1986	1988	1990	1992
Thermal							
Steam	243 477	288 396	358 115	468 692	455 484	444 137	454 009
In-situ Combustion	12 133	10 228	6 445	10 272	6 525	6 090	4 702
TOTAL THERMAL	255 610	298 624	364 560	478 964	462 009	450 227	458 711
Chemical							
Micellar/polymer	930	902	2 832	1 403	1 509	617	254
Polymer	924	2 927	10 232	15 313	20 992	11 219	1 940
Caustic/alkaline	550	580	334	185			
TOTAL CHEMICAL	2 404	4 409	13 398	16 901	22 501	11 836	2 194
Gas							
Hydrocarbon miscible/immiscible	15 448	12 615	14 439	33 767	25 935	55 386	113 072
CO_2 miscible	21 532	21 953	31 300	28 440	64 192	95 591	144 973
CO_2 immiscible		490	702	1 349	420	95	95
Nitrogen miscible/immiscible	2 027	1 400	7 170	18 510	19 050	22 260	22 580
Flue gas miscible/immiscible	35 200	35 200	29 400	26 150	21 400	17 300	11 000
TOTAL GAS	74 207	71 658	83 011	108 216	130 997	190 632	291 720
GRAND TOTAL	332 221	374 691	460 969	604 081	615 507	652 695	752 625

Table 15.11. *USA: EOR project starts, by process, vs average wellhead oil price, years 1980–1989. (Ref. 139)*

	Project starts per year									
	1980	1981	1982	1983	1984	1985	1986	1987	1988	1989
Thermal										
Steam (conventional)	19	37	24	21	22	19	21	8	2	12
(unconventional)	5	11	5	1	1	5	3			
In-situ combustion	11	4	2		1	2				
TOTAL THERMAL	35	52	31	22	24	26	24	8	2	12
Chemical										
Micellar/polymer	20	3	3	1	1	3				
Polymer	19	34	33	53	52	40	36	7	2	2
Caustic/alkaline	13	9		1			3	1		
TOTAL CHEMICAL	52	46	36	55	53	43	39	8	2	2
Gas										
Hydrocarbon gas	2	6	2	8	2	3	1	2	2	
CO_2 miscible	26	29	10	5	9	17	11		1	7
CO_2 immiscible		1	4	17	11	9	20		2	1
Nitrogen	3	5	1	1	1	1				
TOTAL GAS	31	41	17	31	23	30	32	2	5	8
GRAND TOTAL	118	139	84	108	100	99	95	18	9	22
Average oil price at wellhead ($/bbl)	21.59	31.77	28.52	26.19	25.88	24.09	12.51	15.40	12.58	15.85

1. In 1992, daily oil production from EOR in the USA was 752 625 bbl/day, representing 10.5% of the total daily production (7 171 000 bbl/day) of oil and condensate. By comparison, over the same period, some 50–52% were produced by water injection.

The production by EOR in the USA in 1992 breaks down process by process as follows:

Thermal	
Steam drive and steam soak	60.33%
In-situ combustion	0.62%
Total, thermal processes	60.95%
Chemical	
Micellar/polymer	0.03%
Polymer	0.26%
Caustic/alkaline	Nil
Total, chemical processes	0.29%
Gas (Miscible and immiscible)	
Hydrocarbon gas	15.02%
CO_2 miscible	19.27%
CO_2 immiscible	0.01%
Nitrogen	3.00%
Flue gas	1.46%
Total, gas processes	38.76%
GRAND TOTAL	100.00%

2. It is clear that the use of EOR passed through a distinct cycle during the period in question.

   In the 1970s, and even more so during the first half of the 1980s, almost every possible EOR technique was field-tested, in some cases without any particular regard being paid to the limits of their applicability. Chemical processes were expected to have a bright future, and polymer flooding was in frequent use. Fiscal incentives in the form of reduced windfall profit taxes led to some EOR projects being initiated with little regard to the technical issues.

3. As the price of oil started its slow decline in the early 1980s, the technologies already in wide use were fine-tuned, and proper infrastructures for EOR projects were developed by the major oil companies to reduce costs.

4. When the price of oil fell sharply in 1986, the economics of EOR changed drastically. The full impact was not felt until 1987, when project starts dropped by 80 to 90% from the level of previous years.

The effect of the lowering of the oil price can be seen in Table 15.11; the momentum of the infrastructure is more obvious when the situation after 1987 is analysed. In particular:

- before the price drop, polymer flooding was the most frequently used process, with 40 to 50 project starts per year. In 1988 and 1989 it had fallen to only 2 per year.
- not one micellar/polymer project was started after 1985; in fact, this and the caustic/alkaline process currently do not contribute in any sizeable way to oil production by EOR in the USA.
- there were roughly half as many conventional steam project starts as polymer project starts before the oil price drop in 1986, but six times as many in 1989. The actual number of starts did, however, fall appreciably.
- no in-situ combustion projects were started after 1985; this process currently makes only a very minor contribution to EOR production in the USA.
- hydrocarbon gas injection was least affected by the drop in the price of oil. Prior to 1986, there was an average of three project starts a year. This was reduced to two starts per year in 1987 and 1988 but, weighting these figures to take into account the size of the very large North Slope projects started in this period, this particular process stands out quite favourably.
- development of carbon dioxide miscible projects has continued, with seven starts in 1987. The infrastructure that was set up – especially the large pipelines conveying CO_2 from natural deposits to Texas and Oklahoma – is the main reason for the success of this process. Its contribution to EOR production is increasing steadily, as Table 15.10 shows.
- there have been no nitrogen or flue gas project starts since 1985; oil production by nitrogen injection has remained almost constant since 1986, while for flue gas injection it is declining.

5. There is a direct correlation in Table 15.11 between the price of oil and the number of annual EOR project starts. This demonstrates that *in the planning of future EOR projects, economic sensitivity to oil price variations is always a fundamental consideration*. We would therefore expect the EOR processes most widely used throughout the world to be those requiring the lowest capital expenditure and operating expenses: steam injection is a good example of this.

If we extend the study to oil-producing areas outside of the USA, the four with the highest production by EOR in 1992 were[136]: Venezuela, with 234 200 bbl/day (13.60 million m^3/year); the former Soviet Union (FSU), with 228 000 bbl/day (13.23 million m^3/year); Indonesia, with 180 000 bbl/day (10.45 million m^3/year); Canada, with 164 700 bbl/day (9.56 million m^3/year). In 1992, there were also pilot and fieldwide EOR projects under way in Columbia, the Libyan Arab Republic, the People's Republic of China, and Western Europe. In Venezuela, all 38 projects were steam drive, concentrated mainly in the Lake Maracaibo area (Bachaquero, Lagunillas and Tia Juana fields).

Table 15.12 lists the usage of various recovery processes in the former Soviet Union in 1990, with the corresponding oil production figures[136]. Of the projects 36% was based on dilute surfactant injection, and 19% on acid stimulation – not EOR, strictly speaking. The evolution of the number and types of projects for the period between 1976 and 1990 is shown[99] in Table 15.13.

As for Indonesia, the entire production by EOR came from the Duri field, which is under a fieldwide steam drive. In Canada, the figures for 1992 were: 49 miscible

Table 15.12. *Active EOR projects and oil production in the former USSR in 1990. (Ref. 136)*

Process	Active projects	Oil production (bbl/day)
Steam stimulation	19	42 000
In-situ combustion	10	8 000
Hot water injection	9	25 000
Hydrocarbon gas injection	9	14 000
CO_2 injection	1	2 000
Polymer	49	30 000
Alkaline	32	8 000
Acid stimulation	13	8 000
Water-based surfactants	27	22 000
Surfactant compositions (including systematic stimulation)	61	59 000
Microbiological	3	2 000
Others	14	8 000
	247	228 000

Table 15.13. *Active EOR projects in the former USSR, 1976–1990. (Ref. 99)*

EOR process	1976	1980	1987	1990
Thermal	16	21	40	38
Miscible gas	5	5	12	10
Chemical	56	40	128	196
Totals	77	66	180	244

hydrocarbon projects, 6 miscible CO_2, 6 in-situ combustion and 3 steam drives, and 3 polymer floods. Miscible hydrocarbon projects accounted for 83% of the total EOR production. We should also add 90 000 bbl/day (5.22 million m^3/year) of extra-heavy oil extracted by steam injection from the bituminous sands in Alberta. There were 11 active EOR projects in Western Europe in 1992: 7 steam drives and 3 polymer floods in Germany, and 1 polymer flood in France. The busiest year for EOR[30] was 1988, with a total of 19 projects running (9 steam drives, 8 polymer floods, 1 hydrocarbon gas drive and 1 micellar/polymer flood). By country, there were 15 projects in Germany and one in each in France, Great Britain, Italy and The Netherlands.

Finally, Table 15.14 summarises[99] the world's oil production from EOR in 1990, by geographical area and type of process. Note that the total EOR contribution (93.37 million m^3/year) is only 2.66% of the world's total output for the year. The production by process type was as follows: 50.6% thermal, 38.8% miscible gas, and only 10.6% chemical (including stimulation by acid, and dilute surfactant solutions, as for the former Soviet Union).

Although it is not possible to make predictions with any great certainty (even the best crystal balls are at present rather hazy), the evidence of this review of the recent history of EOR methods suggests a logical path which future developments in improved and enhanced recovery might follow:

1. The reaction of the industrialised countries to the 1973 and 1979–1980 oil crises was energy conservation and the partial replacement of oil by other energy sources, such as coal and nuclear power.

   There are still, however, a number of uses (road and air transport, petrochemicals, consumption by developing countries, which are not replacing oil by other fuels), where *"the only substitute for oil is oil"*, to quote a well-known slogan. The forecast[60] in Table 15.15 is interesting in this respect. During the 1980 oil crisis, roughly one half of the world's oil consumption was potentially replaceable with other fuels, while in 2000 it is predicted that only about 25% will be replaceable. *For the remaining 75% there will still be no substitute for oil.* On the strength of these predictions, there exists, therefore, a real incentive to intensify both exploration activity to locate new reserves, and the use of improved and enhanced oil recovery methods in existing fields.

Table 15.14. *World oil production by EOR in 1990. (Ref. 99)*

Country	Oil production (millions m^3/year)			
	EOR process			Total
	Thermal	Miscible	Chemical	
USA	26.13	11.06	0.69	37.88
Canada	0.46	7.37	0.99	8.82
Venezuela	6.27	0.64	Zero	6.91
USSR (former)	4.53	0.87	8.12	13.52
Others (/estimated)	9.87[a]	16.25[b]	0.12	26.24
TOTALS	47.26	36.19	9.92	93.37

[a] Mainly from the Duri field (Indonesia)

[b] Mainly from the Hassi Messaoud field (Algeria) and Intisar field (Libya)

Table 15.15. *Forecasted trend of world oil consumption between 1980 and 2000 [millions of t/year oil-equivalent (MToe/yr)]. (Ref. 60)*

Use	1980	1985	2000
– Transport	1 160	1 200	1 650–1 850
– Petrochemicals	155	165	275–350
– Fuels, developing countries	230	250	350–450
– NON-REPLACEABLE OIL	1 545	1 615	2 275–2 650
– Fuels, industrialised countries	1 315	1 030	575–650
– Other uses	140	130	120–200
– REPLACEABLE OIL	1 455	1 160	695–850
– Total oil consumption	3 000	2 775	2 970–3 500
– NON-REPLACEABLE OIL, %	51.5	58.2	75.7–76.6

2. We could reasonably expect significant expansion in the use of EOR to occur in the near future in those heavily industrialised countries which are also oil-producers in the process of depleting their reserves by conventional recovery methods.

   The situation is illustrated in Table 15.16. The data refer to 1989, the last year in which the production figures per country were not affected by the Gulf War following the invasion of Kuwait in August 1990.

   We might expect a big increase in the proliferation of EOR projects in the USA, Canada and what was the USSR. The opposite can be said for the UK: most of its fields are offshore, and EOR entails a high level of investment (in particular, for the drilling of the infill wells necessary to achieve the close well spacings required), and heavy operating expenses.

   Countries like Kuwait, Saudi Arabia, Abu Dhabi, Iraq and Iran, on the other hand, have such vast reserves that they will be able to sustain their current output levels for at least a century. We should not expect them, therefore, to turn to EOR in the near to medium term.

3. The major obstacle to the more widespread use of EOR is, quite simply, cost[4]. Some rough cost estimates for different processes are provided in Table 15.8.

   A number of changes in the prevailing economic climate would have to take place before any prospect of a substantial increase in EOR usage can be expected:

   - A decrease in the investment required to implement an EOR project, and the subsequent operating expenses. It has been suggested[103], for instance, that producing wells should not be plugged at abandonment: they could then be put back on production as and when EOR was under way in the field. This measure would obviate the need to drill replacement wells.

   - An increase in the price of oil in real terms, sufficient to cover the cost of retrieving the oil by enhanced recovery (running costs, interest payments and taxation).

   These two requirements are difficult to reconcile. While it should be possible to reduce costs through innovation and improved technology, a rise in the real price of oil would inevitably feed back to the project itself in the form of an increase in costs [drilling, surface equipment, fuel for steam injection, natural gas and carbon dioxide, polymers and surfactants (both derived from oil), labour, etc.]

Table 15.16. *Statistics for the top 20 oil-producing countries of the world (1989). (Ref. 98)*

	Reserves (10^6 m^3)	$R = \dfrac{\text{proven reserves in 1989}}{\text{oil production in 1989}}$	
		Annual production (10^6 m^3)	R
a) *Countries with R > 100*			
1: Kuwait	15 260	93.0	164.1
2: Saudi Arabia	41 350	282.3	146.5
3: Abu Dhabi	8 640	85.3	101.3
TOTAL	65 250	460.6	141.7
World total (%):	44.0	13.5	
b) *Countries with 75 ≤ R < 100*			
4: Iraq	15 900	159.6	99.6
5: Venezuela	9 620	110.2	87.3
TOTAL	25 520	269.8	94.6
World total (%):	17.2	7.9	
c) *Countries with 50 ≤ R < 75*			
6: Iran	9 938	165.9	59.9
7: Mexico	8 265	150.4	55.0
TOTAL	18 203	316.3	57.5
World total (%):	12.3	9.3	
d) *Countries with 25 ≤ R < 50*			
8: Libya	3 625	79.1	45.8
9: Neutral zone	777	21.5	36.1
10: Nigeria	2 655	93.1	28.5
TOTAL	7 057	193.7	36.4
World total (%):	4.8	5.7	
e) *Countries with 15 ≤ R < 25*			
11: Indonesia	1 903	81.8	23.3
12: Algeria	1 468	65.0	22.6
13: P.R. of China	3 419	159.5	21.4
14: Norway	1 755	89.3	19.7
15: India	691	37.5	18.4
TOTAL	9 236	433.1	21.3
World total (%):	6.2	12.7	
f) *Countries with R < 15*			
16: Canada	1 076	78.2	13.8
17: Egypt	684	50.7	13.5
18: USSR (former)	9 202	704.5	13.1
19: USA	4 186	442.9	9.5
20: UK	605	99.7	6.1
TOTAL	15 753	1,376.0	11.4
World total (%):	10.6	40.5	
GRAND TOTALS:	141 289	3,049.5	46.3
World total (%):	95.2	89.7	

When, and at just what oil price, the equation would be balanced, is difficult to predict. Figure 15.67 is an example of a forecast published by one of the major multinationals[16].

The diagram predicts a period of EOR development, followed by production of extra-heavy oils (from the Athabasca tar sands of Canada, and the Faja

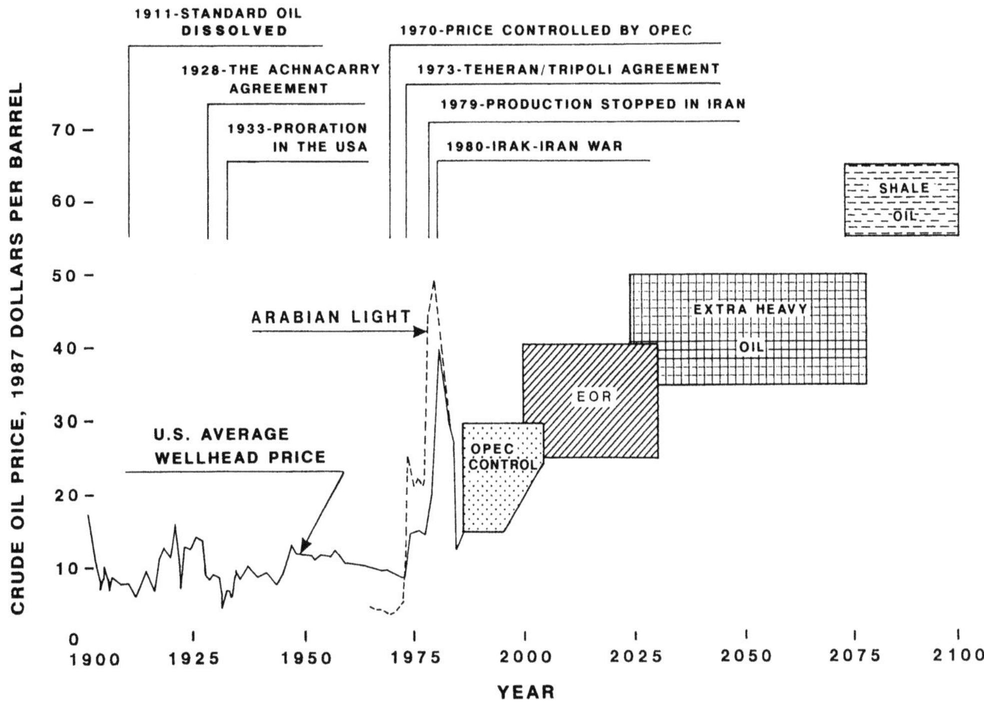

Fig. 15.67. *Historical trends in the price of crude since the beginning of the century, and a forecast*[16] *by Chevron for the period 1990–2100*

Petrolifera in the Orinoco, Venezuela), and then the exploitation of bituminous shales.

In addition to this forecast, the diagram illustrates the historical trend of the oil price in the US from 1900 to the present, and highlights the major events which have affected the price over a period of almost 90 years.

4. The major oil companies will continue research into new processes and the improvement of existing ones, in anticipation of the right combination of technical and economic conditions for an upturn in the use of EOR.

Oil companies in the USA and Western Europe have made a strong commitment to the study of internal reservoir structure. They have recognised that the characterisation of the internal structure is a fundamental prerequisite for the effective application of any improved or enhanced recovery method.

References

1. Aalund LR (1988) EOR projects decline, but CO_2 pushes up production. Oil Gas J April 18: 33–37
2. Anonymous (1988) Energy focus-average oil well production rates. J Pet Technol Feb: 168
3. Anonymous (1988) Reserves up worldwide and outside OPEC. Oil Gas J Dec. 26: 43–47
4. Various authors (1987) The economic complexity of enhanced recovery. Round Table Discussion, Proc 12th World Pet Cong vol 3. Graham & Trotman, London, pp 337–385
5. Bailey RE, Curtis LB (1984) Enhanced oil recovery. National Petroleum Council, Washington, DC
6. Barbe JA, Schnoebelen DJ (1987) Quantitative analysis of infill performance: Robertson Clearfork Unit. J Pet Technol Dec: 1593–1651; Trans, AIME 283

7. Barber AH, Jr, George CJ, Stiles LH, Thompson BB (1983) Infill drilling to increase reserves – actual experience in nine fields in Texas, Oklahoma and Illinois. J Pet Technol Aug: 1530–1538; Trans, AIME 275

8. Bella E, Bilgeri D, Causin E, Chierici GL, Gili V, Mirabelli G, Sozzi I (1982) Piropo heavy-oil accumulation, Adriatic Sea, Italy – study of an exploitation process. Proc 2nd Eur Symp on Enhanced oil recovery, (Paris, Nov. 8–10), pp 477–486

9. Bond DC (ed) (1978) Determination of residual oil saturation. Interstate Oil Compact Commission, IOCC, Oklahoma City

10. Boy de la Tour X, Gadon JL, Lacour JJ (1986) New oil: what's in the future? Energy Explor Exploit 6: 401–495

11. Brigham WE (1985) A field experiment of improved steam drive with in situ foaming – final report. US Department of Energy, Rep DOE/SF/11445-4

12. Brigham WE, Abbaszadeh - Degani M (1987) Tracer testing for reservoir description. J Pet Technol May: 519–527; Trans, AIME 283

13. Carlson LO (1988) Performance of Hawkins Field Unit under gas-drive-pressure maintenance operations and development of an enhanced oil recovery project. Paper SPE 17324 presented at the SPE/DOE Enhanced oil recovery Symp, Tulsa, April 17–20

14. Causin E, Chierici GL, Erba M, Mirabelli G, Turriani C (1982) Study of EOR processes – Cortemaggiore Oilfield; Italy. Proc 18th Congr, Hungarian Nat Soc for Mining and Metallurgy, Siòfok, Hungary, Sept 10–12, Paper C-33

15. Chang MM, Maerefat NL, Tomutsa L, Honarpour MM (1988) Evaluation and comparison of residual oil saturation determination techniques. SPE Formation Evaluation, March, pp 251–269

16. Chevron Organization (1987) World energy outlook

17. Chien Chu (1982) State-of-the-art review of fireflood field projects. J Pet Technol Jan: 19–36; Trans, AIME 273

18. Chien Chu (1985) State-of-the-art review of steamflood field projects. J Pet Technol Oct: 1887–1902; Trans, AIME 279

19. Chierici GL (1960) Spiazzamento del greggio di giacimento a mezzo di idrocarburi miscibili – Fenomeni di "mass transfer" al fronte di miscelazione. Alta tecnologia chimica, Accademia Nazionale dei Lincei, pp 303–326

20. Chierici GL (1980) Enhanced oil recovery processes – a state-of-the-art review. Agip SpA, San Donato Milanese

21. Chierici GL (1990) ARM (Advanced Reservoir Management) vs. EOR. Rev Inst Fr Pét 45 (1): 123–129

22. Chierici GL (1989) Enhanced oil recovery processes: a promise and a challenge. Agip Rev 4 (May): 622–629

23. Chierici GL, Perego M (1987) Ponte Dirillo Field, Italy – a follow-up on a successful CO_2 - enriched natural gas injection process. Proc 4th Eur Symp on Enhanced oil recovery, Hamburg, Germany, Oct 27–29, pp 323–334

24. Chierici GL, Giannone G, Sclocchi G, Terzi L (1981) A wellbore model for two-phase flow in geothermal reservoirs. Paper SPE 10315 presented at the 56th Annu Fall Meet of SPE, San Antonio, Oct 5–7

25. Clancy JP, Gilchrist RE, Kroll DE, MacGregor A (1985) Improved oil and gas recovery using nitrogen – laboratory analysis, cases and economics. Proc 3rd Eur Meet on Improved oil recovery, Rome, April 16–18, vol 1, pp 255–264

26. Coats KH (1978) A highly implicit steamflood model. Soc Pet Eng J Oct: 369–383

27. Coats KH (1980) In-situ combustion model. Soc Pet Eng J Dec: 533–554; Trans, AIME 269

28. Coats KH (1980) An equation of state compositional model. Soc Pet Eng J Oct: 363–376; Trans, AIME, 269

29. Colitti M (1981) Size and distribution of known and undiscovered petroleum resources in the world, with an estimate of future exploration. OPEC Rev 5 (3): 9–65

30. Combe J, Corlay E, Valentin E, Champlon D, Bosio J, de Haan HJ, Hawes RI, Pusch G, Sclocchi G, Stockenhuber F (1989) EOR in Western Europe: status and outlook. Proc 5th Eur Symp. on Improved oil recovery, Budapest, April 25–27, Spec Issue

31. Cutler WW Jr: (1924) Estimation of underground oil reserves by oil-well production curves. Bull USBM 228

32. Deans HA, Majoros S (1980) Single-well chemical tracer test simulator handbook: improved accuracy and interpretation for residual oil saturation measurement. US DOE, Contract No DOE/BC/20006-14

33. de Buyl M, Guidish T, Bell F (1988) Reservoir description from seismic lithologic parameter estimation. J Pet Technol Jan: 75-82

34. de Haan HJ, van Lookeren J (1969) Early results of the first large-scale steam soak project in the Tia Juana Field, western Venezuela. J Pet Technol Jan: 101-110; Trans, AIME 246

35. de Haan HJ, Smart MJ (1983) The present and future role of EOR techniques. OAPEC Enhanced oil recovery seminar, Doha, Qatar, Nov 26-28

36. Dellinger SE, Patton JT, Holbrook ST (1984) CO_2 mobility control. Soc Pet Eng J April: 191-196; Trans, AIME 277

37. Descalzi C, Rognoni A, Cigni M (1988) Synergetic log and core data treatment through cluster analysis: a methodology to improve reservoir description. Paper SPE 17637 presented at the Int Meet on Petroleum engineering, Tianjing, China, Nov. 1-4

38. Dietz DN (1953) A theoretical approach to the problem of encroaching and by-passing edge water. Proc K Ned Akad Wet Ser B 56 (1): 83-92

39. Dumore' JM (1964) Stability considerations in downward miscibile displacements. Soc Pet Eng J Dec: 356-362; Trans, AIME 231

40. Dumore' JM, Schols RS (1974) Drainage capillary pressure functions and the influence of connate water. Soc Pet Eng J Oct: 437-444

41. Falls AH, Lawson JB, Hirasaki G (1988) The role of noncondensable gas in steam foams. J Pet Technol Jan: 95-104

42. Felder RD (1988) Cased-hole logging for evaluating bypassed reserves. J Pet Technol Aug: 963-973

43. Fisher EW (1987) U.S. oil outlook. The University of Texas at Austin, Chair in Mineral Resources

44. Garcia FM (1983) A successful gas-injection project in a heavy oil reservoir. Paper SPE 11988 presented at the 58th Annu Fall Meet of SPE, San Francisco, Oct 5-8

45. Glasø Ø (1985) Generalized minimum miscibility pressure correlation. Soc Pet Eng J Dec: 927-937; Trans, AIME 279

46. Hagoort J (1980) Oil recovery by gravity drainage. Soc Pet Eng J June: 139-150; Trans, AIME 269

47. Hagoort J (1982) The response of interwell tracer tests in watered-out reservoirs. Paper SPE 11131 presented at the 57th Annu Fall Meet of SPE, New Orleans, Sept. 26-29

48. Heller JP, Taber JJ (1986) Improvement of CO_2 flood performance. US Department of Energy, Bartlesville Project Office, Rep DOE/MC/21136-6

49. Holm LW (1987) Evolution of the carbon dioxide flooding process. J Pet Technol Nov: 1377-1442; Trans, AIME 283

50. IOCC (1983) Improved oil recovery. Interstate Oil Compact Commission, Oklahoma City

51. Iverson WP (1988) Crosswell logging for acoustic impedance. J Pet Technol Jan: 75-82

52. Johnston JR (1988) Weeks Island gravity stable CO_2 pilot. Paper SPE 17351 presented at the SPE/DOE Enhanced Oil Recovery Symp, Tulsa, April 17-20

53. Kamal MM, Hegeman PS (1988) New developments in multiple-well testing. SPE Formation evaluation, March: 159-168

54. Keller JM (1987) Present and future trends in production. Proc 12th World Pet Congr, vol 1. John Wiley & Sons, Chichester, pp 183-194

55. King RL, Stiles JH.Jr, Waggoner JN (1970) A reservoir study of the Hawkins Woodbine Field. Paper SPE 2972 presented at the 1970 Fall Meet of SPE, Houston, Oct 5-7

56. Leighton AJ, Wayland JR (1987) Techniques for mapping and diagnosing EOR processes. J Pet Technol Feb: 129-136; Trans, AIME 283

57. Liu Wen-zhang (1987) Pilot steam soak operations in deep wells in China. J Pet Technol Nov: 1441-1448; Trans, AIME 283

58. Matthews CS (1983) Steamflooding. J Pet Technol March: 465-471; Trans, AIME 275

59. McCormick CL, Hester RD (1983) Improved polymers for enhanced oil recovery – synthesis and rheology. US Department of Energy, Bartlesville Project Office, Rep DOE/BC/10321-12 and Rep DOE/BC/10321-16

60. Méo J (1988) R&D in the petroleum industry and its financing. New Technol Explor Exploit Oil Gas Resources, vol 1. Graham & Trotman, London, pp 27-46

61. Mishra S, Brigham WE, Orr FM (1991) Tracer and pressure test analysis for characterization of areally heterogeneous reservoirs. SPE Formation Evaluation, March: 45–54
62. Montadert L (1963) La sédimentologie et l'étude détaillée des hétérogéneités d'un réservoir. Application au gisement d'Hassi Messaoud. Rev Inst Fr Pét 18 (Dec): 258–283
63. Myhill NA, Stegemeier GL (1978) Steam drive correlation and prediction. J Pet Technol Feb: 173–182
64. Needham RB, Doe PH (1987) Polymer flooding review. J Pet Technol Dec: 1503–1507; Trans, AIME 283
65. Nyland E, Dusseaoult MB (1982) Fireflood microseismic monitoring. Results and potential for process control. Paper CIM 82-33-39 presented at the 1982 Annu Meet of the Petroleum Society of CIM, Calgary, Alberta, June 6–9
66. O'Neill N (1988) Fahud field review: a switch from water to gas injection. J Pet Technol May: 609–618
67. Palmer FS, Nute AJ, Peterson RL (1984) Implementation of a gravity-stable miscible CO_2 flood in the 8000 foot sand, Bay St. Elaine Field. J Pet Technol Jan: 101–110; Trans, AIME 277
68. Pope GA, Nelson RC (1978) A chemical flooding compositional simulator. Soc Pet Eng J Oct: 339–354
69. Prats M (1982) Thermal recovery. Monograph volume 7. Society of Petroleum Engineers, Dallas
70. Pullen N, Mathews L, Kirsche W (1986) Techniques to obtain very high resolution 3D-imaging at an Athabasca tar sands thermal pilot. Proc, Conv of Can Soc Explor Geophysicists
71. Schlumberger (1985) Log interpretation charts. Schlumberger Well Services USA
72. Schlumberger (1987) Log interpretation principles/application. Schlumberger Educational Services, Houston
73. Sebastian HM, Wenger RS, Renner TA (1985) Correlation of minimum miscibility pressure for impure CO_2 streams. J Pet Technol Nov: 2076–2082: Trans, AIME 279
74. Shah DO, Schechter RS (eds) (1977) Improved oil recovery by surfactant and polymer flooding. Academic Press, New York
75. Sheely CQ, Jr, Baldwin DE, Jr (1982) Single-well tracer test for evaluating chemical enhanced oil recovery processes. J Pet Technol Aug: 1887–1896; Trans, AIME 273
76. Simon R, Graue DJ (1965) Generalized correlations for predicting solubility, swelling and viscosity behavior of CO_2-crude oil systems. J Pet Technol Jan 102–106; Trans, AIME 234
77. Stalkup FI, Jr (1983) Miscible displacement. Monograph volume 8. Society of Petroleum Engineers, Dallas
78. Stegemeier GL, Laumbach DD, Volek CW (1980) Representing steam processes with vacuum models. Soc Pet Eng J June: 151–174; Trans, AIME 269
79. Stosur GJ, Singer MI, Luhning RW, Yurko WJ (1989) Enhanced oil recovery in North America: status and prospects. Proc 5th Eur Symp on Improved oil recovery, Budapest, April 25–27, pp 433–452
80. Stosur GJ (pers. comm.) Budapest, April 25, 1989
81. Thomas RD, Stosur GJ, Pautz JF (1987) Analysis of trends in U.S.A. enhanced oil recovery projects. Proc 4th Eur Symp on Enhanced oil recovery, Hamburg, Oct 27–29, pp 43–54
82. Thomas EC, Richardson JE, Shannon MT, Williams MR (1987) The scope and perspective of ROS measurement and flood monitoring. J Pet Technol Nov: 1398–1406; Trans, AIME 283
83. Tomich JF, Dalton RL, Jr, Deans HA, Shallenberger LK (1973) Single-well tracer method to measure residual oil saturation. J Pet Technol Feb: 211–218; Trans, AIME 255
84. van Everdingen AF (1978) Dialogue. J Pet Technol Feb: 209–210
85. van Everdingen AF, Kriss HS (1980) A proposal to improve recovery efficiency. J Pet Technol July 1164–1168; Trans, AIME 269
86. van Lookeren J (1983) Calculation methods for linear and radial steam flow in oil reservoirs. Soc Pet Eng J June: 427–439; Trans, AIME 275
87. Wayland JR, Lee DO, Cabe TJ (1987) CSAMT mapping of a Utah tar sand steamflood. J Pet Technol March: 345–352; Trans, AIME 283
88. Weber KJ (1982) Influence of common sedimentary structures on fluid flow in reservoir models. J Pet Technol March: 665–672; Trans, AIME 273
89. White PD (1985) In situ combustion – appraisal and status. J Pet Technol Nov: 1943–1949; Trans, AIME 279
90. Winsor PA (1954) Solvent properties of amphiphillic compounds. Butterworth's, London

91. Witterhold EJ, Kretzschmar JL, Jay KL (1982) The application of crosshole electromagnetic wave measurements to mapping of a steamflood. Paper CIM 82-33-81 presented at the 1982 Annu Meet of the Petroleum Society of CIM, Calgary, Alberta, June 6–9

92. Yellig WF, Metcalfe RS (1980) Determination and prediction of CO_2 minimum miscibility pressure. J Pet Technol Jan: 160–165; Trans, AIME 269

93. Ypma JGJ (1985) Analytical and numerical modelling of immiscible, gravity-stable gas injection into stratified reservoirs. Soc Pet Eng J Aug: 554–564; Trans, AIME 279

94. Ypma JGJ (1985) Compositional effects in gravity-dominated nitrogen displacements. Paper SPE 14416 presented at the 60th Annu Fall Meet of SPE, Las Vegas, Sept 22–25

95. Wu CH, Laughlin BA, Jardon M (1989) Infill drilling enhances waterflood recovery. J Pet Technol Oct: 1088–1095

96. Anonymous (1990) World trends – some song, some refrain. World Oil 211 (8): 19–26

97. Anonymous (1991) Slide seen for non-OPEC oil production. Oil Gas J April 29: 75–76

98. Anonymous (1992) Teamwork, new technology and mature reservoirs. J Pet Technol Jan: 38–40

99. Anonymous (1992) Estimated proven world reserves of crude oil and natural gas. In: World trends – operators abandon the U.S. World Oil 213 (8): 23–28

100. Baird E (1992) New technology in support of improved oil recovery. Paper G1 presented at the Offshore Northern Seas Conf, Stavanger, Aug 25–28

101. Barthel R (1991) The effect of large-scale heterogeneities on the performance of waterdrive reservoirs. Paper SPE 22697 presented at the 66th SPE Annu Technical Conf, Dallas, Oct 6–9

102. Blevins TR (1990) Steamflooding in the U.S.: a status report. J Pet Technol May: 548–554

103. Brashear JP, Biglarbigi K, Becker AB, Ray RM (1991) Effect of well abandonments on EOR potential. J Pet Technol Dec: 1496–1501

104. Briggs P, Corrigan T, Fetkovich M, Gouilloud M, Lo Tien-when, Paulsson B (1992) Trends in reservoir management. Oilfield Rev Jan: 8–24

105. Britt LK, Jones JR, Pardini RE, Plum GL (1991) Reservoir description by interference testing in the Clayton field. J Pet Technol May: 524–577; Trans, AIME 291

106. Bunn GF, Wittmann MJ, Morgan WD, Curnutt RC (1991) Distributed pressure measurements allow early quantification of reservoir dynamics in the Jene field. SPE Formation Evaluation, March, pp 55–62

107. Castanier LM, Brigham WE (1991) An evaluation of field projects of steam with additives. SPE Reservoir Eng Feb: 62–68; Trans, AIME 291

108. Chierici GL (1992) Economically improving oil recovery by advanced reservoir management. J Pet Sci Eng 8: 205–219

109. Corvi P, Heffer K, King P, Tyson S, Verly G, Ehlig-Economides C, Le Nir I, Ronen S, Schulz P, Corbett P, Lewis J, Pickup G, Ringrose P, Guérillot D, Montadert L, Ravenne C, Haldorsen H, Hewett T (1992) Reservoir characterization using expert knowledge, data and statistics. Oilfield Rev Jan: 25–29

110. Corwith JR, Mengel F (1990) Applications of pulsed-neutron capture logs in Ekofisk area fields. SPE Formation Evaluation, March: 16–22

111. Ekrann S (1992) An analysis of gravity-segregated piston-like displacement in stratified reservoirs. SPE Reservoir Eng Feb: 143–148

112. Fjerstad PA, Flaate SA, Hagen J, Pollen S (1992) Long-term production testing improves reservoir characterization in the Oseberg field. J Pet Technol April: 476–485; Trans, AIME 293

113. Freedman R, Rouault GF, (1989) Remaining-oil determination using nuclear magnetism logging. SPE Formation Evaluation, June: 121–130; Trans, AIME 287

114. Gangle FJ, Weyland HV, Lassiter JC, Velth EJ, Garner TA (1992) Light-oil steam-drive pilot test at NPR-1, Elk Hills, California. SPE Reservoir Eng Aug: 315–320

115. Greaves RJ, Beydoun WB, Spies BR (1991) New dimensions in geophysics for reservoir monitoring. SPE Formation Evaluation, June: 141–150; Trans, AIME 291

116. Henriquez A, Tyler KJ, Hurst A (1990) Characterization of a fluvial sedimentology for reservoir simulation modeling. SPE Formation Evaluation, Sept: 211–216; Trans, AIME 289

117. Hirasaki GJ (1974) Pulse tests and other early transient pressure analyses for in-situ estimation of vertical permeability. Soc Pet Eng J Feb: 75–90; Trans, AIME 257

118. Holz MH, Ruppel SC, Hocott CR (1992) Integrated geologic and engineering determination of oil-reserve-growth potential in carbonate reservoirs. J Pet Technol Nov: 1250–1257; Trans, AIME 293

119. Honarpour M, Szpakiewics M, Sharma B, Chang M, Schatzingen R, Jackson S, Tomutza L, Maerafat N (1989) Integrated reservoir assessment and characterization – final report. Rep NIPER-390 (May), IIT Res Inst, Natl Inst Petrol Energy Res, Bartlesville, OK

120. Jacquard P (1991) Improved oil recovery in the global energy perspective. Proc, 6th Eur Symp Improved oil recovery, Stavanger, May 21–23, Plenary Conf

121. Justice JH, Vassiliou AA, Mathisen ME, Throyer AH, Cunningham PS, (1991) Acoustic tomography for improved oil recovery. Proc, 6th Eur Symp Improved oil recovery, Stavanger, May 21–23, vol 1, pp 213–221

122. Kaneda R, Saeedi J, Ayestaran LC (1991) Interpretation of a pulse test in a layered reservoir. SPE Formation Evaluation, Dec: 453–462

123. Kimminau SJ, Plasek RE (1992) The design of pulsed-neutron reservoir-monitoring programs. SPE Formation Evaluation, March: 93–98

124. Kittridge MG, Lake LW, Lucia FJ, Fogg GE (1990) Outcrop/subsurface comparisons of heterogeneity in the San Andres formation. SPE Formation Evaluation, Sept: 233–240

125. Lau HC, Borchardt JK (1991) Improved steam-foam formulations: concepts and laboratory results. SPE Reservoir Eng Nov: 470–476

126. Lo, Tien-when, Inderwiesen PL (1992) Reservoir characterization with cross-well tomography: a case study in the Midway Sunset field, California. Paper SPE 22336 presented at the 3rd Int Meet on Petroleum engineering, Beijing, March 24–27

127. Maerker JM, Gale WW (1992) Surfactant flood design for Loudon. SPE Reservoir Eng Feb: 36–44; Trans, AIME 293

128. Martind FD, Taber JJ (1992) Carbon dioxide flooding. J Pet Technol April: 396–400

129. Massé PJ, Gosney TC, Long DL (1991) Use of pulsed-neutron-capture logs to identify steam breakthrough. SPE Formation Evaluation, Sept: 319–326

130. Masters CD, Root DH, Attanasi ED (1992) World resources of crude oil and natural gas. Proc, 13th World Petrol Congr, vol 4. John Wiley & Sons, Chichester, pp 51–64

131. McCool CS, Green AD, Willhite GP (1991) Permeability reduction mechanism involved in insitu gelation of a polyacrylamide/chromium (VI)/thiourea system. SPE Reservoir Eng Feb: 77–83; Trans, AIME 291

132. McGuire PL, Moritz AL, Jr., (1992) Compositional simulation and performance analysis of the Prudhoe Bay miscible gas project. SPE Reservoir Eng Aug: 329–334; Trans, AIME 293

133. McKeon DC, Scott HD, Olesen J-R, Patton GL, Mitchell RJ (1991) Improved oxygen-activation method for determining water flow behind casing. SPE Formation Evaluation, Sept: 334–342

134. Merritt MB, Groce JF (1992) A case history of the Hanford San Andres miscible CO_2 project. J Pet Technol Aug: 924–929

135. Moritis G (1990) CO_2 and HC injection lead EOR production increase. Oil Gas J April 23: 49–82

136. Moritis G (1992) EOR increases 24% worldwide; claims 10% of U.S. production. Oil Gas J April 20: 51–79

137. Nitzberg KE, Broman WH (1992) Improved reservoir characterization from waterflood tracer movement, Northwest Fault Block, Prudhoe Bay, Alaska. SPE Formation Evaluation, Sept: 228–234

138. Øvreberg O, Damsleth E, Haldorsen HH (1992) Putting error bars on reservoir engineering forecasts. J Pet Technol June: 732–738; Trans, AIME 293

139. Pautz JF, Thomas RD (1991) Applications of EOR technology in field projects – 1990 update. Report NIPER-513, IIT Res Inst, Natl Inst Petrol Energy Res, Bartlesville, OK

140. Ray RM (1990) Producing unrecovered mobile oil: evaluation of potential economically recoverable reserves in Texas, Oklahoma and New Mexico. Report DOE/BC/1400-2 by ICF Resource Inc. and the Bureau of Economic Geology of the University of Texas at Austin

141. Raza SH (1990) Data acquisition and analysis: foundational to efficient reservoir management. Paper SPE 20749 presented at the 65th SPE Annu Technical Conf, New Orleans, Sept 23–26

142. Richardson JE, Wyman RE, Jorden JR, Mitchell FR (1973) Methods for determining residual oil with pulsed neutron capture logs. J Pet Technol May: 593–608; Trans, AIME 255

143. Richardson JG, Sangree JB, Sneider RM (1989) Oil recovery by gravity segregation. J Pet Technol June: 581–582

144. Saleri NG, Toronyi RM, Snyder DE (1992) Data and data hierarchy. J Pet Technol Dec: 1286–1293; Trans, AIME 293

145. Slatt RM, Hopkins GL (1990) Scaling geologic reservoir description to engineering needs. J Pet Technol Feb: 202–210

146. Stein MH, Frey DD, Walker RD, Pariani GJ (1992) Slaughter Estate Unit CO_2 flood: comparison between pilot and field-scale performance. J Pet Technol Sept: 1026–1032

147. Stiles LH, Magruder JB (1992) Reservoir management in the Means San Andres Unit. J Pet Technol April: 469–475

148. Strom ET, Paul JM, Phelps CH, Sampath K (1991) A new biopolymer for high temperature profile control: part 1. Laboratory testing. SPE Reservoir Eng, Aug: 360–364

149. Subroto K (1991) OPEC's perspective. Agip Rev 12 (May-July): 16–23

150. Sydansk RD (1990) A newly developed cromium(III) gel technology. SPE Reservoir Eng Aug: 346–352

151. Thakur GC (1991) Waterflood surveillance techniques – a reservoir management approach. J Pet Technol Oct: 1180–1188

152. Thomas LK, Dixon TN, Pierson RG, Hermansen H (1991) Ekofisk nitrogen injection. SPE Formation Evaluation, June: 151–160

153. Trice ML, Jr., Dawe BA (1992) Reservoir management practices. J Pet Technol Dec: 1296–1349

154. Watfa M, Pollock D (1990) Through-tubing gas saturation monitoring. Paper 21317 presented at the SPE Abu Dhabi Petroleum Conf, Abu Dhabi, May 15–16

155. York SD, Peng CP, Joslin TH (1992) Reservoir management of Valhall field, Norway. J Pet Technol Aug: 918–923; Trans, AIME 293

156. Zaki Abou Zeid, El Bishlawi SH (1990) 3D seismic interpretation in Jarn Yaphour field, Abu Dhabi. J Pet Technol June: 700–707; Trans, AIME 293

157. Dobecki TL (1992) Alternative technologies. In: Sheriff RE (ed) Reservoir geophysics. Society of Exploration Geophysicisists, Tulsa, OK, pp 335–343

158. Gray S, Lines L (1992) Cross-borehole tomographic migration. J Seismic Explor 1: 315–324

159. Greaves RJ (1992) Three-dimensional seismic monitoring of an enhanced oil recovery process. In: Sheriff RE (ed) Reservoir geophysics. Society of Exploration Geophysicists, Tulsa, OK, pp 309–320

160. Hardage BA (1992) Cross well seismology and reverse VSP. Geophysical Press, Castenau-le-Nez, France

161. Johnstad SE, Uden RC, Dunlop KNB (1993) Seismic reservoir monitoring over the Oseberg field. First Break 11 (5): 177–185

162. Justice JH (1992) Geophysical methods for reservoir surveillance. In: Sheriff RE (ed) Reservoir geophysics. Society of Exploration Geophysicists, Tulsa, OK, pp 281–284

163. Justice JH, Vassiliou AA, Mathisen ME, Singh S, Cunningham PS, Hutt PR (1992) Acoustic tomography in reservoir surveillance. In: Sheriff RE (ed) Reservoir Geophysics. Society of Exploration Geophysicists, Tulsa, OK, pp 321–334

164. Justice JH, Mathisen ME, Vassiliou AA, Shiao I, Alameddine BR, Guinzy NJ (1993) Crosswell seismic tomography in improved oil recovery. First Break 11 (6): 229–239

165. Matthews L (1992) 3-D seismic monitoring of an in-situ thermal process: Athabasca Canada. In: Sheriff RE (ed) Reservoir geophysics. Society of Exploration Geophysicists, Tulsa, OK, pp 301–308

166. Wilt MJ, Morrison HF, Becker A, Lee KH (1991) Cross-borehole and surface-to-borehole electromagnetic induction for reservoir characterization. Report DOE/BC/91002253, Lawrence Livermore National Laboratory, Livermore, CA

167. Tinker GE, Turpening R, Chon Y-T (1993) Connectivity mapping improves understanding of reservoir continuity. Oil Gas J (Dec. 20): 87–91

EXERCISES

Exercise 15.1

Derive the equations for the calculation of the percentage of mobile oil produced at gas breakthrough, in the case of *vertical* displacement by gas under *non*-miscible conditions.

Solution

Consider an oil reservoir with a gas cap, or crestal gas injection (Fig. E15/1.1), where oil is produced through non-miscible displacement by gas.

Except in the immediate vicinity of the well, the displacement will be in the vertical (z) direction, with co-current flow of oil and gas.

The total area of the GOC is A.

We will derive the theory making the following assumptions:

1 reservoir pressure is always above, or at least equal to, the bubble point p_b of the oil,
2 the reservoir rock is homogeneous over its entire thickness, with a vertical permeability k_v,
3 production from the reservoir is at a constant total rate q_t (under reservoir conditions) throughout its entire life,
4 because of the low interfacial tension between oil and gas, capillary pressure $P_{c,og}$ will be ignored, as will its vertical gradient $\partial P_{c,og}/\partial z$,
5 since $\mu_g \ll \mu_o$, we can ignore the term:

$$\frac{\mu_g}{\mu_o}\frac{k_{ro}}{k_{rg}}$$

Subject to these assumptions, Eq. (11.47a) becomes:

$$f_g = 1 - k_v\frac{k_{ro}}{\mu_o}\frac{A}{q_t}g\left(\rho_o - \rho_g\right) \tag{11.47c}$$

$$f_o = 1 - f_g = k_v\frac{k_{ro}}{\mu_o}\frac{A}{q_t}g\left(\rho_o - \rho_g\right). \tag{15/1.1a}$$

Since, by definition, $k_{ro,iw} = 1.0$, we have:

$$N_{gv} = \frac{k_v g(\rho_o - \rho_g)}{u_z\mu_o}. \tag{15.11b}$$

We can therefore write:

$$f_o = N_{gv}k_{ro}(S_o), \tag{15/1.1b}$$

where: $u_z = \dfrac{q_t}{A}$.

We next calculate the velocity:

$$v(S_o) = \frac{dz(S_o)}{dt}$$

at which any surface at oil saturation S_o moves vertically through the porous medium. To do this, we will use Buckley and Leverett's theory, introduced in Sect. 11.3.2. Consider two horizontal surfaces A_1 and A_2, a distance dz apart. The mass of oil entering in time dt across the upper surface A_1, of area A, will be:

$$m_1 = u_z f_o \rho_o A\, dt.$$

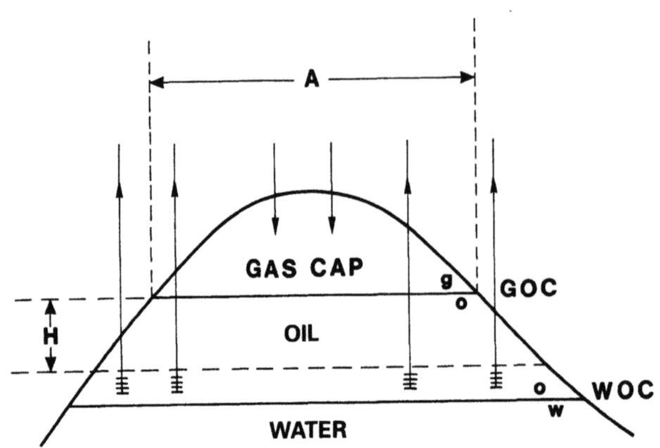

Fig. E15/1.1

The mass of oil leaving in time dt across the lower surface A_2, also of area A, will be:

$$m_2 = u_z \rho_o A (f_o + \frac{\partial f_o}{\partial z} \, dz) \, dt,$$

where $u_z = const.$, and the variation of ρ_o between A_1 and A_2 is considered to be negligible over the small pressure range involved.

The mass of oil between the two surfaces therefore varies in time dt as:

$$dm = m_1 - m_2 = -u_z \rho_o A \frac{\partial f_o}{\partial z} \, dz \, dt. \tag{15/1.2}$$

If ϕ is the porosity of the rock, the mass of oil between A_1 and A_2 at any time is:

$$m = (\phi A \, dz) \rho_o S_o,$$

so that its variation in time dt becomes:

$$dm = \frac{\partial m}{\partial t} \, dt = \phi A \rho_o \frac{\partial S_o}{\partial t} dz \, dt. \tag{15/1.3}$$

Equating Eqs. (15/1.2) and (15/1.3), we get:

$$\frac{u_z}{\phi} \frac{\partial f_o}{\partial z} + \frac{\partial S_o}{\partial t} = 0 \tag{15/1.4a}$$

or, since $f_o = f_o(S_o)$:

$$\frac{u_z}{\phi} \frac{d f_o}{d S_o} \frac{\partial S_o}{\partial z} + \frac{\partial S_o}{\partial t} = 0 \tag{15/1.4b}$$

Since $S_o = S_o(z, t)$, its differential can be expanded as:

$$dS_o = \frac{\partial S_o}{\partial z} \, dz + \frac{\partial S_o}{\partial t} dt \tag{15/1.5}$$

dS_o will of course be zero along any surface of constant S_o.

Therefore, *for* $S_o = const.$, Eq. (15/1.5) becomes:

$$\left(\frac{dz}{dt} \right)_{S_o} \frac{\partial S_o}{\partial z} + \frac{\partial S_o}{\partial t} = 0. \tag{15/1.6}$$

Substituting from Eq. (15/1.4b), we have:

$$\left(\frac{dz}{dt} \right)_{S_o} = v(S_o) = \frac{u_z}{\phi} \left(\frac{d f_o}{d S_o} \right)_{S_o} . \tag{15/1.7a}$$

or, because of Eq. (15/1.1b):

$$v(S_o) = N_{gv} \frac{u_z}{\phi} \left(\frac{d k_{ro}}{d S_o} \right)_{S_o} \tag{15/1.7b}$$

This is the velocity at which a surface of constant oil saturation S_o advances in the porous medium.

Note that $v(S_o)$ is the *real* velocity of advance $(dz/dt)_{S_o}$, while $u_z = q_t/A$ is the *Darcy velocity* of the displacement process.

Since $z(S_o) = 0$ at time $t = 0$, after a time Δt from the start of the displacement the surface of saturation S_o will be at a depth $z(S_o)$ relative to the initial GOC:

$$z(S_o) = N_{gv} \frac{u_z}{\phi} \left(\frac{d k_{ro}}{d S_o} \right)_{S_o} \Delta t. \tag{15/1.8a}$$

With the following definitions:

H : thickness of the oil column, from the original GOC down to the average depth of the tops of the perforated intervals in all the wells.

S_{iw} : irreducible water saturation

$S_{or,g}$: residual oil saturation after gas displacement ($S_{or,g} \approx 0$).

we have:

$$S_o^* = \frac{S_o - S_{or,g}}{1 - S_{iw} - S_{or,g}} \tag{15/1.9a}$$

$$z_{\mathrm{d}} = \frac{z}{H} \tag{15/1.9b}$$

$$t_{\mathrm{d}} = \frac{k_{\mathrm{v}} g (\rho_{\mathrm{o}} - \rho_{\mathrm{g}})}{[\phi (1 - S_{\mathrm{iw}} - S_{\mathrm{or,g}}) H] \mu_{\mathrm{o}}} t \tag{15/1.9c}$$

Note that $\phi (1 - S_{\mathrm{iw}} - S_{\mathrm{or,g}}) H$ is the volume, under reservoir conditions, of *mobile* oil per unit area of gas/oil contact. Since $S_{\mathrm{or,g}} \approx 0$, this is almost the same as the volume of resources per unit area of GOC, under reservoir conditions, between the contact and depth H.

Substituting from Eqs. (15/1.9), Eq. (15/1.8a) becomes:

$$z_{\mathrm{D}}(S_{\mathrm{o}}^*) = \left(\frac{\mathrm{d}k_{\mathrm{ro}}}{\mathrm{d}S_{\mathrm{o}}^*} \right)_{S_{\mathrm{o}}^*} \Delta t_{\mathrm{D}} \tag{15/1.8b}$$

Let $S_{\mathrm{o,f}}^*$ be the normalised oil saturation at the gas/oil displacement front. Then for any value of $S_{\mathrm{o,e}}^* \geq S_{\mathrm{o,f}}^*$, the time $t_{\mathrm{D}}(S_{\mathrm{o,e}}^*)$ required for the saturation $S_{\mathrm{o,e}}^*$ to travel the entire thickness of the oil column ($z = H$; $z_{\mathrm{D}} = 1$) will be:

$$1 = \left(\frac{\mathrm{d}k_{\mathrm{ro}}}{\mathrm{d}S_{\mathrm{o}}^*} \right)_{S_{\mathrm{o,e}}^*} t_{\mathrm{D}}(S_{\mathrm{o,e}}^*), \tag{15/1.10a}$$

from which:

$$t_{\mathrm{D}}(S_{\mathrm{o,e}}^*) = \left(\frac{\mathrm{d}k_{\mathrm{ro}}}{\mathrm{d}S_{\mathrm{o}}^*} \right)_{S_{\mathrm{o,e}}^*}^{-1}. \tag{15/1.10b}$$

k_{ro} can be expressed by a Corey-type equation [see Eq. (3.56a)]:

$$k_{\mathrm{ro}} = \left(\frac{S_{\mathrm{o}} - S_{\mathrm{or,g}}}{1 - S_{\mathrm{iw}} - S_{\mathrm{or,g}}} \right)^n = (S_{\mathrm{o}}^*)^n. \tag{15/1.11}$$

Therefore, from Eq. (15/1.10b):

$$t_{\mathrm{D}}(S_{\mathrm{o,e}}^*) = \left[n \left(S_{\mathrm{o,e}}^* \right)^{n-1} \right]^{-1}, \tag{15/1.12a}$$

which can also be written as:

$$S_{\mathrm{o,e}}^* = \left[n t_{\mathrm{D}} \left(S_{\mathrm{o,e}}^* \right) \right]^{-1/(n-1)}. \tag{15/1.12b}$$

This is the normalised saturation reaching the top of the perforated interval at time t_{D}.

The volume of oil N_{p} produced up to this time is the difference between the volume of *mobile* oil initially in the reservoir, and the volume of *mobile* oil remaining. Referring to Fig. E15/1.2, we can write this as:

$$N_{\mathrm{p}} = \phi A H \left(1 - S_{\mathrm{iw}} - S_{\mathrm{or,g}} \right) - \phi A \int_0^{S_{\mathrm{o,e}} - S_{\mathrm{or,g}}} (H - z) \, \mathrm{d}S_{\mathrm{o}}. \tag{15/1.13a}$$

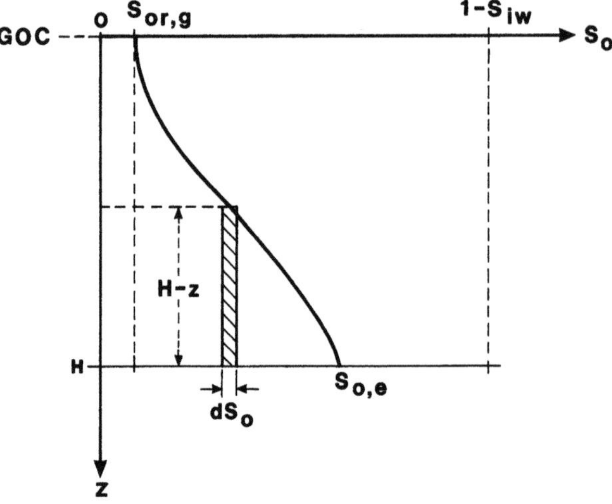

Fig. E15/1.2

We now define the dimensionless term:

$$N_p^* = \frac{N_p}{\phi A H (1 - S_{iw} - S_{or,g})}.$$ (15/1.14)

N_p^* represents the fraction of the *mobile* oil produced up to the time in question. In practice, since $S_{or,g} \approx 0$, N_p^* is effectively the produced fraction of the initial resources between the GOC and the average depth of the tops of the perforated intervals.

We can substitute from Eqs. (15/1.14) and (15/1.8b) so that Eq. (15/1.13a) becomes:

$$
\begin{aligned}
N_p^* &= 1 - \int_0^{S_{o,e}^*} (1 - z_D) \, dS_o^* \\
&= 1 - S_{o,e}^* + \int_0^{S_{o,e}^*} \Delta t_D \frac{dk_{ro}}{dS_o^*} \, dS_o^* \\
&= 1 - S_{o,e}^* + (k_{ro})_{S_{o,e}^*} \, t_D(S_{o,e}^*),
\end{aligned}
$$ (15/1.13b)

which, from Eqs. (15/1.12b) and (15/1.10b), becomes in turn:

$$N_p^* = 1 - \frac{n-1}{n}(n t_D)^{-1/(n-1)}$$ (15/1.15)

With Eq. (15/1.15) we can calculate the fraction of the mobile oil produced up to time t_D, at a constant rate q_t.

For $t_D \le t_D(S_{o,f}^*)$ (prior to gas breakthrough in the producing wells) $f_o = 1$, so:

$$N_p = q_t t$$ (15/1.16a)

and, from Eq. (15/1.9c):

$$\frac{N_p}{\phi A H (1 - S_{iw} - S_{or,g})} = \frac{u_z \mu_o}{k_v g (\rho_o - \rho_g)} t_D,$$ (15/1.16b)

so that:

$$N_p^* = \frac{t_D}{N_{gv}}.$$ (15/1.16c)

If $t_{D,BT}$ is the moment of gas breakthrough $[t_{D,BT} = t_D(S_{o,f}^*]$, we can summarise the conclusions as follows:

for $t_D \le t_{D,BT}$:

$$N_p^* = \frac{t_D}{N_{gv}},$$ (15/1.16c)

for $t_D \ge t_{D,BT}$:

$$N_p^* = 1 - \frac{n-1}{n}(n t_D)^{-1/(n-1)}$$ (15/1.15)

The value of N_p^* at the moment of gas breakthrough $[N_p^*(t_{D,BT}); S_o^* = S_{o,f}^*)$ must of course satisfy both Eqs. (15/1.16c) and (15/1.15). *It therefore corresponds to the intersection of the two curves described by these equations*

In practice, we plot the curves for $N_p^* = f(n, t_D)$ defined by Eq. (15/1.15), for different values of n, and the curves $N_p^*(N_{gv}, t_D)$ defined by Eq. (15/1.16c), for different values of N_{gv}, as in Figs. E15/1.3 and 15/1.4.

Reading off the points of intersection we can construct the curves of:

$$N_p^* \text{ at gas breakthrough at the wells } = f(N_{gv})$$

for different values of the Corey exponent n, as shown in Fig. 15.14.

We can draw the following conclusions from these diagrams:

1. *Even with no miscibility between the oil and the gas*, very high recovery factors can be achieved. For example, for $N_{gv} = 10$ and $n = 3$, we have $N_p^* \approx 0.9$. In other words, with $N_{gv} = 10$ and $n = 3$, when gas breaks through at the wells *90% of the oil present in the reservoir above the top of the perforations has been recovered*.

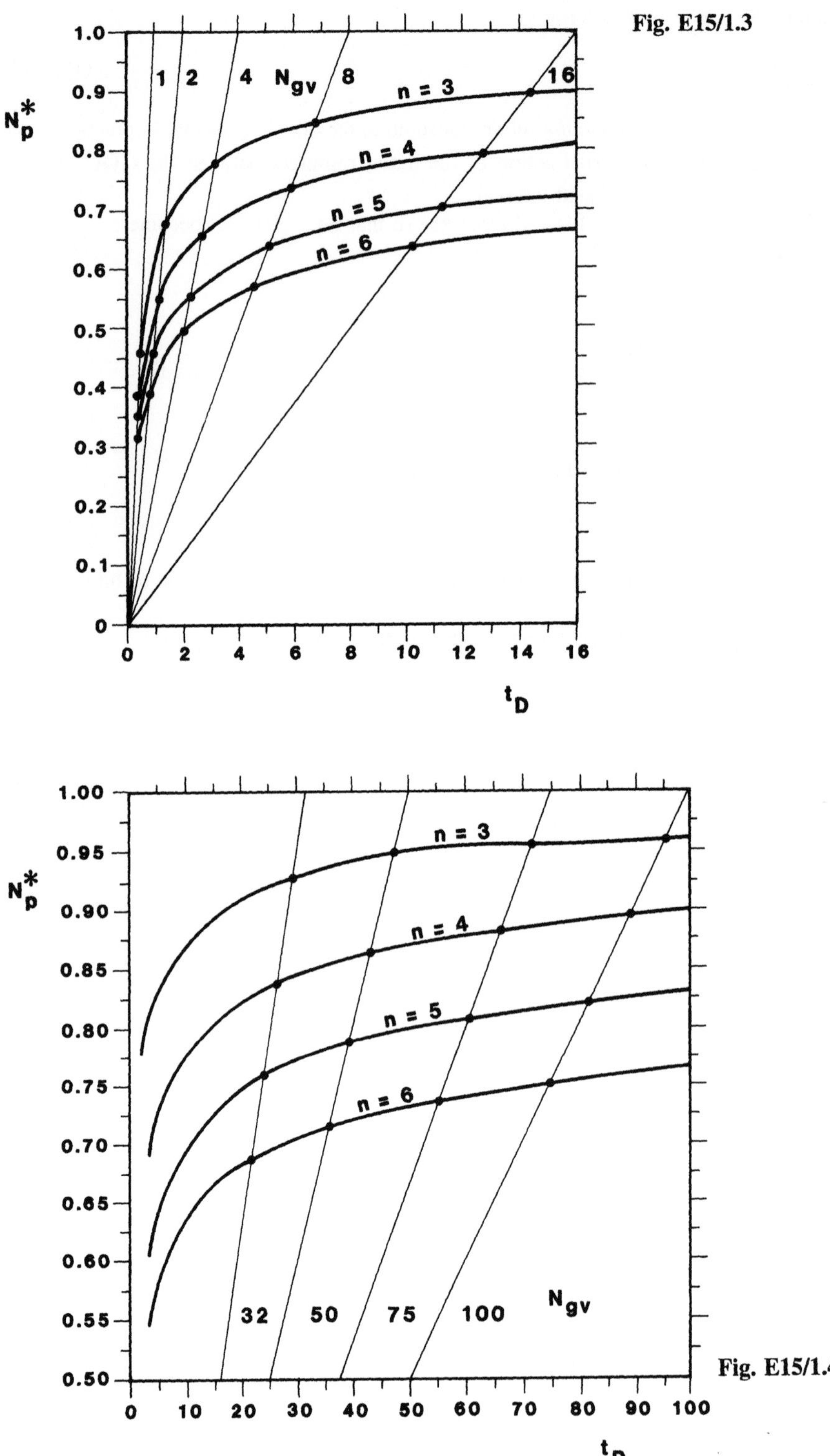

The overall recovery factor $E_{R,o}$ will depend on what fraction of the total reserves the volume above the perforations represents.

If, in addition to a gas cap, there is also an aquifer, the depth of the perforations must be high enough above the WOC to avoid water coning.

The excellent recovery factors achieved in "closed" reservoirs (those not in contact with an aquifer, and which can therefore be perforated right to the bottom of the layer) produced by crestal gas injection, can be explained as a displacement at high N_{gv} (*gravity-dominated* conditions), without needing to postulate miscibility between the oil and the gas (which is very rarely attained).

2. In order that a high N_{gv} should be consistent with a commercially viable rate of production, two conditions must be met [Eq. (15.11b)]:
   - a high rock *vertical* permeability k_v,
   - a low oil viscosity μ_o.

   The presence of intercalations of shale lenses which reduce the vertical permeability (Sect. 13.16.5) and, worse still, the presence of continuous low permeability strata, are highly unfavourable for the use of non-miscible gas to displace oil.

   A high oil viscosity is also undesirable.

3. Figure 15.14 demonstrates that for a given N_{gv}, the value of N_p^* at breakthrough is larger, the smaller n is. The Corey exponent depends on the pore structure, and cannot be modified from the outside. Referring to Eq. (15/1.15), for $n = 1$ we would get $N_p^* = 1.0$ for any value of $t_D > 0$.

   In rocks with secondary porosity (fractures, vugs), n is very close to 1, which explains the high recovery factors for oil produced by crestal injection in this type of reservoir.

   In order to predict N_p^* at gas breakthrough, n must be known precisely. It can be derived from values of $k_{ro}(S_o)$, $S_{or,g}$ and S_{iw} measured on cores.

   From Eq. (15/1.11) we have, in fact:

$$\log k_{ro}(S_o) = n \log \left(\frac{S_o - S_{or,g}}{1 - S_{iw} - S_{or,g}} \right) \tag{15/1.17}$$

which can be solved for n by simple linear regression on a set of data pairs ($\log k_{ro}$, $\log S_o^*$).

The centrifuge technique described by Hagoort[46] is recommended for the determination of the $k_{ro}(S_o)$ curve, and particularly for $S_{or,g}$.

Note that neither the gas relative permeability $k_{rg}(S_o)$, nor the gas viscosity μ_g, figure in this discussion. This is because the term μ_g / k_{rg} can be assumed to be practically zero, μ_g being much smaller than μ_o.

In this context, therefore, the common production laboratory practice of determining both the $k_{ro}(S_o)$ and $k_{rg}(S_o)$ curves is a waste of time.

$\Diamond \ \Diamond \ \Diamond$

Author Index

 MIX
Papier aus verantwortungsvollen Quellen
Paper from responsible sources
FSC® C105338

If you have any concerns about our products,
you can contact us on
ProductSafety@springernature.com

In case Publisher is established outside the EU,
the EU authorized representative is:
**Springer Nature Customer Service Center GmbH
Europaplatz 3, 69115 Heidelberg, Germany**

Printed by Libri Plureos GmbH
in Hamburg, Germany